Smart Nanomaterials Technology

Series Editors

Azamal Husen, Wolaita Sodo University, Wolaita, Ethiopia

Mohammad Jawaid, Department of Chemical and Petroleum Engineering, United Arab Emirates University, Al Ain, United Arab Emirates

Nanotechnology is a rapidly growing scientific field and has attracted a great interest over the last few years because of its abundant applications in different fields like biology, physics and chemistry. This science deals with the production of minute particles called nanomaterials having dimensions between 1 and 100 nm which may serve as building blocks for various physical and biological systems. On the other hand, there is the class of smart materials where the material that can stimuli by external factors and results a new kind of functional properties. The combination of these two classes forms a new class of smart nanomaterials, which produces unique functional material properties and a great opportunity to larger span of application. Smart nanomaterials have been employed by researchers to use it effectively in agricultural production, soil improvement, disease management, energy and environment, medical science, pharmaceuticals, engineering, food, animal husbandry and forestry sectors.

This book series in Smart Nanomaterials Technology (SNT) publishes cutting edge research monographs and edited volumes by top researchers and professionals across the globe. This book series aims to comprehensively cover topics in the fabrication, synthesis and application of these materials for applications in the following fields:

- Energy Systems—Renewable energy, energy storage (supercapacitors and electrochemical cells), hydrogen storage, photocatalytic water splitting for hydrogen production
- Biomedical—controlled release of drugs, treatment of various diseases, biosensors
- Agricultural—agricultural production, soil improvement, disease management, animal feed, egg, milk and meat production/processing
- Forestry—wood preservation, protection, disease management
- Environment—wastewater treatment, separation of hazardous contaminants from wastewater, indoor air filters

Review Process

The proposal for each volume is reviewed by the main editor and/or the advisory board. The chapters in each volume are individually reviewed single blind by expert reviewers (at least two reviews per chapter) and the main editor.

Ethics Statement for this series can be found in the Springer standard guidelines here—https://www.springer.com/us/authors-editors/journal-author/journal-author-helpdesk/before-you-start/before-you-start/1330#c14214

Leonardo Fernandes Fraceto · Deepranjan Sarkar ·
S. Rakesh · Rahul Datta · Sachidanand Singh
Editors

Emerging Nanotechnologies for Agroecosystem Management

Springer

Editors
Leonardo Fernandes Fraceto
Institute of Science and Technology
São Paulo State University
Sorocaba, São Paulo, Brazil

S. Rakesh
Department of Resilient Farming Systems
International Crops Research Institute for the Semi-Arid Tropics
Hyderabad, Telangana, India

Sachidanand Singh
Department of Biotechnology, School of Energy Technology
Pandit Deendayal Energy University
Gandhinagar, Gujarat, India

Deepranjan Sarkar
Department of Food, Agriculture and Natural Resources, School of Environment, Resources and Development
Asian Institute of Technology
Khlong Luang, Pathum Thani, Thailand

Rahul Datta
Department of Geology and Pedology
Mendel University in Brno
Brno, Czech Republic

ISSN 3004-8273 ISSN 3004-8281 (electronic)
Smart Nanomaterials Technology
ISBN 978-981-95-0186-1 ISBN 978-981-95-0187-8 (eBook)
https://doi.org/10.1007/978-981-95-0187-8

This Springer imprint is published by the registered company Springer Nature Singapore Pte Ltd.
The registered company address is: 152 Beach Road, #21-01/04 Gateway East, Singapore 189721, Singapore

Preface

Global food production faces several challenges, including climate change, population growth, resource scarcity, pollution, and economic instability. As the world moves toward 2050, innovative solutions are key to securing a stable and sustainable food future. Nanotechnology can potentially enhance agricultural inputs' efficiency and significantly boost food productivity. The delivery mechanisms of nanoparticles have attracted attention from across the world. Our book has emphasized the transformative potential of nanotechnology by highlighting its recent applications.

Nanomaterials enhance soil fertility, stimulate plant growth, provide crop protection, facilitate environmental remediation, and improve crop productivity. The synthesis of nanoparticles using green synthesis methods offers an eco-friendly approach that minimizes toxic byproducts and utilizes renewable resources. Nanofertilizers enhance nutrient absorption in plants and enable site-specific nutrient delivery. Nanomaterials play a crucial role in water remediation, enhancing irrigation management by improving water quality and availability for agricultural use. Nanosensors are emerging as a real-time monitoring tool in agriculture for detecting biotic and abiotic stresses. The chapter on regulatory aspects and risk assessment of nanoplatforms for agricultural uses focuses on the safe application of nanotechnology in enhancing crop production while protecting consumer health and the environment.

This book represents a significant contribution to the field, drawing insights and perspectives from diverse contributors across various regions of the globe. We extend our heartfelt gratitude to all contributors for their invaluable input and support. The compilation will provide valuable insights for students (undergraduate and postgraduate), experts, stakeholders, policymakers, and other enthusiastic readers.

Sorocaba, Brazil — Leonardo Fernandes Fraceto
Khlong Luang, Thailand — Deepranjan Sarkar
Hyderabad, India — S. Rakesh
Brno, Czech Republic — Rahul Datta
Gandhinagar, India — Sachidanand Singh

Contents

Sensor-Based Approaches

Environmental Cleanup

Risk Assessment and Regulation

Editors and Contributors

About the Editors

Leonardo Fernandes Fraceto earned a bachelor's degree in Chemistry (1997), a master's degree in Functional and Molecular Biology (2000), and a Ph.D. in Functional and Molecular Biology (2003) from the State University of Campinas, Brazil. He is currently an Associate Professor at the Institute of Science and Technology, São Paulo State University (UNESP), Campus Sorocaba, where he teaches in the Environmental Engineering undergraduate program and the postgraduate program in Environmental Sciences. He also served as Coordinator of the Postgraduate Program in Environmental Sciences at UNESP–Sorocaba from 2012 to 2016. His research expertise spans chemistry and biochemistry, with a strong emphasis on environmental nanotechnology and the application of nanotechnology in agriculture. Dr. Fraceto has developed nanocarrier systems for herbicides and other bioactive compounds for pest management in agriculture, resulting in numerous research publications and patents. He continues to actively work on the development of nanocarrier systems for the encapsulation of compounds with bioactive properties.

Deepranjan Sarkar is working as a Postdoctoral Research Fellow at the Department of Food, Agriculture and Natural Resources, School of Environment, Resources and Development, Asian Institute of Technology (AIT), Thailand. Before joining AIT, he served as an Assistant Professor (Soil Science) at the Integral University, Lucknow, India, and also as a Consultant at the International Rice Research Institute (IRRI), India. He received M.Sc. (Ag.) in Soil Science and Agricultural Chemistry degree from Uttar Banga Krishi Viswavidyalaya and Ph.D. from Banaras Hindu University, India. His research interests include soil fertility, plant nutrition, regenerative agriculture, carbon sequestration, climate change, and UNSDGs. During his Ph.D. program, he performed a comprehensive assessment of bio-priming technology in the Middle Gangetic Plains of India and presented his research at several national and international conferences. Dr. Sarkar is involved in teaching soil science courses at UG and PG levels. Currently, he is working on prioritizing adaptation options to address expected future climate challenges in agroecosystems. He has received prestigious awards from BHU, IIT-KGP, TWAS, and the Soil Science Society of America. He is a Life Member/Member of AMI-India, ISCA-Kolkata, IUCN, etc. He has authored more than 30 peer-reviewed publications. He serves on the Editorial Board of Discover Agriculture (Springer Nature) and Modern Agriculture (Wiley), is a Review Editor in Frontiers journals, and regularly reviews manuscripts for journals published by Springer, Elsevier, Taylor & Francis, Wiley, and others.

S. Rakesh is currently serving as an Associate Scientist in the Department of Resilient Farming Systems at the International Crops Research Institute for the Semi-Arid Tropics (ICRISAT), Hyderabad, India. He previously worked as a Research Associate at ICAR-National Academy of Agricultural Research Management (NAARM), Hyderabad (2019-2023) and as a Senior Research Fellow in the Australian (ACIAR) funded project "Sustainable & Resilient Farming System Intensification (SRFSI)" from 2015 to 2019. His Ph.D. research focused on carbon dynamics and conservation agricultural practices. Dr. Rakesh has received several awards, including the Best Thesis Award (2019),

Young Scientist Award (2021), and Young Researcher Award (2023). Currently, he is working on research projects specifically in the area of regenerative agriculture and sustainable crop production through climate adaptation technologies. Dr. Rakesh is a member of several professional societies and organizations like SEE-Agra, ISCA-Kolkata, COBACAS, IUCN, etc. He is actively involved in delivering invited lectures at conferences and Universities and participating in extension works with numerous Farmer's Clubs. He has published more than 50 full length research in reputed international and national journals, with an H-index of 16. Dr. Rakesh has served as editor of two books. He has served as editor of two books, acts as a reviewer for journals published by Elsevier, Springer, and Frontiers, and is an Editorial Board Member of Discover Agriculture (Springer).

Rahul Datta is an Assistant Professor of Soil Science at Mendel University, Brno, Czech Republic, recognized for his notable contributions to sustainable agriculture. His research addresses critical challenges such as drought stress, heavy metal toxicity, and greenhouse gas emissions, significantly enhancing the quality and yield of crops like wheat, maize, barley, rice, mangoes, spinach, and cotton. With over 200 peer-reviewed publications and more than 10,000 citations worldwide, including acknowledgments from institutions like the USDA, Dr. Datta has earned an H-index of 54, underscoring the impact of his work. His achievements also earned him a place in the Top 2% of Scientists as per Stanford University Ranking of World Scientists, recognizing his global influence in the field. Dr. Datta has authored 10 books, contributed to 20 book chapters, and holds 6 patents, further solidifying his leadership in agricultural science. In addition to his research, Dr. Datta holds several editorial roles. He is an Academic Editor for the Turkish Journal of Agriculture and Forestry, serves on the Editorial Board of BMC Plant Biology, and is an Editor for Open Agriculture (De Gruyter) and Phyton-International Journal of Experimental Botany. As a member of the Soil Science Society of America, Dr. Datta remains at the forefront of advancing soil science and sustainable agriculture through his research, publications, and editorial contributions.

Sachidanand Singh is currently working as an Associate Professor and Head in the Department of Biotechnology, School of Energy Technology, Pandit Deendayal Energy University, Gandhinagar, Gujarat, India. He has an engineering background, having earned a B.Tech in Biotechnology and an M.Tech in Bioinformatics, followed by a PhD, and completed postdoctoral research at The Ohio State University, College of Pharmacy, USA. He has more than 14 years of experience in teaching, administration, and research. He has previously served as Professor and Head Internal Quality Assurance Cell and Dean of Allied Sciences at Sankalchand Patel University, Gujarat, Dean of Biotechnology and Agriculture Sciences, NAAC SSR Head, and Research Coordinator at Shri Ramswaroop Memorial University. Work experience includes Karunya University, Shri Ramswaroop Memorial University, Vignan University, and Sankalchand Patel University. His research area includes Plant Biotechnology, metagenomics, Systems Biology, Network Biology, Micro RNA-based studies, Expression Data Analytics, Next-generation Sequencing, and Statistical Genetics. His present focus also includes understanding the Canonical and non-canonical pathways of miRNA Biogenesis, Dynamics, and mechanisms of miRNA-mediated Gene Regulation. He teaches different subjects, especially Genomics and Proteomics, Microarray and Image Analysis, Drug Design, and Molecular Modeling for graduate and postgraduate students. He has been involved in various assignments given by the University authority from time to time, such as Online teaching platform, Research Initiatives, Biogas Plant, NAAC, NBA, and other council approvals. He has published more than 60 research and review articles in peer-reviewed journals, 9 Patents published among which 4 are granted. He has completed one SERB, Govt. of India-funded project, and two University research grant projects. He has received several awards in his area of research and for his teaching strategies. He has organized different conferences and workshops and received Govt. funds for the same. He has shared his area of expertise with different colleges and universities as an invited speaker at different conferences and workshops. He has guided 36 B.Tech and 10 M.Tech projects. He has awarded 3 PhD students under him in the field of Herbal products, drug Design, and integration with network biology.

Contributors

Hajra Abid Department of Botany, University of Poonch Rawalakot, Rawalakot, Pakistan

Khawaja Shafique Ahmad Department of Botany, University of Poonch Rawalakot, Rawalakot, Pakistan

Md. Abdullah Al Sabbir Department of Plant Pathology, Gazipur Agricultural University, Gazipur, Bangladesh

Majed Al-Shaeri Department of Biological Sciences, Faculty of Sciences, King Abdulaziz University, Jeddah, Saudi Arabia

Mohammad Omaish Ansari Center of Nanotechnology, King Abdulaziz University, Jeddah, Saudi Arabia

Raviraj Barot Department of Biotechnology, School of Energy and Technology, Pandit Deendayal Energy University, Gandhinagar, Gujarat, India

Anuradha Beniwal Department of Chemistry, CCS Haryana Agricultural University, Hisar, Haryana, India

Indrani Bhattacharya Department of Biotechnology, Parul Institute of Applied Science, Parul University, Vadodara, Gujarat, India

Suvankar Kumar Biswas Department of Plant Pathology, Gazipur Agricultural University, Gazipur, Bangladesh

Munny Chinyo Department of Agriculture, Integral Institute of Agricultural Science and Technology, Integral University, Lucknow, India

Richa Das Department of Biotechnology, Parul Institute of Applied Science, Parul University, Vadodara, Gujarat, India

Diego Maroso da Silva Food Engineering Pos-Graduate Program, URI, Erechim, RS, Brazil

Rahul Datta Department of Geology and Pedology, Faculty of Forestry and Wood Technology, Mendel University in Brno, Brno, Czech Republic

Nitish R. Dave Department of Science and Technology (GoG), Gujarat State Biotechnology Mission, Gandhinagar, Gujarat, India

Mukul Kumar Deka Department of Entomology, Assam Agricultural University, Jorhat, Assam, India

Adriano Buzutti de Siqueira Department of Chemistry, National Institute of Science and Technology in Nanotechnology for Sustainable Agriculture, INCT-NanoAgro, Federal University of Mato Grosso-UFMT, Cuiabá, MT, Brazil

Jéssica de Souza Rodrigues Institute of Science and Technology, São Paulo State University (UNESP), Sorocaba, SP, Brazil

Faria Fatima Department of Agriculture, Integral Institute of Agricultural Science and Technology, Integral University, Lucknow, India

Humayra Ferdus Department of Plant Pathology, Gazipur Agricultural University, Gazipur, Bangladesh

Leonardo Fernandes Fraceto Institute of Science and Technology, São Paulo State University (UNESP), Sorocaba, SP, Brazil

Felipe Franco de Oliveira Institute of Science and Technology, São Paulo State University (UNESP), Sorocaba, SP, Brazil

Sabuj Ganguly BSM&TC (BTSSO), Central Silk Board, Baripada, Odisha, India

Dania Habib Department of Botany, University of Poonch Rawalakot, Rawalakot, Pakistan

Khalid Habib Department of Agriculture, Integral Institute of Agricultural Science and Technology, Integral University, Lucknow, India

Md. Motaher Hossain Department of Plant Pathology, Gazipur Agricultural University, Gazipur, Bangladesh

Lara Jadhav Department of Microbiology, Parul Institute of Applied Science, Parul University, Vadodara, Gujarat, India

Juliana Steffens Food Engineering Pos-Graduate Program, URI, Erechim, RS, Brazil

Eraldo Antonio Bonfatti Júnior Faculty of Agricultural Sciences, São Paulo State University, Botucatu, SP, Brazil

Simran Kakkar Department of Chemistry, CCS Haryana Agricultural University, Hisar, Haryana, India

R. Karthik Department of Entomology, Assam Agricultural University, Jorhat, Assam, India

Sana Khalid Pakistan Council of Scientific and Industrial Research Laboratories Complex, Lahore, Pakistan

Noopur Khare Bhai Gurdas Institute of Engineering and Technology, Sangrur, Punjab, India;
Shri Ramswaroop Memorial University, Barabanki, Uttar Pradesh, India

Pragati Khare Department of Pharmacy, BPS Educational Institution of Pharmacy, Etah, Uttar Pradesh, India

Elaine Cristine Lengowski Department of Forest Engineering, Federal University of Mato Grosso-UFMT, Cuiabá, MT, Brazil;

Department of Chemical Engineering, University of Waterloo, Waterloo, ON, Canada

R. Loganathan Division of Entomology, ICAR-Indian Agricultural Research Institute, New Delhi, India

Jhones Luiz de Oliveira Institute of Science and Technology, São Paulo State University (UNESP), Sorocaba, SP, Brazil;
B. Nano Soluções Tecnológicas Ltda, São Miguel Arcanjo, São Paulo, Brazil

George Mangse Department of Biotechnology, Faculty of Sciences, Nile University of Nigeria, Abuja, Nigeria

Abdul Mazeed Department of Agriculture, Integral Institute of Agricultural Science and Technology, Integral University, Lucknow, India

Ansar Mehmood Department of Botany, University of Poonch Rawalakot, Rawalakot, Pakistan

Shweta Meshram Department of Plant Pathology, School of Agriculture, Lovely Professional University, Phagwara, Punjab, India

Monika Moond Department of Chemistry, CCS Haryana Agricultural University, Hisar, Haryana, India

Mahabuba Mostafa Department of Plant Pathology, Gazipur Agricultural University, Gazipur, Bangladesh

Giovanna Moura Silva Institute of Science and Technology, São Paulo State University (UNESP), Sorocaba, SP, Brazil

Djadjiti Namla Department of Biological Sciences, Faculty of Sciences, King Abdulaziz University, Jeddah, Saudi Arabia;
Department of Biotechnology, Faculty of Sciences, Nile University of Nigeria, Abuja, Nigeria

Rabia Nazir Pakistan Council of Scientific and Industrial Research Laboratories Complex, Lahore, Pakistan

Saiqa Nazir Department of Botany, University of Poonch Rawalakot, Rawalakot, Pakistan

Mohammad Oves Center of Excellence in Environmental Studies, King Abdulaziz University, Jeddah, Saudi Arabia

Meenakshi Rana Department of Plant Pathology, School of Agriculture, Lovely Professional University, Phagwara, Punjab, India

Jyoti Rani Department of Chemistry, CCS Haryana Agricultural University, Hisar, Haryana, India

Seema Sangwan Department of Microbiology, CCS Haryana Agricultural University, Hisar, Haryana, India

Rajni Kant Sharma Department of Chemistry, CCS Haryana Agricultural University, Hisar, Haryana, India

Muhammad Asif Shehzad Institute of Plant Breeding and Biotechnology, Muhammad Nawaz Shareef University of Agriculture, Multan, Pakistan

Avinash Singh Department of Biotechnology, Axis Institute of Higher Education, Kanpur, India

Indrajeet Singh Department of Biotechnology, Faculty of Engineering and Technology, Rama University, Kanpur, Uttar Pradesh, India

Jay Prakash Singh Department of Plant Pathology, SMMTD College, Balia, Uttar Pradesh, India

Nisha Singh Department of Biotechnology, School of Energy and Technology, Pandit Deendayal Energy University, Gandhinagar, Gujarat, India

Sachidanand Singh Department of Biotechnology, School of Energy and Technology, Pandit Deendayal Energy University, Gandhinagar, Gujarat, India

Shivam Singh Krishi Vigyan Kendra Baghpat, Sardar Vallabhbhai Patel University of Agriculture and Technology, Meerut, Uttar Pradesh, India

Sushila Singh Department of Chemistry, CCS Haryana Agricultural University, Hisar, Haryana, India

P. Smriti Rao Department of Agriculture, Integral Institute of Agricultural Science and Technology, Integral University, Lucknow, India

Seweta Srivastava Department of Plant Pathology, School of Agriculture, Lovely Professional University, Phagwara, Punjab, India

Clarice Steffens Food Engineering Pos-Graduate Program, URI, Erechim, RS, Brazil

Vanessa Takeshita Institute of Science and Technology, São Paulo State University (UNESP), Sorocaba, SP, Brazil

Ailton José Terezo Department of Chemistry, National Institute of Science and Technology in Nanotechnology for Sustainable Agriculture, INCT-NanoAgro, Federal University of Mato Grosso-UFMT, Cuiabá, MT, Brazil

Tushar Thakur Department of Biotechnology, Parul Institute of Applied Science, Parul University, Vadodara, Gujarat, India

Estefânia Vangelie Ramos Campos Institute of Science and Technology, São Paulo State University (UNESP), Sorocaba, SP, Brazil;
B. Nano Soluções Tecnológicas Ltda, São Miguel Arcanjo, São Paulo, Brazil

Gustavo Vínicios Munhoz-Garcia Center of Nuclear Energy in Agriculture, University of São Paulo, Piracicaba, SP, Brazil

Manaal Zahera Department of Biotechnology, Era University, Lucknow, India

Synthesis

Synthesis and Applications of Nanomaterials Derived from Biowastes

Adriano Buzutti de Siqueira, Elaine Cristine Lengowski, Eraldo Antonio Bonfatti Júnior, and Ailton José Terezo

Abstract Circular economy initiatives focus on technologies related to reuse, recycling, and remanufacturing, which aim to recover waste and prevent environmental pollution resulting from improper disposal. One green approach within this framework involves synthesizing nanoparticles using biowastes like stems, leaves, seeds, pulp, stubble, bagasse, wood forest, flours, cereals, animal remains, manure, algal, and microorganisms. This approach is recognized for its low energy consumption and cost-effective production. The composition of the raw materials influences the nanoparticles' final application, which spans various fields and can even be utilized within agriculture itself. Nanobiowastes can be grouped into two principal types: inorganic (nanosilica and nanometallics), and organic (nanocellulose, nanolignin, nanocarbons, and carbon dots) nanoparticles. Green synthesis approaches for nanoparticles focus on several key factors: (i) employing eco-friendly reducing agents, (ii) selecting a solvent, (iii) employing non-toxic and biocompatible materials for synthesis, (iv) ensuring facile and low-energy processing, and (v) enabling scalability of synthesis methods. Indeed, the steps taken to produce nanomaterials using biowaste aim to establish a sustainable and valuable circular economy. This approach holds immense potential for fostering economic and social growth, especially in countries where food security is of paramount importance for the future. By harnessing biowaste to create nanomaterials, we not only mitigate waste but also generate valuable resources that can be utilized in various sectors, thereby contributing to a more resilient and sustainable future. This chapter aims to delve

A. B. de Siqueira (✉) · A. J. Terezo
Department of Chemistry, National Institute of Science and Technology in Nanotechnology for Sustainable Agriculture, INCT-NanoAgro, Federal University of Mato Grosso-UFMT, Cuiabá, MT, Brazil
e-mail: adriano.siqueira@ufmt.br

E. C. Lengowski
Department of Forest Engineering, Federal University of Mato Grosso-UFMT, Cuiabá, MT, Brazil

Department of Chemical Engineering, University of Waterloo, Waterloo, ON, Canada

E. A. B. Júnior
Faculty of Agricultural Sciences, São Paulo State University, Botucatu, SP, Brazil

L. F. Fraceto et al. (eds.), *Emerging Nanotechnologies for Agroecosystem Management*, Smart Nanomaterials Technology, https://doi.org/10.1007/978-981-95-0187-8_1

into the properties of biowastes, exploring their utilization as bionanomaterials and the myriads of potential applications they offer.

Keyword Nanocarbon · Silica · Nanocellulose · Nanolignin · Agriculture

1 Introduction

Climate issues, the greenhouse effect, and food security are interconnected challenges that are essential for ensuring our survival. With the increasing global population, there is a surge in the use of vehicles, gas generators, food consumption, and consequently, waste generation. This scenario appears to perpetuate a harmful cycle. The dwindling of natural resources, coupled with population growth and environmental apprehensions, has spurred heightened interest in the advancement of biomass nanomaterials. The structural versatility of bionanoparticles prove invaluable in comprehending their applications, thereby offering potential solutions to mitigate the depletion of natural resources. Additionally, bionanomaterials hold promise in facilitating sustainable agriculture by providing essential inputs [161].

Biodegradable waste, or biowaste, is increasingly recognized for its significant implications for the economic, social, and environmental future, particularly concerning food security, see Fig. 1. Utilizing waste extends beyond industrial and agricultural sectors, encompassing waste from domestic and public sources as well [222].

Nanotechnology: The small size of particles results in physicochemical properties that differ from those displayed by bulk materials of the same composition. Nanomaterials encompass a variety of substances derived from metals, non-metals, semimetals, and organic compounds (with organic components being predominant). Additionally, they include other biomaterials like chitosan, alginate, cellulose, starch, silica, and clays, all possessing at least one dimension less than 100 nm. Nanotechnology involves the manipulation and reorganization of functional components at the scale of molecules and atoms. In contemporary times, nanomaterials have emerged as a crucial component in material science, showcasing advancements and pioneering trends in research and practical applications aimed at addressing scientific challenges in a meaningful manner. The concept of engineering nanomaterials serves as a cornerstone, unlocking numerous opportunities for the development of nanotools that facilitate the transmission of information within the nanoworld [161].

Nanomaterials find diverse applications in agriculture, incorporating plant growth, nutrient enhancement, weed management, and protection from both biotic and abiotic stresses. These materials are presented in diverse forms, such as nanotubes, nanofibers, nanopowders, nanoclusters, nanosheets, and nanoparticles [199]. Nanotechnology has contributed to the reuse of waste generated in the field, even after harvest, with the majority of nanowaste having important functions in the planting itself, preventing diseases, pest attacks and even improving

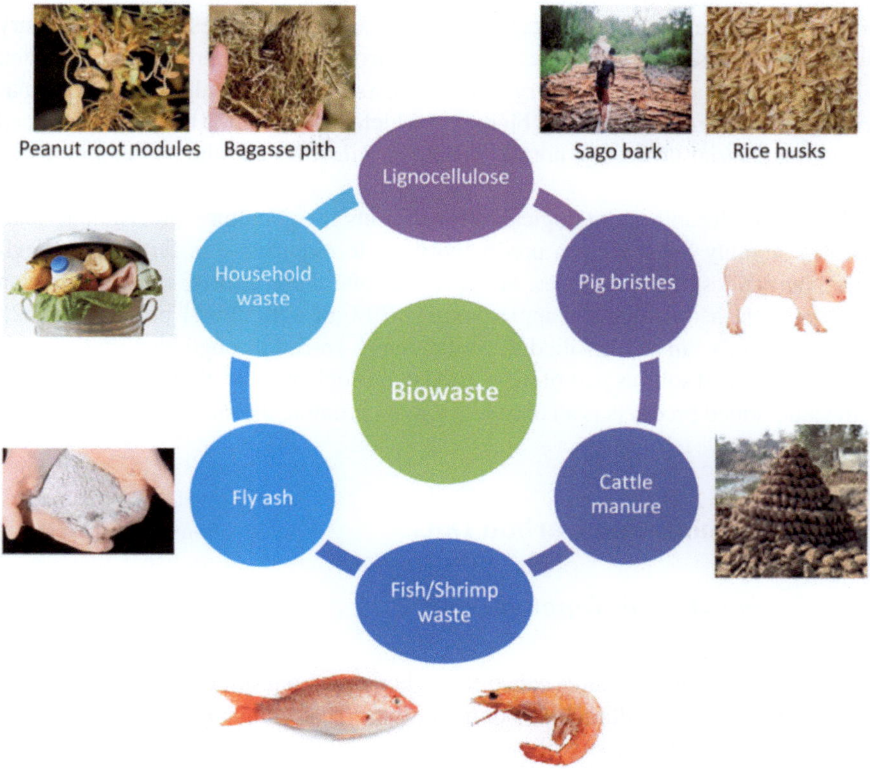

Fig. 1 Frequently used and easily obtainable biowaste materials. Reproduced with permission [222] ROYAL SOCIETY OF CHEMISTRY and Copyright Clearance Center

plant productivity, whether as bionanostimulants, whether as bionanophotosensitizers. Another advantage of using nanomaterials is the reduction in doses of agrochemicals, providing better action against pests and lower impacts on the environment. Nanobiowaste can serve as biosensors, indirectly aiding in the optimal utilization of agrochemicals and the control of environmental contaminants, whether in water or other environments potentially affected by pesticides. Following harvest, various food items, including vegetables, fruits, or grains, can be safeguarded using bionanowastes [199].

Agro-industrial waste refers to the byproducts resulting from the manufacturing, distribution, and sales processes of products such as fruits, vegetables, meat, dairy products, and more [62]. Agriculture stands as one of the essential commodities for the economic development and social well-being of countries. Simultaneously, the farming systems integrated into crop production play a pivotal role in meeting the eco-friendly requirements of sustainable agriculture. Traditional farming practices can affect the surrounding environment by releasing harmful water pollutants such as salts, sediments, pesticides, and fertilizers (such phosphorus and nitrates). The primary biowaste originates directly from agricultural activities, while secondary waste encompasses the byproducts generated from industrial processes associated with agricultural activities. Tertiary residues are a subset of agricultural-based biomaterials, often referred to as "forest biomass," which includes wood materials produced through industrial processes, notably by wood industries and the extraction of tree residues [102].

Despite noticeable technological advancements in agriculture, precision agriculture has notably boosted crop productivity while requiring less fertilizers, fungicides, herbicides, and pesticides. However, despite these efforts, waste generation remains high, accounting for approximately 30% of agricultural production [62]. These materials can be classified as contaminants spread throughout surface water, groundwater, and soils as part of the natural hydrological cycle. Converting biowaste into value-added products is an excellent way to promote the circular economy [62].

2 Nanocarbons and Carbon Dots

2.1 Carbon-Based Nanomaterials

Carbon atoms exhibit a unique capability to bond with one another and with various other elements in a multitude of configurations. This unique property contributes to the ever-growing diversity of carbon compounds. Graphite, diamond, and C-60 stand out as the three most renowned and extensively researched elemental forms, drawing significant attention in the realm of nanomaterial studies. As material science progresses and characterization techniques become more refined, intriguing, and distinctive properties emerge, particularly attributed to the sp^2 bond prevalent in graphite's layered structure and the planar layers of diamond. Moreover, the bonding of sp^3 carbons results in highly organized materials, often crystalline in nature, showcasing the precision and structural integrity inherent to these forms [161]. Carbon atoms' sp^2 and sp^3 hybridizations are key factors in determining the properties of these materials. In graphite, sp2 hybridization fosters the formation of durable π bonds within the graphene layers, which supports electrical conductivity and stable carbon–carbon bonds. In diamond, the sp^3 hybridization results in a highly organized and rigid structure, giving rise to its remarkable hardness and thermal conductivity [105, 171].

Carbon-based nanomaterials (CN) also include porous carbon nanonetworks (3D) nanodiamonds and activated carbon nanoparticles, carbon nanofibers, nanoscrolls, graphene (2D), carbon nanotubes (1D), and fullerenes (0D) and newest carbon dots, see Fig. 2 [228]. Interest in nanocarbons has surged, particularly with sustainable advancements enabling their production from biowaste-derived routes [222]. Functional carbon-based materials are particularly intriguing due to their eco-friendliness compared to other materials such as metals, phosphates, sulfonates, and others.

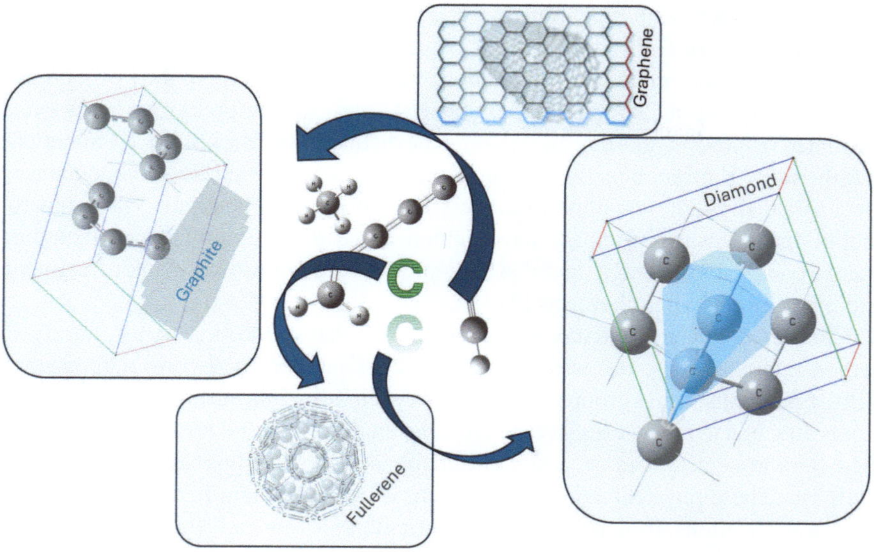

Fig. 2 Nanomaterials based of carbon in different forms

Carbon nanotubes: Can be differentiated into single-walled carbon nanotubes (SWCNTs) and multi-walled carbon nanotubes (MWCNTs), with the classification determined by the number of graphene layers [228]. Needle-shaped graphene layers possess a wide array of mechanical and physicochemical properties. Carbon nanotubes offer the advantage of facile penetration into cellular components, thereby enhancing their utility as carriers for drugs, agrochemicals, bioimaging tools, cancer immunotherapy and in phototherapy applications [26].

Fullerenes: Are carbon compounds characterized by a spherical structure and unique surface properties, including no polarity. Among these, C-60 and C-70 molecules are widely utilized, prepared through various techniques for both manufacturing and functionalization of the carbon cages. The hollow spaces within these cages can host different atoms and small molecules. These characteristics make fullerenes promising candidates for various tumor treatment modalities, including photodynamic therapy, radiography, chemotherapy utilizing imaging techniques, and other controlled release systems [228].

Graphene: Represents a two-dimensional carbon framework, often called 2D graphitic, characterized by a monolayer of carbon atoms. This structure, reminiscent of graphite, has carbon–carbon bonds arranged in a honeycomb pattern, providing it with excellent thermal and electrical conductivity and minimal thermal expansion. Graphene is extensively studied as a carrier for drugs and genes, owing to its exceptionally high total surface area, which is approximately four times greater than that of other nanomaterials investigated for nanomedicine applications [148, 228].

3D-CNT (Three-dimensional porous carbon nanonetworks): Materials characterized by their unique morphology, consisting of intricate three-dimensional network structures with numerous meso- and micropores [222]. The fabrication of 3-D porous structures by integrating 2-D graphene sheets have sparked significant interest due to their broad range of applications, primarily owing to their exceptional porous configuration. Integrating 2D materials into 3D configurations helps avoid unwanted restacking, thereby maintaining the distinct properties of the 2D materials. Additionally, it offers benefits like an expanded accessible surface area, enhanced mechanical strength, and appropriateness for substrate applications. These macroscopic devices with free space have demonstrated diverse applications including energy conversion, storage, and dissipation, as well as in sensors, water treatment, and tissue engineering platforms [214].

Carbon dot: Carbon nanoparticles (0D), the latest addition to the family of carbon nanomaterials, primarily consist of a sp^2 and/or sp^3 hybrid carbon core, with organic functional groups containing oxygen, nitrogen, sulfur, and phosphorus atoms attached to the outermost carbons. This unique structural arrangement renders them luminescent, non-toxic, easily dispersible in water, and typically sized between 2–10 nm. These physical–chemical characteristics make them highly favorable for diverse applications, from healthcare and agriculture to more advanced technological fields such as electronics [29, 169]. Due to its expanding, dynamic, and diversity of applications, it will be better detailed in the following subsections.

2.2 *Nanocarbons from Biowaste*

Green synthesis fundamentally relies on renewable materials, avoids promoting contaminants, and is cost-effective and energy efficient. While leaves, flowers, and roots are traditionally used as renewable sources of raw materials for nanocarbons, the potential extends far beyond these plant parts. However, nanocarbon can also be derived from other forms of plant biomass, such as shells, grains, peels, and stems [42]. Biowaste is an abundant and biodegradable material, though it presents as a complex and heterogeneous bioorganic matrix. It is sourced from diverse origins, including organic domestic waste, perennial grasses, agricultural residues, fisheries, poultry farming, livestock rearing, forestry, and related sectors [96].

Table 1 provides an overview of various studies exploring the synthesis of nanocarbons from biowaste materials. It highlights some potential applications envisioned for these nanocarbons in the respective research articles. It is pertinent to mention that while the table showcases specific applications, these nanocarbons often possess multifunctional properties, suggesting a broader range of potential uses.

The use of biowaste in the synthesis of nanocarbons offers cost advantages for raw materials compared to other carbon sources. Additionally, organic compounds, such as humic acid found in peat, soil, coal, natural water sediments (rivers, lakes, sea), and other decomposition products of plant and animal remains, can be utilized. Humic acid has been tested as a precursor for c-dot using hydrothermal synthesis at 180°C for 5 h. The nanocarbon produced through this one-step, green, and easily controlled process exhibited a quantum yield c.a 5% [185].

Table 1 Biowaste used for nanocarbons and their main properties and applications

Biowaste	Size/nm	Nanocarbon	Application	Main properties	References
Animals	2.6–20	C-dot, N,S-C-dot, 3D porous network carbon, N-CND	Cell imaging, sensor, Battery, Supercapacitor	Luminescence, Electrical conductivity	[30, 52, 80, 192]
Human	2–30	FN-CQD, CND, C-dot, N,S-C-dot, NCND	Bioimaging, display, sensor, photocatalytic activities	Luminescence	[52, 198]
Fish scale	2	N–C-dot	Hypoclorite sensing	Luminescence	[219]
Fruit	10	3D carbon nanosheets	Supercapacitor	Electrical conductivity	[230]
Forest	2–9	C-dot, Carbon nanosheet	Cel imaging, sensor, supercapacitor	Luminescence, Electrical conductivity	[118]
Flowers	9.4	Graphene nanosheet, C-dot	Supercapacitor, sensor	Electrical conductivity, luminescence	[218]
Peels	3–6	C-dot, 3D-porous carbon	Sensor, Supercapacitor, and CO_2 capture	Electrochemical, porous surface, luminescence	[222]
Leaves	2.6–6, 20–40	Porous carbon nanoparticle	Photocatalyst, Sensor Supercapacitor	Electrical conductivity	[86, 120, 213]
Husks	3–6, 606	CNP, Graphene quantum dots, 3D porous network carbon, Ag-Graphene oxide	Supercapacitor, sensor and cell imaging	Electrical conductivity, luminescence, Electrostatic interaction	[43, 222]
Sago Bark (bagasse)	65	Carbon nanosphere, Nanoporous activated carbon	Supercapacitor	Electrical conductivity	[92, 222]
Polymer	2, 22–43	C-dot, Carbon nanotube (MWCNT)	LED, Catalytic activity	Luminescence, Catalytic surface	[233] [215]
Grounds	8.9, 20	N,S-C-dot, carbon nanosheets	Supercapacitor, Sensor	Electrochemical, Luminescence	[115, 222]
Fruits	–	Activated nanocarbon	Adsorbent	Porous surface	[222]

(continued)

Table 1 (continued)

Biowaste	Size/nm	Nanocarbon	Application	Main properties	References
Sawdust	20–50	Tubular graphitic carbon	Catalyst	Electrochemical	
Stem		C-dot	Cell imaging, sensor	Luminescence	

Carbon-based functional materials are interesting because they are more eco-friendly when compared to other materials, such as metallic, phosphates, sulfonates, and others. The transformation of biomass into nanocarbons increases the effectiveness of waste material utilization—currently incinerated, randomly dumped in landfills, or applied in agriculture without considering environmental contamination—but also reduces the cost of preparing nanomaterials [231].

Among the advantages of using biomass and advancements in the heating properties of nanocarbon, C-dots have stood out for their unique characteristics, particularly their non-toxicity to living beings and minimal environmental impact. Improper waste disposal remains a major environmental threat, leading to soil and water contamination. Sewage sludge and agribusiness effluents pose a significant challenge. Current treatment methods are often expensive, and reuse remains limited. The potential for reusing this waste is substantial. Applying green chemistry principles, domestic sewage sludge was converted into C-dots using hydrothermal oven treatment and microwave radiation. Both techniques produced C-dots with a standard size of around 10 nm. These C-dots exhibit notable properties—when exposed to 360 nm light, they emit light at a peak wavelength of 430 nm [169]. This synthesis methodology was also evaluated using swine farming effluent, producing C-dots smaller than 6 nm with emissions in the range of 400 to 550 nm.

2.3 *Newest Nanocarbon: Carbon Dots*

C-dots are the latest group of nanomaterials, characterized by their dimensions of less than 10 nm, luminescence, non-toxicity, biocompatibility, and high affinity for polar solvents like water. Their synthesis methods not only include previously mentioned techniques but can also be adapted with new materials depending on the desired application [117].

The development of various luminescent C-dots has led to the creation of materials with distinct characteristics (Fig. 3), allowing them to be classified into carbon quantum dots (CQDs), carbon nanodots (CNDs), carbon nitride quantum dots (CNQDs), graphene quantum dots (GQDs), and carbonized polymer dots (CPDs). These differences are mainly related to their dimensionality, 2-dimension (GQDs/CNQDs) and 0-dimension (CNDs/CQDs), crystallinity, crystalline (GQDs/CNQDs) and non-crystalline (CQDs/CNDs), hybridization, sp^2 domain (GQDs/CNQDs),

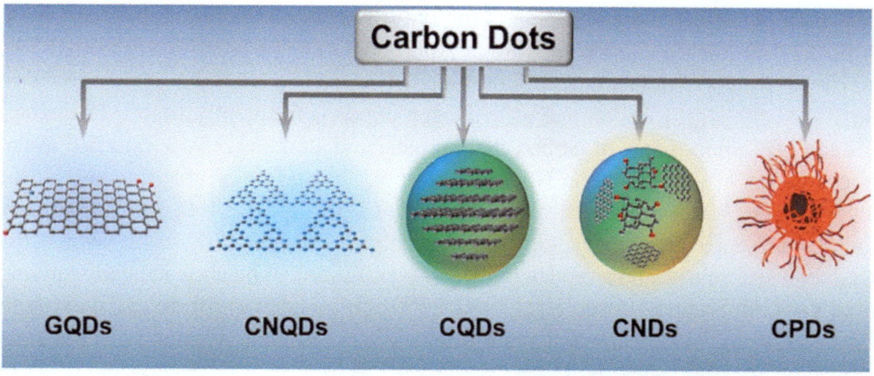

Fig. 3 Classification of Carbon dots. Reproduced with permission [117] Elsevier and Copyright Clearance Center

sp^2 and sp^3 mixture (CQDs/CNDs) and morphology with graphitic layers-GQDs/CNQDs and spheric or quasi-spheric (CNDs/CQDs) [117].

Compared to other types of nanocarbons, CNDs are distinguished by their dimensionality (0D), smaller than 100 nm in all dimensions. This distinction imparts significant physicochemical properties, such as the quantum confinement effect of electrons. This property, indicated by their tiny size, results in electron behavior like that in a single molecule [150].

Doping, passivation, and functionalization are commonly employed in the methods for altering and improving C-dots. Whether by conventional methods (resistance or flame) or microwave radiation, heating remains the most cost-effective, efficient, and controllable synthesis approach. However, the waste materials' complexity complicates the use of open-system heating due to the release of harmful gases and the difficulty in transforming organic matter into carbon nanomaterials. Therefore, inert closed systems capable of withstanding high pressures (up to 3 MPa–22 MPa) are necessary, as the temperatures often exceed 150 ºC, reaching up to 300 ºC. These synthesis processes rely on bond breaking, reorganization, and often the condensation of nanostructures [150, 222].

After synthesizing carbon nanoparticles, it is crucial to analyze their physicochemical properties, including crystallinity, shape, internal structure, and size. Microscopy techniques, primarily Transmission Electron Microscopy (TEM), are employed to gather information about nanoparticle morphology, size, and crystalline structure. The different sizes revealed by TEM were a minimum of 2 nm [130] and maximum 19 nm [58], and some hybrid C-dots can be measured until 50 nm [21]. The C-dots quasi-spherical particles of 1.4 nm, 2 nm, 3 nm, 4.2 nm, 4.6 nm, 4.8 nm, 5 nm, 6.7 nm, 9.57 nm, 11.6 nm, and 19 nm were produced by lychee waste [172], pomelo peel [130], betel leaves [18], coriander leaves [191], black tea [191], natural gas soot [77], kerosene soot [211], pomelo [172], soybean residual [1] spent black tea [1], silkworm chrysalis [58], respectively.

C-dots' external side is a region commonly functionalized by amines, carbonyl, carboxylic, amides, and other functional groups. This can be investigated using FTIR spectroscopy. As C-dots are mostly prepared in aqueous media, the stronger band in the 3060–3600 cm^{-1} region due to hydrogen bonds is frequently found. Therefore, NH and OH components of carboxylic, amines, and alcohols must be studied in the other regions of FTIR spectra. Normally, carboxylic acid or carboxylate anions are confirmed at C-dots by the presence of bands at 1700–1550, 1400, and 1100 cm^{-1}. Amides and ester groups at C-dot show the highest bands, 1700 cm^{-1}, which the bands can differentiate observed between 800–1200 cm^{-1}. The aliphatic and aromatic groups of C-dots can be observed at 3060–2900 cm^{-1}, and fingerprint region between 1500–1000 cm^{-1} [3]. The investigation with FTIR can be correlated with previous knowledge of FTIR spectra of precursor materials used in C-dot synthesis. In some cases, it should be correlated with XPS analysis, UV–Vis Spectra, or elemental analysis.

UV–Vis spectroscopy has proven valuable for assessing the electronic transitions of C–C bond types, indicating their hybridization states such as sp^2 and sp^3 carbon hybrids. Maximum wavelengths ($\lfloor_{max}$) evaluated to abortion at 200–300 nm corresponds to □ − □* electronic transition of aromatic domains in the inner C-dot [227, 237]. Additionally, UV–Vis spectra can reveal n −□* electronic transitions occur at 300–350 nm due to functionalization with groups like CN, CO, and CS on the surface, or doping in the inner part of C-dot with N, B, or P [50]. The band gap of nanocarbon can be estimated using the Tauc plot method and found values to around 2 eV [147]. The nanomaterials semiconductors with bandgap with a range of 1,5–2.5 eV are important to solar energy photocatalytic processes [201].

XPS analysis of C-dot is necessary to improve the components and composition details. The literature has been showing mainly deconvoluted spectra of carbon (C1s), nitrogen (N1s), oxygen (O1s), and sulfur (S2p). The binding energy C1s occurs at 284.8 eV (C–C), 284.5 eV (C = C), 286.0 eV (C–O or C–N), 288–293 eV(C = O), corresponding at 288.5 eV to O–C = O component. The oxygen can be investigated with binding energy around 530 eV, as the deconvoluted O1s spectrum shows peaks at 530.3 eV (C–OH), 531.5 (–C = O), 532.5 eV (C–O–C). Chemical states of N compounds are very investigated to C-dots as amine group ($C–NH_2$ – 397.9–399.3 eV), pyridinic nitrogen groups (400.2 eV), and nitrates (Higher 405 eV). In addition to the study of the presence of sulfur trough C1s, it is possible to investigate the chemical states of S (S2p) by binding energy between 160–170 eV (S–H/S–C = 162.3 eV; C–SO = 164.3 eV; $–SO_3$ = 167.5 eV) [50].

C-dot luminescence is typically studied using photoluminescence (PL) spectroscopy, which reveals that different C-dots exhibit unique light absorption and emission processes specific to each nanomaterial. Normally, C-dot emission bands between 300–500 nm correspond to the following electronic transitions of carbon with carbon (4.1 eV), graphitic nitrogen (3.5 eV), pyridine nitrogen (3.0 eV), amino nitrogen (2.8 eV), and carbonyl groups (2.5 eV) [6]. These transitions are associated with energy transfer due to the interaction between carbon nuclei and surface states. This is a significant advantage for applying these materials, as it allows for their modulation and better configuration, improving their usability.

Biomass offers various compositions and components, which result in different types of C-dots. The primary effect observed in C-dots is quantum confinement, characterized by a redshift in emission wavelength with increasing particle size, like in quantum dots. Some studies have demonstrated that the emission wavelength of C-dots varies with their particle size. In a controlled study, it was observed that C-dots measuring 1 nm emit in the UV region (350 nm), Emission in the visible range (400–700 nm) is characteristic of particles up to 3 nm, while those greater than 3 nm typically emit in the near-infrared spectrum (800 nm) [114]. The luminescence of C-dots does not depend solely on quantum confinement. It can be affected by doping and the types of functional groups involved, the degree of surface oxidation, intermolecular interactions, internal carbon defects, and external factors such as pH and solvent polarity.

C-dots' significance is often linked to their light-emitting efficiency, quantified by their quantum yield (QY), see Table 2.

The International Union of Pure and Applied Chemistry (IUPAC) describes quantum yield as the ratio between the number of emitted and absorbed photons. Consequently, the QY of a substance is the fraction of molecules that release a photon after direct excitation. However, measuring the absolute luminescence requires techniques that may not be readily available. Therefore, it is more common to determine relative quantum yield (QY) using molecules with known luminescence properties. The choice of standard luminescent molecules depends on the similarity of their absorption and excitation wavelengths to those of the C-dots being studied, as represented by Eq. 1 [166].

Table 2 Quantum Yield (QY) of C-dots from some biowaste with QY > 10%

Biowaste to C-dot	L_{exc}/nm	L_{emis}/nm	QY/%	References
Carboxymethylcellulose	350	465.5	44	[90]
Soybean Residual	360	440	28.6	[1]
Lychee Waste	365	443	23.5	[172]
Betel Leaves	360	428	12	[18]
Onion Waste	370	460	28	[25]
Bombyx Mori Silk	360	425	61.1	[125]
Borassus Flabellifer Flowers	320	403	14	[145]
Candle Soot	295	509	21.7	[152]
Kerosene Soot	360	460	10.2	[211]
Silkworm Chrysalis	350	420	46	[58]
Pomelo Peel	330	410	17.3	[233]
Spent Black Tea	390	465	17.5	[1]
Borassus Flabellifer	310	395	19.4	[146]
Corncob	350	450	53.5	[91]

$$QY^S = \frac{F^s f_{St_s}{}^2}{F^{st} f_s \eta_{St}{}^2} QY^{St} \tag{1}$$

where QY^s and QY^{St} represent the photoluminescence quantum yield of the sample and that of the standard, respectively; F^s and F^{St} area the integrated intensities of sample and standard PL spectra, respectively; f_x (f_s *and* f_{St}) are the absorption factor of sample and standard, the fraction of the light incident on the sample that is absorbed ($f_x = 1 - 10^{-Ax}$, where A = absorbance); the refractive indices of the sample and reference solution are I $_s$ and I $_{St}$, respectively.

2.4 Main Approaches to Synthesis of Carbon Dots

Several alternative proportional synthesis methods support the fabrication of nanoparticles (NPs) with diverse structures and sizes, utilizing biowaste as the primary raw material. While these methods yield high-quality NPs with intriguing characteristics for various applications, they often encounter limitations in synthesis yield. Importantly, these techniques are tailored to tackle environmental challenges, ensuring sustainable and eco-friendly NP production [161].

Synthesis methods are broadly classified into two main categories: top-down and bottom-up techniques. Each of these categories includes various submethods based on their operational conditions. In the top-down approach, materials such as graphite, activated carbon, carbon nanotubes, and nanodiamonds are broken down into smaller carbon nanomaterials, including C-dots. Techniques in this category include laser ablation, arc discharge, chemical oxidation, liquid-phase exfoliation, microfluidization, electron beam lithography, and electrochemical processes. Top-down methods often involve complex procedures and sophisticated equipment, resulting in low quantum yields and challenges in controlling particle size and morphology. In contrast, the bottom-up approach involves creating C-dots from smaller precursor molecules, such as carbohydrates, organic acids (e.g., citrates), and natural substances. This method typically involves carbonization and dehydration through thermal treatment, and is generally more straightforward. The bottom-up approach allows greater control over the composition, physical properties, size, and reaction conditions. Common techniques within this category include hydrothermal synthesis, solvothermal synthesis, direct pyrolysis, microwave-assisted synthesis, aldol condensation, plasma jet synthèsis, magnetic hyperthermia, and soft template methods [16, 229].

Solvothermal strategies have become more popular due to their ease of operation, versatility in functionalization, and ability to enhance the quantum yield of C-dots. Additionally, these methods accommodate a wide range of precursors, including biomass. A reorganization process of the carbon atoms by high temperature and high vapor pressure [117]. Tools like chemometrics and machine learning in this synthesis method can help optimize synthesis temperatures and times, whether using conventional oven heating or microwave radiation. Microwave radiation offers the advantage of shorter synthesis times [117, 135].

Some prior knowledge of the synthesis process enables control over modifications in the outermost part of the C-dots and the insertion of heteroatoms into the core, surface of the nanocarbon enhancing their photophysical and photochemical performance. This level of control broadens the potential applications of C-dots derived from biomass [16].

A straightforward one-pot hydrothermal method was utilized to convert dragon fruit peel into two valuable products: solid biochar and a C-dot solution. Even though millions of tons of dragon fruit are harvested yearly in tropical and subtropical areas, the peel is frequently discarded, resulting in a considerable waste of valuable biomass (). The biochar adsorbent showed remarkable ability to adsorb cationic dyes, while the C-dots displayed distinctive fluorescence quenching effects when exposed to Fe (III). These findings indicate that the C-dots might be useful for applications such as cellular imaging and detecting Fe^{3+} within cells. The conversion process was conducted via hydrothermal treatment at 180 °C for 5 h. Following this, the biochar was separated from the C-dots by centrifugation. After dialysis, a C-dot solution of 0.2 mg mL^{-1} was obtained [200].

2.5 *Carbon Dots for Agricultural Applications*

C-dots' applications span various fields due to their remarkable structural advantages and optical properties. From replacing quantum dots in electronic devices to advancements in medicine, agriculture, and food. Even though these applications have been widely studied, this chapter will not cover biomedical and food applications such as photodynamic and photothermal therapy, biosensing, bioimaging, or antiviral and antibacterial functions. Even the limitation in applying solid C-dots due to fluorescence quenching, caused by excessive energy transfer between solid carbon dots, has been overcome by incorporating functional clusters on the surfaces of C-dots [132].

In agriculture, C-dots are utilized in several ways, including as sensors, controlled release systems, fungicides and pesticides, biostimulants, and photosynthesis photosensitizers. As sensors, C-dots are primarily used to monitor pH levels, metals, and pesticides. Their advantages include ease of handling, high sensitivity, excellent selectivity in some cases, non-toxicity, and environmental safety. Real-time monitoring of crops, crop development, insect pest activity, field conditions, disease activity, and environmental stressors is only possible using nanosensors. Their application in agriculture can also help mitigate environmental pollution and decrease product expenses by ensuring the precise use of necessary inputs [90].

Based on their optical properties, C-dots are primarily used in applications involving quenching their fluorescence. Innovatively C-dots have been effectively used as "signal off" fluorescent probes for the selective and sensitive detection of metal ions. These probes offer linear detection ranges from 0 to 3000 $nmolL^{-1}$, depending on the specific ion being evaluated, with detection limits ranging from 1.4 to 2350 $nmolL^{-1}$ [91, 227]. C-dot obtained by dried purple perilla was used as sensor of Ag^+ in aqueous solution containing 25 ions (Co^{2+}, Ag^+, Hg^{2+}, Ni^{2+}, Mg^{2+}, Ca^{2+}, Sr^{2+}, Li^+, Pb^{2+}, Zn^{2+}, Cd^{2+}, Cu^{2+}, Cr^{3+}, Fe^{3+}, Fe^{2+}, Mn^{2+}, Al^{3+}, Na^+, K^+, $NO_3{}^-$, Cl^-, $CO_3{}^{2-}$, S^{2-}, $SO_4{}^{2-}$, $SO_3{}^{2-}$), showing selectivity and sensibility with a detection limit 1.4 $nmolL^{-1}$ [240].

Exploring the chemical interaction between C-dot functional groups and metal ions reveals additional applications for soil treatment. This interaction is believed to be key to unlocking C-dots' possibilities to increase the solubility of poorly soluble phosphates, a major challenge in agriculture. A 30-day experiment demonstrated the effectiveness of C-dot treatment. Soil treated with C-dots exhibited a notable enhancement in the availability of calcium phosphate (106.5%) and copper phosphate (189.1%). These remarkable results suggest that C-dots can significantly improve soil fertility by enhancing phosphate accessibility for plants [70].

Zinc combined with C-dots have been tested to verify their viability in the controlled release of nanofertilizers. The Zn system with C-dots demonstrated a slow release of nanofertilizers, resulting in a 79% increase in zinc uptake in the wheat plants studied, compared to commercial fertilizer use [11]. C-dots are emerging as a promising tool to enhance fertilizer and pesticide efficiency. This growing field of research is exploring how these nanomaterials can optimize nutrient delivery and targeted pest control.

Some studies show C-dots as promising method for detecting pesticides. They act as tiny fluorescent sensors. In the presence of certain pesticides, like atrazine, chlorpyrifos, lindane, and tetradiphon, the fluorescence of the C-dots is quenched, indicating the presence of the contaminant. Interestingly, some pesticides, like imidacloprid, have the opposite effect, causing the fluorescence to increase. This "turn-on" mechanism provides another way to identify specific pesticides [93]. Pesticides are primarily detected through turn-off methods, demonstrating that C-dots can sense pesticides via changes in their PL emission spectra [93]. The determination of glyphosate, herbicide with an important role in agriculture, was evaluated using C-dots, alone and in combination with gold or silver nanoparticles. The results highlighted the substantial potential of these devices for application in precision agriculture [3].

Prolonged contact with nanoparticles can substantially impact plant biology, affecting growth, development, and stress responses. The direct use of nanocarbons in plants depends on their effects and translocation capacity, influenced by factors such as concentration, functional groups, exposure time, size, and shape. Several hypotheses in the literature describe the translocation pathways of nanocarbons of varying sizes, explaining how they move through the vascular bundle from root to shoot. The uptake of C-dots from roots to leaves through the vascular channels was confirmed using SEM and confocal microscopy. Nanocarbons applied to soil can penetrate the plant system through roots via symplastic pathways and subsequently move to the plant's aerial parts. When nanocarbons are suspended or dispersed in water or a growth medium, they are readily absorbed by the plant and essential nutrients through xylem vessels. Effective surface functionalization reduces their hardness and toxic effects. Treated plants showed improved productivity, evidenced by an increased fruit yield per plant [34].

The application of biowaste-derived C-dots in agriculture sector is already a reality. Therefore, developing these technologies to support agricultural regions is crucial, fostering a circular economy in these areas.

3 Nanocellulose and Nanolignin

3.1 Natural Fibers Waste for Producing Nanostructures

Lignocellulosic biomass is regarded as the most abundant biopolymer on the planet, with global annual production estimated at approximately 1.5×10^{12} tons [64]. This biomass encompasses a range of sources, including agricultural residues like palm residues, empty fruit bunches, straw, bagasse, corncobs, Nile rose, and stover [84, 138, 139, 184, 208], (2) forest residues, such as branches, unwanted stems, and fallen leaves [208], and (3) industrial residues, including discarded paper and decommissioned wood [17]. The extensive availability of lignocellulosic residues and their carbon sequestration potential have led to their use in over 200 fields. These include construction materials, medium-strength composites, adhesives, packaging, coatings, dental fillings, implants, biofuels, bioplastics, biocomposites, and drug delivery systems [143].

Natural fibers have been extensively utilized for various applications due to their unique hierarchical structure, which endows them with exceptional mechanical, thermal, and biocompatible properties. Understanding the hierarchical organization of these fibers is crucial for optimizing their performance in different applications. The cell wall of natural fibers comprises a blend of cellulose, lignin, hemicelluloses, and extractives. It contains layers arranged by cellulose microfibrils, alternating between crystalline and amorphous regions [48]. These microfibrils interweave with amorphous polymers like hemicelluloses and lignin, resulting in a biocomposite with notable mechanical strength [57]. The composition of cellulose, hemicellulose, and lignin in lignocellulosic biomass differs based on the source of the waste. Generally, agricultural waste contains about 5–15% lignin, 25–50% hemicellulose, and 35–50% cellulose. Figure 4 provides a schematic representation of lignocellulosic biomass, emphasizing the roles of cellulose, lignin, and hemicellulose. Cellulose is the primary component of plant cell walls and constitutes the major portion of natural fibers. This linear polysaccharide is formed from β-D-glucose units linked through β-1,4-glycosidic bonds. In the cell wall, cellulose molecules form microfibrils aligned parallel to each other, providing structural integrity and mechanical strength to the fiber [59]. The β bonds enable cellulose molecules to form straight chains, capable of organizing into strong fibrous and crystalline structures [217].

Factors including the plant species, tissue type, and developmental phase influence the arrangement of these elements within the cell wall. Typically, cellulose

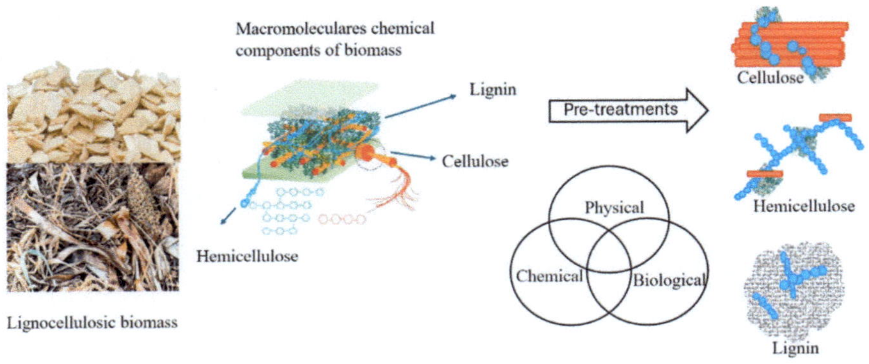

Fig. 4 Schematic structure of lignocellulosic biomass

microfibrils serve as the framework onto which other elements are layered, while hemicellulose and lignin occupy the gaps between these microfibrils. This layered structure enhances the mechanical resilience, flexibility, and other characteristics of natural fibers. A profound grasp of the composition and organization of cell wall constituents is imperative for customizing fiber attributes and crafting eco-friendly materials suited to diverse applications.

Fibers from different plants exhibit varying shapes and dimensions based on their sources. Typically, these fibers range from 1 to 3 mm in length and have a 20–50 μm diameter. Upon reduction to nanoscale dimensions, their diameters decrease to less than 100 nm, representing a reduction in average diameter by approximately 1000 times compared to the original size. In this altered state, they possess the potential for enhanced properties, although not all aspects of these improvements are thoroughly understood [8]. Elemental fibril forms the fundamental structure found in various natural sources of cellulose, including wood, cotton, ramie, jute, and bacterial cellulose [44].

Another polysaccharide found in the cell wall is hemicellulose, along with cellulose. It is a composite polymer of various sugars such as mannose, xylose, galactose, and arabinose. Hemicellulose molecules are branched and interact with cellulose microfibrils, enhancing the overall architecture and flexibility of the cell wall. Additionally, hemicellulose acts as a cementing material between cellulose fibers [217].

Lignin is a term used to denote phenolic polymers that, in combination with cellulose and hemicelluloses, make up the cellular matrix of plants [68]. In plant biomass, it makes up about 15 to 40% (mm^{-1}) of the overall dry mass [163]. Lignin is the most abundant phenolic polymer worldwide, with its core unit being phenyl propane (C6-C3 or C9) (Fig. 5). The carbon atoms of the propane chain are designated using Greek letters, with the letter α (alpha) designated for the carbon linked to the phenyl group and the others by the letters β (beta) and γ (gamma) sequentially, whereas in the aromatic ring, the carbon atoms are numbered starting from the one linked to the propane chain [217].

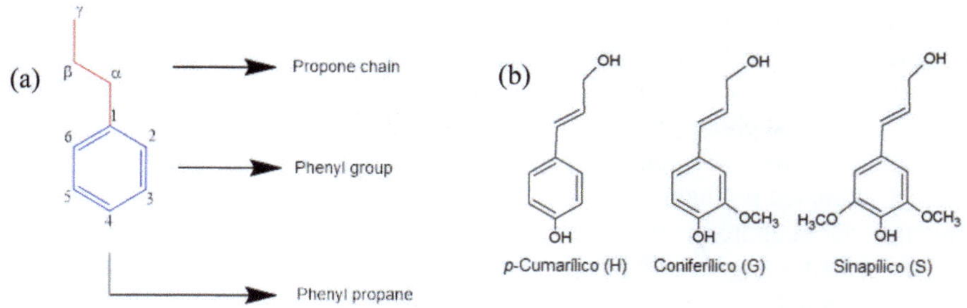

Fig. 5 Structural representation of Phenylpropane, **a** basic structure of lignin and **b** Lignin precursor alcohols

Lignin is derived from three types of hydroxycinnamyl alcohol subunits (monolignols): p-coumaryl alcohol (H), coniferyl alcohol (G), and sinapyl alcohol (S), each differing structurally by the extension of the methyl radical at positions 3′ or 3′–5′ on the phenolic rings (Fig. 5) [47]. Different covalent bonding patterns between these precursor molecules contribute to the intricate structure of lignin.

In lignin, functional groups are responsible for the high reactivity, which is higher than that of carbohydrates [221] and knowledge about them is necessary to have solid and reliable information on the chemical structure of lignin. Several approaches can be employed to separate or reduce lignin in biomass, the chemical reagents used in them can chemically modify this compound, to the point that, after separation, the chemical structure is different from that of lignin in a natural environment. Traditionally considered a product of low commercial value, lignin has become a precursor for high-value products and is increasingly gaining economic interest. The cellulose and paper sector are the main global source of lignin, with annual production of around 50 to 70 million tons [157]. By 2030, this quantity is expected to rise by 225 million tons, with lignin from cellulosic pulp production adding to the residual lignin from biofuel production and other plant biomass processing activities [23, 157].

Despite all this interest and potential, fewer than 2% of the lignin produced in the cellulose and paper sector is directed to produce new products [23]. In addition to waste from cellulose and paper production, agricultural waste is a source of lignocellulosic residue and its use to obtain lignin is a low-cost option [140]. The most common major agricultural residues include straw and crop residues from cereals such as rice, wheat, barley, and corn, and also rice husks and residues from grain crops such as soybeans and beans [178, 187]. Other residues frequently found are sugar cane bagasse, corn residues, including cobs and leaves, and oil crop residues, such as sunflower seeds and cotton bagasse [36, 106].

Obtaining lignin products is only possible through fractionation of lignocellulosic biomass, in which lignin is separated from carbohydrates and other plant wall compounds [225]. The fractionation of lignocellulosic biomass is not simple and often involves combining biomass pre-treatments with more selective treatments [35, 239]. Pre-treatments can be mechanical, chemical (acid pre-treatments, alkaline pre-treatments), thermal and physical chemical (organosolv pre-treatment, steam explosion, ammonia fiber explosion, CO_2 explosion, liquid hot water, ionic liquid approach) [140]. In addition to these, there are pre-treatments that combine advanced techniques to improve their efficiency, including ultrasonography, microwaves, nanoparticles, and photolysis-activated carbon [140].

Kraft lignin, from the pulp and paper sector, is the most produced type of lignin, with an estimated production of 55 million tons per year, followed by soda lignin, which produces approximately six million tons per year, and to a lesser extent has lignosulphonates, organosolv lignin are used [24]. Kraft lignin and lignosulphonates may have limited use due to the presence of sulfur in their molecular structure, while soda lignin and organosolv lignin are sulfur-free and have a similar molecular structure to the original lignin. In addition to these types, Milled Wood Lignin (MWL), Cellulolytic Enzyme Lignin (CEL), Enzymatic Mild Acidolysis Lignin (EMAL), Pyrolysis Lignin (PL), and Steam Explosion Lignin (SEL) are produced on a smaller scale [158]. Table 3 shows the composition of some natural fibers.

Table 3 Composition of some lignocellulosic fibers [2]

Composition of some lignocellulosic fibers (%)				
Biofiber	Source	Cellulose	Hemicellulose	Lignin
Wood	Hardwood	43–47	25–35	16–24
	Softwood	40–44	25–29	25–31
Non-wood	Bagasse	40	30	20
	Coir	32–43	10–20	43–49
	Com cobs	45	35	15
	Com stalks	35	25	35
	Cotton	95	2	1
	Oil Palm fiber	50	30	17
	Flax (retted)	71	221	2
	Flax (unretted)	663	12	3
	Hemp	70	22	6
	Henequen	78	4–8	13
	Istle	73	4–8	17
	Jute	71	14	13
	Kenaf	36	21	18
	Ramie	76	17	1
	Sisal	73	14	11
	Sunn	80	10	6

3.2 *Nanocellulose*

Nanocellulose, a nanoscale cellulose material derived from renewable sources such as wood pulp, agricultural or industrial residues, or bacterial cellulose, has emerged as a promising nanomaterial with diverse applications in industries ranging from engineering to sustainable materials; see Table 4.

Table 4 Nanocellulose forms from different biowastes

Biowaste	Nanocellulose	References
Grape pomace	CNC	[45]
Pineapple residues	CNC and CNF	[155]
Rice husks and straw	CNC	[183]
Coffee husks	CNF	[183]
Turmeric waste	CNF	[85]
Sugar cane	CNC and CNF	[38]
Cornhusk	NCF	[224]
Hemp stalk	CNF and CNC	[131, 149]
Potato peels	CNP	[164]
Pea pod residue	MFC and CNC	[99]
Tomato waste	MFC and CNC	[100]
Banana stem and peel	CNC and CNF	[206]
Eggplant	MFC and CNC	[22]
Oil palm	CNF	[232]
Coconut shells	CNF and CNC	[15]
Sunflower stalks	CNF	[63]
Bamboo	CNC	[39]
Cotton	CNC	[41]
Spruce bark	CNC	[110]
Sisal	CMF	[177]
Forestry residues	CNC	[142]
Garlic straw	CNC	[95]
Mango	CNC and CNF	[20, 79]
Cornbob residue	CNC and CNF	[124]
Weat staw	CNC e CNF	[28, 176]
Kenaf	CNC	[206]
Soy hulls	CNF	[61]
Arecanut husk	CNF	[94]

CNF—cellulose nanofibrillated, CNC—Cellulose nanocrystal, CMF—Cellulose microfibrilated, CNP—Cellulose nanoparticle

Various terms have been used to describe cellulose nanomaterials, sometimes leading to confusion or misinterpretation. These terms include microfibrillated cellulose, nanofibers, nanofibrils, nanocrystals, nanofibrillated cellulose, and microfibrils [56]. Nanocellulose encompasses this diverse group of cellulosic nanomaterials, which, despite sharing common characteristics, exhibit variations in physical and chemical properties influenced by the source material and preparation method. Consequently, nanocellulose (NC) can be categorized based on attributes such as size, shape, and manufacturing technique, with a wide array of applications [56]. There are currently two standards that classify nanocellulose according to its characteristics, the "Standard Terms and Their Definitions for Cellulose Nanomaterials "made by Technical Association of Pulp and Paper Industry (Technical Association of Pulp and Paper Industry 2013) and ISO, in "Nanotechnologies—Standard terms and their definition for nanomaterial cellulose" (International Organization for Standardization 2011). In this paper we use cellulose nanocrystals (CNCs) for chemical methods of production and or cellulose nanofibrils (CNFs) for mechanical methods.

Nanocellulose can be produced through several methods, including mechanical disintegration, chemical treatments, and enzymatic hydrolysis. Mechanical methods open and break down the cellulose fibers into nanoscale dimensions [78]. CNFs are commonly produced through mechanical techniques such as high-pressure homogenizing [78], grinding [112], high-pressure microfluidization [167], high intensity ultrasonication [113], steam explosion [167], and recently extruder as used [168].

A standard production method includes acid hydrolysis, followed by centrifugation, sonication, washing, and dialysis to create a suspension, which is then dried [75]. The primary approach for producing CNCs involves rigorous acid hydrolysis under carefully controlled temperature, concentration, agitation, and exposure time. These CNCs exhibit high crystallinity, occasionally surpassing 90% [129]. The physical properties of CNCs are influenced by the cellulose source, concentration and type of acid used, reaction time, and temperature [75]. For instance, when using sulfuric acid (50–70% w/v) at temperatures ranging from 20 to 70 °C, with agitation at 500 rpm throughout 0.5 to 6 h, esterification can occur, causing the formation of "cellulose sulfate." This modification results within a negatively charged layer on the cellulose crystallites. Longer hydrolysis times and elevated temperatures yield shorter nanocrystals with increased high crystallinity (approximately 80%), surface charge, and reduced polydispersity [88].

Enzymatic hydrolysis involves using enzymes to break down or alter cellulose fibers [197]. Typically, this process operates under mild conditions facilitated by enzymes, although it necessitates extended exposure times [103]. Combining enzymatic hydrolysis with other methods could potentially resolve this issue.

3.3 Nanolignin

Lignin serves as a substantial renewable carbon resource, with the biorefinery aiming to leverage this phenolic polymer for more than just energy applications

Table 5 Nanolignin preparation techniques and the average nanolignin particle diameter

Preparation technique	Type of lignin	DMNP/nm	References
Acid precipitation	Kraft and soda	48–375	[49]
Sonication	Soda	50–500	[32]
Homogenization	Kraft and soda	20–100	[32]
Chemical modification	Kraft, soda, sulfonated, and organosolv	15–800	[241]
Enzyme modification	Kraft	10–100	[223]
Dissolution and reprecipitation	Kraft and organosolv	10–100	[128]
Wet milling	Kraft	20–100	[194]
Drying	Kraft, soda, and organosolv	~30	[4]
Self-assembly	Soda and sulfonated	~100	[32]
Electrospinning	Kraft, soda, and organosolv	60–3000	[203]
Freeze drying and thermal stabilization	Soda	~100	[83]
Carbonization	Wheat straw soda lignin	25–150	[73]

[69]. Potential products derived from lignin encompass a wide range, including adhesives, composites, carbon fibers, plastics, thermoplastic elastomers, polymers, polyurethane foams, polymeric membranes, absorbent materials, emulsifiers, antioxidant compounds, dispersants, flocculants, automotive brakes, battery plates, pharmaceutical products, and epoxy resins, as well as various fuels and chemicals that are currently sourced from petroleum [97]. Furthermore, nanotechnology expands the use of lignin when it is reduced to the nanometric scale [238]. Table 5 shows some nanolignin preparation techniques and the average nanolignin particle diameter.

3.4 *Use of Nanocellulose and Nanolignin in Agricultural Applications*

In recent years, nanocellulose has demonstrated excellent performance across various applications, such as water [98] and wastewater treatment [162], papermaking [151] energy production [51], and biomedical engineering [126]. Various polysaccharides, with their renewable origins and special properties, are emerging at the bioadsorbents' forefront. The application of nanotechnology can further improve their adsorption performance [134]. Specifically, nanocellulose is a new and highly developed biobased adsorbent with numerous additional properties. For agricultural applications, nanocellulose can be found as enhanced-efficiency fertilizers (EEFs) through nanoencapsulation, slow-release hydrogels, stimuli-sensitive hydrogels, controlled release formulations (CRFs) of pesticides, water conditioning of soils, bioremediation of water, reduction of post-harvest food waste and production of composites that can be used in greenhouses that use plastics [153].

The abundant availability, high aspect ratio, hydrophilic characteristics, and adaptable properties of nanocellulose have made it a prime candidate for surface modification aimed at targeted adsorption and/or flocculation of water pollutants [153]. Hydrogels are created and then dried using techniques such as freeze-drying, supercritical drying, and evaporation to form three-dimensional structures—referred to as aerogels, xerogels, and cryogels—with exceptionally low density ($<100\ mg{\cdot}cm^{-3}$), high porosity (80–99.8%), and significant surface area (up to $1000\ m2{\cdot}g^{-1}$) [195]. Nanocomposites also have a wide range of applications, including the catalytic modification of organic pollutants, filtration of drinking water, detection of waterborne pathogens and organic contaminants, adsorption for separating oil from water, and high-efficiency energy conservation devices. Among these adsorbents, surface functionality plays a crucial role in selectively adsorbing specific types of contaminants [179]. The use of nanocellulose in water decontamination filters also showed efficiency, either due to its affinity with the nanocellulose surface or the size of the pore formed [153]

In the controlled release of fertilizers, nanocellulose has hydrophilicity and mechanical properties that make it promising. The encapsulation of NPK (nitrogen, phosphorus, and potassium) has already been tested using cellulose nanofibrils (CNFs) obtained from sugarcane bagasse [141] and cellulose nanocrystals (CNCs) produced from natural hemp stalks as a reinforcement for polyvinyl alcohol (PVA) shells [101] extending the release time of the fertilizer. Liu et al. [127] developed hydrogels using cellulose nanofibrils (CNFs) from bamboo plants, combined with sodium alginate and PVA, to create a structure that encapsulated NPK fertilizer, resulting in a reduced release of up to 15% on the first day and no more than 75% after 30 days.

To minimize the influence of environmental conditions such as UV radiation, temperature fluctuations, and pH levels, [216] developed a pH-sensitive urea-controlled release system using cellulose nanofibrils (CNFs) derived from sodium alginate, eucalyptus pulp and MIL-100 (Fe) as a metal–organic framework. Under basic pH conditions, this system substantially reduced urea release and enhanced wheat plants' germination rate and chlorophyll content by a factor of two compared to the control. In another work, [122] incorporated N-vinyl caprolactam (NVCL) into the previous system as a temperature-sensitive monomer. They showcased how this system effectively enhances urea release at elevated temperatures, leading to notable improvements in wheat plant germination rates, photosynthetic activity, and chlorophyll content.

Nanocellulose can act as a controlled pesticide release agent where the formulation of sugarcane bagasse CNF with novel martial-based diacylhydrazines reduced the proliferation of the fungus *Fusarium oxysporum* [116]. For insects, temephos solutions with Hibiscus cannabinus CNF increased the time of action of the insecticide and the influence on the death rate of *Aedes aegypti* larvae [154].

Many crops are sensitive to water stress, and nanocellulose can zhanretain water [31, 234]. Some tests have already been carried out using cucumber [31] sesame and radish seeds [27].

Nanolignin shows considerable potential to satisfy the demand for agrochemicals, as it has chemical properties suitable for some uses in agriculture, is a natural product, and is an abundant co-product pulp and paper industry and the sugar and alcohol sectors. Within the array of potential uses, the slow and controlled release of chemical compounds is the most studied for application in agriculture. The use of lignin in plant nutrition involves increasing accessibility and promoting the slow and controlled release of nutrients [5]. When lignin is reduced on a micro/nano scale for this purpose, it brings advantages such as the use of a lower dosage of fertilizer, labor savings, gains in workplace safety and lower environmental impact [175]. Furthermore, the production of slow and controlled release fertilizers is very expensive when compared to soluble fertilizers [46], and the use of a natural and renewable product can make this product cheaper [175].

The suitability of nanolignin for the use of slow and controlled release of chemical compounds comes from the characteristic of being able to form hollow nanospheres due to the self-structuring behavior of this molecule [175] and because it is the fraction of lignocellulosic biomass that is most resistant to biodegradation, remaining in the soil for longer.

Urea is one of the main fertilizers used to increase the availability of N in the soil, this fertilizer has a very rapid release of between 3 and 5 days, with the combination of urea and nanolignin this interval increases to 15 days [109]. Added to these results, the release of nitrogen from urea through a lignin-clay nanocomposite was half controlled and efficient than other commercial slow-release fertilizers (SRFs) [236].

Additionally slow-release potential, nanolignin can help with soil fertility because, depending on its extraction method, it can increase soil pH and improve the soil's cation exchange capacity [174]. And, even without encapsulation, lignin can elevate the concentration of macro and micronutrients in the soil, as its structure may contain phosphorus (P), potassium (K), and sulfur (S) [165], requiring care in dosages to avoid soil salinization.

Nanolignin has great potential to encapsulate chemical substances that are biocides, in the case of agriculture herbicides and insecticides [32]. Cis-jasmone [210], isoproturon, imidacloprid and cyromazine [66], cupric pesticides [67], and ivermectin [119] are examples of biocides used in agriculture with chemical compatibility with lignin and nanolignin.

Intrinsically, lignin has antioxidant activity and antimicrobial [220]. These characteristics can be useful in future plant pathology studies and in the development of nano lignin-based pesticides.

4 Inorganic Nanoparticles

The synthesis of inorganic nanoparticles (INPs) from wastes could be categorized into two routes. The first is the nonbiogenic route, nanoparticles (NPs) are directly derived from wastes. An example of this is the production of silica nanoparticles [193]. And second is the Biogenic route, biowastes can mediate the process, acting as a reactant that transforms added chemicals into nanoparticles. An example includes chemical reduction of silver ions by phytochemicals present in plant extracts [226] or microbial activity [9]. In the biogenic route, nanoparticle formation typically follows bottom-up methods. For INPs prepared through a nonbiogenic route derived from biowastes, both bottom-up and top-down methods are present. Various preparation methods can synthesize the INPs, including thermal decomposition (combustion and pyrolysis), microwave heating, sonochemical, ball milling, sol–gel, hydrothermal, mechanochemical, precipitation, solid-state reaction, laser ablation, and supercritical techniques. In this section, we discuss biogenic INPs directly derived from biowastes focusing on their use in plant growth and protection, with emphasis on calcium- and silicon-based nanomaterials.

4.1 Calcium-Based Nanoparticles

Calcium-based nanoparticles (CaNPs), such as calcium oxide (CaO), calcium carbonate ($CaCO_3$), and hydroxyapatite ($HA - Ca_{10}(PO_4)_6(OH)_2$), are derived from biowastes including eggshell, seashell, oyster shell, shrimp shell, fish bone, and cuttlebone. Of these, eggshells have gained notable attention owing to their widespread availability, low cost, and composition including 1% $MgCO_3$, 94% $CaCO_3$, 1% $Ca_3(PO_4)_2$, and 4% of organic compounds [121]. Table 6 presents a compilation of literature from 2019 onward, detailing various sources and preparation procedures for CaNPs. The nanoparticle's size heavily depends on the synthesis conditions, ranging from the nano to micro scales. Hence, synthesizing CaNPs derived from biowaste represents a promising strategy with diverse applications, with special attention in biomedical, catalysis, and agriculture.

4.2 Silicon-Based Nanoparticles

Silicon-based nanoparticles (SiNPs) have applications in several important economic sectors. In industrial applications as rheological modifiers, thermal insulators, encapsulation, and release of drugs and chemical products. Additionally, SiNPs have received notable attention because of their potential uses in diverse fields including solar cells [181], biomedical applications [209], water treatment [65], construction [7], health and beauty care[55], catalysis [136], sensing [159], agriculture [133].

Table 6 CaNPs prepared by different methods and biowastes

CaNPs	Biowaste	Preparation method	Size/nm	References
CaO	Eggshell	Sol–gel	24	[144]
$CaCO_3$	Eggshell	Ultrasound	10–280	[72]
CaO	Eggshell	Sol–gel	19–45	[89]
CaO	Oyster shell *Crassostrea spp*	Thermal decomposition	68 ± 11	[137]
HA	Eggshell Leaf extract *Muntingia calabura*	Thermal decomposition	30 (crystallite size)	[212]
CaO	Eggshell	Sol–gel	–	[74]
$CaCO_3$	Eggshell	Ball milling	<300	[173]
$Ca(OH)_2$	Eggshell	Ultrasound hydrolysis	50–200	[108]
HA	Eggshell	Hydrothermal	84–142	[204]
CaO	Oyster shell	Sol–gel	55	[53]
CaO	Oyster shell *Crassostrea gasar*	Sol–gel	50	[53]
HA	Buffalo bone	Hydrothermal	57–423	[12]
HA	Eggshell	Chemical precipitation	100	[10]
Apatite	Cooked bone	Supercritical	<100	[234]
HA	Eggshell	Microwave heating (MW)	10–80	[71]
$CaCO_3$	Eggshell	Wet ball milling	50–100	[82]
HA	Animal wastes		15–50	[205]
CaO	Eggshell	Sol–gel	100–300	[76]
HA	Cuttlefish bone Eggshell	Mechanochemical	16–170	[60]

SiNPs can be obtained from different wastes, mainly agricultural ones like rice husk [54], but also sugarcane bagasse [180], corn cob [24], straws [189] and leaves [37]. The worldwide estimate of ride husk production is about 120 million tons annually [33]. The concentration of silica in rice husk is about 12%. Nowadays, rice husk is predominantly employed as a boiler fuel and the rice husk ash (RHA) will have about 85–90% of silica [207]. The classic method of obtaining silica from agricultural waste involves an acid pre-treating to remove metal oxide impurities and calcination at 600 to 800 °C. SiNPs can be synthesized by adjusting the method's aging time, temperature, and chemical procedures. Liou [123] produced SiNPs through the thermal degradation of rice husks under non-isothermal conditions in an oxidizing atmosphere. The procedure included pre-treatment with HCl, followed by combustion reactions at temperatures ranging from 300 to 1000 K, resulting in nanoparticles with an average size of 60 nm. Le et al. [111] prepared SiNPs via a sol–gel method using a surfactant template and a sodium silicate intermediate derived from rice husk ash. The resulting SiNPs were non-crystalline and uniformly sized, with an average particle diameter of 3 nm. SiNPs syntheses from sugarcane ash were described by Rovani et al. [170]. The method includes preparing a sodium silicate solution, from which silicon nanoparticles (SiNPs) are formed through hydrolysis and condensation reactions using sulfuric acid in a biphasic system in the presence of cetyltrimethylammonium bromide (CTAB). The resulting silica particles were highly pure and approximately 20 nm in size. Bindar et al. [37] produced SiNPs from bamboo leaves by first washing them with HCl, followed by combustion at 700 °C. Afterward, sodium silicate was obtained by extracting silica with NaOH solution, then precipitated with HCl during the sol–gel process, and subsequently dried. The resulting SiNPs were amorphous, highly pure, and distributed in sizes ranging from nano to micrometers. Corn cob is an interesting alternative source to produce silicon-based nanoparticles. Shim et al. [186] utilized corn cob ash calcinated at 700 °C to obtain SiNPs. The process involved dissolving the ash in NaOH, followed by gel precipitation with HCl and aging to produce xerogel. This method yielded amorphous nanoparticles measuring approximately 60 nm.

4.3 CaNPs and SiNPs Applications in Agriculture

A variety of INPs has potential applications in agriculture as nanopesticides, nanofertilizers, control release agents, and sensors [199, 242]. Among these, those directly derived from biowaste are advantageous because they are considered environmentally friendly, such as CaNPs and SiNPs.

The use of conventional Ca and nano-Ca was evaluated for increasing the carotenoid content in nectarines, thereby improving their quality and economic value [243]. While both forms of calcium enhanced the carotenoid content, the effect of nano-Ca was significantly greater. Nectarine trees were sprayed with solutions of Ca conventional and Nano-Ca (5L per tree), respectively. In addition to increasing carotenoids in the fruit, nano-Ca also enhanced calcium content and calmodulin activity in the peels, flesh, and leaves. Higher concentrations of calcium can contribute to improved boron absorption in plants. Interestingly, boron uptake isn't solely dependent on the amount of boron available in the soil. This suggests that calcium plays a role in facilitating the plant's ability to take up boron, even when soil levels are less than optimal. Sunflower plants grown in alkaline and acidic soils were examined with boron and calcium oxide nanoparticles measuring 84 nm [19]. In addition to enhancing plant nutrition, nanocalcium was also effective in pest control. The comparison of effectiveness in spraying nanocalcium carbonate and colloidal calcium carbonate on Tankan citrus (*Citrus tankan*) leaves. Both approaches successfully controlled California red scale (*Aonidiella aurantii*) and oriental fruit flies (*Bactrocera dorsalis*). In comparison, nanocalcium carbonate demonstrated superior performance over colloidal calcium in enhancing fertilization and in preventing oviposition punctures from oriental fruit flies on Indian jujube. Improved pest control may be associated with the enhanced absorption of calcium nanoparticles, which demonstrated an increase of up to 13 times in field trials [81]. Cucumber plants exhibited significant development when treated with nanocalcium and urea. Compared to the control, the nanofertilizer notably enhanced plant growth, total leaf count, and average leaf size. These nanoparticles enhanced nitrogen availability and stimulated the production of auxins, which increased activity in the cambium, promoting cell division and elongation. This readiness of essential materials, such as amino acids and certain enzymes, aids in the plant's growth and development processes [13]. Other nanocalcium used to agricultural are nanohydroxyapatite. The nano-$Ca_{10}(PO_4)_6(OH)_2$ was prepared by calcined eggshell powder and phosphoric acid. After 10 h of grinding, the resulting nanomaterial consisted of particles around 100 nm in size with irregular shapes. The material served as an alternative phosphorus nutrient for rice cultivation. The lower solubility of nanohydroxyapatite results in an extended retention time of phosphorus in water, providing a more sustained release. This ensures that phosphorus remains readily available for rice plants to absorb throughout their growth cycle [156].

Even as nanocalcium, the nanosilicas are notable for their exceptional properties, including biocompatibility, stability, adjustable pore size, high surface area, and surface reactivity. New technologies in nanosilica have led to an increased use of this micronutrient in plants. Nanosilica is important for various plant development and growth functions, and it also enhances defense against diseases, environmental stress, and soil stabilization [183]. The improved absorption of nanosilica by plants underscores a key advantage of nanofertilizers over traditional fertilizers, plant roots absorb that in silicic acid ($Si(OH)_4$), carried through the xylem to the stems and leaves. This superior uptake increases efficiency, especially under challenging biotic and abiotic stress conditions. Furthermore, the enhanced absorption of nanonutrients allows for a management approach with lower production costs and reduced environmental impact, as smaller quantities of nutrients are required. Monocotyledonous plants (monocots), such as those in the Poaceae (grasses) and Cyperaceae (sedges) families, are known for their impressive silicon (Si) accumulation capabilities, incorporating up to 10% of Si into their total mass. In contrast, most dicotyledonous plants (dicots) tend to accumulate significantly less Si. While dicot shoots generally have low Si levels, a recent greenhouse experiment involving strawberries (a perennial dicot) showed promising results. The study demonstrated that silicon fertilization led to increased vegetative growth and a higher fruit dry weight [202]. Nanosilicas, with sizes ranging from 20–40 nm, were evaluated for their effects on maize growth. Nanosilica concentrations between 5 and 20 kg ha^{-1} were compared to commercial silica in the same concentration. These were applied to soil prepared for tests with maize seeds (Zea mays L.). Among various concentrations tested, nanosilica applied at a rate of 15 kg ha^{-1} emerged as the most effective. This specific concentration proved superior to both other nanosilica doses and commercially available silica products [160, 196]. The improvement in the use of nanosilica has been demonstrated by several parameters compared to the control, including a 112.7% increase in lignin (offers disease resistance) in leaves, a 1.83% increase in carbon assimilation (capturing CO_2), and a 48.3% increase in the transpiration ratio [14]. Other authors showed foliar application of nanosilica in rice resulted in improved plant height, grain number, panicle length, and chlorophyll levels relative to the control. The increase and growth parameters indicate the foliar application can facilitate leaf penetration [104]. Nanosilica treatment in tomato plantations led to quicker germination, greater seed vigor, and improved fresh and dry weight, as well as enhanced chlorophyll, carotenoid, and photosynthetic pigment concentrations [188]. Nanosilica is playing an increasingly pivotal role in the advancement of 'smart fertilizers. These innovative fertilizers have the capability to deliver nutrients in a controlled or slow-release manner. Integrating nanosilica with other nanoformulations can further enhance this controlled release, creating a powerful synergy. Smart fertilizers have been employed to enhance agronomic fertilizer efficiency while reducing environmental impact and costs. As a result, several innovative fertilizers have been developed [202].

5 Concluding Remarks

Nanobiowastes refer to nanomaterials obtained from agricultural waste, sewage sludge, and industrial processes utilizing biomaterial feedstock. Nanotechnology has revolutionized our society, influencing everything from daily life and communication to future planning. While traditional nanomaterials have played a crucial role in advancements, some concerns remain. Production costs can be high, and certain materials pose potential toxicity risks to both people and the environment. Employing biowaste in the production of novel nanomaterials not only lessens the ecological impact of waste management but also drives social and economic advancement in the area. Nanomaterials derived from agricultural waste represent a promising avenue in this regard due to their alignment with sustainability goals. These nanomaterials offer biodegradable properties, allowing them to be broken down by natural processes over time, thereby reducing environmental impact and minimizing waste accumulation.

Nanoengineering applied to biowaste allows the production of materials that can be reused within the region generating the biowaste, thus promoting a sustainable circular economy. The synthesis of nanobiowastes involves various variables, highlighting the importance of studying preparation methods for each new material. Many synthesis methods indicated in the literature are excellent starting points for evaluating the use of waste to produce nanomaterials. One-pot methods and pyrolysis are preferable due to their cost-effectiveness and simplicity, facilitating on-farm production or small manufacturing plants to serve microregions. These efforts are crucial for ensuring the subsistence of small farmers, improving productivity with reduced costs, and promoting circularity in the rural economy. Studies indicate that the use of biobased nanomaterials can increase crop productivity by up to 30%.

A thorough SWOT [S = strengths, W = weaknesses, O = opportunities, and T = threats] analysis providing insights into the strengths and challenges of the study, as shown in Fig. 6.

Implementing a circular economy in agricultural regions enhances their socioeconomic development by fostering local growth and sustainable transport, as biowaste

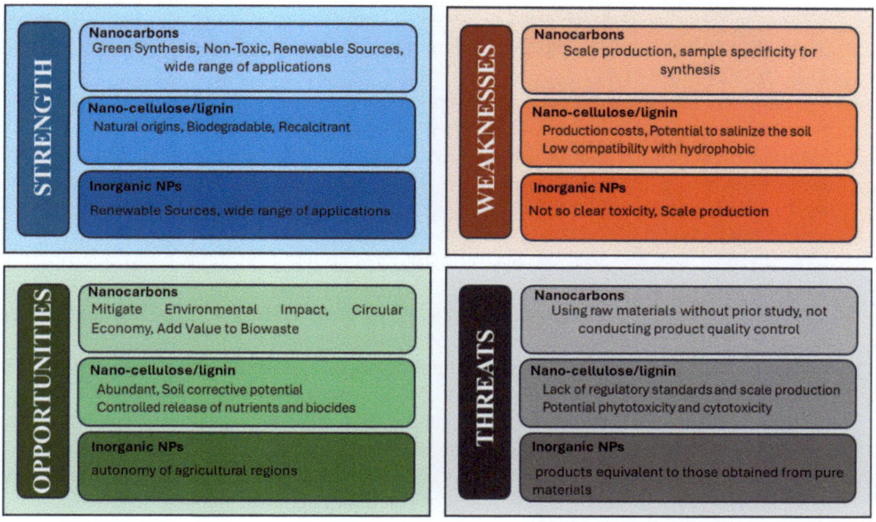

Fig. 6 SWOT Analysis for transforming biowaste into nanocarbons, nanocellulose/lignina and inorganic NPs

and its products do not travel long distances. Using reusable organic materials, such as C-dots, fits within the concept of organic agriculture, boosting crop productivity without compromising organic certification. The phrase "Hence nature what belongs to it" aptly represents the application of biowaste nanomaterials in agriculture. The United Nations has incorporated sustainable agriculture as a crucial component in its 2030 Agenda for Sustainable Development, recognizing its importance in achieving targeted food production goals.

Indeed, the synthesis of advanced nanomaterials represents a remarkable gift from nature with immense potential to benefit humanity and future generations. With careful and responsible utilization, these nanomaterials can revolutionize various fields, including medicine, electronics, energy, environmental remediation, and precision agriculture. Researchers can develop innovative solutions to address pressing challenges and improve the quality of life by harnessing their unique properties- such as high surface area, quantum effects, and tunable characteristics.

References

1. Abbas A, Abbas S, Tabish TA et al (2021) Role of precursor microstructure in the development of graphene quantum dots from biomass. J Environ Chem Eng 9:106154. https://doi.org/10.1016/j.jece.2021.106154
2. Abdul Khalil HPS, Bhat AH, Ireana Yusra AF (2012) Green composites from sustainable cellulose nanofibrils: a review. Carbohydr Polym 87:963–979. https://doi.org/10.1016/J.CARBPOL.2011.08.078

3. Aghoutane Y, Burhan H, Sen F, et al (2024) Glyphosate detection via a nanomaterial-enhanced electrochemical molecularly imprinted polymer sensor. J Anal Sci Technol 15. https://doi.org/10.1186/s40543-024-00417-5
4. Ago M, Tardy BL, Wang L et al (2017) Supramolecular assemblies of lignin into nano- and microparticles. MRS Bull 42:371–378. https://doi.org/10.1557/mrs.2017.88
5. Ahmad UM, Ji N, Li H et al (2021) Can lignin be transformed into agrochemicals? Recent advances in the agricultural applications of lignin. Ind Crops Prod 170:113646. https://doi.org/10.1016/j.indcrop.2021.113646
6. Ai L, Yang Y, Wang B et al (2021) Insights into photoluminescence mechanisms of carbon dots: advances and perspectives. Sci Bull (Beijing) 66:839–856. https://doi.org/10.1016/j.scib.2020.12.015
7. Ainomugisha S, Matovu M, Manga M (2024) Application of green agro-based nanoparticles in cement-based construction materials: a systematic review. J Build Eng 87
8. Aitomäki Y, Oksman K (2014) Reinforcing efficiency of nanocellulose in polymers. React Funct Polym 85:151–156. https://doi.org/10.1016/j.reactfunctpolym.2014.08.010
9. Alborés S, Estevez MB, Raffaelli S, Sanguiñedo P (2023) Microbial silver nanoparticles: synthesis, characterization and applications. In: Applications of silver nanoparticles
10. Alhasan HS, Alahmadi N, Yasin SA, et al (2022) Low-cost and eco-friendly hydroxyapatite nanoparticles derived from eggshell waste for cephalexin removal. Separations 9. https://doi.org/10.3390/separations9010010
11. Alikhani M, Mirbolook A, Sadeghi J, Lakzian A (2023) Improving wheat growth and nutrient uptake in calcareous soil: a novel approach with carbon dots as a slow-release zinc fertilizer. J Soil Sci Plant Nutr 23:6232–6245. https://doi.org/10.1007/s42729-023-01480-4
12. Alsharif SA, Badran MI, Moustafa MH, et al (2023) Hydrothermal extraction and physicochemical characterization of biogenic hydroxyapatite nanoparticles from buffalo waste bones for in vivo xenograft in experimental rats. Sci Rep 13. https://doi.org/10.1038/s41598-023-43989-9
13. Alyasiri YI, Bhiah KM (2021) Effect of traditional and nano calcium fertilizer and nitrogen on cucumber plant growth 8:2602–2611
14. Amin M, Juita N, Asnawi (2023) Application of nano silica fertilizer in agricultural sustainability (a review). In: IOP Conference series: earth and environmental science
15. Arun R, Shruthy R, Preetha R, Sreejit V (2022) Biodegradable nano composite reinforced with cellulose nano fiber from coconut industry waste for replacing synthetic plastic food packaging. Chemosphere 291:132786. https://doi.org/10.1016/J.CHEMOSPHERE.2021.132786
16. Aryamol KS, Kanagaraj K, Nangan S et al (2024) Recent advances of carbon pathways for sustainable environment development. Environ Res 250:118513. https://doi.org/10.1016/j.envres.2024.118513
17. Ashori A (2008) Municipal solid waste as a source of lignocellulosic fiber and plastic for composite industries. Polym Plast Technol Eng 47:741–744. https://doi.org/10.1080/03602550802188565
18. Atchudan R, Edison TNJI, Perumal S et al (2019) Betel-derived nitrogen-doped multicolor carbon dots for environmental and biological applications. J Mol Liq 296:111817. https://doi.org/10.1016/j.molliq.2019.111817
19. Aydogan B, Taskin MB, Gunes A (2023) Effect of synthetic nano-calcium borate as an alternative boron fertilizer on the growth and boron utilization of sunflower plant. Commun Soil Sci Plant Anal 54:2205–2214. https://doi.org/10.1080/00103624.2023.2211619
20. Azeredo HMC, Mattoso LHC, Wood D, et al (2009) Nanocomposite edible films from mango puree reinforced with cellulose nanofibers. J Food Sci 74. https://doi.org/10.1111/J.1750-3841.2009.01186.X
21. Azizi M, Pakravan A, Valizadeh H et al (2024) Fabrication and characterization smart gold-polymer nanostructure as promising theranostic agent for dual-imaging and chemo-photothermal therapy of cancer: an in vitro study. J Photochem Photobiol A Chem 450:115459. https://doi.org/10.1016/j.jphotochem.2024.115459

22. Bahloul A, Kassab Z, El Bouchti M et al (2021) Micro- and nano-structures of cellulose from eggplant plant (Solanum melongena L) agricultural residue. Carbohydr Polym 253:117311. https://doi.org/10.1016/J.CARBPOL.2020.117311
23. Bajwa DS, Pourhashem G, Ullah AH, Bajwa SG (2019) A concise review of current lignin production, applications, products and their environmental impact. Ind Crops Prod 139:111526. https://doi.org/10.1016/j.indcrop.2019.111526
24. Balasubramaniam T, Bakar AHA, Uda MNA, et al (2020) Potential of Syntesized Silica Nanoparticles (Si-NPs) using corn cob for arsenic heavy metal removal. In: IOP conference series: materials science and engineering
25. Bandi R, Gangapuram BR, Dadigala R et al (2016) Facile and green synthesis of fluorescent carbon dots from onion waste and their potential applications as sensor and multicolour imaging agents. RSC Adv 6:28633–28639. https://doi.org/10.1039/c6ra01669c
26. Banihashemi K, Amirmozafari N, Mehregan I, et al (2021) Antibacterial effect of carbon nanotube containing chemical compounds on drug-resistant isolates of acinetobacter baumannii. Iran J Microbiol 13:112–120. https://doi.org/10.18502/ijm.v13i1.5501
27. Barajas-Ledesma RM, Patti AF, Wong VNL et al (2020) Engineering nanocellulose superabsorbent structure by controlling the drying rate. Colloids Surf A Physicochem Eng Asp 600:124943. https://doi.org/10.1016/j.colsurfa.2020.124943
28. Barbash VA, Yaschenko OV, Shniruk OM (2017) Preparation and properties of nanocellulose from organosolv straw pulp. Nanoscale Res Lett 12:1–8. https://doi.org/10.1186/S11671-017-2001-4/FIGURES/9
29. Barot RB, Gawande ND, Omprabha S, Kaushal C, Ghosh S, Saha J, Bhatia D, Sankaranarayanan S (2024) Novel carbon nanoparticles derived from Bougainvillea modulate vegetative growth via auxin–cytokinin signaling in Arabidopsis. Chem Papers 78(8):4733–4750. https://doi.org/10.1007/s11696-024-03421-6
30. Başkaya SK, Tahta B, Uruş S et al (2024) Multifunctional B, N, P, and S-doped fluorescent carbon quantum dot synthesis from pigeon manure: highly effective Hg (II) sensor and fluorescent ink properties. Biomass Convers Biorefin 14:1089–1103. https://doi.org/10.1007/s13399-022-03017-8
31. Bauli CR, Lima GF, de Souza AG et al (2021) Eco-friendly carboxymethyl cellulose hydrogels filled with nanocellulose or nanoclays for agriculture applications as soil conditioning and nutrient carrier and their impact on cucumber growing. Colloids Surf A Physicochem Eng Asp 623:126771. https://doi.org/10.1016/j.colsurfa.2021.126771
32. Beisl S, Miltner A, Friedl A (2017) Lignin from micro- to nanosize: production methods. Int J Mol Sci 18:1244. https://doi.org/10.3390/ijms18061244
33. Bhardwaj AK, Srivastav AL, Dwivedi K, Sillanpää M (2024) Green and sustainable approaches using wastes for the production of multifunctional nanomaterials
34. Bhati A, Gunture TKM et al (2018) Exploration of nano carbons in relevance to plant systems. New J Chem 42:16411–16427. https://doi.org/10.1039/c8nj03642j
35. Bhutto AW, Qureshi K, Harijan K et al (2017) Insight into progress in pre-treatment of lignocellulosic biomass. Energy 122:724–745. https://doi.org/10.1016/j.energy.2017.01.005
36. Bhuyan N, Narzari R, Gogoi L, et al (2020) Valorization of agricultural wastes for multidimensional use. In: Current developments in biotechnology and bioengineering. Elsevier, pp 41–78
37. Bindar Y, Ramli Y, Steven S, Restiawaty E (2024) Optimization of purity and yield of amorphous bio-silica nanoparticles synthesized from bamboo leaves. Can J Chem Eng 102. https://doi.org/10.1002/cjce.25148
38. Bondancia TJ, Batista G, De Aguiar J et al (2022) Cellulose nanocrystals from sugar cane bagasse using organic and/or inorganic acids: techno-economic analysis and life cycle assessment. ACS Sustain Chem Eng 10:4660–4676. https://doi.org/10.1021/ACSSUSCHEMENG.2C00061/ASSET/IMAGES/LARGE/SC2C00061_0007.JPEG
39. Borkotoky SS, Dhar P, Katiyar V (2018) Biodegradable poly (lactic acid)/Cellulose nanocrystals (CNCs) composite microcellular foam: effect of nanofillers on foam cellular morphology, thermal and wettability behavior. Int J Biol Macromol 106:433–446. https://doi.org/10.1016/J.IJBIOMAC.2017.08.036

40. Burpo FJ, Mitropoulos AN, Nagelli EA et al (2018) Cellulose nanofiber biotemplated palladium composite aerogels. Molecules 23:1405. https://doi.org/10.3390/molecules23061405
41. Camarero Espinosa S, Kuhnt T, Foster EJ, Weder C (2013) Isolation of thermally stable cellulose nanocrystals by phosphoric acid hydrolysis. Biomacromol 14:1223–1230. https://doi.org/10.1021/BM400219U/SUPPL_FILE/BM400219U_SI_001.PDF
42. Chahal S, Macairan JR, Yousefi N et al (2021) Green synthesis of carbon dots and their applications. RSC Adv 11:25354–25363. https://doi.org/10.1039/d1ra04718c
43. Chen H, Wang G, Chen L, et al (2018) Three-dimensional honeycomb-like porous carbon with both interconnected hierarchical porosity and nitrogen self-doping from cotton seed husk for supercapacitor electrode. Nanomaterials 8. https://doi.org/10.3390/nano8060412
44. Chinga-Carrasco G (2011) Cellulose fibres, nanofibrils and microfibrils: the morphological sequence of MFC components from a plant physiology and fibre technology point of view. Nanoscale Res Lett 6:417. https://doi.org/10.1186/1556-276X-6-417
45. Coelho CCS, Michelin M, Cerqueira MA et al (2018) Cellulose nanocrystals from grape pomace: production, properties and cytotoxicity assessment. Carbohydr Polym 192:327–336. https://doi.org/10.1016/J.CARBPOL.2018.03.023
46. Cunha FL, Nieri EM, Santos JA dos, et al (2021) Uso dos adubos de liberação lenta no setor florestal. Pesqui Florest Bras 41. https://doi.org/10.4336/2021.pfb.41e201902063
47. del Río JC, Rencoret J, Gutiérrez A et al (2020) Lignin monomers from beyond the canonical monolignol biosynthetic pathway: another brick in the wall. ACS Sustain Chem Eng 8:4997–5012. https://doi.org/10.1021/acssuschemeng.0c01109
48. Datta R (2024) Enzymatic degradation of cellulose in soil: a review. Heliyon 10:e24022. https://doi.org/10.1016/j.heliyon.2024.e24022
49. Deng J, Sun S-F, Zhu E-Q et al (2021) Sub-micro and nano-lignin materials: small size and rapid progress. Ind Crops Prod 164:113412. https://doi.org/10.1016/j.indcrop.2021.113412
50. Dhanush C, Sethuraman MG (2021) Independent hydrothermal synthesis of the undoped, nitrogen, boron and sulphur doped biogenic carbon nanodots and their potential application in the catalytic chemo-reduction of Alizarine yellow R azo dye. Spectrochim Acta A Mol Biomol Spectrosc 260:119920. https://doi.org/10.1016/j.saa.2021.119920
51. Du X, Zhang Z, Liu W, Deng Y (2017) Nanocellulose-based conductive materials and their emerging applications in energy devices-a review. Nano Energy 35:299–320. https://doi.org/10.1016/J.NANOEN.2017.04.001
52. Dubey P (2023) An overview on animal/human biomass-derived carbon dots for optical sensing and bioimaging applications. RSC Adv 13:35088–35126. https://doi.org/10.1039/d3ra06976a
53. Eddy NO, Odiongenyi AO, Garg R, et al (2023) Quantum and experimental investigation of the application of Crassostrea gasar (mangrove oyster) shell–based CaO nanoparticles as adsorbent and photocatalyst for the removal of procaine penicillin from aqueous solution. Environ Sci Pollut Res 30. https://doi.org/10.1007/s11356-023-26868-8
54. Elnazer WY, Farrag TE, Beheary MS, Mansour RA (2024) Efficient adsorption of amoxicillin onto silica nanoparticles synthesized from rice husks. Desalination Water Treat 317. https://doi.org/10.1016/j.dwt.2024.100086
55. Ezhilarasan D, Revathi D, Raghunandhakumar S, et al (2021) Frontier and perspective outlook on agrowaste nanoparticles for healthcare and environment. In: Agri-waste and microbes for production of sustainable nanomaterials
56. Farooq A, Patoary MK, Zhang M et al (2020) Cellulose from sources to nanocellulose and an overview of synthesis and properties of nanocellulose/zinc oxide nanocomposite materials. Int J Biol Macromol 154:1050–1073. https://doi.org/10.1016/J.IJBIOMAC.2020.03.163
57. Fatma S, Hameed A, Noman M, et al (2018) Lignocellulosic biomass: a sustainable bioenergy source for future. Protein Pept Lett 25. https://doi.org/10.2174/0929866525666180122144504
58. Feng J, Wang WJ, Hai X et al (2016) Green preparation of nitrogen-doped carbon dots derived from silkworm chrysalis for cell imaging. J Mater Chem B 4:387–393. https://doi.org/10.1039/c5tb01999k

59. Feng Y-H, Cheng T-Y, Yang W-G et al (2018) Characteristics and environmentally friendly extraction of cellulose nanofibrils from sugarcane bagasse. Ind Crops Prod 111:285–291. https://doi.org/10.1016/j.indcrop.2017.10.041
60. Ferro AC, Guedes M (2019) Mechanochemical synthesis of hydroxyapatite using cuttlefish bone and chicken eggshell as calcium precursors. Mater Sci Eng C 97. https://doi.org/10.1016/j.msec.2018.11.083
61. Flauzino Neto WP, Silvério HA, Dantas NO, Pasquini D (2013) Extraction and characterization of celulose nanocrystals from agroindustrial residue-soy hulls. Ind Crop Prod 42:480–488. https://doi.org/10.1016/j.indcrop.2012.06.041
62. Flores-Contreras EA, González-González RB, Pablo Pizaña-Aranda JJ, et al (2024) Agricultural waste as a sustainable source for nanoparticle synthesis and their antimicrobial properties for food preservation. Front Nanotechnol 6
63. Fortunati E, Luzi F, Jiménez A et al (2016) Revalorization of sunflower stalks as novel sources of cellulose nanofibrils and nanocrystals and their effect on wheat gluten bionanocomposite properties. Carbohydr Polym 149:357–368. https://doi.org/10.1016/J.CARBPOL.2016.04.120
64. Fu Y, Yang L, Zhang M, et al (2022) Recent advances in cellulose-based polymer electrolytes. Cellulose 29:17 29:8997–9034. https://doi.org/10.1007/S10570-022-04834-W
65. Garg VK, Kumar N, Anand A (2023) Green synthesis of nanomaterials for the removal of emerging water pollutants. In: Role of green chemistry in ecosystem restoration to achieve environmental sustainability
66. Garrido-Herrera FJ, Daza-Fernández I, González-Pradas E, Fernández-Pérez M (2009) Lignin-based formulations to prevent pesticides pollution. J Hazard Mater 168:220–225. https://doi.org/10.1016/j.jhazmat.2009.02.019
67. Gazzurelli C, Migliori A, Mazzeo PP et al (2020) Making agriculture more sustainable: an environmentally friendly approach to the synthesis of Lignin@Cu pesticides. ACS Sustain Chem Eng 8:14886–14895. https://doi.org/10.1021/acssuschemeng.0c04645
68. Gibson LJ (2012) The hierarchical structure and mechanics of plant materials. J R Soc Interface 9:2749–2766. https://doi.org/10.1098/rsif.2012.0341
69. Gillet S, Aguedo M, Petitjean L et al (2017) Lignin transformations for high value applications: towards targeted modifications using green chemistry. Green Chem 19:4200–4233. https://doi.org/10.1039/C7GC01479A
70. Gogoi J, Sarma TK, Chowdhury D (2023) Carbon dots as solubilizing agent for insoluble phosphates in enhancing soil fertility. ChemistrySelect 8. https://doi.org/10.1002/slct.202204695
71. Goh KW, Wong YH, Ramesh S, et al (2021) Effect of pH on the properties of eggshell-derived hydroxyapatite bioceramic synthesized by wet chemical method assisted by microwave irradiation. Ceram Int 47. https://doi.org/10.1016/j.ceramint.2020.12.009
72. Gómez-Alvarez LM, Zapata Montoya JE (2024) Effect of fortification with $CaCO_3$ nanoparticles obtained from eggshell on the physical and sensory characteristics of three food matrices. Heliyon 10. https://doi.org/10.1016/j.heliyon.2024.e24442
73. Gonugunta P, Vivekanandhan S, Mohanty AK, Misra M (2012) A study on synthesis and characterization of biobased carbon nanoparticles from lignin. World J Nano Sci Eng 02:148–153. https://doi.org/10.4236/wjnse.2012.23019
74. Gopinath S, Raja K, Manonmani V, et al (2023) CaO Nanoparticles seed invigoration on germination and seedling growth of rice var. TRY 3. Int J Environ Clim Change 13. https://doi.org/10.9734/ijecc/2023/v13i102877
75. Habibi Y, Lucia LA, Rojas OJ (2010) Cellulose nanocrystals: chemistry, self-assembly, and applications. Chem Rev 110:3479–3500. https://doi.org/10.1021/CR900339W/ASSET/IMAGES/MEDIUM/CR-2009-00339W_0022.GIF
76. Habte L, Shiferaw N, Mulatu D, et al (2019) Synthesis of nano-calcium oxide fromwaste eggshell by sol-gel method. Sustainability (Switzerland) 11. https://doi.org/10.3390/su11113196

77. Han L, Ghosh D, Chen W et al (2009) Nanosized carbon particles from natural gas soot. Chem Mater 21:2803–2809. https://doi.org/10.1021/cm900709w
78. Henriksson M, Henriksson G, Berglund LA, Lindström T (2007) An environmentally friendly method for enzyme-assisted preparation of microfibrillated cellulose (MFC) nanofibers. Eur Polym J 43:3434–3441. https://doi.org/10.1016/j.eurpolymj.2007.05.038
79. Henrique MA, Silvério HA, Flauzino Neto WP, Pasquini D (2013) Valorization of an agro-industrial waste, mango seed, by the extraction and characterization of its cellulose nanocrystals. J Environ Manage 121:202–209. https://doi.org/10.1016/J.JENVMAN.2013.02.054
80. Horst FH, Rodrigues CV da S, Carvalho PHPR, et al (2021) From cow manure to bioactive carbon dots: a light-up probe for bioimaging investigations, glucose detection and potential immunotherapy agent for melanoma skin cancer. RSC Adv 11:6346–6352. https://doi.org/10.1039/d0ra10859f
81. Hua KH, Wang HC, Chung RS, Hsu JC (2015) Calcium carbonate nanoparticles can enhance plant nutrition and insect pest tolerance. J Pestic Sci 40:208–213. https://doi.org/10.1584/jpestics.D15-025
82. Huang X, Dong K, Liu L, et al (2020) Physicochemical and structural characteristics of nano eggshell calcium prepared by wet ball milling. LWT 131. https://doi.org/10.1016/j.lwt.2020.109721
83. Hussin MH, Appaturi JN, Poh NE et al (2022) A recent advancement on preparation, characterization and application of nanolignin. Int J Biol Macromol 200:303–326. https://doi.org/10.1016/j.ijbiomac.2022.01.007
84. Ibrahim MM, Moustafa H, El Rahman ENA et al (2020) Reinforcement of starch based biodegradable composite using nile rose residues. J Market Res 9:6160–6171. https://doi.org/10.1016/J.JMRT.2020.04.018
85. Ilangovan M, Guna V, Hu C et al (2018) Curcuma longa L. plant residue as a source for natural cellulose fibers with antimicrobial activity. Ind Crops Prod 112:556–560. https://doi.org/10.1016/J.INDCROP.2017.12.042
86. Inayat A, Albalawi K, Rehman A ur, et al (2023) Tunable synthesis of carbon quantum dots from the biomass of spent tea leaves as supercapacitor electrode. Mater Today Commun 34:105479. https://doi.org/10.1016/j.mtcomm.2023.105479
87. International Organization for Standardization (2011) ISO/TS 80004–4:2011 nanotechnologies—vocabulary—Part 4: nanostructured materials. Geneva
88. Ioelovich M (2012) Optimal conditions for isolation of nanocrystalline cellulose particles 2:9–13. https://doi.org/10.5923/j.nn.20120202.03
89. Ismael E, Fahim KM, Ghorab SMO, et al (2024) Sustainable recycling of poultry eggshell waste for the synthesis of calcium oxide nanoparticles and evaluating its antibacterial potency against food-borne pathogens. J Adv Vet Res 14
90. Issa MA, Abidin ZZ, Sobri S, et al (2019) Facile synthesis of nitrogen-doped carbon dots from lignocellulosic waste. Nanomaterials 9. https://doi.org/10.3390/nano9101500
91. Jagannathan M, Dhinasekaran D, Soundharraj P et al (2021) Green synthesis of white light emitting carbon quantum dots: fabrication of white fluorescent film and optical sensor applications. J Hazard Mater 416:125091. https://doi.org/10.1016/j.jhazmat.2021.125091
92. Jain A, Tripathi SK (2015) Nano-porous activated carbon from sugarcane waste for supercapacitor application. J Energy Storage 4:121–127. https://doi.org/10.1016/j.est.2015.09.010
93. John BK, Abraham T, Mathew B (2022) A review on characterization techniques for carbon quantum dots and their applications in agrochemical residue detection. J Fluoresc 32:449–471. https://doi.org/10.1007/s10895-021-02852-8
94. Julie Chandra CS, George N, Narayanankutty SK (2016) Isolation and characterization of cellulose nanofibrils from arecanut husk fibre. Carbohydr Polym 142:158–166. https://doi.org/10.1016/j.carbpol.2016.01.015
95. Kallel F, Bettaieb F, Khiari R et al (2016) Isolation and structural characterization of cellulose nanocrystals extracted from garlic straw residues. Ind Crops Prod 87:287–296. https://doi.org/10.1016/J.INDCROP.2016.04.060

96. Kang C, Huang Y, Yang H et al (2020) A review of carbon dots produced from biomass wastes. Nanomaterials 10:1–24. https://doi.org/10.3390/nano10112316
97. Karagoz P, Khiawjan S, Marques MPC et al (2023) Pharmaceutical applications of lignin-derived chemicals and lignin-based materials: linking lignin source and processing with clinical indication. Biomass Convers Biorefin. https://doi.org/10.1007/s13399-023-03745-5
98. Karim Z, Claudpierre S, Grahn M et al (2016) Nanocellulose based functional membranes for water cleaning: tailoring of mechanical properties, porosity and metal ion capture. J Memb Sci 514:418–428. https://doi.org/10.1016/J.MEMSCI.2016.05.018
99. Kassab Z, Abdellaoui Y, Salim MH, El Achaby M (2020) Cellulosic materials from pea (Pisum Sativum) and broad beans (Vicia Faba) pods agro-industrial residues. Mater Lett 280:128539. https://doi.org/10.1016/J.MATLET.2020.128539
100. Kassab Z, Kassem I, Hannache H et al (2020) Tomato plant residue as new renewable source for cellulose production: extraction of cellulose nanocrystals with different surface functionalities. Cellulose 27:4287–4303. https://doi.org/10.1007/S10570-020-03097-7/FIGURES/9
101. Kassem I, Ablouh EH, El Bouchtaoui FZ et al (2021) Cellulose nanocrystals-filled poly (vinyl alcohol) nanocomposites as waterborne coating materials of NPK fertilizer with slow release and water retention properties. Int J Biol Macromol 189:1029–1042. https://doi.org/10.1016/J.IJBIOMAC.2021.08.093
102. Kaur M, Mehta A, Bhardwaj KK, Gupta R (2020) Bionanomaterials from agricultural wastes. Adv Struct Mater 126:243–260. https://doi.org/10.1007/978-981-15-3560-4_10
103. Khan MN, Hassan T, Rehman N et al (2023) Isolation of nanocellulose from Dodonaea viscosa plant: structural and thermal properties. Zeitschrift fur Physikalische Chemie 237:1807–1819. https://doi.org/10.1515/ZPCH-2023-0308/ASSET/GRAPHIC/J_ZPCH-2023-0308_FIG_007.JPG
104. Kheir AMS, Abouelsoud HM, Hafez EM, Ali OAM (2019) Integrated effect of nano-Zn, nano-Si, and drainage using crop straw–filled ditches on saline sodic soil properties and rice productivity. Arab J Geosci 12:0–7. https://doi.org/10.1007/s12517-019-4653-0
105. Kościk I, Jankowski D, Jagusiak A (2022) Carbon nanomaterials for theranostic use. C-J Carb Res 8. https://doi.org/10.3390/c8010003
106. Koul B, Yakoob M, Shah MP (2022) Agricultural waste management strategies for environmental sustainability. Environ Res 206:112285. https://doi.org/10.1016/j.envres.2021.112285
107. Kuswandi B, Damayanti F, Jayus J et al (2015) Simple and low-cost on-package sticker sensor based on litmus paper for real-time monitoring of beef freshness. J Math Fundam Sci 47:236–251. https://doi.org/10.5614/j.math.fund.sci.2015.47.3.2
108. Lanzón M, Madrid-Mendoza JA, Navarro-Moreno D, García-Vera VE (2023) Use of eggshell waste: a green and effective method for the synthesis of pure calcium hydroxide suspensions. Constr Build Mater 377. https://doi.org/10.1016/j.conbuildmat.2023.131106
109. Latha M, Subramanian KS, Sundara Sharmila DJ et al (2023) Urea–lignin/chitosan nanocomposite as slow-release nanofertilizer. ACS Agric Sci Technol 3:463–476. https://doi.org/10.1021/acsagscitech.2c00334
110. Le Normand M, Moriana R, Ek M (2014) Isolation and characterization of cellulose nanocrystals from spruce bark in a biorefinery perspective. Carbohydr Polym 111:979–987. https://doi.org/10.1016/J.CARBPOL.2014.04.092
111. Le VH, Thuc CNH, Thuc HH (2013) Synthesis of silica nanoparticles from Vietnamese rice husk by sol–gel method. Nanoscale Res Lett 8. https://doi.org/10.1186/1556-276x-8-58
112. Lengowski EC, Bonfatti Júnior EA, Simon L et al (2020) Different degree of fibrillation: strategy to reduce permeability in nanocellulose-starch films. Cellulose 27:10855–10872. https://doi.org/10.1007/s10570-020-03232-4
113. Levanič J, Svedström K, Liljeström V et al (2022) Cellulose fiber and nanofibril characteristics in a continuous sono-assisted process for production of TEMPO-oxidized nanofibrillated cellulose. Cellulose 29:9121–9142. https://doi.org/10.1007/s10570-022-04845-7

114. Li H, He X, Kang Z et al (2010) Water-soluble fluorescent carbon quantum dots and photocatalyst design. Angewandte Chemie International Edition 49:4430–4434. https://doi.org/10.1002/anie.200906154
115. Li J, Xu O, Zhu X (2021) A facile green and one-pot synthesis of grape seed-derived carbon quantum dots as a fluorescence probe for Cu(ii) and ascorbic acid. RSC Adv 11:34107–34116. https://doi.org/10.1039/d1ra05656e
116. Li Q, Wu Y, Fang R et al (2021) Application of nanocellulose as particle stabilizer in food pickering emulsion: scope, merits and challenges. Trends Food Sci Technol 110:573–583. https://doi.org/10.1016/J.TIFS.2021.02.027
117. Li S, Li L, Tu H et al (2021) The development of carbon dots: from the perspective of materials chemistry. Mater Today 51:188–207. https://doi.org/10.1016/j.mattod.2021.07.028
118. Li Y, Wang G, Wei T et al (2016) Nitrogen and sulfur co-doped porous carbon nanosheets derived from willow catkin for supercapacitors. Nano Energy 19:165–175. https://doi.org/10.1016/j.nanoen.2015.10.038
119. Li Y, Zhou M, Pang Y, Qiu X (2017) Lignin-based microsphere: preparation and performance on encapsulating the pesticide avermectin. ACS Sustain Chem Eng 5:3321–3328. https://doi.org/10.1021/acssuschemeng.6b03180
120. Liang YM, Yang H, Zhou B et al (2022) Waste tobacco leaves derived carbon dots for tetracycline detection: improving quantitative accuracy with the aid of chemometric model. Anal Chim Acta 1191:339269. https://doi.org/10.1016/j.aca.2021.339269
121. Li-Chan ECY, Kim HO (2007) Structure and chemical compositions of eggs. In: Egg bioscience and biotechnology
122. Lin X, Guo L, Shaghaleh H et al (2021) A TEMPO-oxidized cellulose nanofibers/MOFs hydrogel with temperature and pH responsiveness for fertilizers slow-release. Int J Biol Macromol 191:483–491. https://doi.org/10.1016/J.IJBIOMAC.2021.09.075
123. Liou TH (2004) Preparation and characterization of nano-structured silica from rice husk. Mater Sci Eng A 364. https://doi.org/10.1016/j.msea.2003.08.045
124. Liu C, Li B, Du H et al (2016) Properties of nanocellulose isolated from corncob residue using sulfuric acid, formic acid, oxidative and mechanical methods. Carbohydr Polym 151:716–724. https://doi.org/10.1016/J.CARBPOL.2016.06.025
125. Liu H, Zhang Y, Liu JH et al (2017) Preparation of nitrogen-doped carbon dots with high quantum yield from: Bombyx mori silk for Fe(III) ions detection. RSC Adv 7:50584–50590. https://doi.org/10.1039/c7ra10130a
126. Liu J, Willför S, Mihranyan A (2017) On importance of impurities, potential leachables and extractables in algal nanocellulose for biomedical use. Carbohydr Polym 172:11–19. https://doi.org/10.1016/J.CARBPOL.2017.05.002
127. Liu S, Wu Q, Sun X et al (2021) Novel alginate-cellulose nanofiber-poly(vinyl alcohol) hydrogels for carrying and delivering nitrogen, phosphorus and potassium chemicals. Int J Biol Macromol 172:330–340. https://doi.org/10.1016/J.IJBIOMAC.2021.01.063
128. Low LE, Teh KC, Siva SP et al (2021) Lignin nanoparticles: the next green nanoreinforcer with wide opportunity. Environ Nanotechnol Monit Manag 15:100398. https://doi.org/10.1016/j.enmm.2020.100398
129. Lu P, Hsieh Y-L (2010) Preparation and properties of cellulose nanocrystals: rods, spheres, and network. Carbohydr Polym 82:329–336. https://doi.org/10.1016/j.carbpol.2010.04.073
130. Lu W, Qin X, Liu S et al (2012) Economical, green synthesis of fluorescent carbon nanoparticles and their use as probes for sensitive and selective detection of mercury(II) ions. Anal Chem 84:5351–5357. https://doi.org/10.1021/ac3007939
131. Luzi F, Fortunati E, Jiménez A et al (2016) Production and characterization of PLA_PBS biodegradable blends reinforced with cellulose nanocrystals extracted from hemp fibres. Ind Crops Prod 93:276–289. https://doi.org/10.1016/J.INDCROP.2016.01.045
132. Ma X, Li J (2024) Preparation of multicolor carbon quantum dots by hydrothermal method and their functionalization applications. J Lumin 266:120296. https://doi.org/10.1016/j.jlumin.2023.120296

133. Mahawar L, Ramasamy KP, Suhel M, et al (2023) Silicon nanoparticles: comprehensive review on biogenic synthesis and applications in agriculture. Environ Res 232. https://doi.org/10.1016/j.envres.2023.116292
134. Mahfoudhi N, Boufi S (2016) Nanocellulose as a millennium material with enhancing adsorption capacities. Biodegrad Biobased Polym Environ Biomed Appl 349–383. https://doi.org/10.1002/9781119117360.CH10
135. Malar CG, Muthulingam S, Murugesan M et al (2022) A comprehensive review of the importance of thermal activation in the production of carbon dots and the potential for their use in the bioenergy industry. J Therm Anal Calorim 148:505–516. https://doi.org/10.1007/s10973-022-11687-9
136. Malpani SK, Goyal D (2023) Green agricultural waste derived silica nanoparticles as catalyst support materials. In: Advances in green and sustainable nanomaterials: applications in energy, biomedicine, agriculture, and environmental science
137. Maravilla BG, Gan NMC, Guinanao DYB, et al (2024) Photocatalytic degradation of methyl orange in water using oyster shell-derived calcium oxide nanoparticles. Bioresour Technol Rep 25. https://doi.org/10.1016/j.biteb.2024.101813
138. Mehanny S, Darwish L, Ibrahim H et al (2016) High-content lignocellulosic fibers reinforcing starch-based biodegradable composites: properties and applications. Compos Renew Sustain Mater. https://doi.org/10.5772/65262
139. Mehanny S, Farag M, Rashad RM, Elsayed H (2013) Fabrication and characterization of starch based bagasse fiber composite. ASME international mechanical engineering congress and exposition. Proceedings (IMECE) 3:1345–1353. https://doi.org/10.1115/IMECE2012-86265
140. Meraj A, Singh SP, Jawaid M et al (2023) A review on eco-friendly isolation of lignin by natural deep eutectic solvents from agricultural wastes. J Polym Environ 31:3283–3316. https://doi.org/10.1007/s10924-023-02817-x
141. Messa LL, Faez R (2020) Spray-dried chitosan/nanocellulose microparticles: synergistic effects for the sustained release of NPK fertilizer. Cellulose 27:10077–10093. https://doi.org/10.1007/S10570-020-03482-2/FIGURES/9
142. Moriana R, Vilaplana F, Ek M (2016) Cellulose nanocrystals from forest residues as reinforcing agents for composites: a study from macro- to nano-dimensions. Carbohydr Polym 139:139–149. https://doi.org/10.1016/J.CARBPOL.2015.12.020
143. Mujtaba M, Fernandes Fraceto L, Fazeli M et al (2023) Lignocellulosic biomass from agricultural waste to the circular economy: a review with focus on biofuels, biocomposites and bioplastics. J Clean Prod 402:136815. https://doi.org/10.1016/j.jclepro.2023.136815
144. Muleta WS, Denboba SM, Bayu AB (2024) Corncob-supported calcium oxide nanoparticles from hen eggshells for cadmium (Cd-II) removal from aqueous solutions; synthesis and characterization. Heliyon 10. https://doi.org/10.1016/j.heliyon.2024.e27767
145. Murugan N, Sundramoorthy AK (2018) Green synthesis of fluorescent carbon dots from Borassus flabellifer flowers for label-free highly selective and sensitive detection of Fe3+ ions. New J Chem 42:13297–13307. https://doi.org/10.1039/c8nj01894d
146. Nagaraj M, Ramalingam S, Murugan C et al (2022) Detection of Fe3+ ions in aqueous environment using fluorescent carbon quantum dots synthesized from endosperm of Borassus flabellifer. Environ Res 212:113273. https://doi.org/10.1016/j.envres.2022.113273
147. Nugroho D, Wannakan K, Nanan S, Benchawattananon R (2024) The Synthesis of carbon dots/ /zincoxide (CDs/ZnO-H400) by using hydrothermal methods for degradation of ofloxacin antibiotics and reactive red azo dye (RR141). Sci Rep 14:1–14. https://doi.org/10.1038/s41598-024-53083-3
148. Osman AI, Zhang Y, Farghali M, et al (2024) Synthesis of green nanoparticles for energy, biomedical, environmental, agricultural, and food applications: a review. Springer International Publishing
149. Pacaphol K, Seraypheap K, Aht-Ong D (2019) Development and application of nanofibrillated cellulose coating for shelf life extension of fresh-cut vegetable during postharvest storage. Carbohydr Polym 224:115167. https://doi.org/10.1016/J.CARBPOL.2019.115167

150. Palacio-Vergara M, Álvarez-Gómez M, Gallego J, López D (2023) Biomass solvothermal treatment methodologies to obtain carbon quantum dots: a systematic review. Talanta Open 8:1–12. https://doi.org/10.1016/j.talo.2023.100244
151. Park NM, Koo JB, Oh JY et al (2017) Electroluminescent nanocellulose paper. Mater Lett 196:12–15. https://doi.org/10.1016/J.MATLET.2017.03.003
152. Patidar R, Rebary B, Bhadu GR, Paul P (2016) Fluorescent carbon nanoparticles as label-free recognizer of Hg2+ and Fe3+ through effective fluorescence quenching in aqueous media. J Lumin 173:243–249. https://doi.org/10.1016/j.jlumin.2015.12.051
153. Peña-Ortiz M, Rincón E, Serrano L, García A (2023) Recent nanocelullose applications for sustainable agriculture—a review, pp 185–213
154. Pengiran H, Kamaldin J, Leo BF, Ahmad H (2021) Larvicidal efficacy of temephos impregnated onto kenaf cellulose nanofibre to control Aedes aegypti (Diptera: Culicidae) larvae. Trop Biomed 38:453–461. https://doi.org/10.47665/tb.38.3.088
155. Phosanam A, Moreira J, Adhikari B et al (2023) Stabilization of ginger essential oil Pickering emulsions by pineapple cellulose nanocrystals. Curr Res Food Sci 7:100575. https://doi.org/10.1016/j.crfs.2023.100575
156. Pohshna C, Mailapalli DR (2023) Characterization and testing of nanohydroxyapatite synthesized from eggshells as a phosphorus source for rice crops. J Soil Sci Plant Nutr 23:4491–4504. https://doi.org/10.1007/s42729-023-01366-5
157. Poletto M (2018) Lignin-trends and applications. InTech, London
158. Pompêu GCS, Pasquini D (2023) Extraction of lignin and modifications. In: Thomas S, Hosur M, Pasquini D, Chirayil JD (eds) Handbook of biomass. Springer Nature Singapore, Singapore, pp 1–35
159. Prabha S, Durgalakshmi D, Rajendran S, Lichtfouse E (2021) Plant-derived silica nanoparticles and composites for biosensors, bioimaging, drug delivery and supercapacitors: a review. Environ Chem Lett 19
160. Prajapati D, Patani A, Jain T, Patel A, Singh S (2023) ROS responsive silica nanoparticles for controlled and targeted drug delivery. In: Chawla S, Singh S, Husen A (eds) Smart nanomaterials targeting pathological hypoxia. Smart nanomaterials technology. Springer, Singapore
161. Prasad R, Kumar M, Kumar V (2017) Nanotechnology: an agricultural paradigm
162. Putro JN, Kurniawan A, Ismadji S, Ju YH (2017) Nanocellulose based biosorbents for wastewater treatment: study of isotherm, kinetic, thermodynamic and reusability. Environ Nanotechnol Monit Manag 8:134–149. https://doi.org/10.1016/J.ENMM.2017.07.002
163. Ralph J, Lapierre C, Boerjan W (2019) Lignin structure and its engineering. Curr Opin Biotechnol 56:240–249. https://doi.org/10.1016/j.copbio.2019.02.019
164. Ramesh S, Radhakrishnan P (2019) Cellulose nanoparticles from agro-industrial waste for the development of active packaging. Appl Surf Sci 484:1274–1281. https://doi.org/10.1016/J.APSUSC.2019.04.003
165. Rangan A, Manjula MV, Satyanarayana KG, Menon R (2016) Lignin/nanolignin and their biodegradable composites. In: Kalia S (ed) Biodegradable green composites. Wiley, Hoboken, pp 167–198
166. Resch-Genger U, Rurack K (2013) Determination of the photoluminescence quantum yield of dilute dye solutions (IUPAC Technical Report). Pure Appl Chem 85:2005–2026. https://doi.org/10.1351/PAC-REP-12-03-03
167. Rojas J, Bedoya M, Ciro Y (2015) Current trends in the production of cellulose nanoparticles and nanocomposites for biomedical applications. In: Cellulose-fundamental aspects and current trends. InTech
168. Rol F, Karakashov B, Nechyporchuk O et al (2017) Pilot-scale twin screw extrusion and chemical pretreatment as an energy-efficient method for the production of nanofibrillated cellulose at high solid content. ACS Sustain Chem Eng 5:6524–6531. https://doi.org/10.1021/acssuschemeng.7b00630
169. Rossi BL, Andrade CMB, Therézio EM, et al (2022) Carbon quantum dots : an environmentally friendly and valued approach to sludge disposal, pp 1–14. https://doi.org/10.3389/fchem.2022.858323

170. Rovani S, Santos JJ, Corio P, Fungaro DA (2018) Highly pure silica nanoparticles with high adsorption capacity obtained from sugarcane waste ash. ACS Omega 3. https://doi.org/10.1021/acsomega.8b00092
171. Ru Y, Sui L, Song H et al (2021) Rational design of multicolor-emitting chiral carbonized polymer dots for full-color and white circularly polarized luminescence. Angewandte Chemie International Edition 60:14091–14099. https://doi.org/10.1002/anie.202103336
172. Sahoo NK, Jana GC, Aktara MN et al (2020) Carbon dots derived from lychee waste: application for Fe3+ ions sensing in real water and multicolor cell imaging of skin melanoma cells. Mater Sci Eng C 108:110429. https://doi.org/10.1016/j.msec.2019.110429
173. Sahu BK, Sharma S, Kaur K, et al (2023) Farm waste-eggshell nanoparticles constitute gel for safe navigation of probiotic across the stomach. Mater Today Commun 34. https://doi.org/10.1016/j.mtcomm.2022.104876
174. Sajjadi M, Ahmadpoor F, Nasrollahzadeh M, Ghafuri H (2021) Lignin-derived (nano)materials for environmental pollution remediation: current challenges and future perspectives. Int J Biol Macromol 178:394–423. https://doi.org/10.1016/j.ijbiomac.2021.02.165
175. Sameni J, Jaffer SA, Tjong J, Sain M (2020) Advanced applications for lignin micro- and nano-based materials. Curr For Rep 6:159–171. https://doi.org/10.1007/s40725-020-00117-4
176. Sánchez R, Espinosa E, Domínguez-Robles J et al (2016) Isolation and characterization of lignocellulose nanofibers from different wheat straw pulps. Int J Biol Macromol 92:1025–1033. https://doi.org/10.1016/j.ijbiomac.2016.08.019
177. Santos RPO, Rodrigues BVM, Ramires EC et al (2015) Bio-based materials from the electrospinning of lignocellulosic sisal fibers and recycled PET. Ind Crops Prod 72:69–76. https://doi.org/10.1016/J.INDCROP.2015.01.024
178. Sarkar S, Skalicky M, Hossain A et al (2020) Management of crop residues for improving input use efficiency and agricultural sustainability. Sustainability 12:9808. https://doi.org/10.3390/su12239808
179. Sellergren B, Hall AJ (2012) Molecularly imprinted polymers. Supramol Chem. https://doi.org/10.1002/9780470661345.SMC137
180. September LA, Kheswa N, Seroka NS, Khotseng L (2023) Green synthesis of silica and silicon from agricultural residue sugarcane bagasse ash-a mini review. RSC Adv 13
181. Shahbaz A, Ayaz M, Bin Khalid U, Liaqat L (2023) Sustainable synthesis of silica nanoparticles from agricultural waste and its utilization in modern technology: a review. Energy Sour Part A Recover Util Environ Eff 45
182. Sharma B, Kumawat KC, Tiwari S, et al (2023) Silicon and plant nutrition—dynamics, mechanisms of transport and role of silicon solubilizer microbiomes in sustainable agriculture: a review. Pedosphere 33. https://doi.org/10.1016/j.pedsph.2022.11.004
183. Sharma N, Allardyce BJ, Rajkhowa R, Agrawal R (2023) Rice straw-derived cellulose: a comparative study of various pre-treatment technologies and its conversion to nanofibres. Sci Rep 13:1–12. https://doi.org/10.1038/s41598-023-43535-7
184. Sherif M, Lamis D, Abd El Haleem M El H, et al (2019) Effect of glue and temperatures on mechanical properties of starch-based biodegradable composites reinforced with bagasse fibers. Int J Biotechnol Biomater Eng 1. https://doi.org/10.35840/2633-8920/8802
185. Shi W, Fan H, Ai S, Zhu L (2015) Preparation of fluorescent graphene quantum dots from humic acid for bioimaging application. New J Chem 39:7054–7059. https://doi.org/10.1039/c5nj00760g
186. Shim J, Velmurugan P, Oh BT (2015) Extraction and physical characterization of amorphous silica made from corn cob ash at variable pH conditions via sol gel processing. J Ind Eng Chem 30. https://doi.org/10.1016/j.jiec.2015.05.029
187. Shinde R, Shahi DK, Mahapatra P et al (2022) Management of crop residues with special reference to the on-farm utilization methods: a review. Ind Crops Prod 181:114772. https://doi.org/10.1016/j.indcrop.2022.114772
188. Siddiqui MH, Al-Whaibi MH (2014) Role of nano-SiO2 in germination of tomato (Lycopersicum esculentum seeds Mill.). Saudi J Biol Sci 21:13–17. https://doi.org/10.1016/j.sjbs.2013.04.005

189. Singh G, Arora H, Hariprasad P, Sharma S (2023) Development of clove oil based nanoencapsulated biopesticide employing mesoporous nanosilica synthesized from paddy straw via bioinspired sol-gel route. Environ Res 220. https://doi.org/10.1016/j.envres.2022.115208
190. Singh Y, Sharma S, Kumar U et al (2024) Strategies for economic utilization of rice straw residues into value-added by-products and prevention of environmental pollution. Sci Total Environ 906:167714. https://doi.org/10.1016/j.scitotenv.2023.167714
191. Song P, Zhang L, Long H et al (2017) A multianalyte fluorescent carbon dots sensing system constructed based on specific recognition of Fe(III) ions. RSC Adv 7:28637–28646. https://doi.org/10.1039/c7ra04122e
192. Song Y, Qi N, Li K et al (2022) Green fluorescent nanomaterials for rapid detection of chromium and iron ions: wool keratinbased carbon quantum dots. RSC Adv 12:8108–8118. https://doi.org/10.1039/d2ra00529h
193. Sinamo S, Tarigan S, Gea S, Putra DP (2022) Utilization of silica nanoparticles from rice husks for improving the mechanical properties of dental materials: a literature review. Bioscientia Medicina J Biomed Transl Res 6. https://doi.org/10.37275/bsm.v6i14.662
194. Stygar Lopes M, Elita Carneiro M, Sulato de Andrade A, Potulski DC (2017) HIDRÓLISE ÁCIDA PARA PRODUÇÃO DE NANO LIGNINA EM PÓ. BIOFIX Sci J 3:41. https://doi.org/10.5380/biofix.v3i1.56180
195. Sun Y, Chu Y, Wu W, Xiao H (2021) Nanocellulose-based lightweight porous materials: a review. Carbohydr Polym 255:117489. https://doi.org/10.1016/j.carbpol.2020.117489
196. Suriyaprabha R, Karunakaran G, Yuvakkumar R, et al (2012) Growth and physiological responses of maize (Zea mays L.) to porous silica nanoparticles in soil. J Nanoparticle Res 14. https://doi.org/10.1007/S11051-012-1294-6
197. Szymańska-Chargot M, Cieśla J, Pękala P et al (2022) The influence of high-intensity ultrasonication on properties of cellulose produced from the hop stems, the byproduct of the hop cones production. Molecules 27:2624. https://doi.org/10.3390/MOLECULES27092624/S1
198. Tai JY, Leong KH, Saravanan P et al (2021) Facile green synthesis of fingernails derived carbon quantum dots for Cu2+sensing and photodegradation of 2,4-dichlorophenol. J Environ Chem Eng 9:104622. https://doi.org/10.1016/j.jece.2020.104622
199. Tamta S, Vimal V, Verma S et al (2024) Recent development of nanobiomaterials in sustainable agriculture and agrowaste management. Biocatal Agric Biotechnol 56:103050. https://doi.org/10.1016/j.bcab.2024.103050
200. Tan J, Zhou L (2024) Two birds with one stone: One-pot hydrothermal treatment of dragon fruit peel to simultaneously prepare efficient biochar adsorbent and fluorescent carbon quantum dots. Ind Crops Prod 210:118086. https://doi.org/10.1016/j.indcrop.2024.118086
201. Targhan H, Rezaei A, Aliabadi A et al (2024) Photocatalytic removal of imidacloprid pesticide from wastewater using CdS QDs passivated by CQDs containing thiol groups. Sci Rep 14:1–17. https://doi.org/10.1038/s41598-023-49972-8
202. Tayade R, Ghimire A, Khan W, et al (2022) Silicon as a smart fertilizer for sustainability and crop improvement. Biomolecules 12. https://doi.org/10.3390/biom12081027
203. Tian G, Zhong X, Wu X, Wang Z (2021) Self-Assembly preparation of nano-lignin/cationic polyacrylamide complexes. Polymers (Basel) 13:1726. https://doi.org/10.3390/polym13111726
204. Torres-Mansilla A, Álvarez-Lloret P, Fernández-Penas R, et al (2023) Hydrothermal transformation of eggshell calcium carbonate into apatite micro-nanoparticles: cytocompatibility and osteoinductive properties. Nanomaterials 13. https://doi.org/10.3390/nano13162299
205. U. Tosun G, Sakhno Y, Jaisi DP (2021) Synthesis of hydroxyapatite nanoparticles from phosphorus recovered from animal wastes. ACS Sustain Chem Eng 9. https://doi.org/10.1021/acssuschemeng.1c01006
206. Ul-Islam S, Shalla AH, Khan SA (2022) Handbook of biomass valorization for industrial applications. Handb Biomass Valorization Ind Appl 1–560. https://doi.org/10.1002/9781119818816
207. Ullal VN, Shivaramu HT, Aveen KP (2022) Rice husk ash utilization, composition and properties: a brief review. J Metals Mater Miner 32. https://doi.org/10.55713/JMMM.V32I4.1544

208. Väisänen T, Haapala A, Lappalainen R, Tomppo L (2016) Utilization of agricultural and forest industry waste and residues in natural fiber-polymer composites: a review. Waste Manag 54:62–73. https://doi.org/10.1016/J.WASMAN.2016.04.037
209. Varshney S, Nigam A, Pawar SJ, Mishra N (2022) An overview on biomedical applications of versatile silica nanoparticles, synthesized via several chemical and biological routes: a review. Phosphorus Sulfur Silicon Relat Elem 197
210. Vaz Júnior S (2018) Estudo da aplicação de lignina kraft em liberação controlada para manejo integrado de pragas. Embrapa Agroenergia, Brasília
211. Venkatesan S, Mariadoss AJ, Kathiravan A, Muthupandian A (2019) Fuel waste to fluorescent carbon dots and its multifarious applications. Sens Actuators B Chem 282:972–983. https://doi.org/10.1016/j.snb.2018.11.144
212. Vinayagam R, Kandati S, Murugesan G, et al (2023) Bioinspiration synthesis of hydroxyapatite nanoparticles using eggshells as a calcium source: evaluation of Congo red dye adsorption potential. J Mater Res Technol 22. https://doi.org/10.1016/j.jmrt.2022.11.093
213. Wang C, Xu J, Zhang R, Zhao W (2022) Facile and low-energy-consumption synthesis of dual-functional carbon dots from Cornus walteri leaves for detection of p-nitrophenol and photocatalytic degradation of dyes. Colloids Surf A Physicochem Eng Asp 640:128351. https://doi.org/10.1016/j.colsurfa.2022.128351
214. Wang C, Yao Y, Li J, Yamauchi Y (2022) Metal-organic frameworks: a robust platform for creating nanoarchitectured carbon materials. Acc Mater Res 3:426–438. https://doi.org/10.1021/accountsmr.1c00258
215. Wang R, Li S, Huang H et al (2023) Preparation of carbon dots from PET waste by one-step hydrothermal method and its application in light blocking films and LEDs. J Fluoresc 33:1305–1315. https://doi.org/10.1007/s10895-022-03132-9
216. Wang Y, Shaghaleh H, Hamoud YA et al (2021) Synthesis of a pH-responsive nano-cellulose/sodium alginate/MOFs hydrogel and its application in the regulation of water and N-fertilizer. Int J Biol Macromol 187:262–271. https://doi.org/10.1016/J.IJBIOMAC.2021.07.154
217. Wastowski AD (2018) Química da Madeira. Interciência, Rio de Janeiro
218. Wen Y, Rufford TE, Hulicova-Jurcakova D et al (2016) Structure control of nitrogen-rich graphene nanosheets using hydrothermal treatment and formaldehyde polymerization for supercapacitors. ACS Appl Mater Interfaces 8:18051–18059. https://doi.org/10.1021/acsami.6b04572
219. Wu G, Feng M, Zhan H (2015) Generation of nitrogen-doped photoluminescent carbonaceous nanodots via the hydrothermal treatment of fish scales for the detection of hypochlorite. RSC Adv 5:44636–44641. https://doi.org/10.1039/c5ra04989j
220. Xing Q, Buono P, Ruch D et al (2019) Biodegradable UV-blocking films through core-shell lignin–melanin nanoparticles in poly(butylene adipate- *co* -terephthalate). ACS Sustain Chem Eng 7:4147–4157. https://doi.org/10.1021/acssuschemeng.8b05755
221. Xu C, Ferdosian F (2017) Conversion of lignin into bio-based chemicals and materials. Springer, Berlin Heidelberg, Berlin, Heidelberg
222. Xu C, Nasrollahzadeh M, Selva M et al (2019) Waste-to-wealth: biowaste valorization into valuable bio(nano)materials. Chem Soc Rev 48:4791–4822. https://doi.org/10.1039/c8cs00543e
223. Xu E, Yu H, Wu W et al (2024) Preparation of high antioxidant nanolignin and its application in cosmetics. Int J Biol Macromol 272:132635. https://doi.org/10.1016/j.ijbiomac.2024.132635
224. Xu J, Krietemeyer EF, Boddu VM et al (2018) Production and characterization of cellulose nanofibril (CNF) from agricultural waste corn stover. Carbohydr Polym 192:202–207. https://doi.org/10.1016/J.CARBPOL.2018.03.017
225. Xu J, Li C, Dai L et al (2020) Biomass fractionation and lignin fractionation towards lignin valorization. Chemsuschem 13:4284–4295. https://doi.org/10.1002/cssc.202001491
226. Yadav M, Dwibedi V, Sharma S, George N (2022) Biogenic silica nanoparticles from agro-waste: properties, mechanism of extraction and applications in environmental sustainability. J Environ Chem Eng 10. https://doi.org/10.1016/j.jece.2022.108550

227. Yang X, Guo Y, Liang S et al (2020) Preparation of sulfur-doped carbon quantum dots from lignin as a sensor to detect Sudan i in an acidic environment. J Mater Chem B 8:10788–10796. https://doi.org/10.1039/d0tb00125b
228. Yasir M, Mishra R, Tripathi AS, et al (2024) Theranostics: a multifaceted approach utilizing nano-biomaterials. Discover Nano 19. https://doi.org/10.1186/s11671-024-03979-w
229. Younis MR, He G, Lin J, Huang P (2020) Recent advances on graphene quantum dots for bioimaging applications 8:1–25. https://doi.org/10.3389/fchem.2020.00424
230. Yu D, Chen C, Zhao G et al (2018) Biowaste-derived hierarchical porous carbon nanosheets for ultrahigh power density supercapacitors. Chemsuschem 11:1678–1685. https://doi.org/10.1002/cssc.201800202
231. Yuan L, Shao C, Zhang Q et al (2024) Biomass-derived carbon dots as emerging visual platforms for fluorescent sensing. Environ Res 251:118610. https://doi.org/10.1016/j.envres.2024.118610
232. Zaini LH, Gindl-Altmutter W, Gusenbauer C et al (2024) Nanofibrils from oil palm trunk: effect of delignification and fibrillation technique. J Wood Sci 70:19. https://doi.org/10.1186/s10086-024-02133-5
233. Zhang D, Zhang F, Liao Y, et al (2022a) Carbon quantum dots from pomelo peel as fluorescence probes for "Turn-Off–On" high-sensitivity detection of Fe3+ and L-Cysteine. Molecules 27. https://doi.org/10.3390/molecules27134099
234. Zhang H, Yang M, Luan Q et al (2017) Cellulose anionic hydrogels based on cellulose nanofibers as natural stimulants for seed germination and seedling growth. J Agric Food Chem 65:3785–3791. https://doi.org/10.1021/ACS.JAFC.6B05815/ASSET/IMAGES/LARGE/JF-2016-05815X_0007.JPEG
235. Zhang H, Zhang C, Zheng Q, Zeng X (2022b) Extraction of calcium phosphate nanoparticles from cooked bone wastes by supercritical water. Waste Biomass Valorization 13. https://doi.org/10.1007/s12649-021-01667-9
236. Zhang S, Fu X, Tong Z et al (2020) Lignin-clay nanohybrid biocomposite-based double-layer coating materials for controllable-release fertilizer. ACS Sustain Chem Eng 8:18957–18965. https://doi.org/10.1021/acssuschemeng.0c06472
237. Zhang X, Jiang M, Niu N et al (2018) Natural-product-derived carbon dots: from natural products to functional materials. Chemsuschem 11:11–24. https://doi.org/10.1002/cssc.201701847
238. Zhao W, Simmons B, Singh S et al (2016) From lignin association to nano-/micro-particle preparation: extracting higher value of lignin. Green Chem 18:5693–5700. https://doi.org/10.1039/C6GC01813K
239. Zhao X, Li S, Wu R, Liu D (2017) Organosolv fractionating pre-treatment of lignocellulosic biomass for efficient enzymatic saccharification: chemistry, kinetics, and substrate structures. Biofuels Bioprod Biorefin 11:567–590. https://doi.org/10.1002/bbb.1768
240. Zhao X, Liao S, Wang L et al (2019) Facile green and one-pot synthesis of purple perilla derived carbon quantum dot as a fluorescent sensor for silver ion. Talanta 201:1–8. https://doi.org/10.1016/j.talanta.2019.03.095
241. Zhao X, Zhang Z, Pang J, Su L (2022) Study on the preparation of epoxy resin materials from nano-lignin polyols. Ind Crops Prod 185:115158. https://doi.org/10.1016/j.indcrop.2022.115158
242. Zhou XQ, Hayat Z, Zhang DD, et al (2023) Zinc oxide nanoparticles: synthesis, characterization, modification, and applications in food and agriculture. Processes 11
243. Zhu M, Ma W, Deng J et al (2023) Effect of nano-calcium on carotenoid and anthocyanin contents of nectarine fruit. Plant Physiol Biochem 204:108088. https://doi.org/10.1016/j.plaphy.2023.108088

Plant-Based Synthesis of Metal Nanoparticles for Managing Agroecosystem

Rabia Nazir and Sana Khalid

Abstract The plant-mediated synthesis of nanoparticles (NPs) is a simple, economical, environment-friendly, easily scalable, time-efficient, low-cost, low-temperature, and safe approach for preparing metal-based nanoparticles. Various plant parts, such as bark, flowers, leaves, and roots are commonly used in this process. These NPs have been successfully utilized for improved seed germination, shoot and root growth, shoot and root biomass, and shoot branching in plants. High levels of chlorophyll, protein, reducing sugar, phenols, etc. have also been observed in plants treated with these NPs. In addition, they have been found effective as pesticides and are able to tackle many fungal, bacterial, and pest-borne infections. Furthermore, they have shown promise in mitigating the adverse effects of environmental stressors such as drought and salinity.

Keywords Nanoparticles · Plant-mediated synthesis · Priming agents · Stress tolerance · Nano-pesticides

1 Introduction

Rapid population increase is demanding an increase in agricultural productivity while maintaining food quality and security. This together with biotic and abiotic stresses is putting enormous stress on the agricultural economy [32, 42]. Scientists are driving different pathways to enhance agricultural production and designing new agrochemicals that can effectively tackle multiple issues faced by crops during their life cycle. Moreover, there is also a need to focus on the environmental implications of these agrochemicals [31, 77]. Hence, nanotechnology has arisen as a domain that can be helpful in designing such chemicals that can promote sustainability and maintain a balance between agriculture and agroecosystems [93].

Nanotechnology revolution has brought with it novel developments in almost every field of life ranging from simple household items to hi-tech developments. In the

R. Nazir (✉) · S. Khalid
Pakistan Council of Scientific and Industrial Research Laboratories Complex, Lahore, Pakistan
e-mail: rabiapcsir@yahoo.com

L. F. Fraceto et al. (eds.), *Emerging Nanotechnologies for Agroecosystem Management*, Smart Nanomaterials Technology, https://doi.org/10.1007/978-981-95-0187-8_2

field of agriculture, multitudinous benefits have been offered by nanotechnology like food security, food quality, agrochemicals, monitoring, water and soil reclamation, etc. [63]. The nanomaterials especially metal and their metal oxides have found applications in the field of soil amelioration [16, 49], plant disease [44] and stress management [4]. Moreover, these can be used as fertilizers [46, 47], plant growth regulators [39], priming agents [2] etc.

Nanomaterials with their tailor-made properties that are unique to them and this is because of these value additions that these nanomaterials have brought revolution in the agrochemical's domain. These materials offer several benefits like reduced dose application, enhanced uptake nutrient efficacy, minimization of nutrient losses, targeted delivery of active agents to reduce negative impacts on plants and environment, improved pest management, better adhesion and penetration, etc. [16]. Despite these advantages, commercialization of nanomaterials and widespread use are still in their infancy, as few products have made it to the market [24].

Chemical and physical synthetic approaches [18] have been used to prepare these metal-based nanoparticles (NPs) but owing to their hazardous nature green synthesis mode is preferred [90]. Plants are readily available in ecosystem and also in diverse nature. These diverse nature plants carry in them different combinations of phytochemicals that facilitates the synthesis of nanomaterials with variable sizes, properties, shapes surface area, and versatility [29]. Hence these nanomaterials together with their remarkable characteristics find applications in diverse domains of agriculture. This chapter emphasizes on impact and applications of plant-mediated synthesized NPs in promoting agroeconomics.

2 Plant-Mediated Synthesis of Nanoparticles

Nanoparticles synthesize by different routes is a constantly expanding field. Generally, three approaches have been in practice i.e. physical, chemical, and biological. Recently, NPs' synthesis by employing green approaches has gained considerable attention because of being certain advantages i.e. simple process, economic, environmental-friendly approach, easily scalable, time-efficient, low-cost, low-temperature, and safe as it avoids use of toxic chemicals [19, 58, 92] (Fig. 1). Among green synthesis that incorporated use of microbes and plants, plant-mediated synthesis is deemed best owing to being less tedious as it doesn't require dealing with microbes and related protocols of purification, isolation, and maintenance [21, 62]. Plant-mediated synthesis (PMS is a bottom-up approach in which plant extracts that contain phytochemical agents like hydroquinone, phenols, alkaloids, flavonoids, alcohols, amino acids, amines, terpenoids, carboxylic acids, etc. that play multiple roles as reducing, capping and stabilizing agents [58, 88] (Fig. 1). For the purpose of PMS of NPs, different parts of plants have been employed as raw materials i.e. roots, fruits, barks, seeds, stem, flowers, leaves, and peels (Fig. 1).

PMS involves three phases i.e.

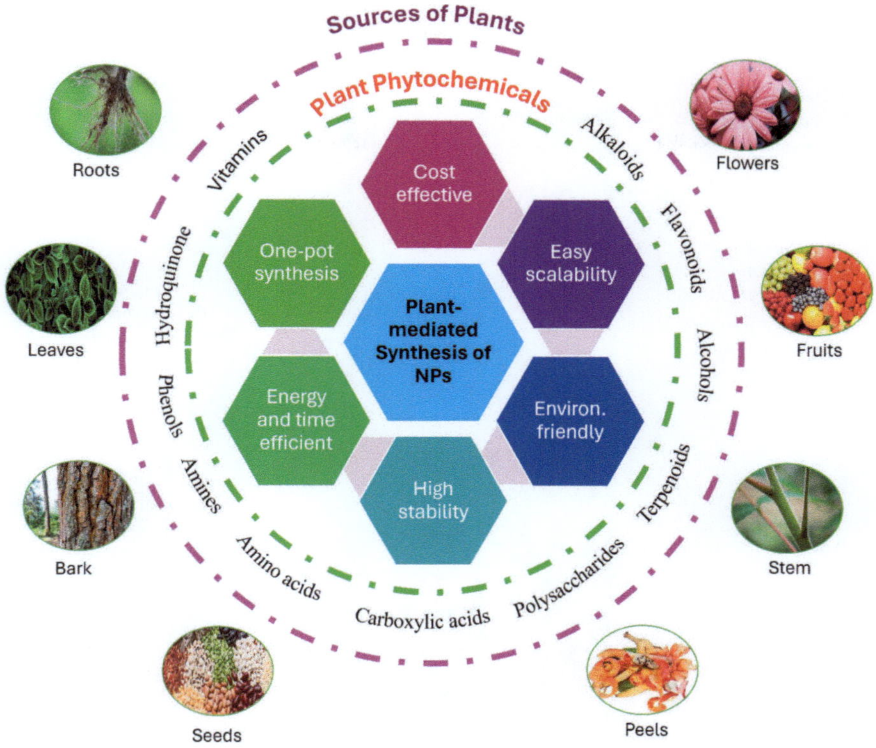

Fig. 1 Advantages (inner circle) of plant-mediated synthesis using different parts of plants (outer circle) containing multiple phytochemicals (middle circle) responsible for NPs synthesis

(a) *Initial phase* which consists of attaining the precursors like metal salts and plant extracts
(b) *Activation phase* deals with metal ions reduction followed by the formation of NPs and nucleation center generation.
(c) *Growth phase* is demarcated by the formation of larger particles from the fusion of small particles.
(d) *Termination phase* defines the nanoparticles' shape and size. In this final stage, phytochemicals in the plants promote NPs stabilization [64] (Fig. 2).

The PMS is influenced by many factors as described below:

(a) pH of solution—it impacts generation of nucleation centers, reaction rates, size, shape and aggregation of NPs.
(b) Temperature—increase in reaction temperature promotes formation of nucleation centers, enhancing reaction rates and reduction in required plant extract concentration. The reaction yield, size, shape, particle growth, and distribution of NPs is also impacted by changes in reaction temperature.
(c) Reaction time—the length of reaction defines NPs' size and shape as well as reaction yields.

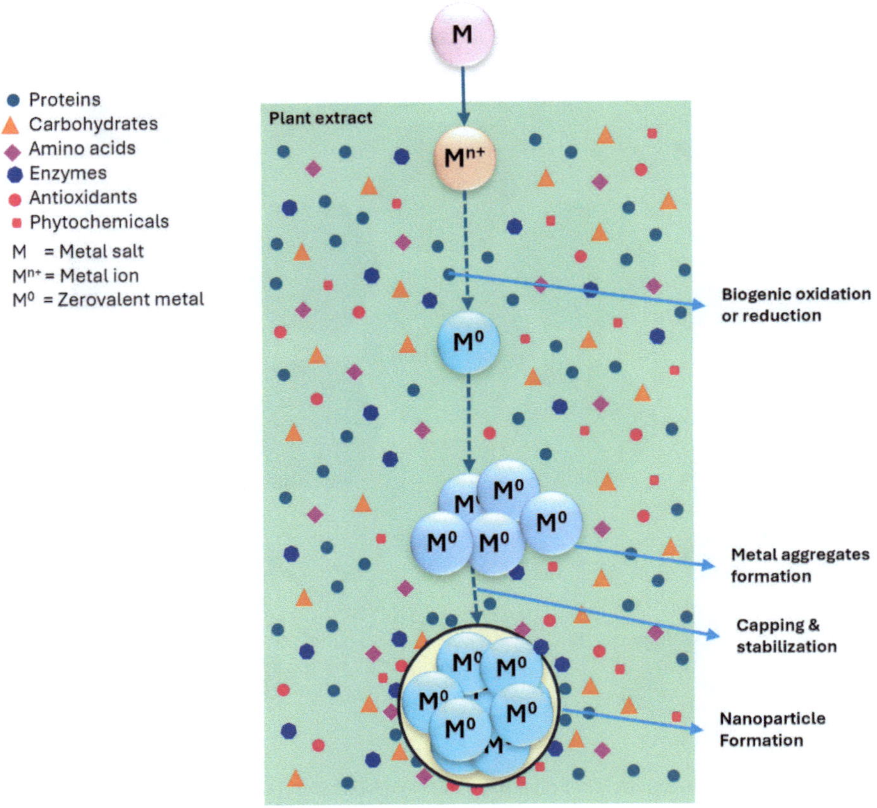

Fig. 2 Proposed mechanism of PMS NPs

(d) Plant extract & biomass dosage—right amount of plant extract and biomass dosage is decisive for defining the NPs' size, shape, and yield. The plant extract concentration to be utilized is itself dependent on plant species, extraction solvent, and extraction temperature used.
(e) Metal ion concentration—it affects the NPs size, reaction time, and yield [19, 64, 88, 92].

Two pathways are suggested for movement of NPs into the plant i.e.

(a) Apoplastic route—movement of water outside plasma membranes through extracellular spaces and cell walls themselves. Root central cylinders as well as vascular tissues are approached by NPs using this route by employing radial movement within the plant [38]
(b) Symplastic route—movement of water through plasmodesmata via cytoplasm.

3 Role of Nanoparticles as Fertilizers and Priming Agents

There are several nanparticles (metal and metal oxides) prepared by using green synthesis route and used as proming agents and fertilziers (Table 1)

3.1 Iron Oxide Nanoparticles

Iron oxide NPs are used as priming agents to promote early germination in the seeds (Table 1). Fe_2O_3 NPs (20–40 nm) synthesized using fruit extract of Cornelian cherry were applied to the barley plant seeds in 10–100 mg/L concentrations and compared the results with bulk counterpart. It was indicated that NPs significantly stimulated the barley plant germination by promoting root biomass (42%) in contrast to bulk iron oxide NP which only resulted in 18% biomass increase [73]. Lower dosages (5 mg/L) of iron oxide NPs, synthesized via *Artemisia afra* leaf extract, were found suitable for early germination of the carrot as well as spinach seeds leading to increase in root (73 & 42 mm) and shoot (9 & 27 mm) lengths, respectively [58]. Biogenic iron oxide NPs when used in 20 mg/L dosage improved both rice seed germination and its vigor as compared to ferrous sulfate. Enhancement in α-amylase activity, reactive oxygen species, soluble sugar levels, and antioxidant activities of the plant were also recorded for NP-treated seeds in contrast to ferrous sulfate as well as control [2].

The blend of PMS iron oxide NPs and chitosan loaded with NPK (500, 60, 400 mg/L) was applied for corn seed treatment and results were compared with the control. No substantial change was recorded in contrast to the control for germination rate, shoot and root lengths (~78%, 70 cm, 36 cm) at dose rate of 100 ppm (~77%, 68 cm, 32 cm) while much less rates and lengths were observed at dose rate of 10ppm (~55%, 55 cm, 25 cm). Contrary to that, there was almost 45% increase in the index-speed of germination owing to enhanced permeation of NPs in the seed. It is proposed that iron oxide NPs seed treatment caused inactivation of metabolic pathways in seeds hence lowering the germination rate while enhanced speed of germination was facilitated by speedy access to water facilitated by iron oxide NPs [89].

Soil application of γ-Fe_2O_3 NPs (10 mg/kg of soil) promotes early sprouting (10 days) and high sprouting percentage (45%) as compared to the control (27 days; 30%) and simple $FeSO_4$ treatment (20 days; 66%). Increase in the NPs dosage from 10 to 50 mg/kg resulted in an increase in days to appearance of first leaves in mulberry by 3 days but enhanced the sprouting percentage to 73%. In addition to that, the iron oxide NPs treatment also facilitates increase in root & shoot length, branching and biomass, number of leaves/plants and leaf length & breadth as compared to the control and iron salt treatment. Higher chlorophyll, protein, total & reducing sugar, total phenol, and flavonoid contents was also observed with introduction of iron oxide NPs [38]. The dose-based induction of green and chemically synthesized FeO NPs and the resulting plant response were compared with the direct application of $FeCl_3$.

Shoot length of maize seedlings registered positive impact on induction of green synthesized FeO NP at dosage of 50 mg/L while same concentration of chemically synthesized FeO NP induces toxicity effect. Further increase in the concentrations of both the NPs caused undesirable impacts on plant growth. In the case of $FeCl_3$, high toxicity was registered by the plants leading to root and shoot lengths even smaller than the control. Leaf area and number of leaves showed −37% and −50% reduction in case of chemically synthesized NPs while PMS NPS showed 13 and 26% improvement, respectively. In contrast to chemically synthesized NPs, PMS NPs enhanced total chlorophyll content, nitrate content and plant anti-oxidative activity [37].

3.2 Zinc Oxide Nanoparticles

ZnO NPs prepared from pea extract were applied in different concentrations (0.2–1000 mg/L) to treat maize seeds. Early sprouting (2–3 days) was observed in all the treatments as compared to the control that took 4–5 days with much less germination % i.e. 68%. On other hand seeds treated with 700–1000 mg/L NPs showed maximum germination of 92%. Furthermore, treated seeds depicted enhanced root and shoot lengths as well as improved fresh and dry seedling weights. Comparison of NP treatments with zinc sulfate treated seeds showed better germination (19%), shoot length (55%), root length (37.5%), leaf area (33%) and cobs per plants (33%) [57]. Studies conducted on Mung bean seeds and Neem leaves procured ZnO NPs, showed 53% enhancement in growth when 200 mg/L NPs concentration was used as compared to the control. The application of NPs also resulted in increased root and shoot growth achieving maximum growth at 100 and 200 mg/L [85]. Similar studies performed on wheat seed showed that employing higher concentrations of ZnO induces toxic effects to the plant. Maximum germination i.e. 100% as compared to control was observed at 600 mg/L after which further increase to 900 mg/L reduce the germination rate to 70%. The root and shoot lengths were also recorded indicating that highest values were observed at 150 mg/L while the maximum fresh root and shoot weights were attained at 900 mg/L [87]. Biogenic ZnO NPs have augmented seed vigor index in case of black mustard and *Pennisetum glaucum* seeds [65, 66] and growth parameters and yield of tomato plants [13]. The role of Zn NPs is suggested to improve plant metabolism, reduce seed dormancy, and improve rapid water uptake via cell wall into the seed kernel consequently promoting early seed germination. Furthermore, improved shoot and root length was attributed to zinc being essential micronutrient imperative for plant metabolic processes. Improved chlorophyll a & b, and carotenoid contents were also recorded in maize plants grown from NPs treated seeds at concentration levels of 800, 800, and 1000 mg/L owing to the important role of Zn in chlorophyll synthesis. In comparison to untreated seeds, there is a significant increase in sugar contents of the plants with an increase in NPs dosage attaining maximum value of 455 μg/g at 400 mg/L, after which it starts decreasing. Maximum protein (17.1 μg/g) and proline (343 μg/g) contents were recorded at 900 mg/L.

Evaluation of chlorophyll, phenolic content, proline, and malondialdehyde (MDA) contents of the plants grown from treatment of Zn NPs and micro $ZnSO_4$ showed enhancement in values by 19, 2, 36, and 160%, respectively while sugar values of the two treatments vary insignificantly. There is also an appreciable increase in the zinc and starch content of the seeds treated with NPs in contrast to salt treated ones. The difference was attributed to higher translocation and mobility of Zn NPs as compared to the salt [57]. The foliar application of these ZnO NPs is also effective in enhancing the growth of tree saplings. In a study carried out over a period of 6 months showed noteworthy increase in heights of the plants treated with 30 mg/l ZnO NPs which follows the order: *Alstonia scholaris* (~95 cm) > *Pongamia pinnata* (~80 cm) > *Azadirachta indica* (~65 cm) [22].

3.3 Silver Nanoparticles

Impact of soaking seeds of wheat and tomato in silver NPs attained by Rosemary plant leaves extract showed varied behaviors for the two plants (Table 1). Ag NPs reduced wheat seed germination (48%) and seedling dry biomass (0.026 g/plant) in contrast to control (57%; 0.028 g/plant). The wheat plants also showed suppression in shoot and root lengths of Ag NPs treated seeds as compared to untreated ones. In contrast to these results, similar treatment of Ag NPs to tomato seeds have resulted in enhanced shoot and root length. In wheat plants, Ag NPs seed treatment resulted in higher photosynthetic pigments but for tomato chlorophyll a content drops appreciably. The seed treatment also inhibited antioxidant enzymes (catalase (CAT) and peroxidase) as well as soluble proteins in wheat plants while in tomato plant both antioxidant activity and soluble protein contents are increased significantly [28].

4 Role of Nanoparticles as Nano-Pesticides

Nanomaterials as novel pesticidal agents are playing a remarkable role in crop protection, especially in the control and management of pests as they have high efficacy with less environmental effects. The introduction of nanomaterials as novel pesticidal agents have reformed the pest management practices owing to their high efficacy and low environmental impacts. Nano-pesticides offer sustainable release and improved penetration over conventional pesticides, leading to increased efficacy against target pests. For this purpose, various plant-based metal NPs as nano-pesticides are researched to determine their efficacy against different pests (Table 2).

Ag NPs have been extensively explored for their effectiveness against various pests. Danish et al. [23] applied green-synthesized (GS) Ag NPs at three stages in plants injected with *Meloidogyne incognita* (famous for root-knot disease), i.e., NPs application a week prior to nematode inoculation (T1), simultaneous induction

Table 1 Plant-mediated nanoparticles with application as fertilizers and priming agents

NPs	Plant & part used	Particle size (nm)	Experimental crop	Application	Refs
Fe_2O_3	Fruit extract of Cornelian cherry	20–40	Barley seedlings	Enhancement in root and stem biomass (Seed soaking)	[73]
Fe_2O_3 + chitosan + NPK	*Moringa oleifera* leaves extract	3	Corn seeds	Seed germination	[89]
γ-Fe_2O_3	*Camellia sinensis* leaves extract	11.98	Mulberry stem cutting	Sprouting, Growth (Soil application)	[38]
FeO	*Withania coagulans*	–	Maize	Seedling growth (Hydroponic application)	[37]
FeO	*Artemisia afra* leaf extract	15–20	Spinach & Tomato seeds	Seed germination (Seed soaking)	[58]
FeO	*Cassia occidentalis L.* flower extract	–	Rice seed	Seed germination (Seed soaking)	[2]
ZnO	Leaves extract of *Azadirachta indica*	50	Mung bean seeds	Seed germination (Seed soaking)	[85]
ZnO	Scallions onion peels extract	40–100	Wheat seeds	Seed germination (Seed soaking)	[87]
ZnO	*Citrus reticulate* peels extract	8.89–8.62	*Brassica nigra* (black mustard) seeds	Seed germination (Seed soaking)	[65]
ZnO	Leaves extract of *Azadirachta indica*	50	Tomato seeds and plants	Seed germination and plant growth (Seed soaking)	[13]
ZnO	Leaves extract of *Calotropis gigantea*	20	*Azadirachta indica, Pongamia pinnata, Alstonia scholaris* saplings	Seedling growth (Foliar application)	[22]
ZnO	pea (*Pisum sativum* Linn)	79.8	*Zea mays*	Seed germination and seedling growth (Seed soaking)	[57]
ZnO	*Syzygium cumini* leaves extract	16.4	*Pennisetum glaucum*	Seed germination (Seed soaking)	[66]
ZnO	*Cassia renigera* leaf extract	28–112	Cauliflower	Seed germination (Seed soaking)	[61]
Ag	Rosemary leaves extract	17	Wheat and tomato seeds	Seed germination (Seed soaking)	[28]

Table 2 Plant-mediated meal nanoparticles with their application against common pests

Biogenic metal NPs	Plants	Target pest	Effects	Refs
Ni-Pd	*Cocos nucifera* coir dust	*C. maculate* insect & eggs (isolates)	Pesticidal and ovicidal activity	[26]
Ag	*Senna siamea* leaves	*M. incognita* (seedlings inoculation)		[23]
Ag	*Myriostachya wightiana* leaves	*S. oryzae*, *T. castaneum*, *R. dominica*	Mortality; 47.4% (*S. oryzae*), 55.2% (*T. castaneum*), and 52.8% (*R. dominica*) at 150 μg concentration after 1 day of exposure	[91]
Ag	*Annona raticulata* leaves	*S. oryzae*	Insect repellence (90%) after 5 h of exposure, and rice grain weight loss (0.13 g)	[51]
Ag	*Datura stramonium* leaves	*Trogoderma granarium*	Mortality of 41.4% at dosage of 300 ppm after 96 h, and insect repellence of 67.89% after 24 h	[33]
Ag	*Moringa oleifera* leaves	*S. oryzae*	100% mortality after 9 days of exposure, and improvements in seed germination (88%), length of seedling (30.95%), seed biomass (2.42 g), seed weight loss (0.00%) and vigor index (2723)	[69]
SiO_2	Cinnamon oil	*C. cephalonica*	Reduction in pupae development (45.0%), decrease in pupae duration (5.5%), emergence of adult insects (0%), adult insect life span (0%), and adult mortality (100%)	[15]
FeO	*Trigonella foenum-graecum* leaves	*Tuta absoluta*	Fatality of 72% pest larvae after 96 h at 100 μg mL^{-1} concentration (LD_{50}, 147.32 μg mL^{-1})	[68]
ZnO	*Ricinus communis* leaves	*T. granarium*, *T. castaneum*	Mortality; 49.5% (*T. granarium*), 66.3% (*T. castaneum*)	[35]
Cu/Zn	*Prosopis juliflora* leaves	*P. solenopsis*	> 30% insect fatality, 50% decrease in cell viability of pest	[52]

(continued)

Table 2 (continued)

Biogenic metal NPs	Plants	Target pest	Effects	Refs
Cu, Cu/Zn, Cu/Mn	*Capsicum chinensis* (Habanero pepper)	*T. castaneum, P. solenopsis*	Mortality; 63–80% (*T. castaneum*) for Cu, Cu/Zn, Cu/Mn NPs, and 20–28% (*P. solenopsis*) for Cu and Cu/Zn NPs	[10]
Cu/Zn	*Crotaloria longirostrata* leaves	*P. solenopsis*	65% mortality	[53]
TiO_2	*Juniperus phoenica*	*S. oryzae*	85% fatality, increased levels of acetylcholine esterase causing continuous stimulation in synaptic nerves of insect and leading to paralysis, low levels of total soluble CAT and protein (TSP), showing oxidative stress and protein break down in the body of pest	[36]

of both (T2) and inoculation one week before NPs introduction (T3). There was a noteworthy reduction in plant growth attributes in the *M. incognita* diseased plants. In all the cases, 50 mg/L NPs exposure leads to almost the same shoot and root lengths as well as fresh and dry weights of shoot and root as was observed in healthy plants (control). Above this concentration, considerable decrease in these attributes was recorded probably owing to toxicity induced by NPs. Similar observations were recorded for number of branches and umbels, seed yield, and oil content. Evaluation of photosynthetic pigments (chlorophyll a & b, Carotenoid), biochemical activity (nitrate reductase, carbonic anhydrase), and leaf N and protein contents indicated significant reduction in infected plants but the introduction of Ag NP in all three cases have resulted in overcoming this biotic stress enabling improved concentrations. Antioxidant enzymes such as peroxidase (POS), catalase (CAT), superoxide dismutase (SOD), and ascorbate peroxidase (APX) registered an increased in infected plants. The application of Ag NPs appreciably improved these levels enabling plant healthy growth especially in T3. The NP treatment amicably reduced galls and eggs masses count in infested plants owing to appreciable reduction in the population of nematodes in soil and roots, with best results achieved in case of T3. Sankar and Abideen [76] determined the insecticidal activity of mangroves (*Avivennia marina*) leaves aqueous extract derived Ag and Pb NPs by using variable concentrations against *Sitophilus oryzae*, resulting in notably high insecticidal efficacy (100%) as compared to aqueous extract (30%). However, Pb NPs exhibited better performance than Ag NPs as 100% fatalities against stored grain pest for Pb and Ag NPs were achieved after 72 and 96 h of exposure, respectively. In another study, Ag NPs

(52.4 nm and rod-like morphology) fabricated using *Euphorbia prostrate* (prostrate spurge) were applied against *S. oryzae* and caused 100% mortality after one week of exposure, whereas in case of aqueous extract and $AgNO_3$ solution almost similar results were attained after two weeks of treatment. The LC_{50} values for $AgNO_3$ solution, aqueous extract and phyto-fabricated Ag NPs against *S. oryzae* were recorded as 247.9, 213.3 and 44.7 mg kg^{-1} [97]. When applied at dose of 1 $\mu g\ mL^{-1}$, Ag NPs fabricated from rosemary achieved maximum fatality rate of 100% for *Sitophilus granaries* (*L.*) after 120 h, whereas cardamom and Laura-derived Ag NPs exhibited similar pesticidal effectiveness after 144 h of treatment [71]. Ag NPs synthesized from neem leaves (*Nerium oleander*) registered potent insecticidal activity against *Callosobruchus maculatus* and *Tribolium castaneum*, leading to reduction in number of eggs produced/female insect, and adult insect emergence (71.66 and 68.33%) for Ag NPs as compared to control (i.e., 90 and 95%) against *C. maculatus* and *T. castaneum*, respectively [12]. Shahzadi et al. reported that treating *T. castaneum* with *Ricinus communis* and *Citrus paradise* oil-based Ag NPs when applied at dose of 300 mg/L resulted in larvicidal activity of 41.4 and 38.8% after 96 h, and insect repellent efficacies of 67.9 and 48.3% after 24 h, respectively [81]. *Syzygium aromaticum* (clove) methanolic (1%) and acetone (4%) extract-mediated Ag NPs presented notable insect repellent properties (71 and 66%) against *T. castaneum and Corcyra cephalonica*, respectively, and larvicidal effectiveness (100%) by acetone (4%) extract-mediated Ag NPs against Bombyx mori after 2 days [80]. Eucalyptus aqueous extract-mediated Ag NPs (4250 mg/L) were found lethal for different stages (egg, crawler and adult female insect) of citrus mealybug (*Planococcus citri*), causing mortality rates of 76.7, 93.6 and 66.3%, respectively after 3 days of exposure with respective LD_{50} values calculated as 1248, 1296 and 2236 mg/L [8]. Ag NPs obtained from *Cameline sativa* leaf extract at concentration of 500 mg/L presented 42.7 and 36.2% of insect repellence after 36 h, and 60.1 and 42.6% of fatality after 72 h against *Orzyaephilus surinamensis* and *S. granaries*, respectively [72]. Ag NPs prepared using *Citrus sinensis* (*L.*) (Sweet orange) peels extract was assessed for their insecticidal activity against *Tribolium confusum*. The findings of the study indicated that plant extract by itself did not induce toxicity against target pest. However, fabricated NPs demonstrated remarkably high insecticidal activity (77–83%) when compared to commercial $AgNO_3$ (40–75%) [79]. Moreover, treatment of *Callosobruchus chinensis* infested soybean seeds with curry leaf extract derived Ag NPs (PS) and standard Ag NPs (SS) at two different concentrations (35 and 70 mg/L) in solution: seed ratio of 0.3:1 was done to evaluate the insecticidal efficacy of green fabricated Ag NPs in terms of fatality, seeds damage and reduction in seeds weight. The complete mortality (100%) was noticed with treatment SS-70 mg/L after one week of exposure, while treatment PS-70 mg/L depicted 100% fatality after exposure of two weeks. The minimum soybean seeds damage was observed with SS-70 mg/L (38.7%) followed by PS-70 mg/L (42.0%), and minimum reduction in seeds weight was noted for SS-70 mg/L (0.78%) followed by PS-70 mg/L (0.96%) after two weeks of exposure [95].

Comparative insecticidal effectiveness against *Rhyzopertha dominica* was evaluated utilizing *Euclyptus globulus* leaf extract (300, 600, 900, 1200, 1500, and

1800 mg/L) and biosynthesized ZnO NPs (100, 200, 300, 400, 500, and 600 mg/L) for treatment period of 3,7,11 and 15 days. After 15 days of exposure, insecticidal activity showed higher mortality rate of 80.5% for ZnO NPs (600 mg/L) than *E. globulus* extract (i.e., 62.5%) when used in 1800 ppm concentration. The LD_{50} values for ZnO NPs and plant extract were 202.11 and 1043.06 mg/L, respectively. Post-treatment inhibitory effects on the growth of *R. dominica* were 87.0 and 75.7% by NPs and extract, respectively after one month [83]. *Pongamia pinnata* leaf extract-mediated ZnO NPs induced insecticidal effects against *C. maculatus*, leading to significant reduction in fecundity and number of eggs hatching of target pest. The findings of the study depicted that ZnO NPs amicably slow down the larvae-pupae and adult developmental period at dose of 25 μg/mL and caused complete mortality (100%) of *C. maculatus* with LC_{50} value recorded as 10.85 μg/mL. Biogenic ZnO NPs also interfered with several digestive enzymes in the midgut of *C. maculatus*, resulting in remarkable reduction in the activity of α and β-glycosidase, lipase, α-amylase and cysteine protease [50]. Similar findings were presented by Shayaraj et al., where pungam oil-based Au NPs substantially lengthened the larvae-pupae and adult lifespan of *Pericallia ricini* [75].

Likewise, the remarkable reduction in cell viability of adult insects (*T. castaneum*) after treatment with *Acacia cornigera* and *Annona purpurea* leaves aqueous extract-based Cu NPs was noted up to 95 and 90%, respectively after 6 days of exposure. The Annona purpurea and Acacia cornigera derived Cu NPs also exhibited mortality rate of 76.6 and 90% against *T. castaneum*, respectively [86]. Ghidan et al. studied the potential of *Punica granatum* peels derived CuO NPs as an effective insecticidal agent against green peach aphid (*Myzus persicae*). The biogenic CuO NPs (8000 μg/mL) successfully registered mortality rate of 86% third and fourth stage juvenile instars [30]. CuO NPs (<50 nm) were phyto-fabricated by using *Orobunche aegyptica* (holoparasitic plant) stem extract were applied as an effective nematocidal agent against *M. incognita.* Under in vitro conditions, treatment with Cu NPs (800 μg/ mL) induced high mortality (91.5%) of second stage nymphal instar of *M. incognita* after 16 h of exposure. Biosynthesized Cu NPs also shoed ovicidal activity causing significant suppression in eggs hatchability (83%) at 400 μg/mL concentration [6]. *Pluchea sericea* and *Prosopis juliflora* leaves extract-based CuO NPs (average size, 68.1 and 33.8 nm, respectively) when applied in concentration of 300 μg/mL demonstrated mortality rate (14 and 20%) and cell viability (30 and 38%), respectively against *Phenacoccus solenopsis* after 96 h of treatment [48]. In another study, *Achillea fragrantissima*-derived CuO NPs displayed notable molluscicidal effects against *Pirenella conica* than plant extract with LD_{50} values noted as 5.65 and 279.53 mg/L, respectively [7].

Nopal cactus fruit extract-derived cadmium sulfide (CdS) NPs were applied in different doses (0.005–0.04%) to *Spodoptera litura* larvae, and their pesticidal effectiveness and oral administration effects on various biochemical, physiological, and oxidative stress profile were studied in concentration-dependent manner. The application of 0.04% dosage level resulted in 46.66% of larvae fatality, 25.28% of decreased reproducibility, 22.74% of low fertility, 27.78% of pupal and adult insect malformation, and 43.87% of inhibitory effects on adult emergence. The significant reduction in SOD, CAT, GAT, and increase in malondialdehyde (MAD) levels was observed during 48–72 h of NPs exposure which is suggestive of potent toxicity, oxidative stress, and substantial damage to pest developmental and reproductive systems, respectively [41]. For *M. persicae, Artimisia herba-alba* leaf extract-mediated iron sulfide (FeS) NPs showed highest mortality (100%) at 800 mg/L concentration after 72 h of treatment with LD_{50} values registered as 251 and 302 mg/L for initial and late juvenile instars, respectively [14]. Titanium dioxide (TiO_2) NPs obtained by using *Syzygium cumini* leaf extract were applied at variable doses (125, 250, 375, and 500 mg/L) for evaluating their nematocidal activity against *M. incognita* eggs and second-stage nymphs under in vitro and pot experimental conditions. Lowest percentage eggs hatchability and highest mortality was noted with increase in NPs dosage levels. At highest concentration (500 mg/L), mortality rate of 69.23% was achieved (LC_{50}, 232.4 mg/L) after 48 h of treatment. In case of pot experiment, second-stage nymphs inoculated carrot seeds showed considerable improvements in plants' physiological and pathological profile (i.e., increase in fresh root and shoot weight, root and shoot length, chlorophyll content, carotenoid levels, and suppression in root-knot disease index etc.) after 5 days of treatment with NPs [43].

Pd–Ni NPs prepared by using coir waste were checked for their pesticidal and ovicidal efficacy against *C. maculatus* beetle and eggs and comparing the results with Azadirachtin as standard drug. At 20 µg/mL concentrations, NP depicted comparable mortality rates (92.9%) than the drug (99.01%) while in case of ovicidal activity the percentage of eggs that didn't hatch increased with increase in exposure time as well as NPs dose. At 1.25 mg/L with exposure levels of 24 h and 72 h the ovicidal activity registered increased from 62 to 86% while at 20 mg/L the activity rose to 82 and 100% [26].

5 Role of Nanoparticles Under Stress Conditions

Plants are prone to different types of abiotic and biotic stresses that influence the plant growth and hence yield. The share of abiotic and biotic stresses in overall losses suffered by agriculture productivity is 50 and 30%, respectfully. These stresses impact negatively on plants' morphology, physiology and biochemistry [5] (Fig. 3). Nanoparticles are frequently used to overcome these stresses in plants and enhance yield. In this regard, different NPs (Table 3) are employed as detailed below.

5.1 Drought

Drought, one of the key abiotic stress, hampers plant growth and seed germination by limiting water availability. This also causes shrinkage of xylem and phloem, inhibits physiological functioning, affects photosynthetic efficiency and damage root structure. This further leads to reduction in membrane permeability, cell wall thickening,

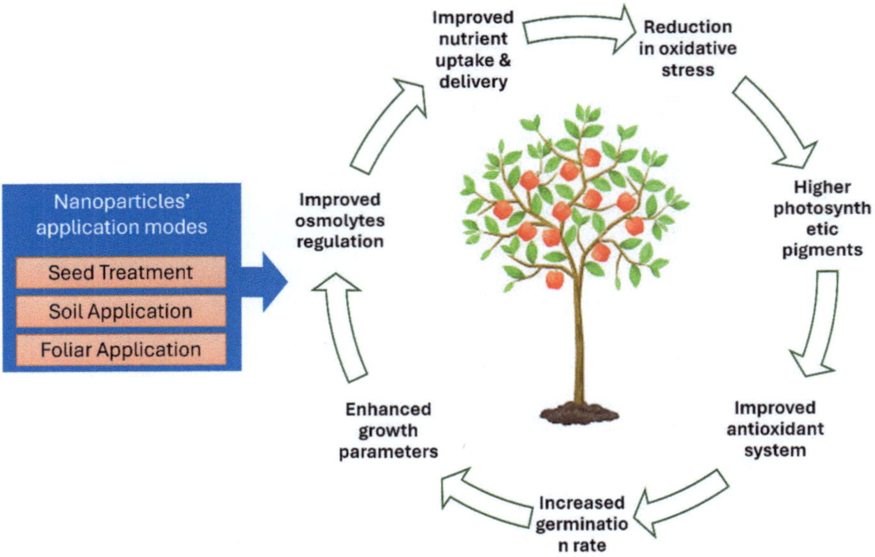

Fig. 3 Nanoparticles role in mitigating impacts of drought stress in plants [1, 42, 94]

enhancement in reactive oxygen species, damaging of DNA, proteins, and lipids. Plants overcome these stressful conditions by undergoing morphological changes as well as other processes, but crop yield remains compromised. Various strategies have been employed but use of NPs have sprouted as an effective strategy to overcome the negative influences of droughts in plants. They have registered to improve plant tolerance, enzymatic and non-enzymatic activities, photosynthetic activity, pollen germination and prevent oxidative damage [4, 60, 70] (Fig. 3).

Fe_3O_4 NPs prepared by using agro-waste, when applied to the soya bean subjected to drought stress, resulted in enhanced chlorophyll content up to 6 folds by just employing 0.03 mg dose while the control samples exhibit chlorosis [17]. Seed treatment with biogenic Fe_2O_3 (dose 10 mg/L) resulted in 70% plant height, 55% fresh weight and 34% dry weight increase of the plants grown under water deprived conditions as compared to the well-watered ones (control). Fe_2O_3 primed seeds registered enhancement in total chlorophyll (160%) while the untreated ones showed 81% reduction. The treatment not also prevent epidermal damage and vascular bundles deformation that is visible under drought conditions but also helps in maintaining the same anatomical structure as was depicted by well-watered plants. Micronutrients (Mg, Si, P and K) transport and distribution that was significantly hampered in the water-deficit plants was also improved in plants grown from treated seeds leading to enhancement in 119% Mg, 100% Si, 284% P and 218% K. Blue spot appearance on leaves, evidenced in the drought stress plants owing to oxidative damage, was significantly reduced by the Fe_2O_3 treatment owing to suppression in reductive oxidative species production. The NP treatment also helps in reduction of osmolytes (proline and soluble sugars) that are generally enhanced in water-deficit conditions in the

Table 3 Plant-mediated nanoparticles synthesized using different plants parts with their application in stress management

NPs	Plant and part used	Particle size (nm)	Experimental crop	Application	Category	Refs
Fe_3O_4	Extract of *Punica granatum* and *Citrus limetta* peels and *Curcuma longa* and *Carica papaya* leaves	1.26	Soya bean	Drought stress (Soil application)	Drought stress tolerance	[17]
α-Fe_2O_3	*Aspalathus linearis* leaves	3–31	Sorghum seeds	Drought stress (Seed soaking)	Drought stress tolerance	[60]
Iron oxide	*Prosopis cineraria leaves extract*	42.6	Wheat	Drought stress (Seed Priming, foliar application at seedling development stage and tillering stage)	Drought stress	[84]
Fe_2O_3	*Polyalthia longifolia* leaves extract	25	*Trigonella foenum-graecum*	Drought stress (Soil application)	Drought stress	[20]
ZnO	*Coleus forskohlii* leaf extract	62–91	Tomato	Stress conditions (foliar spray)	Drought stress	[25]
ZnO	Mint leaves	Spherical 9.4, rod shaped 15.2 length and 3.5 width	Broad beans	Stress conditions (Foliar spray)	Salt stress	[55]
ZnO	*Agathosma betulina*	27.5	*Sorghum bicolor*	Seed priming	Salt stress	[67]
Zn	leaves of *Sorghum bicolor*	17	Okra	Foliar application	Salt stress	[96]
Ag + melatonin	–	–	*E. uniflora*	Drought stress (seed sowing)	Drought stress	[45]

(continued)

Table 3 (continued)

NPs	Plant and part used	Particle size (nm)	Experimental crop	Application	Category	Refs
Ag	leaf extract of *Hibiscus sabdarifa*	8	Roselle	Seed treatment	Salt stress	[74]
Ag	*Ochradenus arabicus* plant extract	9–30	*Maerua oblongifolia*	Nodal segments growing media	Salt stress	[82]
Ag-Au	*Mentha piperita* leaves	–	*Mentha piperita*	Rhizome treatment	Salt stress	[11]
TiO_2	*Buddleja asiatica* leaves	30–111	Wheat	Salt stress (Foliar application)	Slat stress	[56]
Se	Barley leaves	50–200	Barley	Salt stress (Nutrient solution application)	Salt stress	[34]
Se	Grape extract	55.8	Mustard plants	Salt stress (Hydroponic media application)	Salt stress	[78]
Se	*Allium sativum* bud extract	50–150	Wheat	Drought stress (Foliar application)	Drought stress tolerance	[40]
Se	Garlic powder	100	Sesame	Seed priming, Foliar, Seed priming & foliar	Salt stress	[3]

plants. Drop in contents of soluble sugar and proline was recorded by 57 and 64% in treated seeds grown under water stress condition [60]. Drought resilience in different wheat crop varieties was also affected by introduction of phyto-synthesized iron oxide NPs at three different stages i.e. seed priming, and foliar applications at seedling development and tillering stage. In all the wheat varieties, the seed treatment recorded a reduction in germination time, days to 50% heading and days to 50% anthesis as compared to the untreated seeds when 100 mg/L concentration were employed. At this concentration there is also an ample increase in germination percentage. Below these concentrations there is a significant increase in days to germination, 50% heading and 50% anthesis and reduction in germination percentage. Moreover, treatments (100 mg/L of iron oxide) facilitated enhanced plant heights, no. of grains per spike, grain yield and 1000 grain weight [84]. Biosynthesized Fe_2O_3 NPs are also found effective in regulating nitrogen metabolism and its biological fixation that consequently ascertain survival of plant in drought stress conditions. Stressed plants have enhanced levels of proline and soluble sugar which were effectively reduced by use of NP in two different varieties of fenugreek to 22–31 and 16–22%, respectively. Use of NPs also facilitated improvement in amino acids as well as proteins levels owing to nitrogen regulation. Nitrogen assimilation in plants is imperative for plant health and growth which is facilitated by high nitrate reductase, glutamate synthase, glutamate synthetase, and glutamate dehydrogenase activities. In plants the nitrate is converted to nitrite which consequently transforms to ammonium ion by utilization of nitrate and nitrate reductase, respectively. In stressed plants these activities are hampered leading to incremental hype in nitrate and ammonium ions and decrease in nitrite levels in untreated plants with an increase in days to stress exposure. The high values of nitrate and ammonium ions were considerably reduced in treated plants with exponential increase in nitrite levels pointing to enhancement in nitrate reductase activity and suppression in nitrite reductase activity and hence reducing ammonium ion-induced toxicity in plants. In addition to that exposure to NPs also improved glutamate synthase, glutamate synthetase and glutamate dehydrogenase activities [20].

Dose–response studies related to Selenium NPs effect on plant's ability to fight drought stress showed that lower concentrations are effective while higher one's impact negatively. Foliar application of biosynthesized Selenium NPs on wheat varieties (drought tolerant (V_1) and drought susceptible (V_2)) have inculcated positive impact on plant height, shoot & root length, shoot & root fresh weights, shoot & root dried weights, leaf area, leaf number and leaf length when compared with the untreated stressed growth conditions. In comparison to the drought stressed V_1 & V_2 plants there is an increase of 9.1, 12.9; 34.9, 26.1; 34.1, 25.3; 13.6, 16.2; 7.8, 8.9; 12.4, 30.6; 25.2, 17.8; 22.3, 52.1; 44.2, 38.3, and 22.6, 18.1% in the parameter's studies, respectively [40].

ZnO NPs prepared by using PMS were able to strengthen the plants grown under four different water stressed conditions in field capacity (FC) i.e. control (S1), mild (S2), moderate (S3), severe (S4) stress. The introduction of 50 mg/L NPs resulted in increased shoot fresh weight by 18.7, 23.2, 50.6 and 88.4% in S1, S2, S3, and S4 cases, respectively. In comparison to S1, root fresh weight rose by 39.9, 55.7, and 50.4% in treatments S2, S3, and S4. Drought conditions promote higher levels of ROS and lower levels of antioxidants in the plants, pointing to its oxidative stress. ZnO NPs application also considerably helped in fighting the oxidative stress by reducing the generation of ROS (i.e., MDA and H_2O_2) and enhancing the production of antioxidant content (i.e., Ascorbic acid, free phenols) under all stress conditions. The significant enhancement in antioxidant enzymes (superoxide dismutase (SOD), CAT, and ascorbate peroxidase (APX)) was also facilitated by NP treatment [25].

Stress tolerance role of silver NPs have shown to improve plant attributes by improvising nutrient uptake and enzymatic processes. It also enhances the attachment of good bacteria to the plant roots that facilitates plant growth [9]. Priming of *E. uniflora* seeds with green-synthesized Ag NPs when loaded with melatonin can act as bio stimulant for dealing with drought stressed plants. The introduction of NP improved germination %, time, and vigor and facilities shoot elongation. It also promotes 100% production of CAT, APX and proline dehydrogenase and balancing of ROS in the stressed plant [45].

5.2 Salinity

Salinity of the soil is regarded as accumulation of excessive salts which can inhibit crop growth leading to its death. This high salt places the plants in salt stress condition and is responsible for decrease in soil productivity and hence crop yields. Approximately 2) and 50% of cultivated and irrigated lands, respectively, are prone to salt stress posing world food production at high risk. Major ions responsible for this stress are sodium, potassium, calcium, chloride, and nitrate. Salinity affects soil's microbial count, osmotic stress, and reduction in water potential which consequently affects plants in multiple ways. These effects include malfunctioning in metabolic and nutritional processes, reduction in cell growth, membrane restructuring, oxidative and water stress, ion toxicity, reduction in nitrogen fixation, etc. (Fig. 4) [5, 54]. Nanoparticles loading can be helpful in developing resistance to salt stress in plants and promoting growth. NPs based on their nature, size, reactivity, and dose inculcate different responses in the plants promoting their morphological, biochemical and physiological well-being [27] (Fig. 4).

Salt stress tolerance of Roselle plant were studies in presence of different levels of salt concentrations (0–180 mM) and seed treatment with phyto-synthesized Ag NPs (0–90 mg/L). The Germination rate decreased with enhancement in salinity, control plants at 45 mM salinity showed germination percentage of 74.6 at 45 mM salt concentration while under normal conditions the germination percentage was 94%. The decrease in seedling growth parameters and relative water content of

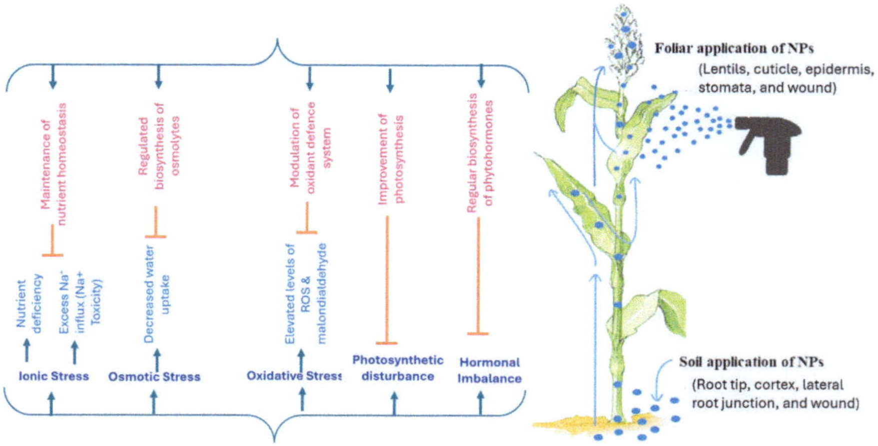

Fig. 4 Role of nanoparticles in overcoming salinity stress in plants (reproduced from [27])

plant was also registered under salinity conditions. The seed treatment with Ag NPs was able to increase the germination percentage, plant height and relative water content. In addition to that the antioxidant activity (CAT, APX, peroxidase, and SOD) that was hampered by salt stress was successfully activated by use of Ag NP activation of the antioxidant system. Enhancement in proline, soluble sugars, and carbohydrates contents were also noticed in salt stressed plants but application of NPs has considerably improved these contents enabling osmo-protection. Concentrations of Flavonoid and anthocyanin were appreciably improved in plants grown from treated seeds [74]. Similarly, introduction of Se NPs biosynthesized using grapes extract have improved seed germination, plant root-shoot lengths and fresh and dry weights. Increase in water content %, chlorophyll, carbohydrates and proteins were also suggested by use of Se NPs [78]. Seed priming with phyto-synthesized ZnO improved *Sorghum bicolor* growth traits, biomass, oxidative stress markers (MDA, H_2O_2), enzyme activities (SOD, CAT and APX) and osmoregulation (proline, and soluble sugars) [67].

Foliar treatment of green-synthesized (GS) ZnO and Zn NPs have also induced positive impact on the salt stressed plants that otherwise suffer from reduction in plant height, shoot and roots lengths, fresh and dry weights owing to osmotic stress and water shortage. The dosage of 50 mg/L ZnO and 0.3% Zn NPs marked positive impacts on shoot & root lengths and fresh shoot & root weights of the plants in contrast to unstressed plants [55, 96]. The foliar treatment with ZnO NPs considerably improved levels of chlorophyll a & b (4.8, 2.1 mg/g), carotenoids (0.9 mg/g), and total pigments (7.9 mg/g) in the stressed plants as compared to the affected plants (2.1, 0.9, 0.7, 3.7 mg/g). The treatment improved secondary metabolites, soluble sugar, amino acids, glycine betaine, proline, glycine, and antioxidant contents (non-enzymatic (glutathione, anthocyanin, ascorbic acid, α-tocopherol) and enzymatic (peroxidase (POX), CAT, polyphenol oxidase (PPO))). As compared to the plants grown normally, MDA and H_2O_2 levels that rose to 42 and 75% in the stressed plants were reduced to about 17 and 17.5% on ZnO NPs exposure. Electrolytes leakage from plasma membrane, caused by high salinity, decreases almost 30.7% on NPs treatment in stressed *Vicia faba* plants [55]. High salt stress negatively affects ions (N, P, K, Zn, Ca, NH_4, and NO_3) uptake owing to high Na and Cl concentrations [11, 55]. NPs exposure can also enhance ions uptake in affected plants by reducing the Na and Cl ions' translocation. Anomalous behavior was observed in case of Fe that showed declined levels in plants exposed to salt content and a further reduction when exposed to NPs [55]. A comparison of chemically (Ch) and GS-Zinc revealed better salt-tolerant effect of GS-Zn causing improved plant shoot/ root lengths, dry weights, fresh weights, and antioxidant activities [96].

Application of GS-Ag NPs to nodal segments of *M. oblongifolia* exposed to very high saline concentrations (200 mM NaCl) didn't show significant effect but at lower exposure levels the NPs play positive role in managing plant growth attributes. Positive impact of these NP at all salt exposure levels was observed in handling appropriate levels of photosynthetic pigments, total sugar, proline, MDA, H_2O_2, enzymatic and non-enzymatic antioxidants [82]. Rhizome treatment with green-synthesized Au–Ag NPs in the case of *Mentha piperita* facilitated recovery of plants prone to salt stress at vegetative state while the application of silver chloride and gold chloride showed reduction in growth parameters [11]. Correspondingly, Se NPs synthesized from Barley showed positive impact in recovery of salt stressed barley seedlings grown in hydroponic settings. The harvested plants when analyzed for Se content, showed its higher levels in both stressed and normally grown plants [34].

Higher doses of NPs can induce adverse impacts in plants with respect to stress tolerance as was the case in ZnO NPs [55] and TiO_2 [56]. Application of TiO_2 in higher concentrations (>40 mg/L) have caused reduction in the plant length, plant, root and shoot fresh and dry weights and number of leaves per plant in both wheat stress tolerant and non-tolerant varieties. Lower concentrations of biogenic TiO_2 (40 mg/L) have enhanced the physiological parameters of the plants in addition to chlorophyl, membrane stability index, and relative water content [56].

6 Conclusion

The plant-mediated approach for the synthesis of NPs is an environment-friendly approach that outperforms their bulk counterparts. These biogenic metallic NPs are found to be effective as priming agents as well as fertilizers positively impacting plants germination, root and shoot lengths and biomass, chlorophyll content, etc. Under stress conditions, such as drought or salinity, these NPs enhance both enzymatic and non-enzymatic activities, improve nutrient uptake, and boost antioxidant responses in plants. The pesticidal impact of these green-synthesized NPs has recorded high mortality rates with a significant reduction in fecundity and number of egg hatchings of target pests. Overall, plant-mediated NPs are a viable and sustainable option, capable of transforming the agricultural sector through their simple, cost-effective synthesis and wide range of applications.

References

1. Abasi F, Raja NI, Mashwani ZUR, Amjad MS, Ehsan M, Mustafa N, Haroon M, Proćków J (2022) Biogenic silver nanoparticles as a stress alleviator in plants: a mechanistic overview. Molecules 27(11):3378
2. Afzal S, Sharma D, Singh NK (2021) Eco-friendly synthesis of phytochemical-capped iron oxide nanoparticles as nano-priming agent for boosting seed germination in rice (*Oryza sativa L.*). Environ Sci Pollut Res 28(30):40275–40287. https://doi.org/10.1007/s11356-020-12056-5
3. Ahmad I, Younas Z, Mashwani Z-U-R, Raja NI, Akram A (2023) Phytomediated selenium nanoparticles improved physio-morphological, antioxidant, and oil bioactive compounds of sesame under induced biotic stress. ACS Omega 8(3):3354–3366. https://doi.org/10.1021/acsomega.2c07084
4. Ahmad J, Qamar S, Kausar N, Qureshi MI (2020) Nanoparticles: the magic bullets in mitigating drought stress in plants. In Hakeem K, Pirzadah T (eds) Nanobiotechnology in agriculture: an approach towards sustainability. Springer, Cham, pp 145–161

5. Kumar A, Verma J (2018) Does plant-microbe interaction confer stress tolerance in plants: a review? Microbiol Res 207:41–45
6. Akhter G, Khan A, Ali SG, Khan TA, Siddiqi KS, Khan HM (2020) Antibacterial and nematicidal properties of biosynthesized Cu nanoparticles using extract of holoparasitic plant. SN Appl Sci 2:1–6
7. Al-Ghabban A, Eldiasty J (2022) Green synthesis of copper oxide nanoparticle by using *Achillea fragrantissima* and *Nigella sativa* extracts and their effects as larvicidal, molluscicidal and antimicrobial agents. Egypt Acad J Biol Sci E Med Entomol Parasitol 14(2):127–148
8. Al-Shammari HI, Al-Zubaidi HK (2016) Toxicity of silver nanoparticles prepared by extract of Eucalyptus sp in some biological aspects of Citrus mealybug Planococcus citri (Risso), Hemiptera: Pseudococcidae. Eur Acad Res IV:7658–7675
9. Alabdallah NM, Hasan MM (2021) Plant-based green synthesis of silver nanoparticles and its effective role in abiotic stress tolerance in crop plants. Saudi J Biol Sci 28(10):5631–5639. https://doi.org/10.1016/j.sjbs.2021.05.081
10. Alfaro-Corres AE, González-Mendoza D, Ruiz-Sánchez E, Ail-Catzin C, Valdez-Salas B, Gutiérrez-Miceli F, Reyes-Ramírez A, Fils Pierre J (2023) Insecticidal activity and physicochemical characterization of nanoparticles from foliar extract of Capsicum chinense. J Renew Mater 11(11):3933–3944
11. Aliakbarpour F, Vaziri A, Mohseni M (2020) Green synthesis of Au-Ag nanoparticles using *Mentha piperita* and effects of Au-Ag alloy nanoparticles on the growth of *Mentha piperita* under salinity stress. J Plant Process Funct 9:47–57
12. Annon MR, Jafar FS (2020) The effectiveness of silver and silica nanoparticles on productivity and adult emergence of T. castaneum and C. maculatus. Paper presented at the Journal of physics: conference series
13. Arshi N, Prashanthi Y, Rao TN, Ahmed F, Kumar S, Oves M (2020) Biosynthesis of ZnO nanostructures using *Azadirachta indica* leaf extract and their effect on seed germination and seedling growth of tomato: an eco-friendly approach. J Nanoelectron Optoelectron 15(11):1412–1422
14. Asoufi HM, Al-Antary TM, Awwad AM (2018) Biosynthesis and characterization of iron sulfide (FeS) nanoparticles and evaluation their aphicidal activity on the green peach aphid *Myzus persicae* (Homoptera: Aphididae). Fresenius Environ Bull 27(11):7767–7775
15. Attia RG, Rizk SA, Hussein MA, Fattah HMA, Khalil MM, Ma'moun SA (2020) Effect of cinnamon oil encapsulated with silica nanoparticles on some biological and biochemical aspects of the rice moth, Corcyra cephalonica (Staint.) (Lepidoptera: Pyralidae). Ann Agric Sci 65(1):1–5
16. Attia TS, Elsheery N (2020) Nanomaterials: scope, applications, and challenges in agriculture and soil reclamation. In Sustainable agriculture reviews 41: nanotechnology for plant growth and development. Springer International Publishing, pp 1–39
17. Azeez A (2016) Phyto synthesis of ferric oxide nano particles amended with agro wastes to treat calcareous soil: a low cost iron supplement. Int J Res Pharmacol Pharmacother 1:105–114
18. Baig N, Kammakakam I, Falath W (2021) Nanomaterials: a review of synthesis methods, properties, recent progress, and challenges. Mater Adv 2(6):1821–1871
19. Berta L, Coman N-A, Rusu A, Tanase C (2021) A review on plant-mediated synthesis of bimetallic nanoparticles, characterisation and their biological applications. Materials 14(24):7677
20. Bisht S, Sharma V, Kumari N (2022) Biosynthesized hematite nanoparticles mitigate drought stress by regulating nitrogen metabolism and biological nitrogen fixation in Trigonella foenum-graecum. Plant Stress 6:100112. https://doi.org/10.1016/j.stress.2022.100112
21. Chandraker SK, Ghosh MK, Lal M, Shukla R (2021) A review on plant-mediated synthesis of silver nanoparticles, their characterization and applications. Nano Express 2(2):022008. https://doi.org/10.1088/2632-959X/ac0355
22. Chaudhuri SK, Malodia L (2017) Biosynthesis of zinc oxide nanoparticles using leaf extract of *Calotropis gigantea*: characterization and its evaluation on tree seedling growth in nursery stage. Appl Nanosci 7(8):501–512

23. Danish M, Altaf M, Robab MI, Shahid M, Manoharadas S, Hussain SA, Shaikh H (2021) Green synthesized silver nanoparticles mitigate biotic stress induced by *Meloidogyne* incognita in *Trachyspermum ammi* (L.) by improving growth, biochemical, and antioxidant enzyme activities. ACS omega 6(17):11389–11403
24. Dwivedi S, Saquib Q, Al-Khedhairy AA, Musarrat J (2016) Understanding the role of nanomaterials in agriculture. Microbial inoculants in sustainable agricultural productivity, vol 2: Functional applications, pp 271–288
25. El-Zohri M, Al-Wadaani NA, Bafeel SO (2021) Foliar sprayed green zinc oxide nanoparticles mitigate drought-induced oxidative stress in tomato. Plants 10(11):2400. https://www.mdpi.com/2223-7747/10/11/2400
26. Elango G, Roopan SM, Al-Dhabi NA, Arasu MV, Dhamodaran KI, Elumalai K (2016) Coir mediated instant synthesis of Ni-Pd nanoparticles and its significance over larvicidal, pesticidal and ovicidal activities. J Mol Liq 223:1249–1255
27. Etesami H, Fatemi H, Rizwan M (2021) Interactions of nanoparticles and salinity stress at physiological, biochemical and molecular levels in plants: a review. Ecotoxicol Environ Saf 225:112769
28. Farghaly FA, Nafady NA (2015) Green synthesis of silver nanoparticles using leaf extract of Rosmarinus officinalis and its effect on tomato and wheat plants. J Agri Sci 7(11):277–287
29. Francis D, Balachandran M, Daniel EC, Kunnel SG (2024) Nanoparticles in healthcare: applications in therapy, diagnosis, and drug delivery. In: Daniel EC (ed) Mater Res Forum LLC
30. Ghidan AY, Al-Antary TM, Awwad AM (2016) Green synthesis of copper oxide nanoparticles using Punica granatum peels extract: effect on green peach Aphid. Environ Nanotechnol Monit Manag 6:95–98
31. Gill R, Naeem M, Ansari AA, Kumar A, Kumar A, Chhikara A, Bremont JFJ, Tuteja N, Gill SS (2022). Agrochemicals in soil and environment: an overview. In: Naeem M, Bremont JFJ, Ansari AA, Gill SS (eds) Agrochemicals in soil and environment: impacts and remediation Singapore. Springer Nature Singapore, pp 3–24
32. Gull A, Lone AA, Wani NUI (2019) Biotic and abiotic stresses in plants. Abiotic Biot Stress Plants 7:1–9
33. Gulzar MU, Zia T, Shahzad M, Ibrahim MU, Ihsan T (2020) Impacts of bio synthesized silver-nanoparticles ($AgNO_3$) and plant oils against *Trogoderma granarium.* GSC Biol Pharm Sci 10(3):010–015
34. Habibi G, Aleyasin Y (2020) Green synthesis of Se nanoparticles and its effect on salt tolerance of barley plants. Int J Nano Dimens 11(2):145–157
35. Haider Z, Ayub AR, Tahir MU, Sajid SA, Khalid M (2020) Efficacy and characterization of synthesized zinc oxide nanoparticles against *Tribolium castaneum* and *Trogoderma granarium.* Int J Sci Res 10(3):131–151
36. Hamdi H, Rashwan RS, Abu-Zaid AA, Gado E, Zarah RK, Sami R, Almatrafi M, Al-Harthi HF, Abushal SA, Al-Zahrani MH (2024) Efficacy of Titanium Dioxide nanoparticles Using *Juniperus phoenicea* in controlling Rice Weevil (*Sitophilus oryzae*) and its effect on the microbial contents and nutritive value of grains. J Biobased Mater Bioenergy 18(6):1002–1009
37. Hasan M, Rafique S, Zafar A, Loomba S, Khan R, Hassan SG, Khan MW, Zahra S, Zia M, Mustafa G (2020) Physiological and anti-oxidative response of biologically and chemically synthesized iron oxide: zea mays a case study. Heliyon 6(8)
38. Haydar MS, Ghosh S, Mandal P (2021) Application of iron oxide nanoparticles as micronutrient fertilizer in mulberry propagation. J Plant Growth Regul:1–21
39. Hermes P-H, Gabriela M-P, Ileana V-R, Fusaro C, Fernando L-V, Mariana M-A, Padilla-Rodríguez C, Fabián F-L (2020) Carbon nanotubes as plant growth regulators: prospects. In: Patra J, Fraceto L, Das G, Campos E (eds) Green nanoparticles: synthesis and biomedical applications. Spriger, Cham, pp 77–115
40. Ikram M, Raja NI, Javed B, Mashwani Z-U-R, Hussain M, Hussain M, Ehsan M, Rafique N, Malik K, Sultana T, Akram A (2020) Foliar applications of bio-fabricated selenium nanoparticles to improve the growth of wheat plants under drought stress. Green Process Synth 9(1):706–714. https://doi.org/10.1515/gps-2020-0067

41. Jameel M, Rauf MA, Khan MT, Farooqi MK, Alam MA, Mashkoor F, Shoeb M, Jeong C (2023) Ingestion and effects of green synthesized cadmium sulphide nanoparticle on Spodoptera Litura as an insecticidal and their antimicrobial and anticancer activities. Pestic Biochem Physiol 190:105332
42. Khalid MF, Iqbal Khan R, Jawaid MZ, Shafqat W, Hussain S, Ahmed T, Rizwan M, Ercisli S, Pop OL, Alina Marc R (2022) Nanoparticles: the plant saviour under abiotic stresses. Nanomaterials 12(21):3915. https://www.mdpi.com/2079-4991/12/21/3915
43. Khan A, Raza A, Hashem A, Dolores Avila-Quezada G, Fathi Abd_Allah E, Ahmad F, Ahmad A (2024) Green fabrication of titanium dioxide nanoparticles via Syzygium cumini leaves extract: characterizations, photocatalytic activity and nematicidal evaluation. Green Chem Lett Rev 17(1):2331063
44. Khandelwal N, Barbole RS, Banerjee SS, Chate GP, Biradar AV, Khandare JJ, Giri AP (2016) Budding trends in integrated pest management using advanced micro-and nano-materials: challenges and perspectives. J Environ Manage 184:157–169
45. Labulo AH, David OA, Terna AD, Omotosho TP, Tanko NS, Hassan I, Oluwole BR, Odebode A (2024) Modulation of physiological and biochemical activities of Eugenia uniflora by green-synthesized silver nanoparticle and melatonin under drought stress. Plant Biotechnol Rep 18(2):289–299. https://doi.org/10.1007/s11816-024-00887-4
46. Lateef A, Nazir R, Jamil N, Alam S, Shah R, Khan MN, Saleem M (2016) Synthesis and characterization of zeolite based nano–composite: an environment friendly slow release fertilizer. Micropor Mesopor Mater 232:174–183
47. Lateef A, Nazir R, Jamil N, Alam S, Shah R, Khan MN, Saleem M (2019) Synthesis and characterization of environmental friendly corncob biochar based nano-composite–a potential slow release nano-fertilizer for sustainable agriculture. Environ Nanotechnol Monit Manag 11:100212
48. León Jimenez E, Valdéz-Salas B, González-Mendoza D, Tzintzun-Camacho O (2019) Biosynthesis and insecticidal activity of Cu-nanoparticles of Pluchea sericea (Nutt) and Prosopis juliflora (Sw) DC. against Phenacoccus solenopsis Tinsley (Hemiptera: Pseudococcidae). Prueba De Galera 2:12–21
49. Liu R (2012) Lal R (2012) Nanoenhanced materials for reclamation of mine lands and other degraded soils: a review. J Nanotechnol 1:461468
50. Malaikozhundan B, Vinodhini J (2018) Nanopesticidal effects of Pongamia pinnata leaf extract coated zinc oxide nanoparticle against the Pulse beetle, Callosobruchus maculatus. Mater Today Commun 14:106–115
51. Malathi S, Rameshkumar G, Rengarajan R, Rajagopal T, Muniasamy S, Ponmanickam P (2019) Phytofabrication of silver nanoparticles using Annona reticulata and assessment of insecticidal and bactericidal activities. J Environ Biol 40(4):626–633
52. Mendez-Trujillo V, Valdez-Salas B, Carrillo-Beltran M, Curiel-Alvarez M, Tzintzun-Camacho O, Ceceña-Duran C, Gonzalez-Mendoza D (2019) Green synthesis of bimetallic nanoparticles from Prosopis juliflora (Sw) DC., and its effect against cotton mealybug, Phenacoccus solenopsis (Hemiptera: Pseudococcidae). Phyton 88(3):269
53. Mendez-Trujillo V, Valdez-Salas B, Curiel-Alvarez M, Beltran-Partida E, Alfaro-Corres A, Ruiz-Sanchez E, Bautista-Trujillo G, Camacho OT, Grimaldo-Juarez O, Ceceña-Duran C (2022) Insecticidal effect of green bimetallic nanoparticles from Crotalaria longirostrata on Cotton Mealybug, Phenacoccus solenopsis. J Renew Mater 10:2543–2552
54. Mitra A, Kataki S, Singh AN, Gaur A, Razafindrabe BHN, Kumar P, Chatterjee S, Gupta DK (2021) Plant stress, acclimation, and adaptation: a review. In: Growth P, Physiology S (eds) Gupta DK, Palma JM. Springer International Publishing, Cham, pp 1–22
55. Mogazy AM, Hanafy RS (2022) Foliar spray of biosynthesized zinc oxide nanoparticles alleviate salinity stress effect on vicia faba plants. J Soil Sci Plant Nut 22(2):2647–2662. https://doi.org/10.1007/s42729-022-00833-9
56. Mustafa N, Raja NI, Ilyas N, Ikram M, Mashwani Z-U-R, Ehsan M (2021) Foliar applications of plant-based titanium dioxide nanoparticles to improve agronomic and physiological attributes of wheat (Triticum aestivum L.) plants under salinity stress. Green Process Synth 10(1):246–257. https://doi.org/10.1515/gps-2021-0025

57. Naseer I, Javad S, Shah AA, Tariq A, Ahmad A (2024) Influence of phyto-mediated zinc oxide nanoparticles on growth of (Zea mays L.). Pak J Bot 56(3):911–923
58. Ndaba B, Roopnarain A, Haripriya R, Maaza M (2022) Biosynthesized metallic nanoparticles as fertilizers: an emerging precision agriculture strategy. J Integr Agric 21(5):1225–1242
59. Ndaba B, Roopnarain A, Vatsha B, Marx S, Maaza M (2022) Synthesis, characterization, and evaluation of Artemisia Afra-mediated iron nanoparticles as a potential nano-priming agent for seed germination. ACS Agr Scie Technol 2(6):1218–1229. https://doi.org/10.1021/acsagscitech.2c00205
60. Ndou N, Rakgotho T, Nkuna M, Doumbia IZ, Mulaudzi T, Ajayi RF (2023) Green synthesis of iron oxide (hematite) nanoparticles and their influence on Sorghum bicolor growth under drought stress. Plants 12(7):1425. https://www.mdpi.com/2223-7747/12/7/1425
61. Parveen K, Kumar N, Sharma D, Ledwani L (2022) Evaluation of efficacy of "Cassia renigera" leaf extract mediated ZnO nanoparticles as nanofertilizer for cauliflower plant. ChemistrySelect 7(23):e202200367
62. Patani A, Patel A, Prajapati D, Khare N, Singh S (2023) Various agriculture crop plant-based bioactive compounds and their use in nanomaterial synthesis and applications. In: Husen A (eds) Nanomaterials from agricultural and horticultural products. Smart nanomaterials technology. Springer, Singapore. https://doi.org/10.1007/978-981-99-3435-5_12
63. Prasad R, Bhattacharyya A, Nguyen QD (2017) Nanotechnology in sustainable agriculture: recent developments, challenges, and perspectives. Front Microbiol 8. https://doi.org/10.3389/fmicb.2017.01014
64. Rafael Á-C, Jesús Ángel A-A (2023). Green synthesis of nanoparticles: a biological approach. In: Kinjal JS (ed) Advances in green chemistry. IntechOpen, Rijeka, pp Ch 5
65. Rafique M, Sohaib M, Tahir R, Bilal Tahir M, Rizwan M (2021) Plant-mediated green synthesis of zinc oxide nanoparticles using peel extract of citrus reticulate for boosting seed germination of Brassica nigra seeds. J Nanosci Nanotechnol 21(6):3573–3579
66. Rafique M, Tahir R, Gillani SSA, Tahir MB, Shakil M, Iqbal T, Abdellahi MO (2022) Plant-mediated green synthesis of zinc oxide nanoparticles from Syzygium cumini for seed germination and wastewater purification. J Environ Anal Chem 102(1):23–38
67. Rakgotho T, Ndou N, Mulaudzi T, Iwuoha E, Mayedwa N, Ajayi RF (2022) Green-synthesized zinc oxide nanoparticles mitigate salt stress in sorghum bicolor. Agriculture 12(5):597. https://www.mdpi.com/2077-0472/12/5/597
68. Ramkumar G, Asokan R, Ramya S, Gayathri G (2021) Characterization of Trigonella foenum-graecum derived iron nanoparticles and its potential pesticidal activity against Tuta absoluta (Lepidoptera). J Cluster Sci 32:1185–1190
69. Rani SS, Justin CGL, Gunasekaran K, Roseleen SSJ (2019) Efficacy of green synthesized silver nanoparticle, plant powders and oil against rice weevil Sitophilus oryzae L.(Coleoptera: Curculionidae) on sorghum seeds. J Pharmacogn Phytochem 8(5):38–42
70. Rasheed A, Li H, Tahir MM, Mahmood A, Nawaz M, Shah AN, Aslam MT, Negm S, Moustafa M, Hassan MU, Wu Z (2022) The role of nanoparticles in plant biochemical, physiological, and molecular responses under drought stress: a review. Front Plant Sci 13. https://doi.org/10.3389/fpls.2022.976179
71. Rashwan RS, Abu-Zaid AA (2018) Evaluation of nutritive values and microbial content of wheat grains treated with some types of nano particles to control Sitophilus granaries (L.). Arab Univ J Agric Sci 26(1):293–301
72. Rehman HU, Majeed B, Farooqi MA, Rasul A, Sagheer M, Ali Q, Akhtar ZR (2021) Green synthesis of silver nitrate nanoparticles from Camelina sativa (L.) and its effect to control insect pests of stored grains. Int J Trop Insect Sci 41:3031–3039
73. Rostamizadeh E, Iranbakhsh A, Majd A, Arbabian S, Mehregan I (2020) Green synthesis of Fe_2O_3 nanoparticles using fruit extract of Cornus mas L. and its growth-promoting roles in Barley. J Nanostructure Chem 10(2):125–130. https://doi.org/10.1007/s40097-020-00335-z
74. Sadat-Hosseini M, Naeimi A, Boroomand N, Aalifar M, Farajpour M (2022) Alleviating the adverse effects of salinity on Roselle plants by green synthesized nanoparticles. Sci Rep 12(1):18165. https://doi.org/10.1038/s41598-022-22903-9

75. Sahayaraj K, Madasamy M, Radhika SA (2016) Insecticidal activity of bio-silver and gold nanoparticles against Pericallia ricini Fab.(Lepidaptera: Archidae). J Biopestic 9(1):63–72
76. Sankar MV, Abideen S (2015) Pesticidal effect of green synthesized silver and lead nanoparticles using *Avicennia marina* against grain storage pest *Sitophilus oryzae*. Int J Nanomater Biostruct 5(3):32–39
77. Sarkar D, Dubey PK, Chaurasiya R, Sankar A, Shikha CN, Ganguly S, Meena VS, Meena SK, Parewa HP, Rakshit A (2021) Organic interventions conferring stress tolerance and crop quality in agroecosystems during the UN decade on ecosystem restoration. Land Degrad Dev 32(17):4797–4816. https://doi.org/10.1002/ldr.4094
78. Sarkar RD, Kalita MC (2022) Green synthesized Se nanoparticle-mediated alleviation of salt stress in field mustard, TS-36 variety. J Biotechnol 359:95–107. https://doi.org/10.1016/j.jbiotec.2022.09.013
79. Sedighi A, Imani S, Moshtaghi Kashanian G, Najafi H, Fathipour Y (2019) Efficiency of green synthesized silver nanoparticles with sweet orange, Citrus sinensis (L.)(Rutaceae, Sapindales) against Tribolium confusum Duval.(Coleoptera, Tenebrionidae). J Agr Sci Techn 21(6):1485–1494
80. Selvaraj P, Angelin PC, Jayaseeli JPR (2019) Insecticidal and antibacterial potential of Syzygium aromaticum (L.) merrill and perry. J Biopest 12(2):191–196
81. Shahzadi K, Bashir F, Nazir K, Seemab S, MaA S, Waheed I, Shehzaib A, Fatima K (2019) Impacts of green synthesized Silver-Nanoparticles against Tribolium castaneum (Coleoptera: Tenebrionidae). Int J Entomol Res 4(6):114–117
82. Shaikhaldein HO, Al-Qurainy F, Nadeem M, Khan S, Tarroum M, Salih AM, Alansi S, Al-Hashimi A, Alfagham A, Alkahtani J (2022) Assessment of the impacts of green synthesized silver nanoparticles on Maerua Oblongifolia shoots under in vitro salt stress. Materials 15(14):4784. https://www.mdpi.com/1996-1944/15/14/4784
83. Siddique MA, Hasan MU, Sagheer M, Sahi ST (2022) Comparative toxic effects of Eucalyptus globulus L. (Myrtales: Myrtaceae) and its green synthesized zinc oxide nanoparticles (ZnONPs) against Rhyzopertha dominica (F.)(Coleoptera: Bostrichidae). Int J Trop Insect Sci 42:1697–1706
84. Singh A, Basnal N, Shukla G, Chaudhary N, Singh S, Gaurav SS (2022) Evaluation of efficacy of Phyto-synthesized iron oxide nanoparticles in contributing drought resilience in wheat (*Triticum aestivum* L.). Nanotechnology 33(48):485101
85. Singh M, Singh J, Sharma D, Kaur B, Rawat M (2018) Plant leaves mediated synthesis of semiconductor ZnO nanoparticles and its application for seed germination. Paper presented at the AIP Conference Proceedings
86. Solorzano Toala R, Gutierrez-Miceli F, Valdez-Salas B, Beltran-Partida E, Gonzalez-Mendoza D, Tzintzun-Camacho O, Grimaldo-Juarez O, Basilio-Cortes A (2024) Biosynthesis of copper nanoparticles from acacia cornigera and annona purpurea and their insecticidal effect against tribolium castaneum. Reactions 5(2):274–284
87. Soltani S, Gacem A, Choudhary N, Yadav VK, Alsaeedi H, Modi S, Patel A, Khan SH, Cabral-Pinto MMS, Yadav KK, Patel A (2023) Scallion peel mediated synthesis of zinc oxide nanoparticles and their applications as nano fertilizer and photocatalyst for removal of organic pollutants from wastewater. Water 15(9):1672. https://www.mdpi.com/2073-4441/15/9/1672
88. Thatyana M, Dube NP, Kemboi D, Manicum A-LE, Mokgalaka-Fleischmann NS, Tembu JV (2023) Advances in Phytonanotechnology: a plant-mediated green synthesis of metal nanoparticles using Phyllanthus plant extracts and their antimicrobial and anticancer applications. Nanomater 13(19):2616. https://www.mdpi.com/2079-4991/13/19/2616
89. Tovar GI, Briceño S, Suarez J, Flores S, González G (2020) Biogenic synthesis of iron oxide nanoparticles using Moringa oleifera and chitosan and its evaluation on corn germination. Environ Nanotechnol Monit Manag 14:100350. https://doi.org/10.1016/j.enmm.2020.100350
90. Ullah A, Lim SI (2022) Plant extract-based synthesis of metallic nanomaterials, their applications, and safety concerns. Biotechnol Bioeng 119(9):2273–2304
91. Vadlapudi V, Amanchy R (2017) Phytofabrication of silver nanoparticles using Myriostachya wightiana as a novel bioresource, and evaluation of their biological activities. Braz Arch Biol Techn 60:e17160329

92. Vijayaraghavan K, Ashokkumar T (2017) Plant-mediated biosynthesis of metallic nanoparticles: a review of literature, factors affecting synthesis, characterization techniques and applications. J Environ Chem Eng 5(5):4866–4883. https://doi.org/10.1016/j.jece.2017.09.026
93. Vurro M, Miguel-Rojas C, Pérez-De-Luque A (2019) Safe nanotechnologies for increasing the effectiveness of environmentally friendly natural agrochemicals. Pest Manag Sci 75(9):2403–2412
94. Wahab A, Batool F, Muhammad M, Zaman W, Mikhlef RM, Naeem M (2023) Current knowledge, research progress, and future prospects of phyto-synthesized nanoparticles interactions with food crops under induced drought stress. Sustainability 15(20):14792. https://www.mdpi.com/2071-1050/15/20/14792
95. Yerragopu PS, Hiregoudar S, Nidoni U, Ramappa K, Sreenivas A, Doddagoudar S (2019) Effect of plant-mediated synthesized silver nanoparticles on pulse beetle, Callosobruchus chinensis (L.). Int J Curr Microbiol Appl Sci 8:1965–1972
96. Zafar S, Hasnain Z, Aslam N, Mumtaz S, Jaafar HZ, Wahab PEM, Qayum M, Ormenisan AN (2021) Impact of Zn nanoparticles synthesized via green and chemical approach on okra (Abelmoschus esculentus L.) growth under salt stress. Sustainability 13(7):3694
97. Zahir AA, Bagavan A, Kamaraj C, Elango G, Rahuman AA (2012) Efficacy of plant-mediated synthesized silver nanoparticles against Sitophilus oryzae. J Biopest 5:95

Uptake and Translocation

Nanoparticles: Uptake, Transport, and Effects on Plant Growth

Richa Das, Tushar Thakur, Avinash Singh, Indrajeet Singh, Lara Jadhav, Indrani Bhattacharya, and Rahul Datta

Abstract Global agricultural systems are facing a number of unprecedented challenges. Nanotechnology is a useful tool for ensuring sustainability and increasing crop productivity in order to attain food security. Due to the unique properties of nanoparticles, such as their large surface area, reactivity, agglomeration, penetration capabilities, size, and structure, nanotechnology has made a wide range of applications in agriculture over time. When nanoparticles (NPs) are present in the soil, they can be taken up by the roots and migrate between and among the cells of the plant tissues after being adsorbed at the root surface. Additionally, through aerial deposition, NPs can be absorbed by plants, primarily via stomata and cuticles. Nanoparticles significantly influence plant growth and development by improving seed germination, facilitating nutrient delivery, enhancing tolerance to biotic and abiotic stresses, and promoting root and shoot growth. This chapter focuses on the types, modes of entry, uptake mechanisms, transport, interactions, and applications of nanoparticles in promoting plant growth.

Keywords Nanoparticle types · Interaction · Root and shoot growth · Nutrient delivery · Stress

---

R. Das (✉) · T. Thakur · I. Bhattacharya
Department of Biotechnology, Parul Institute of Applied Science, Parul University, Vadodara, Gujarat, India
e-mail: dasricha722@gmail.com

A. Singh
Department of Biotechnology, Axis Institute of Higher Education, Kanpur, India

I. Singh
Department of Biotechnology, Faculty of Engineering and Technology, Rama University, Kanpur, Uttar Pradesh, India

L. Jadhav
Department of Microbiology, Parul Institute of Applied Science, Parul University, Vadodara, Gujarat, India

R. Datta
Department of Geology and Pedology, Faculty of Forestry and Wood Technology, Mendel University in Brno, Brno, Czech Republic

L. F. Fraceto et al. (eds.), *Emerging Nanotechnologies for Agroecosystem Management*, Smart Nanomaterials Technology, https://doi.org/10.1007/978-981-95-0187-8_3

1 Introduction

The increasing concerns over food security and sustainable production have spurred significant technical advancements and breakthroughs in agriculture [12]. Continuous improvements in farming techniques like these are crucial if we are to harness the power of both natural and artificial resources to feed a planet that is getting bigger at an alarming rate. When it comes to many agricultural problems, nanotechnology in particular may hold the key to real solutions. Scientists are captivated by the idea of using nanoparticles in order to vanish the gap between large-scale matter and their atomic or molecular structures [58]. Nanoparticles (NPs), typically ranging in size from 1 to 100 nm (nm), exhibit colloidal properties [25]. They possess remarkable attributes—including an extraordinarily specific surface area and high surface-to-volume ratio—which determine their excellent chemical, optical, electrical, and mechanical properties [6]. Nanomaterials are examined from the angles of structural and chemical composition, they are a diverse collection of substances. These materials include carbonaceous substances, quantum dots, lipids, metal oxides, zero-valent metals, dendrimers, and nanopolymers, in a variety of shapes and sizes [26].

The pathways that plants use to transport nanoparticles allow them to enter plants more easily [17]. Plants absorb NPs via micro- and nanoscale pores after they have adsorbed on different plant surfaces. Surface charge, concentration, size, and shape of NPs impact the rate of absorption, which can occur through many pathways in plant bodies [61]. The aerial component assists in the communication between shot surfaces and NPs in the air. Some of the engineered nanoparticles released into the environment could deposit onto plant leaves. Follicle hydathodes and stomata are essential for nanoparticle absorption because plants get involved with NPs in atmosphere through their leaves. Because of its high transport velocity and massive size exclusion limit (more than 10 nm), the stomatal channel is extraordinarily capacitive. Permeability fluctuation, however, makes this pathway quite unpredictable [13]. Applying a solution topically and soaking and drying it multiple times might increase the number of stomata, which can help with absorption in rare cases. After entering the leaf through the stomata or hydathodes, NPs make their way through the cell walls of the parenchymatous palisade, enter the phloem of leaf, and continue their journey to other portions of the plant body. Additional sites for NP deposition include neighbouring cells and the substomatal cavity's cell walls. Aerial components can also be attached to nanomaterials through the bark surfaces, stigma, or cuticle. Nanoparticle deposition in the epicuticular cavities is hastened by the epicuticular structures. The trichomes on the shoot surface also increase the likelihood of NP deposition. Particle concentration and size have a substantial impact on NP absorption and distribution in plant systems. In order to absorb tiny lipophilic NPs, the cuticle's apolar fluid areas can use both polar and apolar absorption pathways. Stomata, hydathodes, and flower stigmas are cuticle-free areas that absorb larger particles [38]. Soil NPs interact with plant root systems, NPs adsorb on roots, which facilitates NP entry into cell walls and subsequent absorption by cells. Most primary roots have suberised endodermis and exodermis, which prevent soil water and solute particles to enter the central vascular cylinder through the apoplastic bypass. Conversely, apoplastic bypass is now within reach thanks to the freshly grown lateral or adventitious roots that reach the cortex. This means the nanoparticles can potentially reach the vascular tissues like xylem through the cortical cells and central vascular cylinder, right where the adventitious roots start. Plant cells that have a negatively charged surface can transfer substances with a negative charge into the apoplast [40]. The nanoparticles as a novel platform in scientific discovery have led to the creation of numerous low-cost nanotech applications that increase plant lifecycle progression, environmental adaptation, and seed germination. Specific nanoparticles, such as titanium dioxide (TiO_2), zinc oxide (ZnO), iron oxide (FeO), multi-walled carbon nanotubes (MWCNTs), ZnFeCu-oxide, and hydroxyfullerenes, have the ability to enhance growth, development, and quality of crops [58]. This chapter explores the types, modes of entry, uptake mechanisms, transport, interactions, and applications of nanoparticles in promoting plant growth.

2 Types of Nanoparticles

Nanoparticles can be categorised on the basis of their attributes, such as dimensions, reactivity, structure, and composition. Particles that do not belong to the category of organic or inorganic substances can be broadly classified as NPs. Furthermore, the category of inorganic NPs encompasses metal and metal oxide NPs, carbon-based NPs, ceramic NPs, and semiconducting NPs. Conversely, organic nanoparticles can be categorised into two groups: polymeric nanoparticles and nanoparticles derived from biomolecules.

2.1 Metal and Metal Oxide NPs

Nanoparticles made of metal oxides or metals can have many different physicochemical characteristics. The mechanical, electrical, optical, chemical, and electromagnetic properties of these materials are superior to those of their larger equivalents. They also have large surface areas, may be easily customised in size and shape, and are surface-modifiable. As a result, nanoparticles have found extensive usage in different fields such as molecular sensing, environmental remediation, medicine, catalysis, and energy generation. They have been created through the utilisation of both top-down and bottom-up techniques. Commonly used metal and metal oxide nanoparticles include Au, Ag, Pt, Pd, Cu, CuO, Ni, NiO, Zn, ZnO, iron oxide, titanium dioxide, and cerium oxide [59].

2.2 Carbon-Based NPs

Restricting carbon nanoparticles (CNPs) to a spherical shape with a diameter smaller than 100 nm is unattainable. Carbon nanoparticles (NPs) encompass all types of carbon nanomaterials. Carbon nanoparticles (NPs) encompass a variety of materials such as fluorescent carbon quantum dots (CQDs), graphene, carbon dots, single-walled and multi-walled CNTs, and other similar substances [16]. Their exceptional mechanical, chemical, thermal, and physical properties have resulted in their extensive application in various fields. Due to their extensive surface area, superb biocompatibility, minimal toxicity, affordable cost, and environmentally benign synthesis methods, CNPs have been extensively utilised in numerous scientific and technological applications. These traits, along with their exceptional optical features, enabled their use in biological imaging, biomedical, photocatalysis, and chemical optical sensing.

2.3 Ceramics NPs

A solid core, created by the use of pressure and/or heat, made up of one or more of the following elements, is one approach to characterise ceramics: (i) both metals and nonmetals, (ii) a combination of at least two nonmetallic elemental solids along with at least one metal, (iii) a combination of at least two nonmetallic elemental solids, and (iv) a combination of an inert substance with two or more nonmetallic elemental solids. Ceramic nanoparticles are mostly composed of metal and metalloid oxides, phosphates, carbides, and carbonates, including silicon, titanium, calcium, and many more. The two main components of ceramic NPs are silica and alumina. Because they are porous, NPs physically protect against degranulation and degradation. Nanophase ceramics can be categorised into three main types: nanoparticles, nanoscaffolds, and nanoclays. Their unique qualities, like their great heat resistance and chemical inertness, allow them to be used in a variety of contexts. Because of their adjustable size, surface functionalisation, porosity, and surface area-to-volume ratio, they are extensively investigated in biomedical applications, especially in prescription delivery.

2.4 Semiconducting NPs and QDs

Quantum dots (QDs), also known as semiconductor nanoparticles, have demonstrated outstanding labelling capabilities for proteins, cells, and DNA [19]. They provide tunable emission spectra in the visible, mid-infrared, near-infrared, and ultraviolet spectral regions. They also have highly manipulable surface characteristics, great photostability, and resistance to photobleaching. The "quantum confinement effect" is another benefit of semiconductor QDs. The size of QDs affects the emission spectra. They are a superior substitute for natural fluorophores, and a variety of parameters, including size, shape, doping, and environmental conditions, affect their optical characteristics. Compared to dyes, QDs provide numerous advantages. Problems with hydrophobicity of certain dyes, photodegradation in response to laser stimulation, and solvent-dependent quantum yields, can be addressed by replacing the dyes in bioassays with QDs. Some well-known examples of semiconductor QDs include cadmium selenide (CdSe), zinc sulphide (ZnS), zinc oxide (ZnO), and mercuric selenide (HgSe). A wide variety of applications make use of semiconductor quantum dots, such as photocatalysis, optical devices, cellular imaging, and detection of trace levels of analytes [30].

2.5 Polymeric NPs

Polymeric nanoparticles, also referred to as colloidal particles, have a size range of 10 nm to 1 μm. Usually, biodegradable and biocompatible polymers are employed in their fabrication. Pharmaceuticals are encapsulated or entrapped by them, which makes them valuable for transporting drugs. The drugs have the potential to undergo chemical or physical adsorption onto the surface. Due to their small dimensions, ability to dissolve in water, lack of toxicity, long shelf life, and remarkable stability, they are highly effective as carriers. They can be classified into two architectural types based on the synthesis technique. (i) Nanospheres: These represent a matrix structure where the medicine is uniformly dispersed. (ii) Nanocapsules: Nanocapsules refer to nanoparticles that contain a drug enclosed within a polymeric membrane, forming a cavity. Polymeric nanoparticles are produced by using hydrophilic polymers that occur naturally, such as proteins and polysaccharides. In many applications, prepolymerised or polymerised synthetic hydrophobic polymers are utilised. There are three main methods commonly used to create polymeric nanoparticles: (i) synthesising them from a dispersion of preformed polymers, (ii) polymerising them from monomers, and (iii) forming them through coacervation or ionic gelation of hydrophilic polymers. Polymeric NPs have excellent control over pharmacokinetics, a wide range of drug delivery capabilities, and can be easily modified on their surface using chemical methods. Notable examples of NPs in this context consist of gelatin, poly (lactic-co-glycolic acid) copolymer, chitosan, polylactic acid, etc. [28]. In addition, several types of NPs can also be coated with these polymers on their surface.

2.6 Biomolecules Derived NPs

Proteins, nucleic acids, lipids, and polysaccharides are examples of biomolecules with distinctive properties that can be used to create NPs. Although these NPs have historically gotten less attention than inorganic NPs, they are currently at the forefront of many applications in research and development. The increasing need for biocompatible and biodegradable nanoparticles has led to the rise in the popularity of biomolecule-derived NPs. Moreover, biological nanoparticles are nonimmunogenic and widely accessible. In addition to serving specific purposes on their own, proteins can conjugate with other inorganic NPs to create unique hybrids between biomolecules and NPs.

3 Modes of Entry, Uptake, and Interaction of Nanoparticles with Plants

Plants can absorb nanoparticles through their leaves, absorb them through their roots, or penetrate them through their stomata (Fig. 1). Numerous variables, including the plant species involved, surface chemistry, nanoparticle size, and shape, influence the mode of entry of NPs into plants [8]. The uptake and transport process of NPs is illustrated in the schematic diagram (Fig. 1). The following are some typical pathways by which plants absorb nanoparticles.

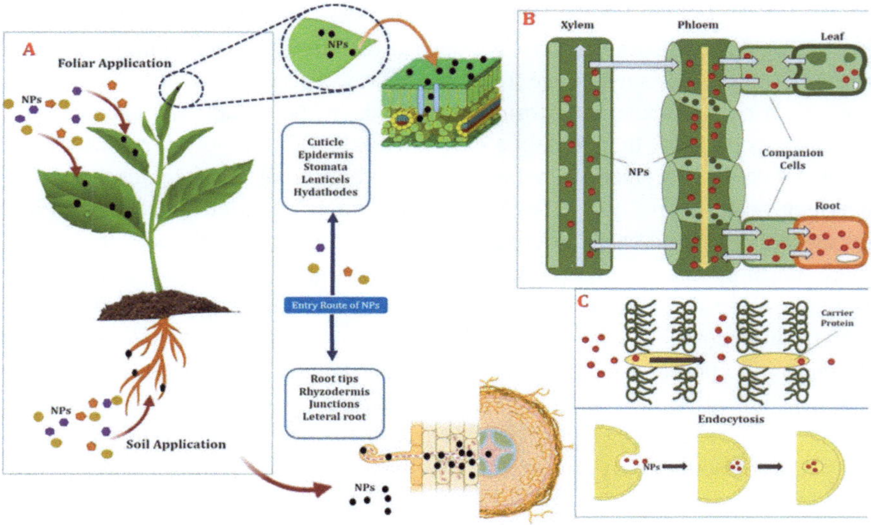

Fig. 1 Modes of entry and interaction pathways of nanoparticles in plants: **a** uptake through roots and leaves; **b** long-distance transport via xylem and phloem; and **c** probable routes of cellular uptake within plant cells

3.1 Root Uptake

Plants Can Absorb Nanoparticles Through Their Roots in a Number of Ways, Including

- Passive Uptake: Nanoparticles can passively enter roots by travelling through the soil with water. The physicochemical parameters of the soil and the size and features of the nanoparticles have a role in this process.
- Active Uptake: Certain nanoparticles could be actively absorbed by plant roots via specialised transporter proteins or mechanisms like endocytosis. The surface biochemistry and integration of nanoparticles can impact this active absorption.

3.2 Foliar Uptake

It is possible for plants to absorb nanoparticles through their leaves, in particular through pores or cuticular holes. When using nanoparticles as aerosols or foliar sprays, this method is more typical. Some factors that affect nanoparticle size and surface charge, and the properties of the leaf can all influence their absorption efficiency through this pathway [66].

3.3 Interactions on Cell Organelles

When absorbed by plant cells, nanoparticles may have interactions with different cell organelles, including nuclei, mitochondria, and chloroplasts. These interactions may have an impact on the metabolism, cellular functions, and general health of the plant.

3.4 Translocation and Distribution

After entering the plant, nanoparticles may use the circulatory system to move and distribute across the various plant tissues. Influence by the features of the NPs and the physiology of the plants, this transport may take place either in the xylem or phloem.

1. Important information on the accumulation and future of various plants can be gleaned from the mobility of NP inside plants. For example, NPs that are transported via the phloem are more commonly found in fruits and grains. Evidence of NPs translocating laterally between several types of cells suggests that translocation is not specific to any one cell type [44]. It is possible for NP-plant interactions to occur both inside and outside of cells. While symplastic transport, which

relies on plasmodesmata, moves water and materials between neighbouring cells' cytoplasms and sieve plates, apoplastic transport happens outside of the plasma membrane, via xylem vessels, cell walls, and extracellular spaces [49, 54].

2. **Apoplast**: This kind of mobility is crucial because it permits NPs to migrate radially within plant tissues [54]. By following the transpiration stream via the xylem, it enables NPs to ascend to the aerial portions of the plant by entering the vascular tissues through the root central cylinder [29, 43]. But after NPs cross the xylem, they arrive at the Casparian strip, which needs to be pierced by endodermal cells via the symplastic pathway [55]. As a result, the Casparian strip which may likewise be utilised for NP delivery under control observes higher accumulation [29, 31].
3. **Symplast**: Water and materials can pass between neighbouring cells through plasmodesmatal structures, sieve plates, and sieve tube elements found in the phloem. This uptake pathway, which is another way for NP entrance, occurs through the symplastic system [49]. Numerous NPs have been reported to translocate via the symplastic process, according to studies. This technique permits the distribution of NPs throughout nonphotosynthetic tissues and organs of plants [46, 65]. Some of the symplastic mechanisms include:
   (a) Endocytosis: NPs may encapsulate themselves inside the plasma membrane as a result of invagination, which is one more way that NPs might be transported [14].
   (b) Pore formation: Damage to the plasma membrane has the potential to cause pore development, which in turn allows NPs to enter the cell. This method of entrance guarantees that the NP can proceed freely into the cytosol without being impeded by any organelles [57, 67].
4. **Carrier proteins**: NPs may be assisted in internalising if they bind to nearby proteins, such as cell membrane proteins [37]. Aquaporins are assumed to be NP transporters inside the cell. However, because of their extremely small pore size, which ranges from 2.8 to 3.4 A°, the rate of absorption and transportation is quite low [68].
5. **Plasmodesmata**: NPs can also enter cells through a different route, which is through specialised structures called plasmodesmata, which are used for transfer between cells. This process aids in NP translocation via the phloem. Nonetheless, for successful translocation, the NPs must already be present in the symplast.
6. **Ion channels**: According to certain studies, ion channels are one of the most likely entrance points for NPs into plants. NP transport across the ion channels is unlikely, nevertheless, due to their 1 nm size. NPs must therefore go through a lot of changes in order to flow through these channels [47, 55].

Although plants may absorb nanoparticles and use them for biotechnology and agriculture, there are worries regarding possible toxicity and buildup in culinary plant sections. To evaluate the environmental impact of nanoparticles and guarantee food safety, it is essential to comprehend the biological processes of their uptake and how they affect plant physiology. The contained nutrients are released by the nanoparticles once they are inside the plant cells. This regulated release makes sure that the necessary nutrients are given to the crop at a pace that corresponds with its requirements. One common method of applying nanoparticles to crops is to sprinkle them on the leaves. Plants then take in the substances via their cuticle or stomata. The leaf epidermis typically has a protective waxy cuticular layer that is mostly made up of pectin, cutin, and wax [66]. It functions as the main natural protective shield to keep nanoparticles from penetrating the leaves and shields them from transpirational loss. However, the waxy stratum corneum layer has two different passages on its surface. Nanoparticles can enter the root column by creating new adsorption surfaces via lateral root growth [42]. The initial contact between nanoparticles and plant roots occurs through adsorption at the root surface. Positively charged nanoparticles are more likely to concentrate inside the root and be more easily absorbed on the root surface because the root's hairs have the ability to discharge chemicals such as mucus or acidic substances, resulting in negative charges on the root surface [31]. According to certain studies, plant roots absorbed nanoparticles and employed the hydrophilic channel as their major mode of cell penetration. Nonetheless, only a few nanoparticles may reach cells through this pathway due to tiny hole size. Plant cells can absorb nanoparticles through different types of mechanisms, including endocytosis, diffusion, and interaction with cell membrane proteins [4].

4 Applications of Nanoparticles

The modelling of strategies for a wide range of low-cost nanotech applications for better germination of seeds, growth of plants, development, and environment acclimation constitutes nanoscience, a new platform for scientific discovery. With the aim to satisfy the projected increase in global food consumption, the introduction of nanomaterials that have been engineered and their applications in the context of agricultural sustainability have drastically transformed world agriculture. These developments are notable for their originality, rapid growth, and scale. The primary goal of commerce in sustainable agriculture is the safeguarding of the environment from contamination, and nanomaterials offer an assurance of improved conservation for plant development and input management. With fewer farming risks, the potential benefits of nanomaterials promote a new green revolution. Nonetheless, there are still a great deal of unknowns regarding the ecotoxicity, allowable limit, and absorption capacity of many nanomaterials. Their ability to feel and recognise environmental circumstances or impairments has been significantly enhanced by the merging of biology and nanotechnology.

Nanotechnology has increasingly transitioned from laboratory-based experimental trials to real-world applications throughout time. The goal of controlled delivery approaches is to achieve complete biological competency while reducing loss and negative effects by releasing a defined amount of required and adequate quantities of agrochemicals throughout time [60]. The agriculture sector is facing an increasing demand to boost food production in order to combat hunger. The high cost, restricted availability, and recurrent scarcity of inorganic fertilisers have had a substantial effect on small-scale crop production. This can be partially attributed to the COVID-19 pandemic epidemic, which has resulted in increased prices for food and oil. Inorganic fertilisers have been widely used in recent years to enhance plant growth and yields. However, in many cases, farms apply more than the recommended amounts, and the excess can easily seep into rivers, contributing to water contamination. From the root's epidermis and endodermis, nanofertilisers enter the xylem vessels and reach the aerial portions of the plant. Furthermore, the plant's phloem and stomata of leaf allow these nutrients in the form of nanoparticles to be transferred to other parts of the plant. There are a number of nanoparticles developed for the enhancement of plant growth and widely used for their unique and specific importance in agricultural science. Some examples of widely used nanoparticles with their general applications in plant regulation are presented in Table 1. Figure 2 illustrates the various effects of nanoparticles on plant growth.

4.1 Role of Nanoparticles in Seed Germination

In addition to boosting seed germination rates, nanoparticles can also improve seedling growth, increase nutrient uptake, and offer defence against environmental stressors and diseases. They can control water intake, stimulate enzymatic activity essential for germination, and function as carriers for nutrition or chemicals that promote growth. But depending on the kind of nanoparticle, how much of it is present, and the kind of plant, the results can differ, therefore using them in agriculture needs to be carefully considered. The methods of conventional seed primer and seed nanopriming differ significantly in that the former uses solutions containing vitamins, hormones, or biopolymers that adsorb on the seeds and produce seed coating, while the latter uses water, or hydropriming. Phase I involves imbibition, during which rapid water absorption causes the transcription process, the production of proteins, and mitochondrial activity in the seed's initial metabolism. Water absorption is restricted during phase II, often known as the lag phase, but metabolism becomes hyperactive due to the creation of enzymes including phytase, endoxylanase, and amylases that are needed for reserve mobilising and embryo growth. Phase III sees the seeds absorb water quickly once more, and the expansion of the embryo culminating in the emergence of radicles. The sole consistent ingredient in nanotechnology are the nanoparticles, which vary depending on the kind of product. Based on their chemical makeup, nanoparticles are categorised into inorganic and biological nanomaterials. Both organic and inorganic nanoparticles are employed as promising coating, pelleting, and seed priming agents. Following are the nanoparticles that are used in seed germination:

(a) **Inorganic Nanoparticles**:

   (i) Manganese oxide nanoparticles: Manganese was known to increase the pace at which seeds germinate. Compared to typical mass or ionic

Table 1 List of nanoparticles with their importance in plant growth

S.no	Nanoparticles	Plants species	Impact	References
1	ZnO	Soybean	Enhanced the rate of seed germination and decreased the weight of remaining seeds	[56]
		Sorghum	Enhanced growth, enhanced crop yield, supplemented edible grains with essential nutrients, and improved nitrogen uptake	[9]
2	Ag	Tomato	Enhanced the percentage of germination and the ratio of germination, as well as the length of the roots, and the fresh weight and dry weight of the seedlings. Four genes (MAPK2, AREB, CRK1 and P5CS) were upregulated, whereas three genes (DDF2, TAS14, and ZFHD1) were downregulated, resulting in the alleviation of salt stress	[5]
		Satureja hortensis L	Higher germination rate and improved growth metrics, such as increased shoot length. Enhanced resistance to salt-induced stress	[39]
3	FeO	Sorghum	Enhanced the rate of photosynthesis, efficiency of photosystem II (PSII), chlorophyll index, relative water content (RWC), and lipid peroxidation	[36]
		Wheat	Elevated levels of antioxidant enzymes activity, photosynthetic pigments and NPK. Enhanced proliferation, greater plant bulk, and greater biomass	[35]
4	TiO_2	*Linum usitatissimum*	Elevated levels of chlorophyll and carotenoids	[2]
		Broad bean	Enhanced plant growth by enhancing the activity of antioxidant enzymes and elevating the concentrations of amino acids, soluble sugars, and other metabolites, therefore aiding in osmoprotection	[1]

(continued)

Table 1 (continued)

S.no	Nanoparticles	Plants species	Impact	References
5	CeO_2	Canola	Improved growth and physiology under salt stress	[50]
		Arabidopsis	Enhanced retention of K^+ in leaf mesophyll, increased chlorophyll concentration, biomass, and photosynthesis	[69]

Fig. 2 Role of nanoparticles in plant growth

manganese compounds, nanoscale manganese has been proven to be less harmful as well as more successful at minimising environmental stressors. Concerning the effects of manganese nanoparticles on agricultural crops from a physiological and toxicological perspective, not much is known. But more investigation is required to figure out the precise impact of manganese-infused seed priming on agricultural productivity [72].

(ii) Carbon nanoparticles: Plant development and seed germination have been said to be promising applications for nanocarbon. Carbon nanoparticles show promise as regulators of plant development and germination of seeds. Under salinity stress, carbon nanoparticles may improve plant post-germination development and seed germination [7].

(iii) Silver nanoparticles: By increasing amylase activity, silver nanopriming may result in a larger concentration of soluble sugars, which would promote the growth of seedlings. The treatment with nanopriming led to an increase in amylase activity and a consequent rise in the number of sugars that were soluble. The osmotic force and water potential in cells decline as sugar concentrations rise. Water can enter into the seeds via osmosis as a result of an increase in the gradient (difference) between the potential for water inside and outside the tissues. The increase in water intake in nanoprimed seeds may potentially be related to the shift in interior osmotic potential brought about by soluble sugars (solutes), given the higher soluble sugar amount and amylase activity [32].

(iv) Zinc oxide nanoparticles: Zinc is necessary for a variety of physiological and biochemical functions in seeds. Numerous studies demonstrate that seed priming with nanoparticles of zinc oxide has a larger zinc content, which raises the rate of development and yield. Crops flourishing in salinised soils may benefit from increased growth and yield thanks to zinc nanoparticles. More research is needed to understand the processes that help zinc oxide nanoparticles mitigate the adverse impacts of salinity in seedlings [48].

(b) **Organic Nanoparticles**:

(i) Cellulose nanofibers: These cellulose nanoparticles are inexpensive, easily processed, biodegradable, recyclable, and compatible with several organic solvents. Biodegradable nanofiber seed coatings made of cellulose and biopolymers. Cellulose nanofiber-based anionic hydrogels for substantial germination of seeds and seedling development [70].

(ii) Chitosan nanoparticles: It possesses strong antibacterial properties, serving as a bioprotectant for rice phytopathogens. In order to fight rice blast and blight disease, chitosan nanoparticles might be employed as an antibacterial agent, per this study. Fertiliser testing is done on nanochitosan that has been loaded with potassium, phosphorus, and nitrogen. A study by [51] showed that copper–chitosan nanoparticles improve the mobilisation of food reserves, like starch, by raising α-amylase levels, which in turn promotes seedling growth.

4.2 Role of Nanoparticles in Root and Shoot Growth

The morphology of a plant's shoots and roots is crucial to the plant's development and growth, and any factor that alters this morphology can have positive or bad effects on the plant. Researchers looked at how the shape of plant roots and shoots was affected by a small number of nanoparticles. Studies have shown that seed germination and the first development of roots and shoots are affected differently by varying concentrations of nanoparticles. Although some nanoparticles can really lengthen plant roots and shoots at low concentrations, they are toxic and stunt plant growth when present in higher concentrations. The plant's ability to absorb and transfer NPs is widely acknowledged. Multiple factors, such as surface charge, particle size, concentration, plant type, and exposure time, impact this process. Nanoparticles are able to traverse the plant system via a variety of apertures, such as root hairs, stomata, and surface fissures on leaves. Nanoparticles, once within a plant, can travel through its systems by diffusion, phloem loading, or bulk flow. The size, shape, and surface characteristics of the nanoparticles, in addition to the pH and presence of any other ions or chemicals in the solution, are some of the numerous variables that could impact the nanoparticles' mobility [66]. Nanoparticles, due to their unique properties compared to bulk materials, offer a novel approach to addressing crop nutritional deficiencies and promoting plant growth and development. One way that nanoparticles can come into touch with plant roots is by adsorption on the root layer. Because root hairs can secrete substances like organic acids or mucus, the root surface has a negative charge. Consequently, positively charged nanoparticles are more likely to collect inside the root and absorb easily on the root surface [31, 73]. The plant's leaf surface and the root epidermis are structurally and functionally similar. The epidermal layer is fully matured on the surface of the plant, but it is not yet present on the root apices or the root surface hairs of the main and secondary (branch) roots. The root epidermal layer is directly touched and penetrated by nanoparticles when they are introduced to this area. The extent to which wheat roots absorb nanoparticles is affected by their size. As an example, wheat roots are capable of absorbing TiO_2 NPs ranging in size from 36 to 140 nm. As the size of the particles increases, their overall absorption decreases. Despite this, according to [29], TiO_2 NPs bigger than 140 nm are not absorbable. Positively charged nanoparticles tend to cluster on the root surface rather than penetrate root tissues, due to electrostatic attraction between the nanoparticles and the negatively charged cell wall [15]. The effect of electrical charges on the uptake of nanoparticles in roots is slightly different from that in leaves. Spots on the surface of leaves called stomata help plants control the exchange of water and gases in their shoots. The size of stomata can vary greatly among plant species, with some having stomata that are 10 to 100 μm in diameter and others that are denser.

Improved root and shoot growth have been reported by several studies. For example, Dutta Gupta [11] observed that bio-synthesised silver (Ag) NPs improved the root and shoot growth of rice seedlings. Similarly, seaweed-based biogenic ZnO NPs were reported to enhance the root length and shoot length of rice [24].

4.3 Role of Nanoparticles in Biotic and Abiotic Stress

Research in the intriguing and quickly expanding field of nanotechnology has produced a number of advancements. Specifically, agriculture-related issues can be effectively resolved with the aid of nanotechnology paving the way for a safe and sustainable agricultural future. Due to its wide range of applications including medicine, drug delivery, energy, poultry production, and the agri-food industry, nanotechnology has attracted a lot of attention in recent years. The existence of a robust ecosystem depends on both biotic and abiotic components, which are intricately interdependent.

a. **Abiotic Stress**: Agricultural production is greatly impacted by abiotic stresses, including salt, drought, extreme heat and cold, and heavy metals [53]. These stresses can also have an impact on the characteristics of plant growth and yield, including fresh and dry weight, leaf area, shoot and root length, and total plant growth [45]. According to [21], applying NPs through seed priming is a suitable strategy to enhance seed germination, root growth, and shoot growth under stress conditions. By applying selenium (Se) NPs, the temperature stress-induced reductions in sorghum yield, seed set, and pollen germination were restored [10]. Under high temperatures, plant-derived AgNPs improved the leaf area, root length, shoot length, root number, fresh weight, and dry weight of wheat [23]. Under salinity stress, TiO_2 NPs have been reported to enhance agronomic traits and boost antioxidant enzyme activity in *Dracocephalum moldavica* [20]. Under drought stress, maghemite (γ-Fe_2O3) NPs have demonstrated to improved agronomic traits in *Brassica napus* [41].
b. **Biotic Stress**: Biological factors, such as pests and diseases, have a significant impact on agricultural yields. Pesticides have been an important tool for farmers in reducing crop failures. The health of humans and the sustainability of ecosystems are jeopardised by the use of these chemicals. New research suggests that nanoparticles have the potential to lessen the prevalence of pests and diseases, which could lessen the impact of crop losses and environmental problems. Silica NPs reduced the populations of pests such as aphid (*Aphis craccivora*), leafminer (*Liriomyza trifolii*), and cotton leafworm (*Spodoptera littoralis*) in faba bean–soybean system [62]. According to [64], *Xanthomonas axonopodis* pv. *malvacearum* and *Xanthomonas campestris* pv. *campestris* are two significant bacterial pathogens affecting crops in the Malvaceae and Brassicaceae families, respectively. Biosynthesised AgNPs derived from cotton plant (*Gossypium hirsutum*) stem extract have demonstrated strong antibacterial activity against these two pathogens [64]. Similarly, [33] reported that AgNPs were effective in inhibiting the mycelial growth of *Botrytis cinerea.* Copper-based NPs are effective in controlling the infection of *Phytophthora infestans* in tomato [18]. Magnesium oxide NPs induced systemic resistance in tomato plants against bacterial wilt (disease) caused by *Ralstonia solanacearum* [22].

4.4 Role of Nanoparticles in Nutrient Delivery in Plants

Given the soil's lack of most macro- and micronutrients, fertilisers are employed in agriculture to meet the nutrient requirements. Fertilisers contribute to around 35–40% of the entire productivity of crops, primarily by directly influencing plant development. The current nutrient utilisation efficiency is very low, as approximately 50 to 70% of the nitrogen provided by conventional fertilisers is wasted [3]. Nitrogen, phosphorus, potassium, calcium, magnesium, and sulphur are essential macronutrients for crop growth. In nanofertilisers, these macronutrients are engineered at the nanoscale to enhance the efficiency of fertilisers. Nitrogen, the main nutrient, is slowly released through urea-coated zeolite chips and is crucial for the growth of all plants. Zeolites are naturally occurring minerals with honeycomb-like crystal structures that can be used to achieve the slow release of nutrients into the environment [34].

5 Conclusion

The use of NPs in plants is an exciting and diverse field of study with potential applications in many areas of agriculture. With the use of nanoparticles, one can increase crop yield, improve plant growth, and lessen environmental stress. Their special qualities allow for accurate control over the distribution of nutrients, focused control over pests, and effective use of resources, all of which support sustainable agricultural methods. Furthermore, nanoparticles have the capacity to clean up contaminated water and soil, which promotes environmental stewardship. However, before these technologies are widely adopted in agriculture, it's important to thoroughly assess their potential risks and impacts on ecosystems, human health, and food safety. Comprehensive evaluations of nanoparticle toxic effects, fate, and solubility in plant structures, as well as their lasting effects on the soil microbiota and ecosystem resilience, must be given top priority in research endeavours.

References

1. Abdel Latef AAH, Srivastava AK, El Sadek MSA, Kordrostami M, Tran LSP (2018) Titanium dioxide nanoparticles improve growth and enhance tolerance of broad bean plants under saline soil conditions. Land Degrad Dev 29:1065–1073
2. Aghdam MTB, Mohammadi H, Ghorbanpour M (2016) Effects of nanoparticulate anatase titanium dioxide on physiological and biochemical performance of *Linum usitatissimum* (Linaceae) under well-watered and drought stress conditions. Braz J Bot 39:139–146
3. Agrahari S, Dubey A (2020) Nanoparticles in plant growth and development. In: Ghorbanpour M, Bhargava P, Varma A, Choudhary D (eds) Biogenic nano-particles and their use in agro-ecosystems. Springer, Singapore, pp 9–37. https://doi.org/10.1007/978-981-15-2985-6_2

4. Ali S, Mehmood A, Khan N (2021) Uptake, translocation, and consequences of nanomaterials on plant growth and stress adaptation. J Nanomater 2021:6677616. https://doi.org/10.1155/2021/6677616
5. Almutairi ZM (2016) Influence of silver nano-particles on the salt resistance of tomato (*Solanum lycopersicum*) during germination. Int J Agric Biol 18:449–457
6. Baig N, Kammakakam I, Falath W (2021) Nanomaterials: a review of synthesis methods, properties, recent progress, and challenges. Mater Adv 2:1821–1871
7. Baz H, Creech M, Chen J, Gong H, Bradford K, Huo H (2020) Water-soluble carbon nanoparticles improve seed germination and post-germination growth of lettuce under salinity stress. Agronomy 10(8):1192
8. Barot RB, Gawande ND, Omprabha S, Kaushal C, Ghosh S, Saha J, Bhatia D, Sankaranarayanan S (2024) Novel carbon nanoparticles derived from Bougainvillea modulate vegetative growth via auxin–cytokinin signaling in Arabidopsis. Chem Pap 78(8):4733–4750
9. Dimkpa CO, Singh U, Bindraban PS, Elmer WH, Gardea-Torresdey JL, White JC (2019) Zinc oxide nanoparticles alleviate drought-induced alterations in sorghum performance, nutrient acquisition, and grain fortification. Sci Total Environ 688:926–934
10. Djanaguiraman M, Belliraj N, Bossmann SH, Prasad PV (2018) High-temperature stress alleviation by selenium nanoparticle treatment in grain sorghum. ACS Omega 3(3):2479–2491
11. Dutta Gupta S, Agarwal A, Pradhan S (2018) Phytostimulatory effect of silver nanoparticles (AgNPs) on rice seedling growth: an insight from antioxidative enzyme activities and gene expression patterns. Ecotoxicol Environ Saf 161:624–633
12. Dwivedi S, Saquib Q, Al-Khedhairy AA, Musarrat J (2016) Understanding the role of nanomaterials in agriculture. Microbial inoculants in sustainable agricultural productivity: Vol. 2: functional applications, pp 271–288
13. Eichert T, Kurtz A, Steiner U, Goldbach HE (2008) Size exclusion limits and lateral heterogeneity of the stomatal foliar uptake pathway for aqueous solutes and water-suspended nanoparticles. Physiol Plant 134:151–160
14. Etxeberria E, Gonzalez P, Bhattacharya P, Sharma P, Ke PC (2016) Determining the size exclusion for nanoparticles in citrus leaves. HortScience 51(6):732–737
15. Francis DV, Abdalla AK, Mahakham W, Sarmah AK, Ahmed ZFR (2024) Interaction of plants and metal nanoparticles: exploring its molecular mechanisms for sustainable agriculture and crop improvement. Environ Int 190:108859. https://doi.org/10.1016/j.envint.2024.108859
16. Gaur M, Misra C, Yadav AB, Swaroop S, Maolmhuaidh FÓ, Bechelany M, Barhoum A (2021) Biomedical applications of carbon nanomaterials: fullerenes, quantum dots, nanotubes, nanofibers, and graphene. Materials 14:5978. https://doi.org/10.3390/ma14205978
17. Ghorbanpour M, Bhargava P, Varma A, Choudhary DK (eds) (2020) Biogenic nano-particles and their use in agro-ecosystems. https://doi.org/10.1007/978-981-15-2985-6
18. Giannousi K, Avramidis I, Dendrinou-Samara C (2013) Synthesis, characterization and evaluation of copper based nanoparticles as agrochemicals against *Phytophthora infestans*. RSC Adv 3(44):21743–21752
19. Gill R, Zayats M, Willner I (2008) Semiconductor quantum dots for bioanalysis. Angew Chem Int Ed 47(40):7602–7625
20. Gohari G, Mohammadi A, Akbari A, Panahirad S, Dadpour MR, Fotopoulos V, Kimura S (2020) Titanium dioxide nanoparticles (TiO_2 NPs) promote growth and ameliorate salinity stress effects on essential oil profile and biochemical attributes of *Dracocephalum moldavica*. Sci Rep 10:912
21. Hassanisaadi M, Barani M, Rahdar A, Heidary M, Thysiadou A, Kyzas GZ (2022) Role of agrochemical-based nanomaterials in plants: biotic and abiotic stress with germination improvement of seeds. Plant Growth Regul 97(2):375–418
22. Imada K, Sakai S, Kajihara H, Tanaka S, Ito S (2016) Magnesium oxide nanoparticles induce systemic resistance in tomato against bacterial wilt disease. Plant Pathol 65(4):551–560
23. Iqbal M, Raja NI, Mashwani ZUR, Hussain M, Ejaz M, Yasmeen F (2019) Effect of silver nanoparticles on growth of wheat under heat stress. Iran J Sci Technol Trans A Sci 43:387–395

24. Itroutwar PD, Govindaraju K, Tamilselvan S, Kannan M, Raja K, Subramanian KS (2020) Seaweed-based biogenic ZnO nanoparticles for improving agro-morphological characteristics of rice (*Oryza sativa* L.). J Plant Growth Regul 39:717–728
25. Jośko I, Oleszczuk P (2013) Manufactured nanomaterials: the connection between environmental fate and toxicity. Crit Rev Environ Sci Technol 43(23):2581–2616
26. Khan MR, Adam V, Rizvi TF, Zhang B, Ahamad F, Jośko I, Zhu Y, Yang M, Mao C (2019) Nanoparticle-plant interactions: two-way traffic. Small 15(37):e1901794. https://doi.org/10.1002/smll.201901794
27. Khan I, Awan SA, Rizwan M, Hassan ZU, Akram MA, Tariq R, Xie W (2022) Nanoparticle's uptake and translocation mechanisms in plants via seed priming, foliar treatment, and root exposure: a review. Environ Sci Pollut Res 29(60):89823–89833
28. Lai P, Daear W, Löbenberg R, Prenner EJ (2014) Overview of the preparation of organic polymeric nanoparticles for drug delivery based on gelatine, chitosan, poly (d, l-lactide-co-glycolic acid) and polyalkylcyanoacrylate. Colloids Surf B 118:154–163
29. Larue C, Laurette J, Herlin-Boime N, Khodja H, Fayard B, Flank AM, Brisset AM, Carriere M (2012) Accumulation, translocation and impact of TiO_2 nanoparticles in wheat (*Triticum aestivum* spp.): influence of diameter and crystal phase. Sci Total Environ 431:197–208
30. Lesiak A, Drzozga K, Cabaj J, Bański M, Malecha K, Podhorodecki A (2019) Optical sensors based on II-VI quantum dots. Nanomaterials 9(2):192. https://doi.org/10.3390/nano9020192
31. Lv J, Christie P, Zhang S (2019) Uptake, translocation, and transformation of metal-based nanoparticles in plants: recent advances and methodological challenges. Environ Sci Nano 6(1):41–59
32. Mahakham W, Sarmah AK, Maensiri S, Theerakulpisut P (2017) Nanopriming technology for enhancing germination and starch metabolism of aged rice seeds using phytosynthesized silver nanoparticles. Sci Rep 7(1):8263
33. Malandrakis AA, Kavroulakis N, Chrysikopoulos CV (2019) Use of copper, silver and zinc nanoparticles against foliar and soil-borne plant pathogens. Sci Total Environ 670:292–299
34. Manjunatha SB, Biradar DP, Aladakatti YR (2016) Nanotechnology and its applications in agriculture: a review. J Farm Sci 29(1):1–13
35. Manzoor N, Ahmed T, Noman M, Shahid M, Nazir MM, Ali L et al (2021) Iron oxide nanoparticles ameliorated the cadmium and salinity stresses in wheat plants, facilitating photosynthetic pigments and restricting cadmium uptake. Sci Total Environ 769:145221. https://doi.org/10.1016/j.scitotenv.2021.145221
36. Maswada HF, Djanaguiraman M, Prasad PVV (2018) Seed treatment with nano-iron (III) oxide enhances germination, seeding growth and salinity tolerance of sorghum. J Agro Crop Sci 204:577–587
37. Nel AE, Mädler L, Velegol D, Xia T, Hoek EM, Somasundaran P, Thompson M (2009) Understanding biophysicochemical interactions at the nano–bio interface. Nat Mater 8(7):543–557
38. Nair R (2016) Effects of nanoparticles on plant growth and development. In: Kole C, Kumar D, Khodakovskaya M (eds) Plant nanotechnology. Springer, Cham, pp 95–118. https://doi.org/10.1007/978-3-319-42154-4_5
39. Nejatzadeh F (2021) Effect of silver nanoparticles on salt tolerance of *Satureja hortensis* L. during *in vitro* and *in vivo* germination tests. Heliyon 7:e05981. https://doi.org/10.1016/j.heliyon.2021.e05981
40. Nowack B, Bucheli TD (2007) Occurrence, behavior and effects of nanoparticles in the environment. Environ Pollut 150:5–22
41. Palmqvist NM, Seisenbaeva GA, Svedlindh P, Kessler VG (2017) Maghemite nanoparticles acts as nanozymes, improving growth and abiotic stress tolerance in *Brassica napus*. Nanoscale Res Lett 12:631. https://doi.org/10.1186/s11671-017-2404-2
42. Peng C, Duan D, Xu C, Chen Y, Sun L, Zhang H, Shi J (2015) Translocation and biotransformation of CuO nanoparticles in rice (*Oryza sativa* L.) plants. Environ Pollut 197:99–107
43. González-Melendi P, Fernández-Pacheco R, Coronado MJ, Corredor E, Testillano PS, Risueño MC, Marquina C, Ibarra MR, Rubiales D, Pérez-de-Luque A (2008) Nanoparticles as smart

treatment-delivery systems in plants: assessment of different techniques of microscopy for their visualization in plant tissues. Ann Bot 101(1):187–195
44. Pate JS, Sharkey PJ, Lewis OAM (1975) Xylem to phloem transfer of solutes in fruiting shoots of legumes, studied by a phloem bleeding technique. Planta 122:11–26
45. Rakshit A, Singh HB, Singh AK, Singh US, Fraceto L (eds) (2020) New frontiers in stress management for durable agriculture. Springer Nature, Singapore
46. Raliya R, Franke C, Chavalmane S, Nair R, Reed N, Biswas P (2016) Quantitative understanding of nanoparticle uptake in watermelon plants. Front Plant Sci 7:1288
47. Rico CM, Majumdar S, Duarte-Gardea M, Peralta-Videa JR, Gardea-Torresdey JL (2011) Interaction of nanoparticles with edible plants and their possible implications in the food chain. J Agric Food Chem 59(8):3485–3498
48. Rizwan M, Ali S, Ali B, Adrees M, Arshad M, Hussain A, Waris AA (2019) Zinc and iron oxide nanoparticles improved the plant growth and reduced the oxidative stress and cadmium concentration in wheat. Chemosphere 214:269–277
49. Roberts A (2003) Plasmodesmata and the control of symplastic transport plant cell environ. Plant Cell Environ 26:103–124
50. Rossi L, Zhang W, Lombardini L, Ma X (2016) The impact of cerium oxide nanoparticles on the salt stress responses of *Brassica napus* L. Environ Pollut 219:28–36
51. Saharan V, Kumaraswamy RV, Choudhary RC, Kumari S, Pal A, Raliya R, Biswas P (2016) Cu-chitosan nanoparticle mediated sustainable approach to enhance seedling growth in maize by mobilizing reserved food. J Agric Food Chem 64(31):6148–6155
52. Sajid M (2022) Nanomaterials: types, properties, recent advances, and toxicity concerns. Curr Opinion Environ Sci Health 25:100319. https://doi.org/10.1016/j.coesh.2021.100319
53. Sarkar D, Pal S, Mehjabeen M, Singh V, Singh S, Pul S, Garg J, Rakshit A, Singh HB (2018) Addressing stresses in agriculture through bio-priming intervention. In: Rakshit A, Singh HB (eds) Advances in seed priming. Springer, Singapore, pp 107–113
54. Sattelmacher B (2001) The apoplast and its significance for plant mineral nutrition. New Phytol 149(2):167–192
55. Schwab F, Zhai G, Kern M, Turner A, Schnoor JL, Wiesner MR (2016) Barriers, pathways and processes for uptake, translocation and accumulation of nanomaterials in plants–critical review. Nanotoxicology 10(3):257–278
56. Sedghi M, Mitra H, Sahar T (2013) Effect of nano zinc oxide on the germination of soybean seeds under drought stress. Ann West Univ Timişoara ser Biol 16:73–78
57. Serag MF, Kaji N, Habuchi S, Bianco A, Baba Y (2013) Nanobiotechnology meets plant cell biology: carbon nanotubes as organelle targeting nanocarriers. RSC Adv 3(15):4856–4862
58. Shang Y, Hasan MK, Ahammed GJ, Li M, Yin H, Zhou J (2019) Applications of nanotechnology in plant growth and crop protection: a review. Molecules (Basel, Switzerland) 24(14):2558. https://doi.org/10.3390/molecules24142558
59. Sharifi R, Vatani A, Sabzi A, Safaei M (2024) A narrative review on application of metal and metal oxide nanoparticles in endodontics. Heliyon 15:e34673. https://doi.org/10.1016/j.heliyon.2024.e34673
60. Shojaei TR, Salleh MAM, Tabatabaei M, Mobli H, Aghbashlo M, Rashid SA, Tan T (2019) Applications of nanotechnology and carbon nanoparticles in agriculture. In: Synthesis, technology and applications of carbon nanomaterials. Elsevier, pp 247–277
61. Tarafdar JC, Xiang Y, Wang W-N, Dong Q, Biswas P (2012) Standardization of size, shape and concentration of nanoparticle for plant application. Appl Biol Res 14:138–144
62. Thabet AF, Boraei HA, Galal OA, El-Samahy MF, Mousa KM, Zhang YZ, Tuda M, Helmy EA, Wen J, Nozaki T (2021) Silica nanoparticles as pesticide against insects of different feeding types and their non-target attraction of predators. Sci Rep 11:14484
63. Tombuloglu H, Ercan I, Alshammari T, Tombuloglu G, Slimani Y, Almessiere M, Baykal A (2020) Incorporation of micro-nutrients (nickel, copper, zinc, and iron) into plant body through nanoparticles. J Soil Sci Plant Nutr 20:1872–1881
64. Vanti GL, Nargund VB, Basavesha KN, Vanarchi R, Kurjogi M, Mulla SI, Tubaki S, Patil RR (2019) Synthesis of *Gossypium hirsutum*-derived silver nanoparticles and their antibacterial

efficacy against plant pathogens. Appl Organomet Chem 33:e4630. https://doi.org/10.1002/aoc.4630
65. Wang P, Lombi E, Zhao FJ, Kopittke PM (2016) Nanotechnology: a new opportunity in plant sciences. Trends Plant Sci 21(8):699–712
66. Wang X, Xie H, Wang P, Yin H (2023) Nanoparticles in plants: uptake, transport and physiological activity in leaf and root. Materials 16(8):3097
67. Wong MH, Giraldo JP, Kwak SY, Koman VB, Sinclair R, Lew TTS, Strano MS (2017) Nitroaromatic detection and infrared communication from wild-type plants using plant nanobionics. Nat Mater 16(2):264–272
68. Wu B, Beitz E (2007) Aquaporins with selectivity for unconventional permeants. Cell Mol Life Sci 64:2413–2421
69. Wu H, Shabala L, Shabala S, Shabala S, Giraldo JP (2018) Hydroxyl radical scavenging by cerium oxide nanoparticles improves *Arabidopsis* salinity tolerance by enhancing leaf mesophyll potassium retention. Environ Sci Nano 5:1567–1583
70. Xu T, Ma C, Aytac Z, Hu X, Ng KW, White JC, Demokritou P (2020) Enhancing agrichemical delivery and seedling development with biodegradable, tunable, biopolymer-based nanofiber seed coatings. ACS Sustain Chem Eng 8(25):9537–9548
71. Yang GW (2007) Laser ablation in liquids: applications in the synthesis of nanocrystals. Prog Mater Sci 52(4):648–698
72. Ye Y, Cota-Ruiz K, Hernández-Viezcas JA, Valdes C, Medina-Velo IA, Turley RS, Gardea-Torresdey JL (2020) Manganese nanoparticles control salinity-modulated molecular responses in *Capsicum annuum* L. through priming: a sustainable approach for agriculture. ACS Sustain Chem Eng 8(3):1427–1436
73. Zhao L, Peralta-Videa JR, Ren M, Varela-Ramirez A, Li C, Hernandez-Viezcas JA, Gardea-Torresdey JL (2012) Transport of Zn in a sandy loam soil treated with ZnO NPs and uptake by corn plants: electron microprobe and confocal microscopy studies. Chem Eng J 184:1–8. https://doi.org/10.1016/j.cej.2012.01.041

Nutrient Delivery

Nanofertilizers for Plant Nutrition

P. Smriti Rao, Faria Fatima, Abdul Mazeed, Munny Chinyo, and Khalid Habib

Abstract In a startling saga of agricultural overabundance, the once-fertile soil has suffered greatly from the pervasive use of synthetic fertilizers in contemporary agricultural practices, which has severely depleted its essential nutrients and irreversibly damaged the vital equilibrium of the surrounding environment. Overuse of these fertilizers degrades soil quality, pollutes the environment, disturbs agrarian ecosystems, and deposits residue on agricultural products. Moreover, between 50 and 70% of the nutrients including potassium, phosphate, and nitrogen are lost from the soil before they can be used. Nanoparticles regulate nutrient delivery, making nanofertilizers more economical and efficient than conventional fertilizers. A new class of fertilizers called nanofertilizers makes use of cutting-edge nanotechnology to offer a potent and eco-friendly approach to crop fertilization. They ensure that nutrients are distributed uniformly over an extended period of time with their regulated nutrient delivery system, providing plants that provide a consistent flow of vital components. Because it requires less fertilizer and less frequent management, the regulated discharge technology works better than standard fertilizers. In agriculture, nanotechnology is primarily utilized to enhance crop yields, reduce losses, and stimulate plants to defend against pests, insects, and environmental challenges. Nanofertilizers have shown promising results in various plant species, whether applied topically or in the soil. Most nanomaterials are precursors of macro- and micronutrients and their nanoscale properties. This chapter explores new approaches to using them as crop growth promoters, as well as how to apply and incorporate them into plant matter. It also discusses future concerns and possible drawbacks for the commercial implementation of nanofertilizers in agriculture.

Keywords Nanofertilizers · Nutrient use efficiency · Slow-release formulations · Soil fertility · Sustainable agriculture

---

P. Smriti Rao (✉) · F. Fatima · A. Mazeed · M. Chinyo · K. Habib
Department of Agriculture, Integral Institute of Agricultural Science and Technology, Integral University, Lucknow, India
e-mail: smritirao19@gmail.com

L. F. Fraceto et al. (eds.), *Emerging Nanotechnologies for Agroecosystem Management*, Smart Nanomaterials Technology, https://doi.org/10.1007/978-981-95-0187-8_4

1 Introduction

Global food security and agricultural sustainability are facing significant difficulties due to climate change, extreme weather, industrialization, and the growing human population [48]. Earth's temperature has increased by about 1% on average over the past century, and future decades are expected to see much more temperature increases. Global warming is also predicted to double the amount of arable land by the end of the decade. Due to social unrest, poverty, food insecurity, water scarcity, and health problems, people are moving from rural to urban areas increasingly frequently. Due to the aforementioned issues, freshwater depletion and declining land fertility have had a significant impact on the world food supply [66]. By 2050, it is anticipated that high dosages of agrochemical and synthetic fertilizers will be used to feed 9.6 billion people [51]. In contemporary agricultural methods, synthetic fertilizers have not been able to improve crop output or plant nutrient uptake, despite the fact that they may increase crop growth [23].

As a result, it is now evident that consistent fertilizer use is essential to preserving soil fertility and raising crop yields. Conventional fertilizers are extensively utilized to replenish essential nutrients in the soil. These consist of monoammonium phosphate, urea, nitrogen, potassium, and phosphorous, among others. Leaching lowers the nutrient utilization efficiency of traditional fertilizers, causing significant financial losses and reduced soil fertility. As nutrients leach from the soil, fertility declines dramatically. The main cause of this is the extremely low-nutrient consumption rates of conventional fertilizers, which are roughly 35–40% for potassium, 18–20% for phosphorus, and 30–35% for nitrogen [42]. One viable way to address these issues is with nanofertilizers due to their more effectiveness and less of an influence on the environment. One might categorize them according to their composition, activity, and nutrient content. Nanoparticles (NPs) play a crucial role in sustainable agriculture, owing to their increased surface area and nanoscale size (1–100 nm). Nanotechnology has enabled various agricultural products, including nanofertilizers, nanoinsecticides, and nanopesticides [43]. Recent studies indicate that nanoparticles can help manage plant pests such as weeds, fungi, and insects [7, 59]. Some researchers suggest that nanofertilizers can deliver nutrients more efficiently to plants, significantly enhancing crop productivity [1].

Because nanofertilizers can meet the alimentary demands of plants deprived of sacrificing agricultural output, they could undergo a radical transformation the way that food is now produced. Figure 1 illustrates how the increased mobility and surface area to volume ratio of these sub-microscopic particles enhance agricultural output and plant nutrient uptake. Thus, by increasing the production of food grains, one way to think of nanofertilizers is to consider them a "smart system of nutrients" that can help lower poverty and hunger. By 2030, achieving sustainable development targets of eradicating hunger, guaranteeing food security, and enhancing the ecological integrity of worldwide agriculture may depend heavily on these innovative nanofertilizers. Because of this, the current chapter concentrates on the urgent requirement for nanofertilizers for sustainable farming methods. It seeks to provide

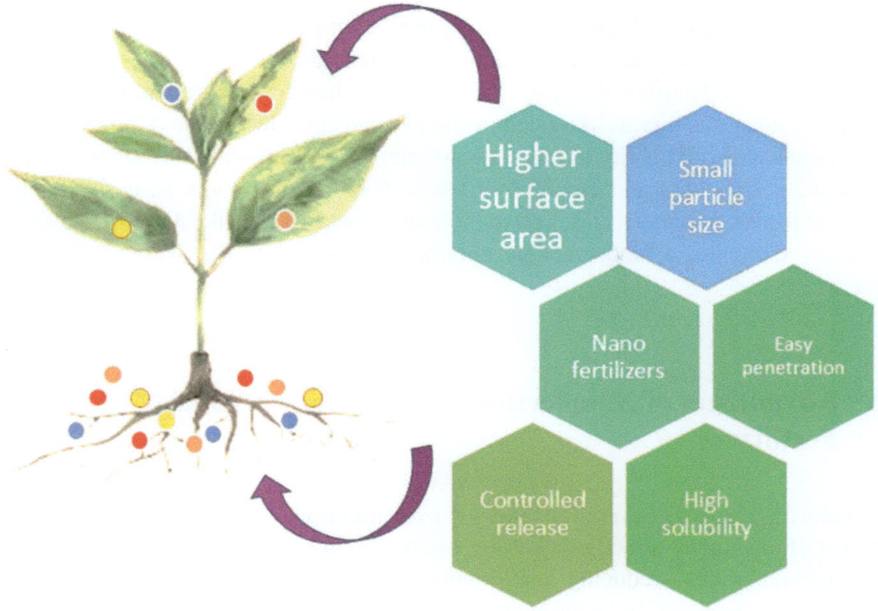

Fig. 1 Behavior and impact of nanofertilizers in soil and plant systems

an in-depth explanation of several kinds of nanofertilizers that are already on the market as well as their possible uses in contemporary farming. We will go over their efficacy, benefits, and drawbacks as well as the materials and techniques for delivering NPs in a targeted and regulated manner. We will also explore the characteristics of a successful nanofertilizer, possible application-related hazards, and the prospects for this developing technology. Through our assessment and comparison of several nanofertilizers, we intend to provide insightful information to farmers, researchers, and other agriculture industry stakeholders.

2 Manufacturing of Nanofertilizers

Nanotechnology involves manipulating materials at the nanoscale to develop advanced applications, including "smart fertilizers" [71]. Nanofertilizers enhance nutrient uptake, restore soil fertility, reduce toxicity, and minimize environmental pollution [65]. These fertilizers contain nanomaterials like silica, iron, zinc oxide, and titanium dioxide, with effectiveness depending on crop type, size, concentration, and chemical properties.

Nanofertilizers release nutrients into the soil through nanoparticle suspensions and can be encapsulated with polymers to prevent nutrient loss [38]. They improve nutrient utilization by using nano-adsorbents and surface-modified cationic or anionic nutrients. There are three encapsulation methods: nanoscale particles/ emulsions, polymer coatings, and nanoporous materials [16].

Nano-enabled nutrient delivery systems, such as hydroxyapatite-based fertilizers, provide essential minerals like calcium and phosphorus. Mesoporous silica nanoparticles enhance crop quality and stress resistance, while carbon-based nanomaterials improve germination and chlorophyll levels. Nanofertilizers can be derived from organic (lipids, polymers, chitosan) or inorganic (metal oxides like AgO, MgO, ZnO) sources. Chitosan, a biodegradable polymer, serves as an effective agrochemical carrier [21].

3 Advantages of Nanofertilizers Over Conventional Fertilizers

Nanofertilizers offer several advantages over traditional fertilizers. They provide higher efficiency by directly delivering essential nutrients to plants while requiring smaller quantities, reducing environmental impact. This approach moreover enhances crop yields by minimizing negative consequences of fertilizers on the environment, as shown in Fig. 2.

Modern agricultural systems are characterized by reduced nutrient use efficiency, diminished productivity, widespread nutrient deficiencies, and a decline in soil biodiversity, largely due to intensive farming practices and the unbalanced use of chemical

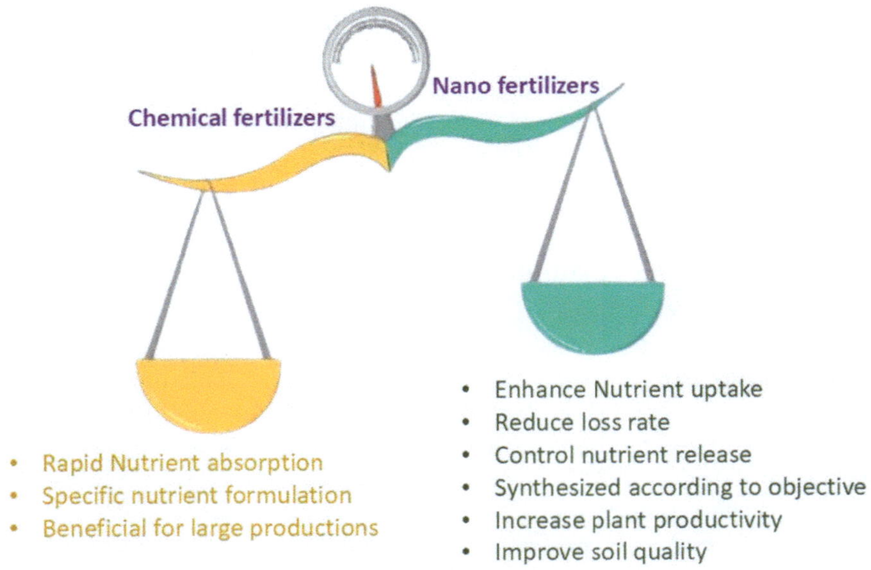

Fig. 2 Advantages of nanofertilizers over chemical fertilizers

fertilizers [49]. Key benefits of nanofertilizers include controlled nutrient release, improved plant absorption, and reduced fertilizer input requirements [20]. Additionally, they enhance the physical and chemical properties of soil while reducing overall fertilizer usage. Due to their high nutrient density, nanofertilizers can be applied in lower amounts than conventional fertilizers, leading to cost savings and a reduced environmental footprint from production and transportation.

Slow-release nanofertilizers can enhance plant growth and productivity by providing a consistent supply of nutrients over an extended period. They also reduce the need for frequent fertilizer applications and minimize nutrient leaching into the environment. Improved nutrient uptake leads to increased plant growth and yield while decreasing environmental nutrient losses. By enhancing fertilizer efficiency, nanofertilizers help reduce the overall environmental impact of agricultural practices. There are various advantages of using nanofertilizers over traditional fertilizers, some of which are mentioned below.

3.1 Increased Surface Area

Nanofertilizers increase plant nutrient uptake since their particles are smaller than 100 nm, which makes them more effective at reaching plants through applied surfaces like soil or leaves [29]. This allows for a wider variety of sites and metabolic activities in plant systems, which increases the production of photosynthate. The effectiveness of nutrient uptake and utilization is enhanced by increased surface area, which also makes nanofertilizers more reactive to other chemicals [55].

3.2 Higher Solubility

Solubility mentions the determined quantity of a substance capable to dissolve in a specified solvent under definite temperature and pressure conditions. Nanofertilizers display higher solubility attributable to their larger surface area and smaller particle size, facilitating dissolution in soil solutions. Their ability to dissolve in various solvents, including water, enhances the availability of previously insoluble soil nutrients to nearby organisms [10]. Studies have shown that nano-hydroxyapatite, a phosphorus-based nanofertilizer, has greater solubility than conventional phosphorus fertilizers [17]. Similarly, nano-zinc oxide is significantly more soluble than its bulk counterpart [46].

3.3 Fertilizer Encapsulation Inside Nanoparticles

Fertilizers coated in NPs increase crop nutrient absorption [61]. One way of encapsulating fertilizers is polymer-based encapsulation, which involves using interfacial and emulsion polymerization to create a polymer matrix enclosing the fertilizer [57]. Adding inorganic materials that can precipitate or create a sol-gel synthesis shell around the fertilizer, such as silica or calcium carbonate, is another method. Another approach is layer-by-layer assembly, where charged polymers, nanoparticles, or other compounds are sequentially applied to the fertilizer surface, forming a multilayered coating [30].

3.4 Minimal Invasion

Nanofertilizers enhance fertilizer uptake and utilization by improving penetration into plant tissues. Their high surface area-to-volume ratio allows easy absorption through plant cell walls and membranes [27]. Functionalization with targeting molecules ensures precise nutrient delivery [73]. This improved absorption reduces fertilizer use, minimizes nutrient leaching, and lowers environmental contamination.

3.5 Slow-Release Formulations

Low concentrations of micronutrients in soils, such zinc or iron, can have a significant impact on crop yields even if they may not be readily available by conventional methods. With the use of nanofertilizer technology, this particular form of specific plant nutrition requirement can be addressed. Nanoparticles enable the development of slow-release formulations that minimize leaching while ensuring optimal plant nutrition during critical growth stages like flowering and fruiting. This targeted nutrient delivery supports reproductive processes and maximizes crop yield.

3.6 High Efficiency of Nutrient Absorption

Nanofertilizers enhance soil nutrient absorption and fertilizer efficiency, reducing the need for fertilizers while boosting crop yield. They also prevent fertilizer leaching. Research shows that zinc oxide nanoparticles significantly improved its assimilation, root extent, and bud dry weight in wheat related to outmoded zinc sulfate fertilization [37]. Similarly, iron oxide nanoparticles increased iron uptake and chlorophyll content in wheat, leading to higher yields. A nitrogen release study found that a nano-urea-HAP mixture released urea 44% more slowly at pH 7 than conventional urea over 60 days [41].

3.7 Frequency of Effective Nutrient Release

The controlled nutrient release in nanofertilizers enhances agricultural productivity and sustainability by ensuring a steady supply of essential nutrients. This reduces nutrient losses, minimizes frequent fertilizer applications, and lowers environmental pollution [45].

3.8 Accelerated Microbiological Activity

The duration of nutrient release significantly impacts agricultural sustainability and productivity. Key factors include NP-microorganism interactions, biofertilizer application time, and nanofertilizer dispersion. Nano biofertilizers, combining natural polymers and microbes, enhance soil fertility, nutrient uptake, and plant growth. Advancing nanofertilizers supports smart fertilization, improving soil health and boosting crop yields [6].

3.9 Higher Soil Absorption

Soil activity measures the biological activity of the soil, which includes the breakdown and cycling of nutrients. These activities, which are fueled by soil microbes like fungi and bacteria, are crucial for preserving healthy soil. Nanofertilizers increase soil activity by efficiently supplying nutrients to the root zone. Nanofertilizers enhance nutrient cycling and organic matter decomposition by increasing the activity of soil microorganisms [78]. Nanofertilizers accelerate the soil's nitrogen mineralization process to maintain soil fertility [22]. They boost the soil's bacterial and fungal activity, which is essential for maintaining the soil ecology [12].

3.10 Enhanced Soil Water-Holding Absorbency

Nanofertilizers enhance organic matter, improve soil structure, and boost water retention, creating a favorable environment for beneficial microbes. With humic acid and clay, they reduce water loss and increase crop yields. Carbon-based and clay nanofertilizers promote soil aggregation, strengthening structure, improving water retention, and reducing erosion [31].

3.11 Nature that is Friendly to the Environment

Nanofertilizers are considerably better for the environment in addition to being more effective. However, nanofertilizers reduce the probability of discharge of nutrients harming the environment and provide more exact control over the nutritional balance of the soil because they release their nutrients gradually over time. Nanofertilizers are extra effectual than traditional fertilizers, requiring less contribution and reducing growers' costs [24]. Their slow nutrient release also lowers labor expenses and application frequency.

3.12 Minimal Production Cost

Because of their targeted delivery, controlled release, and improved nutrient use efficiency, nanofertilizers have the potential to reduce field fertilizer waste and result in cheaper production costs. Because they need less effort, require less fertilizer per application, as well as higher absorption rates than conventional fertilizers, nanofertilizers are usually far less expensive than conventional fertilizers. Moreover, longer soil retention times for nanofertilizers mean fewer applications and less prices [69].

3.13 Meets the Purpose of Precision Agriculture

Precision farming is a method of managing agricultural output that makes use of information technology to monitor and regulate inputs used in crop production, like water, herbicides, and fertilizers. Precision farming techniques could be enhanced by the use of nanofertilizers, which have been made possible by recent developments in nanotechnology. There are a number of possible advantages to precision farming when using nanofertilizers. First off, compared to regular fertilizers, nanofertilizers allow for a more exact application because they may be delivered directly to the plant. As a result, expenses are decreased, less fertilizer is used, and there may be less runoff volume that reaches the environment. Furthermore, targeted nutrient delivery to the plant is possible with nanofertilizers, enabling more accurate crop nutrition management [2].

3.14 Enhances Plant Resistance to Stress

Many environmental stressors have an impact on plants at different stages of their lives. They therefore alter physiological, pharmacological, and genetic pathways to fortify their resistance to environmental stressors at different stages of development. Plants react to such via changing the molecular expression of genes under various abiotic stress conditions. Several studies show that NPs have dose-dependent impacts on the growth and development of plants. In response to various stress situations, plants' defensive mechanisms are triggered by signals from their signaling system, which in turn triggers molecular mechanisms. Plant stress tolerance has been discovered to be improved by nanofertilizers in a number of ways [58].

3.15 Encourages the Growth of Plants

Nitrogen, phosphorus, and potassium are examples of vital nutrients that are made more readily available by NPs and are essential to plants' development and maturation. Furthermore, NPs improve plant use by facilitating the metabolism of these essential elements. Additionally, soil structure, texture, and water retention are all improved with nanofertilizers. By coating the soil's surface with a protective layer, nanofertilizers might lessen the quantity of vital nutrients lost. Furthermore, nanofertilizers may support preservation of the pH equilibrium in the soil, which will benefit vegetal evolution [69].

4 Types of Nanofertilizers

Nanofertilizers contain nanosized particles that enhance plant absorption, boosting crop yields. While they represent advanced technology with agricultural potential, their classification remains inconsistent. Nevertheless, other goods like nanoscale delivery systems and nanobiosensors are also included in certain definitions. The conflicting definitions of nanofertilizers have baffled the scientific community. For example, fertilizers classed as both a form of fertilizer and a subset of nanotechnology are called nanofertilizers. The definition and classification of nanofertilizers are unclear as a result of this uncertainty, which could cause misunderstandings when discussing their application and potential benefits.

Another way to categorize nanofertilizers is by the material they are made of. For instance, metals or polymers are used to make some nanofertilizers, whereas carbon nanotubes are used in the production of others. The characteristics and potential impacts of each form of nanofertilizer on plants vary. They are categorized generally in this evaluation according to the nutrients they contain, the tasks they complete, and their consistency. Knowing the characteristics of nanofertilizer is necessary to choose the best delivery method. Florae can receive foliar, water, or soil application of nanofertilizers. The types of nanofertilizers are also represented in Fig. 3 and Table 1.

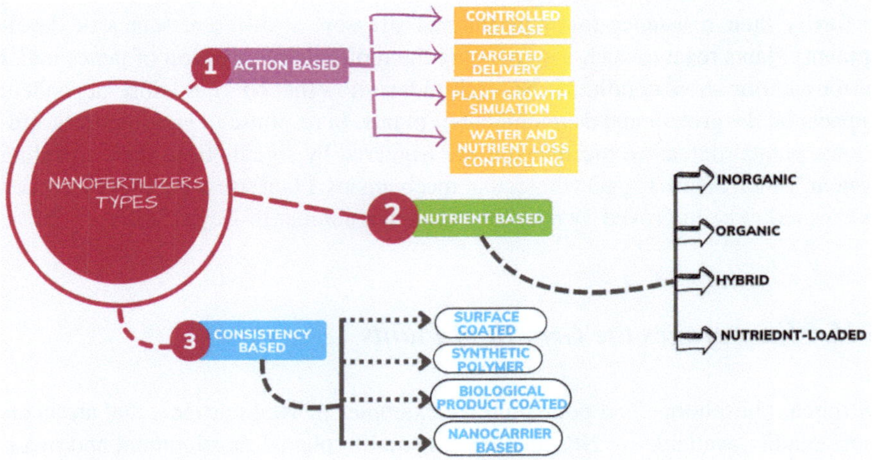

Fig. 3 Types of nanofertilizers

4.1 Action Based

Growth-stimulating, loss-minimizing, controlled-release, and targeted delivery are the four categories into which nanofertilizers fall. These advanced fertilizers enhance nutrient efficiency, regulate release, support targeted delivery, promote vegetal development, and reduce nutrient forfeiture, assembling them ideal for regenerative agroecology.

4.1.1 Targeted Delivery

Nanoaptamers are an innovative fertilizer delivery system revolutionizing agriculture. They target specific soil molecules, efficiently delivering nutrients to plants by binding to hormones and enzymes [32]. Composed of peptides or oligonucleotides, aptamers modify nanofertilizers to release nutrients in response to rhizosphere signals. By attaching fertilizers to nanoparticles like liposomes or gold particles, nanoaptamers enable precise delivery, enhancing nutrient absorption and plant growth.

Target-based nanofertilizers embrace clay-based formulations, nano-coated urea, iron oxide NPs, nanohydroxyapatite, carbon-based transporters, nano-emulsions, and nano-encapsulated micronutrients. These advanced fertilizers improve nutrient uptake, boost yields, and reduce environmental impact. While nanoaptamers show promise, their agricultural applications are still emerging, with potential for future commercialization.

Table 1 Classification of nutrient based nanofertilizer

Nutrient-based nanofertilizer types		Type	Nutrient composition	Application method	Benefits	References
Inorganic nanofertilizer	Macronutrient-based nanofertilizer	Nano nitrogen	Nano-urea or other nitrogenous compounds	Soil application/ foliar spray	Enhances nitrogen-use efficiency, reduces traditional fertilizer usage, and improves crop growth.	[67, 74]
		Nano phosphorus	Nano-phosphates	Soil application/ foliar spray	Increases phosphorus availability, supports root development and enhances yield.	[40, 72]
		Nano-potassium	Nano-potassium particles	Foliar spray	Improves stress tolerance, boosts metabolic activities, and ensures balanced plant growth.	[75], Rostami and Amiri (2022)
	Micronutrient-based nanofertilizer	Micronutrient-based nanofertilizer	ZnO, FeO, CuO nanoparticles	Soil application/ foliar spray	Improves plant health by addressing micronutrient deficiencies, enhancing enzyme activities.	[26], Das and Ghish (2023)
Organic nanofertilizer		Organic nanofertilizer	Organic compounds, plant extracts, or bio-based materials with nanoparticles	Soil application/ foliar spray	Environmentally friendly, improves soil health, and supports sustainable agriculture.	[4, 19]

(continued)

Table 1 (continued)

Nutrient-based nanofertilizer types	Type	Nutrient composition	Application method	Benefits	References
Hybrid nanofertilizer	Hybrid nanofertilizer	Combination of organic and inorganic nanoparticles	Foliar spray, seed coating, or soil	Balanced nutrient release, synergistic benefits from both organic and inorganic components.	[70, 63]
Nutrient loaded nanofertilizer	Nutrient loaded nanofertilizer	Nanoparticles carrying N, P, K, or micronutrients like Zn, Fe, etc.	Soil application/ Foliar spray	High nutrient-use efficiency, precise delivery, reduced fertilizer wastage.	[50, 76]

4.1.2 Controlled Released Nanofertilizers

According to studies, controlled-release nanofertilizers can address issues with conventional fertilizers, including nutrient leaching and inadequate nutrient use [7]. These fertilizers enhance nutrient absorption and mitigate their adverse environmental impacts by controlling the distribution of nutrients [59].

Controlled-release nanofertilizers capture nutrients inside nanoscale carriers made of lipids, polymers, or inert materials [7]. Their nutrient release is inclined by conservational factors alike temperature, pH, and moisture, as well as stimuli-responsive mechanisms alike enzyme-driven or biodegradation processes [1].

Sustainable agriculture benefits from the many advantages that controlled-release nanofertilizers provide. Improved nutrient use efficiency is one of these advantages because the regulated release of nutrients enables targeted and prolonged delivery, which improves plant nutrient uptake and lowers fertilizer application rates [7]. Additionally, they help to mitigate the adverse environmental impacts of agricultural practices. Controlled-release nanofertilizers safeguard aquatic ecosystems by reducing eutrophication and nutrient leaching, which minimizes contamination of water bodies [59]. Controlled-release nanofertilizers have been shown to enhance crop yields by improving nutrient efficiency. Research indicates that their application can boost agricultural productivity and optimize nutrient use. These fertilizers have demonstrated promising results across numerous produces, including wheat, soybeans, maize, and paddy.

Examples of controlled-release nanofertilizers include:

Carbon Based

Carbon, a vital element in all organic materials, plays a key role in biological processes. Nanostructured carbon, such as biochar from plant biomass, benefits plant growth. Carbon-based nanofertilizers offer enhanced nutrient integration, delivery, and complete growth [58]. Among them, carbon nanotubes (CNTs) distinguish out via their distinct chemical characteristics, high aspect ratio, and robustness. Additionally, graphene-based nanofertilizers have shown promise in improving plant growth and nutrient efficiency.

Chitosan Based

Experts in agriculture, food, and medicine all agree that chitosan, a biopolymer, is an excellent option. The shells of arthropods include a deacetylated version of chitin. Chitosan, a cationic polymer, binds with negatively charged molecules, forming complexes that enhance nutrient availability to plants. Its adjustable size makes it highly versatile. Chitosan NPs can act as carriers for the precise release of NPK fertilizers in agriculture. The progression is simple, as enrichers dissolve in the nanoparticle solution and are mixed via magnetic stirring.

Clay Based

Clay-based NPs have a lot of surface area and are reactive to nanolayers; they are capable of producing controlled-release fertilizer (CRF) formulations. The active surface that nano clay offers to several physicochemical and biological processes makes it an essential part of the CRF production process [35]. Organoclays are clay minerals with particle sizes under 2 mm, characterized by a high surface area, hydrogel capacity, charge, and crystalline structure. These properties, apart from their colloidal size, make them ideal for CRF synthesis.

Layer Double Hydroxides

Layered double hydroxides (LDH) are two-dimensional compounds with anionic components intercalated within interlayer gaps, enabling the controlled release of anions to balance their positive charge [56]. Additionally, properties such as anion exchange and thermal stability can be tailored through composition and synthesis methods.

Nanocapsule Based

Fertilizers are delivered and encapsulated in nanocapsules, which are little capsules usually made of either organic or inorganic material. Carbon nanotubes, metal oxides, silica, lipids, and biopolymers are the materials used to make them. Nanocapsules create a protective layer that ensures nutrient stability while gradually releasing nutrients over time. This controlled release enhances fertilizer efficiency and provides crops providing a consistent flow of vital nutrients. This guarantees that the nutrients are given to the crops at the appropriate times. Furthermore, there is less chance of leaching because of the gradual release of nutrients.

Nanogel Based

Nanogels are porous polymer-based materials that gradually release absorbed fertilizers or other substances. Infused with essential nutrients like nitrogen, phosphorus, and potassium, they enhance plant growth while improving soil water retention and reducing erosion. Widely used in fertilizers, water purification, and drug delivery, nanogels offer high nutrient capacity due to their large surface area and are environmentally safe due to their biodegradability and non-toxicity.

Polyurethane Based

Polyurethane-based nanofertilizers utilize polyurethane NPs, composed of organic units linked by urethane bonds, to enhance nutrient delivery. These fertilizers are often synthesized using ring-opening metathesis polymerization, where a monomer is formed from two molecules with double bonds before polymerization. The resulting nutrient-rich polymers release essential elements like potassium, phosphorus, and nitrogen. Their polyurethane matrix ensures controlled nutrient release while shielding the particles from environmental degradation, improving fertilizer efficiency.

Starch Based

The starch nanocrystals used to make starch-based nanofertilizers dissolve readily in water and can be sprayed on plants as an aerosol or liquid. The nanocrystals use a superior renewable energy source to fertilize crops effectively without producing chemical waste.

Zeolite Based

Because of their ability to retain and absorb organic molecules, water, and nutrients, zeolites—a family of aluminosilicate, microporous minerals—have been added to soil for many years. The process of creating zeolite NPs, which are then mixed with different substances to make fertilizer, is the basis for developing zeolite-based nanofertilizers. Due to their small size, these nanofertilizers can more effectively reach the root zone of a crop by penetrating deeper into the soil. Additionally, zeolites act as a reservoir for the slow release and storage of nutrients, providing plants with a more consistent supply of nutrients.

4.1.3 Nanofertilizers that Stimulate Plant Growth

Carbon nanotubes (CNTs), a type of nanofertilizer, interact with plant roots to enhance hormone production and promote growth while enriching the soil with carbon and essential nutrients. Composed of rolled-up carbon atom sheets, CNTs possess unique properties that improve soil structure, increase water retention, and facilitate nutrient absorption. While high concentrations may have adverse effects, low doses of CNTs can benefit water transport, root development, and seed germination without causing phytotoxicity [25]. Beyond soil supplementation, CNTs can serve as a protective seed coating, enhancing water and nutrient uptake while deterring pests. Additionally, their exceptional strength and rigidity may reinforce plant stems and stalks, improving resistance to bending and breakage, ultimately boosting harvests.

4.1.4 Nanofertilizer for Water and Nutrient Loss Control

Nanofertilizers, containing NPs that regulate fertilizer release, enable reduced fertilizer use by producers without sacrificing crop yields. Various strategies remain employed to control nutrient release and minimize desiccation. One way encompasses encapsulating nanofertilizers in a porous milieu for gradual nutrient release, while another modifies their surface to enhance hydrophilicity, improving water retention and reducing evaporation [76]. Nanobeads and nanoemulsions are notable nanofertilizers that help prevent water and nutrient loss from the soil.

Nanobeads

Nanobeads are tiny structures designed to store and gradually release nutrients, reducing water loss in plants. Made from materials like iron, carbon, and other metals, they can penetrate even the smallest soil pores, enhancing fertilization. Additionally, nanobeads can aid in pollution cleanup. One popular nanobead-based fertilizer, NanoFert, contains both macro- and micronutrients [39, 76]. These particles dissolve rapidly in soil, constructing nutrients voluntarily obtainable to vegetal roots. Moreover, NanoFert has low salt content, preventing harm to beneficial soil microbes and preserving soil structure.

Nanoemulsions

Nanoemulsions are microscopic droplets that contain both water-soluble and insoluble nutrients, enabling efficient nutrient delivery to plants. Produced through high-energy mixing, these fertilizers create a stable and uniform blend of liquid and surfactant. The surfactant ensures the droplets remain suspended in the solution for even distribution. With sizes below 100 nanometers—significantly smaller than conventional fertilizer particles—nanoemulsions enhance nutrient uptake, boosting crop health and yield.

4.2 Nutrient-Based Nanofertilizer

With growing universal populace, the culinary desire continues to rise, highlighting the need for sustainable agricultural practices. Nutrient-based nanofertilizers offer an innovative solution by refining nutrient obtainability, absorption, and efficiency in plants.

4.2.1 Nanofertilizers that Are Inorganic

Inorganic nanofertilizers can provide plants with essential nutrients like potassium, phosphate, and nitrogen include metals, metalloids, and non-metallic NPs. These fertilizers may be utilized to increase agricultural yields because they are made to increase plants' efficiency in absorbing nutrients [37]. The material assortment is based on the desired fertilizer potentials. Since, they are personalized to the precise nutritional desires of the florae they are intended for, inorganic nanofertilizers have a special benefit that allows for targeted treatments to increase production. It is anticipated that as more farmers implement precision agriculture techniques, the usage of inorganic nanofertilizers in agriculture would rise. There are two categories of inorganic nanofertilizers, namely, macronutrient and micronutrient-based nanofertilizers.

Macronutrient-Based Nanofertilizers

Macronutrient nanofertilizers enhance plant growth and environmental sustainability. Nitrogen-based variants boost efficiency while reducing emissions and eutrophication. Phosphorus nanofertilizers aid nutrient uptake and soil reclamation, while potassium-based types improve soil properties and leaching resistance. Calcium nanofertilizers enhance disease resistance and crop quality, magnesium-based one's support growth and pest resistance, and sulfur variants provide slow-release nutrition, minimizing soil acidification [66].

Micronutrient-Based Nanofertilizers

Advanced agricultural inputs known as micronutrient-based nanofertilizers use NPs to more efficiently supply vital micronutrients to plants than traditional fertilizers. Tiny levels of micronutrients, such as zinc, titanium, nickel, copper, iron, boron, and nickel, are necessary for plants' growth, development, and overall health.

Plants benefit greatly from micronutrient nanofertilizers, which boost nutrient uptake, yield, and tolerance to both biotic and abiotic stressors.

4.2.2 Organic Nanofertilizers

The NPs in organic nanofertilizers are made of organic ingredients and are intended to progressively release nutrients into the soil. They come from renewable resources, are safe for the environment, and may be able to maintain soil moisture and change pH levels, which will improve plants' ability to absorb essential nutrients. Organic NPs are composed of atoms or molecules and exist in numerous forms, including capsules, polymer conjugates, vesicles, micelles, liposomes, polymersomes, dendrimers, and polymeric NPs. Organic nanofertilizers enhance plant growth and yield by enabling slow nutrient release for better absorption and utilization. They also improve soil structure by enhancing aggregation, porosity, water infiltration, aeration, and root penetration. Notably, liquid fertilizers like Ferbanat and Nanonat represent a new generation of nanobiostimulators derived from organic materials [15].

4.2.3 Hybrid Nanofertilizers

Hybrid nanofertilizers combine conventional and nanotechnology-based fertilizers for boosted nutrient accessibility and controlled release. By integrating nanomaterials within a polymeric matrix, they improve efficiency while minimizing environmental impact. Nanocomposites, formed by dispersing nanofillers in a polymer phase, offer superior mechanical, thermal, and chemical properties, including increased strength, stability, and conductivity. These fertilizers optimize nutrient uptake through nano-fertigation, reducing waste and costs while promoting sustainable agriculture.

4.2.4 Nanofertilizers Packed with Nutrients

Crops can be effectively nourished by nutrient-loaded nanofertilizers, such as nano-porous zeolites. Because zeolite fertilizers are composed of minerals created by the response of volcanic ash and alkaline lake water, they are a great choice for fertilizing and soil amendment. These minerals' honeycomb-like structure and great porosity make it easier for moisture, nutrients, and other substances to be absorbed and retained. Natural zeolite is ground into a fine powder and combined with nitrogen, phosphorus, and potassium to create a nanofertilizer tailored to various crop needs. This process enhances nutrient absorption and retention by increasing the zeolite's surface area, improving soil efficiency.

4.3 Consistency-Based Nanofertilizers

Consistency-based nanofertilizers, including surface-coated and nanocarrier-based technologies, offer a sustainable alternative for agriculture by enhancing nutrient absorption, minimizing loss, and reducing environmental impact. Surface-coated nanofertilizers feature nanoparticles covered with synthetic polymers or biological materials—polyacrylamide improves nitrogen use efficiency, while alginate enhances phosphorus uptake. Nanocarrier-based nanofertilizers, using carriers like mesoporous silica nanoparticles, rally nutrient solubility, constancy, and accessibility, ultimately boosting crop yields.

4.3.1 Nanofertilizers Applied on the Surface

Silica-NPs, iron nanoparticles are examples of surface-coated nanofertilizers that have been covered with nanoscale materials and fertilizers. Surface-coated nanofertilizers are produced by veneer fertilizer particles with nanomaterials alike titanium dioxide, carbon, silver, and gold. Consequently, this raises the rate at which the fertilizer is absorbed, improving plant growth and use efficiency.

4.3.2 Nanofertilizers Coated with Synthetic Polymers

While nanofertilizers efficiently deliver nutrients, their small size can make handling challenging. To address this, researchers have advanced polymer-coated nanofertilizers with a shrill copied polymer layer. This coating improves handling, protects against environmental degradation, and minimizes nitrogen loss, making them ideal for high-value applications. Their nitrogen release follows a complex pattern, controlled by the gradual diffusion of the polymer matrix.

4.3.3 Biological Product Coated Nanofertilizers

Nanofertilizer coatings enhance soil absorption, transport, and nutrient availability, ensuring efficient fertilization. Biological product-coated nanofertilizers, made from organic materials like NPs and humic acid, surpass conventional fertilizers by improving soil fertility, water retention, and microbial activity—ideal for organic farming.

Plant growth regulator-coated nanofertilizers regulate plant growth while supplying essential nutrients. Nanobiofertilizers, combining beneficial microbes and NPs, enhance nutrient absorption, solubilization, and fixation. Their nanosize and high surface area enable efficient nutrient delivery, while coatings of zeolite, chitosan, and polymers ensure sustained nutrient release, optimizing crop fertilization.

4.3.4 Based on Nanocarriers the Use of Nanofertilizers

Nanocarrier-based fertilizers deliver nutrients to plants and soil, enhancing growth, reducing nutrient loss, and improving availability. Creating next-generation nanocarriers to be used in the biofortification. Controlled release capabilities are required to efficiently distribute nutrients and reduce adversative effects on the ecosystem. Nanocarrier-based fertilizers offer aids such as lower environmental impact, boosted nutrient approval, and controlled release. Their use across innumerable crops has publicized better-quality produce and nutrient efficacy.

5 Modes of Nanofertilizer Application

Applying nanofertilizers can be done in a variety of ways, including as foliar spraying, dusting, fogging, emulsion, soil application, seed treatment, etc. Applying the proper type of nanofertilizer is essential for plants to grow as vigorously as they can, and soil and climate both play a role in this. The decision is influenced by soil quality and climate, which have an impact on the absorption of nutrients and utilization. After taking these factors into account, selecting the best course of action can boost agricultural yields, reduce environmental impact, and create more ecologically friendly farming practices as mentioned in Table 2.

5.1 Soil Treatment

Nanofertilizers are applied through fertigation, side-dressing, or broadcasting. Once in the soil, NPs adhere to roots or enter root cells, interacting with plants, microbes, and soil [33]. Their controlled nutrient release enhances growth and productivity [5]. However, challenges include regulatory issues, high costs, and uncertain long-term effect [34]s.

5.2 Seed Nanopriming

Seed priming, a pre-sowing technique, modulates metabolic and signaling pathways to enhance germination and plant growth. Soaking seeds in nanofertilizers can cut fertilizer use by half while preserving high-quality yields. By inflowing seed pores, distributing inside, and prompting growth-promoting plant hormones, nanobiofertilizers work as stimulants to improve germination and growth. By reducing reactive oxygen species and controlling plant growth hormones, using nanofertilizer during seed priming boosts seed germination [52]. Additionally, seed priming increases the expression of several genes during germination, especially those linked to plant resilience, which strengthens the plant's resistance.

5.3 Foliar Spray

Liquid fertilizers are applied directly using this method. Compared to soil treatment, foliar fertilization allows for faster nutrient uptake and quicker replenishment of detected deficits. Studies comparing the administration of nanoparticles to plants by leaf spraying versus soil amendments reveal significant benefits for nanoscale nutrient uptake from foliar application. Foliar spraying turns out to be more economical and

Table 2 Modes of nanofertilizer application

Modes of application	Advantages	Disadvantages	References
Soil treatment	• Increased nutrient use efficiency • Controlled nutrient release • Reduced fertilizer requirement • Environmental sustainability • Improved crop yield and quality	• Potential soil toxicity • Unknown long-term environmental effects • Cost of production • Complex application procedures • Regulatory and safety concerns	[28, 76]
Seed nanopriming	• Enhanced nutrient use efficiency • Better seed germination and early growth • Enhanced stress resistance • Reduced fertilizer input • Less environmental runoff	• Environmental and health concerns • Potential toxicity • High production cost • Uncertain regulatory frameworks • Variable effectiveness	[54, 76]
Foliar spray	• Improved nutrient efficiency • Reduced environmental impact • Targeted delivery • Enhanced uptake speed • Lower dosage requirements	• High production cost • Limited research on long-term effects • Potential toxicity • Risk of leaf phytotoxicity • Complex application requirements	[7, 76]
Aeroponics treatment	• Enhanced nutrient uptake • Precise nutrient delivery • Reduced nutrient loss • Stimulates plant growth • Decreased frequency of application	• Potential toxicity • Cost and accessibility • Environmental and health concerns • Regulatory challenges • Technical requirements	[19]
Hydroponics treatment	• Enhanced nutrient absorption • Controlled nutrient release • Increased plant health and yield • Reduced fertilizer usage • Environmental sustainability	• Challenges in application • High production cost • Potential for nanotoxicity • Environmental and health concerns • Regulatory and market limitations	[7, 77]

effective than soil application for immobilized nutrients like iron, manganese, and copper. The morning and evening spraying timings are the method's main drawback because the stomata are open during these hours. Furthermore, failure to apply fertilizer at the proper dosage may result in plant harm [18].

5.4 Aeroponics Treatment

Using an oxygen mist in the growing chamber, nutrients are made available to horticultural crop plants using the efficient soilless cultivation method known as aeroponics. Here, plants are constantly exposed to nutritional elements in the form of dust on their roots or as foliar sprays [11]. In plants, gas exchange takes place via stomatal apertures. Micronutrients can be replaced with nanofertilizers, which have the potential to be even more advantageous than chemical fertilizers. By 20 to 20%, these nanofertilizers improve nutrient efficiency. Since their size and shape affect how quickly nutrients that are nanosized are absorbed. Larger particles can more easily pass through non-cuticle areas like stomata, stigma, and hydathodes than smaller ones can via the cuticle [53]. One effective method of maximizing nutrient consumption with the least amount of waste could be to use aeroponic treatments in conjunction with nanofertilizers.

5.5 Hydroponics Treatment

Water serves as the main supply channel for micronutrients in hydroponics technology, which eliminates the need for soil. Nanofertilizers can be added to hydroponic systems as stable colloids that contain plant nutrients in nanoparticle form. Additional agrochemicals and micro- and macronutrients in nanoparticle form can be used to treat hydroponic systems [60]. Therefore, approaches based on nanotechnology are showing promise as a way to use nutrients with minimal resource waste [36].

6 Mechanisms of Nanofertilizer Absorption in Plants

The absorption, translocation, and aggregation of nanoparticles are influenced by a variety of factors, including plant age, species, stability, functionalization, and mode of delivery.

Nucleic acids can enter plant tissue from the roots or the top portions. Important variables influencing the captivation of nanofertilizers in plants include shape, extent, and nanoparticles interactions with cell walls. The cell wall's size exclusion barrier, which is between 5 and 20 nm, prevents larger particles from entering plant cells

[37]. Surface-functionalized nanoparticles improve the absorption of nanoparticles by enlarging existing pores or opening new ones in the cell wall.

Additional pathways through which nanoparticles may arrive in vegetal cells are endocytosis, intricate development with membrane transporters, binding to carrier proteins via aquaporin, and ion channels. Encapsulated nutrients are shielded from soil filtration by nanocarriers, which help keep them in the soil surrounding the roots. Encapsulated components can extend their spatial scale by entering the soil network through surface tension, hydrogen bonds, or viscous forces. NPs may arrive at leaves or cuticles by stomata or the base of the trichome for foliar applications. Nanoparticles can move through the circulatory system apoplastically or symplastically after entering the cell through stomatal pores. Plasmodesmata have the ability to move NPs from one cell to another.

NPs have a well-established ability to regulate plant development. But at a certain threshold, the adverse effects begin to show. One of the characteristics of nanoparticles' distinct physio-chemical features is their eco-toxicity and destiny. Almost every type of nanoparticle can be either genotoxic or phytotoxic, depending on how they are structured and functionalized. The availability of nanoparticles in the surrounding soil and plants is directly correlated with phytotoxicity. To comprehend how the NPs interact with environmental elements and epigenetic variables, more study and money are required. It is necessary to have a solid understanding of production processes, characterization and detection, application techniques, dose requirement, their release into the environment, and their risk. To lessen these NP-induced toxicities, the acceptable limit of NPs must be established while taking into account all application mode factors.

7 Present Status

Nanofertilizers are widely accessible in markets for use in modern agricultural techniques; yet, prominent fertilizer businesses have not yet designed agricultural fertilizers [66]. To boost agricultural output, studies on the uptake of nanofertilizers from the soil, their bioavailability, and potential toxicity of quite a few metal oxides (NPs), including TiO_2, Al_2O_3, CeO_2, ZnO, and FeO, have received significant support within the last 10 years. When it comes to addressing climate change, smart agriculture serves as a bridge to the twenty-first century and prioritizes both short- and long-term growth. It seeks to support countries and other functional aspects in guaranteeing the security of vital agricultural functions [44].

Understanding how nanofertilizers affect plants dependent on the growing environment in the farming system has been the focus of recent research. Plant growth metrics that are influenced by nanofertilizers include root systems, photosynthetic rate, fresh and dry matter content differences, plant height, number of leaves per plant, and leaf area. When compared to artificial fertilizers, this leads to careful examination of an increase in output because of a higher rate of photosynthetic translocation to different regions of the plant. Some experts are concentrating on employing nanotechnology (nanofertilizers) to increase the nutrient efficiency of plants [79]. Thorough field experiments must come after the nanofertilizers' strong acceptability for use in investigations pertaining to growth factors. Consequently, supporting the researchers' and decision-makers' recommendations. It would be necessary to apply particular dosages in bulk on the fields without having a major negative impact on the ecosystem [68].

8 Future Prospects

To comprehend the properties of nanofertilizers, their mode of action with regard to the type and mode of administration, and their effects on plants, a great deal of research is necessary. When NF concentrations rise above a particular threshold, they can adversely impact plants, causing them to grow and yield less. Therefore, it is crucial to optimize the dosage quantity of various nanofertilizers for various plants. A careful and systematic analysis of the conceivable healthiness effects, clearance, and safe discarding of nanofertilizers might enhance the development of new applications for these materials [80]. Therefore, it is highly advised that future research emphasizes on determining the dosage for commercial nanofertilizer field applications. The recommendations for nanofertilizers that are specific to farmers should be the main focus of scientific advancement. Additionally, it is expected that the creation and approval of nanofertilizers will boost the industry by acting as a counterbalance to the manufacturing of chemical fertilizers.

This will assist policymakers in determining when and how much nanofertilizer should be applied.

Despite being environmentally and economically viable, the distribution of nanofertilizers as Gen Beta fertilizers for agrarian applications poses challenges. Ensuring the accessibility of raw materials for the commercial amalgamation of nanofertilizers is imperative, in addition to scaling up the process for large-scale distribution.

When these problems are fixed, it will be easy to utilize the right procedures to make commercially viable nanofertilizers. This will open the door to the day when even a small amount of these fertilizers will be used to achieve the desired greater agricultural production [9]. Further research and commercialization of smart nanofertilizers require thorough evaluation. Developing slow-release variants with environmental, energy, and economic benefits would be a major breakthrough. Enhancing manufacturing efficiency and using cost-effective, eco-friendly materials can make nanofertilizer production more viable. Techniques that can accurately and selectively identify and track environmental pollutants in a variety of biological matrices are needed to facilitate effective restoration. Governments and corporations must also actively fund research groups that use a variety of animal models to evaluate the safety of nanomaterials. More effective regulations—which are currently either nonexistent or weak in some countries—will be built upon the findings of this research.

9 Conclusion

The demand for manufacturing more food is growing along with the global population. Nutrient deficiencies are one of the primary causes of less crop output and financial losses. Applications of chemical fertilizers can increase agricultural output, but long-term utilization of it is definitely not a feasible solution. Finding a substitute for chemical fertilizers is a crucial focus in modern agricultural research since it would allow for a more ecologically friendly method of farming.

Nanotechnology has been of great importance in the agriculture sector in recent years. Nanotechnology-based smart fertilizers have been established to boost soil health and agricultural productivity. To ensure continued scientific advancements and commercialization, an integrated research of these nano-based smart fertilizers can be conducted. To make nanofertilizers a profitable venture, this analysis should focus on streamlining production processes and finding low-cost, non-toxic, or biodegradable continuous matrix materials.

This finding has tremendous potential ramifications, especially for enhancing environmental sustainability and energy conservation in global farming practices. For example, slow-release nanofertilizer solutions may increase nutrient uptake by up to 40% while reducing water usage by up to 20%. Furthermore, less chemical-based fertilizer applications will be required when enhanced production techniques are implemented widely in many countries throughout the world, which will lower carbon emissions associated with agricultural activities.

Nanofertilizers are predicted to transform global perspectives on sustainable agricultural methods, resulting in more economical use of resources like land and water and less reliance on dangerous chemicals that over time could degrade our ecosystem. With constant sustenance from communal and private sectors, academia, and agriculturalist assemblies, slow-release nanofertilizers can greatly impact the environment, energy use, and economic growth worldwide.

References

1. Abdelmigid HM, Morsi MM, Hussien NA, Alyamani AA, Alhuthal NA, Albukhaty S (2022) Green synthesis of phosphorous-containing hydroxyapatite nanoparticles (nHAP) as a novel nano-fertilizer: preliminary assessment on pomegranate (*Punica granatum* L.). Nanomaterials 12:1527
2. Achari GA, Kowshik M (2018) Recent developments on nanotechnology in agriculture: plant mineral nutrition, health, and interactions with soil microflora. J Agric Food Chem 66:8647–8661
3. Sustainability A (2023) Agrochemicals 2:296–336. https://doi.org/10.3390/agrochemicals2020019
4. Ahanger MA, Qi M, Huang Z, Xu X, Begum N, Qin C, Zhang L (2021) Improving growth and photosynthetic performance of drought stressed tomato by application of nano-organic fertilizer involves up-regulation of nitrogen, antioxidant and osmolyte metabolism. Ecotoxicol Environ Saf 216:112195

5. Ahmed B, Rizvi A, Ali K, Lee J, Zaidi A, Khan MS, Musarrat J (2021) Nanoparticles in the soil–plant system: a review. Environ Chem Lett 19:1545–1609
6. Akhtar N, Ilyas N, Meraj TA, Pour-Aboughadareh A, Sayyed RZ, Mashwani ZU, Poczai P (2022) Improvement of plant responses by nanobiofertilizer: a step towards sustainable agriculture. Nanomaterials 12:965
7. Ali SS, Darwesh OM, Kornaros M, Al-Tohamy R, Manni A, El-Shanshoury AERR, Sun J (2021) Nano-biofertilizers: synthesis, advantages, and applications. In: Biofertilizers. Woodhead Publishing, pp 359–370
8. Ali SS, Darwesh OM, Kornaros M, Al-Tohamy R, Manni A, El-Shanshoury AERR, Metwally MA, Elsamahy T, Sun J (2021) Nano-biofertilizers: synthesis, advantages, and applications. In: Rakshit A, Meena VS, Parihar M, Singh HB, Singh AK (eds) Biofertilizers: advances in bioinoculants. Woodhead Publishing, Sawston, UK, vol 1, pp 359–370
9. Babu S, Singh RP, Yadav D, Rathore SS, Raj R, Avasthe R, Yadav SK, Das A, Yadav V, Yadav B et al (2021) Nanofertilizers for agricultural and environmental sustainability. Chemosphere 292:133451
10. Butt BZ, Naseer I (2020) Nanofertilizers. In: Nanoagronomy. Springer, Berlin/Heidelberg, Germany, pp 125–152
11. Cui HX, Sun CJ, Liu Q, Jiang J, Gu W (2010) Applications of nanotechnology in agrochemical formulation, perspectives, challenges and strategies. In: Proceedings of the international conference on Nanoagri, Sao Pedro, Brazil 20–25:28–33
12. Das P, Davis K, Penton CR, Westerhoff P, Bi Y (2022) Impacts of graphitic nanofertilizers on nitrogen cycling in a sandy, agricultural soil. J Nanoparticle Res 24:1–21
13. Das SK, Ghosh GK (2023) Development and evaluation of biochar-based secondary and micronutrient enriched slow release nano-fertilizer for reduced nutrient losses. Biomass Convers Biorefinery 13(13):12193–12204
14. De Oliveira JL, Campos EVR, Bakshi M, Abhilash PC, Fraceto LF (2014) Application of nanotechnology for the encapsulation of botanical insecticides for sustainable agriculture: prospects and promises. Biotechnol Adv 32:1550–1561. https://doi.org/10.1016/j.biotechadv.2014.10.010
15. Ekinci M, Dursun A, Yildirim E, Parlakova F (2014) Effects of nanotechnology liquid fertilizers on the plant growth and yield of cucumber (*Cucumis sativus* L.). Acta Scientiarum Polonorum. Hortorum Cultus 13(3)
16. Elnahal ASM, El-Saadony MT, Saad AM, Desoky ESM, El-Tahan AM, Rady MM, AbuQamar SF, El-Tarabily KA (2022) The use of microbial inoculants for biological control, plant growth promotion, and sustainable agriculture: a review. Eur J Plant Pathol 162:759–792. https://doi.org/10.1007/s10658-021-02393-7
17. Elsayed AAA, El-Gohary A, Taha ZK, Farag HM, Hussein MS, AbouAitah K (2022) Hydroxyapatite nanoparticles as novel nano-fertilizer for production of rosemary plants. Sci Hortic 295:110851
18. Fageria NK, Filho MPB, Moreira A, Guimarães CM (2009) Foliar fertilization of crop plants. J Plant Nutr 32:1044–1064
19. Fatima F, Hashim A, Anees S (2021) Efficacy of nanoparticles as nanofertilizer production: a review. Environ Sci Pollut Res 28(2):1292–1303
20. Gade A, Ingle P, Nimbalkar U, Rai M, Raut R, Vedpathak M, Jagtap P, Abd- Elsalam KA (2023) Nanofertilizers: the next generation of agrochemicals for long-term impact on sustainability in farming systems. Agrochemicals 2:257–278. https://doi.org/10.3390/agrochemicals2020017
21. Ghormade V, Deshpande MV, Paknikar KM (2011) Perspectives for nano-biotechnology enabled protection and nutrition of plants. Biotechnol Adv 29:792–803. https://doi.org/10.1016/j.biotechadv.2011.06.007
22. Gupta R (2022) Mineralization of soil carbon, nitrogen, and phosphorus and role of nanofertilizers in soil fertility and plant growth. In: Giri B, Kapoor R, Wu Q, Varma A (eds) Structure and functions of pedosphere. Springer, Singapore, pp 393–409
23. Jakhar AM, Aziz I, Kaleri AR, Hasnain M, Haider G, Ma J, Abideen Z (2022) Nano-fertilizers: a sustainable technology for improving crop nutrition and food security. NanoImpact 27:100411. https://doi.org/10.1016/j.impact.2022.100411

24. Kah M, Beulke S, Tiede K, Hofmann T (2013) Nanopesticides: state of knowledge, environmental fate, and exposure modeling. Crit Rev Environ Sci Technol
25. Kalia A, Sharma SP, Kaur H, Kaur H (2020) Novel nanocomposite-based controlled-release fertilizer and pesticide formulations: prospects and challenges. In: Multifunctional hybrid nanomaterials for sustainable agri-food and ecosystems, pp 99–134
26. Kalra T, Tomar PC, Arora K (2020) Micronutrient encapsulation using nanotechnology: nanofertilizers. Plant Arch 20(2):1748–1753
27. Kottegoda N, Munaweera I, Madusanka N, Karunaratne V (2011) A green slow-release fertilizer composition based on urea-modified hydroxyapatite nanoparticles encapsulated wood. Curr Sci 101:73–78
28. Kumar N, Samota SR, Venkatesh K, Tripathi SC (2023) Global trends in use of nano-fertilizers for crop production: advantages and constraints–a review. Soil Tillage Res 228:105645
29. Liscano J, Wilson C, Norman R, Jr. Slaton N (2000) Zinc availability to rice from seven granular fertilizers. Arkansas agricultural experiment station Fayetteville, Fayetteville, CA, USA, p 963
30. Lvov Y, Ariga K, Ichinose I, Kunitake T (1995) Assembly of multicomponent protein films by means of electrostatic layer-by-layer adsorption. J Am Chem Soc 117:6117–6123
31. Mastronardi E, Tsae P, Zhang X, Monreal C, DeRosa MC (2015) Strategic role of nanotechnology in fertilizers: potential and limitations. In: Nanotechnologies in food and agriculture. Springer, Berlin/Heidelberg, Germany, pp 25–67
32. Mim JJ, Rahman SM, Khan F, Paul D, Das HP, Orny NT, Shuvo MRH, Hossain N (2025) Towards smart agriculture through nano-fertilizer-a review. Mater Today Sustain 30:101100. https://doi.org/10.1016/j.mtsust.2025.101100
33. Miralles P, Church TL, Harris AT (2012) Toxicity, uptake, and translocation of engineered nanomaterials in vascular plants. Environ Sci Technol 46:9224–9239
34. Mishra M, Dashora K, Srivastava A, Fasake VD, Nag RH (2019) Prospects, challenges and need for regulation of nanotechnology with special reference to India. Ecotoxicol Environ Saf 171:677–682
35. Moradi S, Babapoor A, Ghanbarlou S, Kalashgarani MY, Salahshoori I, Seyfaee A (2024) Toward a new generation of fertilizers with the approach of controlled-release fertilizers: a review. J Coat Technol Res 21(1):31–54
36. Nagula S, Ramanjaneyulu AV (2020) Nanofertilizers: the next generation fertilizer. Biotica Res Today 2:905–907
37. Nair R, Varghese SH, Nair BG, Maekawa T, Yoshida Y, Kumar DS (2010) Nanoparticulate material delivery to plants. Plant Sci 179:154–163
38. Nongbet A, Mishra AK, Mohanta YK, Mahanta S, Ray MK, Khan M, Baek KH, Chakrabartty I (2022) Nanofertilizers: a smart and sustainable attribute to modern agriculture. Plants (Basel, Switzerland) 11(19):2587. https://doi.org/10.3390/plants11192587
39. Oancea F (2022) Development of safe nanoagrochemicals—the nanoporous route. Chem Proc 7:68
40. Poudel A, Singh SK, Jiménez-Ballesta R, Jatav SS, Patra A, Pandey A (2023) Effect of nanophosphorus formulation on growth, yield and nutritional quality of wheat under semi-arid climate. Agronomy 13(3):768
41. Prasad TNVKV, Sudhakar P, Sreenivasulu Y, Latha P, Munaswamy V, Reddy KR, Sreeprasad TS, Sajanlal PR, Pradeep T (2012) Effect of nanoscale zinc oxide particles on the germination, growth and yield of peanut. J Plant Nutr 35:905–927
42. Preetha PS, Balakrishnan N (2017) A review of nano fertilizers and their use and functions in soil. Int J Curr Microbiol Appl Sci 6:3117–3133
43. Raguraj S, Wijayathunga WMS, Gunaratne GP, Amali RKA, Priyadarshana G, Sandaruwan C, Karunaratne V, Hettiarachchi LSK, Kottegoda N (2020) Urea–hydroxyapatite nanohybrid as an efficient nutrient source in Camellia sinensis (L.) Kuntze (tea). J Plant Nutr 43:2383–2394
44. Rajput VD, Singh A, Minkina T, Rawat S, Mandzhieva S, Sushkova S, Shuvaeva V, Nazarenko O, Rajput P, Komariah Kandasamy S, Prema RS (2015) Methods of synthesis of nano particles and its applications. J Chem Pharm Res 7:278–285

45. Raliya R, Saharan V, Dimkpa C, Biswas P (2018) Nanofertilizer for precision and sustainable agriculture: current state and future perspectives. J Agric Food Chem 66:6487–6503
46. Reed RB, Ladner DA, Higgins CP, Westerhoff P, Ranville JF (2012) Solubility of nano-zinc oxide in environmentally and biologically important matrices. Environ Toxicol Chem 31:93–99
47. Rostami Ajirloo AA, Amiri E (2022) Effects of nano-potassium fertilizer on yield and water use efficiency of soybean under water deficit conditions (case study: Moghan plain, Iran). Commun Soil Sci Plant Anal 53(12):1542–1551
48. Sarkar D, Kar SK, Chattopadhyay A, Shikha RA, Tripathi VK, Dubey PK, Abhilash PC (2020) Low input sustainable agriculture: a viable climate-smart option for boosting food production in a warming world. Ecol Indic 115:106412
49. Sarkar D, Rakshit A (2023) Amalgamation of farmers' bio-priming knowledge in integrated nutrient management for sustainable management of red cabbage soil under Middle Gangetic Plains, India. Environ Manag 72:382–395
50. Saurabh K, Prakash V, Dubey AK et al (2024) Enhancing sustainability in agriculture with nanofertilizers. Discov Appl Sci 6:559. https://doi.org/10.1007/s42452-024-06267-5
51. Seleiman MF, Almutairi KF, Alotaibi M, Shami A, Alhammad BA, Battaglia ML (2020) Nano-fertilization as an emerging fertilization technique: why can modern agriculture benefit from its use? Plants 10(1):2. https://doi.org/10.3390/plants10010002
52. Sharma B, Tiwari S, Kumawat KC, Cardinale M (2023) Nano-biofertilizers as bio-emerging strategies for sustainable agriculture development: potentiality and their limitations. Sci Total Environ 860:160476
53. Sharma U, Barupal M, Shekhawat NS, Kataria V (2018) Aeroponics for propagation of horticultural plants: an approach for vertical farming. Hortic Int J 2:443–444
54. Singh A, Sengar RS, Sharma R, Rajput P, Singh AK (2021) Nano-priming technology for sustainable agriculture. Biogeosystem Tech 8:79–92
55. Singh A, Singh N, Hussain I, Singh H (2017) Effect of biologically synthesized copper oxide nanoparticles on metabolism and antioxidant activity to the crop plants *Solanum lycopersicum* and *Brassica oleracea* var. botrytis. J Biotechnol 262:11–27
56. Singha Roy A, Kesavan Pillai S, Ray SS (2022) Layered double hydroxides for sustainable agriculture and environment: an overview. ACS Omega 7(24):20428–20440
57. Sinha T, Bhagwatwar P, Krishnamoorthy C, Chidambaram R (2019) Polymer based micro-and nanoencapsulation of agrochemicals. In: Gutiérrez TJ (ed) Polymers for agri-food applications. Springer, Cham, Switzerland, pp 5–28
58. Solanki P, Bhargava A, Chhipa H, Jain N, Panwar J (2015) Nano-fertilizers and their smart delivery system. In: Nanotechnologies in food and agriculture. Springer, Berlin/Heidelberg, Germany, pp 81–101
59. Sood R (2021) Nano-biofertilizers for sustainable agriculture. Just Agric Multidiscip E-Newsl 1:1–6
60. Tarafdar JC (2022) Nanofertilizers and nanobioformulations: blessings for global farming. In: Transforming coastal zone for sustainable food and income security: proceedings of the international symposium of ISCAR on coastal agriculture. Springer International Publishing, Cham, pp 645–659
61. Tarafdar J, Xiang Y, Wang W, Dong Q, Biswas P (2012) Nanoparticle synthesis characterization and application to solve some chronic agricultural problems. Appl Biol Res 14:138–144
62. Tarafdar JC (2022) Nanofertilizers: future for global food production. In: Proceedings of the 2nd Edition of global conference on agriculture and horticulture, Online
63. Tarafder C, Daizy M, Alam MM, Ali MR, Islam MJ, Islam R, Khan MZH (2020) Formulation of a hybrid nanofertilizer for slow and sustainable release of micronutrients. ACS Omega 5(37):23960–23966
64. Toksha B, Sonawale VAM, Vanarase A, Bornare D, Tonde S, Hazra C, Kundu D, Satdive A, Tayde S, Chatterjee A (2021) Nanofertilizers: a review on synthesis and impact of their use on crop yield and environment. Environ Technol Innov 24:101986
65. Tyagi J, Ahmad S, Malik M (2022) Nitrogenous fertilizers: impact on environment sustainability, mitigation strategies, and challenges. Int J Environ Sci Technol. https://doi.org/10.1007/s13762-022-04027-9

66. Usman M, Farooq M, Wakeel A, Nawaz A, Cheema SA, Rehman HU, Ashraf I, Sanaullah M (2020) Nanotechnology in agriculture: current status, challenges and future opportunities. Sci Total Environ 721:137778
67. Verma KK, Song XP, Degu HD, Guo DJ, Joshi A, Huang HR, Li YR (2023) Recent advances in nitrogen and nano-nitrogen fertilizers for sustainable crop production: a mini-review. Chem Biol Technol Agric 10(1):111
68. Verma KK et al (2021) Nano-enabled products: challenges and opportunities for sustainable agriculture. Plants 10:2727
69. Verma KK, Song XP, Joshi A, Tian DD, Rajput VD, Singh M, Arora J, Minkina T, Li YR (2022) Recent trends in nano-fertilizers for sustainable agriculture under climate change for global food security. Nanomaterials 12:173
70. Waheed Z, Anwar A, Sadiqa A, Ahmad A, Intisar A, Javaid A, Kazi M (2024) Advancing sustainable agriculture: metal-doped urea–hydroxyapatite hybrid nanofertilizer for agro-industry. Nanotechnol Rev 13(1):20240107
71. Wang Y, Wang S, Sun J, Dai H, Zhang B, Xiang W, Hu Z, Li P, Yang J, Zhang W (2021) Nanobubbles promote nutrient utilization and plant growth in rice by upregulating nutrient uptake genes and stimulating growth hormone production. Sci Total Environ 800:149627. https://doi.org/10.1016/j.scitotenv.2021.149627
72. Wang P, Das P, Wang L, Zhou J, Deng C, Vera-Reyes I, Wang Y (2025) Prospects of nano phosphorus fertilizers (NPFs) in plant-based agriculture: effects and mechanisms. J Nanopart Res 27(3):60
73. Wang P, Lombi E, Zhao F-J, Kopittke PM (2016) Nanotechnology: a new opportunity in plant sciences. Trends Plant Sci 21:699–712
74. Wang Q, Gao L, Li Y, Shakoor N, Sun Y, Jiang Y, Zhang P (2023) Nano-agriculture and nitrogen cycling: opportunities and challenges for sustainable farming. J Clean Prod 421:138489
75. Wang Q, Shan C, Zhang P, Zhao W, Zhu G, Sun Y, Rui Y (2024) The combination of nanotechnology and potassium: applications in agriculture. Environ Sci Pollut Res 31(2):1890–1906
76. Yadav A, Yadav K, Abd-Elsalam KA (2023) Nanofertilizers: types, delivery and advantages in agricultural sustainability. Agrochemicals 2(2):296–336
77. Yadav S, Sawarni N, Dahiya T, Rana JS, Sharma M, Batra B (2023b) Nanoagriculture: advantages and drawbacks. In: Agricultural and environmental nanotechnology: novel technologies and their ecological impact. Springer Nature Singapore, Singapore, pp 3–42
78. Yaseen R, Ahmed AIS, Omer AM, Agha MKM, Emam TM (2020) Nano-fertilizers: biofabrication, application and biosafety. Nov Res Microbiol J 4:884–900
79. Yomso J, Menon S (2021) Impact of nanofertilizers on growth and yield parameters of rice crop: a review. Pharma Innov J 10:249–253
80. Zahra Z, Habib Z, Hyun H, Shahzad HMA (2022) Overview on recent developments in the design, application, and impacts of nanofertilizers in agriculture. Sustainability 14:9397

Stress Management

Alleviation of Salt, Drought, Nutrient, and Temperature Stresses in Plants Through Nanoparticles

Hajra Abid, Khawaja Shafique Ahmad, Muhammad Asif Shehzad, Ansar Mehmood, Dania Habib, and Saiqa Nazir

Abstract Global agriculture system is under the influence of various environmental challenges due to climate change thereby raising concern to develop novel approaches to overcome these problems. Plants are exposed to a wide range of environmental stresses, i.e., salinity, drought, cold, high temperature, nutrient deficiencies, and heavy metal toxicity. These stresses negatively impact plant physiology, growth, and productivity, ultimately leading to food insecurity and economic losses. Nanotechnology has emerged as a recent, efficient, and promising approach for abiotic stress management in plants and has the potential to revolutionize sustainable agriculture. Therefore, understanding the role of various nanoparticles (NPs) to mitigate abiotic stresses in plants is crucial to improve plant resilience. Due to their unique physicochemical properties, NPs offer effective solutions to improve stress tolerance, combat nutrient deficiencies, and boost plant growth and productivity. Under different abiotic stress conditions, nanoparticles can activate plant defense mechanisms and enhance the activities of antioxidant enzymes. They also improve photosynthetic efficiency, transpiration rate, stomatal conductance, CO_2 assimilation, relative water content, osmolyte accumulation, protein metabolism, nutrient uptake, germination, hormonal signaling pathways, and the expression of stress-responsive genes. This chapter focuses on the potential of various NPs to alleviate different plant stresses and revolutionize agricultural practices to achieve food security and sustainable agriculture.

Keywords Abiotic stress · Nanoparticles · Antioxidants · Plant growth · Stress mitigation

H. Abid · K. S. Ahmad (✉) · A. Mehmood · D. Habib · S. Nazir
Department of Botany, University of Poonch Rawalakot, Rawalakot, Pakistan
e-mail: ahmadks@upr.edu.pk

M. A. Shehzad
Institute of Plant Breeding and Biotechnology, Muhammad Nawaz Shareef University of Agriculture, Multan, Pakistan

L. F. Fraceto et al. (eds.), *Emerging Nanotechnologies for Agroecosystem Management*, Smart Nanomaterials Technology, https://doi.org/10.1007/978-981-95-0187-8_5

1 Introduction

Agricultural production is significantly challenged by abiotic stresses globally, including salinity, drought, heavy metal contamination, and high temperatures. These stresses have prompted investigators to focus on improving the long-term stability of ecological systems to increase crop resilience and survival under adverse situations [91, 118]. Abiotic stress factors affecting modern agricultural productivity globally include drought, extreme temperatures, salinity, waterlogging, ultraviolet (UV) radiations, and heavy metals (HMs) stress [38, 188]. Global warming leading to climate change which has directed to a sharp rise in the intensity and frequency of drought, heat radiations, and other environmental challenges like excess water, salt, and cold [102, 236]. Besides natural factors, human disturbances of the ecosystem, exhibiting through accelerated industrialization, intensive agriculture, extensive mining, growing populations, and rapid urbanization, have exacerbated global warming. This, in turn, contributes indirectly to these environmental constraints and severe toxicity of vital life-sustaining elements [69]).

Environmental stresses including drought, salinity, and extreme temperature must be endured by plants as they are sessile [182, 238]. Salinity is a huge environmental challenge globally, that significantly harms growth and development of crop plants, from seedlings to harvesting stage [57]. This reduces food production globally by about 30% (Machado and [140]). Approximately one billion hectares of land are overwhelmed by salinity worldwide, which is increasing at a rate of 2 million hectares per year [128]. Salinity impacts over 7% of the total area of earth and about 20% of arable land, particularly in arid and semi-arid countries like Egypt [185]. Moreover, it is predicted that over 50% of the arable regions of world will be salinized by 2050 [26].

Drought is one of the consequential environmental stresses that exacerbate other stresses for instance heat, salinity, and pathogen attacks, thereby damaging cultivated plants and reducing agronomic yield [71]. It is also one of the most prevalent, extreme, and persistent severe weather events, along with heat waves and floods, with incidences varying regionally and annually [170]. Heat and drought stress are the major environmental constraints negatively impacting crop yield, with their prevalence enhanced alarmingly due to changing climate. Cold stress impacts numerous processes in plants, i.e., secondary metabolism, defense mechanisms, respiration and the production of proteins and nucleic acids [32]. Furthermore, abiotic factors including salt, heavy metals, and water stress similarly constrain crop production in most of the regions [151, 164]. Scientists are endeavoring to develop innovative approaches to mitigate adverse impacts of abiotic stresses for fulfilling growing global nutrition requirements, improving crop yield and ensure food security and sustainable utilization [5]. Nanotechnology offers promising strategies with immense potential to develop stress resilience in crops and foster sustainable agricultural practices globally [235]. As depicted in Fig. 1, nanoparticle application significantly helps plants to mitigate stress.

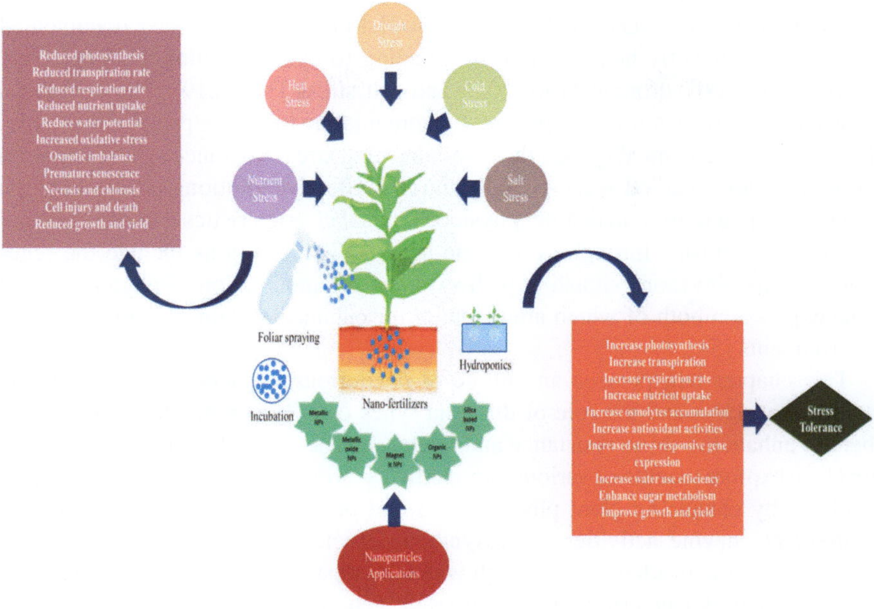

Fig. 1 Stress tolerance mechanism through nanoparticle application in plants

Nanotechnology is a rapidly evolving and aesthetic discipline of science that has prompted numerous innovations [59, 60] and has potential to reform sustainable agriculture [85]. It particularly offers effective solutions to agricultural challenges, contributing to a safe and sustainable development of this sector [187]. NPs are very small size particles which range between 1 to 100 nm [119], utilized to help plants overcome environmental challenges. Various NPs used in agricultural treatments include zinc (Zn), silver (Ag), silicon (Si), copper (Cu), cobalt (Co), magnesium (Mg), manganese (Mn), cesium (Ce), molybdenum (Mo), selenium (Se), zinc oxide (ZnO), copper oxide (CuO), silicon oxide (SiO_2), titanium oxide (TiO_2), iron oxide (FeO, Fe_2O_3), calcium carbonate ($CaCO_3$), magnesium oxide (MgO), aluminum oxide (Al_2O_3, AlO_4, and carbon nanotubes (CNTs [24, 64, 114]). As they have exceptional physiochemical characteristics including more surface area, mobility, reactivity, and specific shapes, NPs have developed as a favorable approach for promoting sustainable agronomic system [107].

NPs can enhance crop production in sustainable agriculture by boosting crop tolerance against environmental constraints; thereby arising current and future production challenges [21, 109]. This development is attained via enzyme activation, increased photosynthetic pigment concentration, improved photosynthetic efficacy, and the regulation of plant microbes [62]. Because they have small size, great solubility, and more plant uptake rate, NPs have developed a widely used latest agricultural tool [21, 53]. The small size of NPs poses valuable characteristics permitting them to pass through the biological membranes in plants more rapidly, thereby alleviating stresses, i.e., salinity, heavy metals, and heat stress [162]. Many researches have documented that NPs significantly help plants in responding to environmental challenges like temperature [100], drought [150, 202], and salt stress [141, 219]. They exhibit a critical role in stress mitigation process by upregulating tolerance proteins and genes that connect biochemical cycles, thereby enhancing stress tolerance. These nanomaterials are often regarded as eco-friendly and cost-effective solutions, possessing high surface-volume ratios and distinct physicochemical characteristics that control plant defense mechanisms through synergistic actions. These actions include the regulation of phytohormone signaling pathways and the modulation of phytohormonal gene expression both of which are essential for enhancing plant growth in adverse environments [58].

This chapter will provide an outline of deleterious impacts of various abiotic stresses on plants and the role of different types of NPs to mitigate these stresses thereby enhancing stress resistance and tolerance. The main objectives of this work are (1) to explore the role of various nanoparticles (NPs) in mitigating abiotic stresses in plants by enhancing plant physiological and biochemical responses, including antioxidant enzyme activities, photosynthetic efficiency, and nutrient uptake, (2) to investigate the mechanisms through which nanoparticles influence plant defense systems, such as triggering stress-responsive gene expression, osmolyte accumulation, and hormonal pathways, under adverse environmental conditions, and (3) to evaluate the potential applications of nanoparticles in revolutionizing agricultural practices for improving plant resilience, productivity, and food security while promoting sustainable agriculture.

2 Effect of Salt, Drought, Nutrient, and Temperature Stress on Plants

Salt stress leads to the accumulation of sodium (Na^+) and chloride (Cl^-) ions in the cytoplasm, triggering severe cellular damage [173]. Salinity impacts all physiological mechanisms of plants, such as photosynthesis, respiration, lipid metabolism, and proteins synthesis [3]. The primary effect of salinity is a decline in the enlargement of leaf surface, which may eventually cease entirely, causing a significant decline in shoot growth [4]. The detrimental consequences of salinity on growth of plants arise from ionic toxicity and osmotic stress [22]. Na^+ and Cl^- toxicity not only led to nutritional imbalances but results in physiological drought as well by dropping the solute potential of soil solutions [116]. This stimulates the enhancement of osmolytes, i.e., proline and soluble sugars, facilitating water absorption [14].

Drought represents one of the most critical environmental challenges to global agronomic yields [106]. Globally, about 70% of the potential damage of agriculture production and efficiency is attributed to water stress, a chronic abiotic stress that impact crops [227]. The detrimental effects of drought are aggravated when combined with other stresses, like salinity [29] or temperature extremes [193, 228].

Low soil moisture impairs nutrient mobility and uptake, leading to cascading effects on metabolic processes, i.e., regulation of hormones and chlorophyll synthesis, along with nutrient acquisition and phenological development [210]. Drought triggers considerable variations in plant morphology and physiology, including decreased photosynthetic activity, transpiration rates, root and shoot growth, osmotic adjustments, overproduction of reactive oxygen species (ROS), and accelerated senescence [12, 71, 99]. In cultivated plants, it diminishes photosynthesis, growth, and biomass production, simultaneously increasing oxidative harm to chloroplasts and protein degradation [13, 123]. Drought mainly interrupts electron transport, photosynthetic pigment synthesis, and the Krebs cycle through enhanced membrane lipid peroxidation and thylakoid degeneration, ultimately impairing the light dependent and independent reactions of photosynthesis [40]. It induces water leakage from cytoplasm, leading to fluctuations in cell volume, vacuole size, changes in membrane permeability, loss of turgor pressure detection, and osmotic shifts within cells [130, 205]. Under such environments, cell contents leak due to membrane damage and changes in lipid composition [39, 146, 222]. Under drought stress, substantial amounts of ROS are produced, involving singlet oxygen (O_2^{-1}), superoxide (O_2^{-}), hydroxyl (OH^{-1}), hydrogen peroxide (H_2O_2), and alkoxy radicals (RO), mainly in the chloroplasts [44, 165] that severely damage plant cells via protein break down, DNA disintegration, and lipid peroxidation eventually causing cell death [146]. The excessive production of ROS damages consistency of membranes and cell functions by disrupting mineral uptake, photosynthesis, and incorporation [81]. Various osmolytes such as amine compounds (e.g., glycine betaine), soluble sugar and proline, exhibit crucial roles in regulating enzyme functions, water contents, and stress indication [56, 231].

Nutrients such as macro and micronutrient deficiencies increase plant susceptibility to abiotic stresses [196]. Nutrient deficiencies can range from mild to acute, with prolonged scarcities leading to suppressed growth and reduced productivity [108]. Nitrogen insufficiency results in leaves yellowing, and minor shoots whereas phosphorus insufficiency leads to dark green leaves with a bronze-purple shade on the lower surfaces and weak branches. The shortage of potassium leads to interveinal chlorosis, crinkling, and ascendant rolling of leaves, and in severe cases, shoot tip death. Calcium deficiency leads to leaf chlorosis with necrotic tips and magnesium deficiency results in thin, brittle leaves that drop prematurely, with shoots appearing normal except the shortage is acute [154]. Nutritional disorders can also exhibit unusual symptoms due to interactions with several living and non-living factors, like water, light, pathogens, pests, and disease [73, 212]. For example, plants experiencing nitrogen deficiency often produce lower concentration of lignin and cellulose, weakening cell walls and making them susceptible to pathogens [197]. Similarly, insufficient potassium levels in plants can impair stomatal closure, facilitating pathogen invasion through the stomata [221].

Plants such as chili plants experience reduced growth, productivity, and flowering under low temperature of below 15 °C [159]. The main concerns of cold stress involve chlorophyll content impairment and photosynthetic disruption, triggering the production of ROS that deteriorate cell membranes. Plants possess defense mechanisms, such as osmotic regulation and antioxidant apparatus, which provide shield from the negative effects of extremely low temperatures [17]. The increased synthesis of osmolytes and antioxidants aids plants in resisting extremely low temperatures [104, 225]. Moreover, research examining the correspondence among physiological and biochemical properties are crucial for dealing chilling stress tolerance [147]. Therefore, plants adjust themselves to climatic fluctuations by altering their morphological and physiological traits [110]. On the other hand, heat stress arises when ambient temperatures exceed a critical threshold, with abrupt rise of 10 to 15 °C [134]. The intensity and time of elevated temperatures determine the extent of their impact on morphology, possibly causing irreversible harms to growth parameters [125]. High-temperature stress exerts both direct and indirect effects, potentially leading to cell death through protein denaturation, lipid degradation, mitochondrial dysfunction, and loss of membrane stability [122]. It typically reduces the efficiency of the photosynthetic machinery, accelerates the plant life cycle, and diminishes overall yield [239].

3 Uptake, Translocation, and Accumulation of NPs in Plants

Generally, NPs have the ability to penetrate into tissues of root and shoot directly due to their extremely small sizes when applied through roots or foliar spray, respectively [138, 175]. The primary route for NPs to enter into plants from soil is via lateral roots, afterward they are transported through the xylem tissues to distribute them in the entire body of plants [103]. Several factors including size, surface properties, charge, functionalization, chemical composition, root secretions, and microorganisms accompanied with roots significantly impact the NP uptake, translocation, and accumulation process [35, 70]. The Van der Waals forces, double-layer forces, Brownian motion, and gravity are factors influencing high mobility of NPs, which is essential for the adhesive characteristics of NPs. Research has demonstrated that NPs can penetrate plants via leaves and roots, subsequently moving through the phloem and xylem, and infiltrating cells, tissues, and organelles, [74, 95]. Plants are regularly in contact with NPs occurring in surrounding environment in air and soil. From air and soil, NPs in liquid form can percolate into plants [35]. Aerosol NPs can travel to the upper portions of plants through pathways, i.e., stomata, direct diffusion, injuries, or through upper parts, unlike NPs in soil [96]. NPs can be applied to plants through seed priming, which results in plants exhibiting altered metabolism than control plants [129].

When NPs applied through leaves, they are primarily absorbed through the stomata and cuticle [233]. It is established that the cuticle can allow entry of NPs up to 5 nm in size [54], whereas the stomatal pore typically ranges from roughly 20–500 nm in size [55]. Known that maximum number of agriculturally applied NPs exceed 5 nm in size, the stomatal uptake is likely the primary route for NPs uptake during foliar spraying. Foliar uptake facilitates the internalization of NPs across a wide size spectrum, from a few nanometers to several hundred nanometers, encompassing elements like cerium oxide (ceria), titanium dioxide (titania), FeO, ZnO, and Ag [46]. Initially, aerosol NPs adhere to the leaves surface, afterward they are being absorbed and enter into epidermal cells. Once they are transported to the epidermal cells, they can move via apoplastic or symplastic routes to various parts [15]. Additionally, the cuticle acts as a crucial protective layer, effectively preventing the entry of NPs exceeding 5 nm in diameter through the leaves surface. The "apoplastic route" refers to the pathway through which small NPs, typically 5–20 nm in size, are absorbed by roots via openings in the epidermal cell walls of roots [132]. Stomatal pores provide an alternative pathway for entry of NPs into plants through foliar uptake [126]. Plant species such as rapeseed, beans, wheat, maize, cucumber, and lettuce have shown evidence of NP translocation via their leaves [46].

Under root application, NP intake typically takes place in immature regions of the underground parts, for example root hairs, root tips, and adjacent root joints, wherever physical barriers (i.e., suberin lamella and casparian strips) are not fully developed [223]. The direct passage through the epidermal cells of root, capillary forces, and osmotic pressure are pathways for entry of smaller size NPs [20]. The microscopic gaps present in partially permeable cell walls of epidermal of roots cells that function as apertures, preventing the pathway of larger NPs. NPs are translocated from root to shoot or shoot to root via xylem and phloem, respectively, after being absorbed by the roots or shoots [139]. The negatively charged NPs are more competently transferred from root to shoot whereas positively charged NPs are attached more tightly to the roots surface through root application [33, 201]. Conversely, positively charged NPs exhibit the highest supply efficacy into apoplastic space and stomata through foliar application [93].

Consequently, it is not amazing that both the characteristics of NPs (i.e., their charge and size) and plant parameters (including species, growth, and developmental phase), effect the immersion and delivery of NPs in plants [211]. Plant parameters as various key factors vary with particular species and developmental phases. These parameters include: a) the area of contact between the plant and NPs, which is based on morphological characters of plants, b) the extent of undeveloped upper and lower parts, that serve as main gateways for NPs, and c) the comparable sizes of pores of membranes, such as those in plasma membrane, cell wall, stomata, cuticle, and the vascular system [48, 138].

4 Role of NPs in Alleviation of Stress in Plants

NPs have shown great prospective to increase water retention and balance water level, thereby enhancing tolerance to environmental stresses in plants [174]. NPs control transpiration rate and gas exchange by influencing anatomy of leaves and promoting stomatal closure [9]. They also protect the photosynthetic apparatus and increase photosynthesis by activating the antioxidant system to recover impairment triggered by ROS in chloroplasts and photosystems. Elevated level of ROS production can be mitigated by upregulating antioxidant systems and enhancing the osmolytes accumulation [11]. Furthermore, they boost the electron transport chain and enhance plants pigments concentration [75, 145]. NPs significantly influence various physiological processes, including stress-related responses, hormone metabolism, and the biosynthesis of osmolytes. They modulate the synthesis of abscisic acid (ABA), ethylene, calcium, and nitric oxide signaling pathways. Moreover, NPs control signal transduction phases in excessive drought and salinity, activating stress-responsive genes as well as adapting plants to thrive under harsh environmental conditions [174]. The NPs are employed greatly following the findings of their potential advantages for plant nutrients and tolerance against stresses.

4.1 Mitigation of Salt Stress

The term "salinity" or "saline soil" represents the soil having water-soluble salts concentration exceeding 4 dS m^{-1}. NPs application has established the prospective to reduce the detrimental effects of salinity by boosting K^+ uptake, alleviating Na^+ ion toxicity, maintaining ionic homeostasis, triggering antioxidant defense mechanisms, enhancing the pigments and compatible solutes accumulation, improving membrane stability, and enhancing gas exchange and water contents of leaf [171]. NPs significantly influence various plant developmental stages, promoting growth and development under salinity condition exerting both positive and negative effects [241]. In addition, they have been observed to profoundly affect the plants' morphological, physiological, and biochemical attributes along with molecular functions when plants uptake [117]. They can modulate plant responses to salinity by influencing antioxidant system, hormonal concentrations, expression of genes, ion homeostasis, and defense system functions [219, 241]. NPs' composition and plant species may affect these interactions, exhibiting either optimistic or pessimistic effects. The critical factors influencing plant responses include the NPs' size, structural composition, and reactivity [161, 211]. Photosynthesis is among the most profoundly impacted process by salinity, can be enhanced by NPs, which increase the synthesis of light-harvesting complexes, leading to improved light absorption and photosynthesis. Several studies have indicated that NPs considerably enhance chlorophyll concentration [20].

NPs are extensively employed to increase resistance against salinity in various species of plants (Table 1). Numerous researches have highlighted the positive impacts of NPs on plants under saline condition [157, 241]. NPs' role in salinity stress alleviation has been studied, for instance, nano-silicon in tomato [23], CeO_2 NPs in *Brassica napus* [178], iron oxide NPs in tomato [79], ZnO NPs in *B. napus*, and *Triticum aestivum* [36, 67], TiO_2 NPs in *T. aestivunm* and *Zea mays* [66, 89], chitosan NPs in *B. juncea* and *Oryza sativa* [27, 179] and AgNPs in tomato crops, *Pennisetum glaucum* [115, 124]. Further research has recognized proliferating effects of silica NPs on lupine and wheat [203], Fe_2O_3 NPs on soybean [80], Fe_3O_4 NPs on alfalfa and soybean [97] and Si NPs on *B. napus* [94]. They play a pivotal role

Table 1 Role of NPs in mitigation of salt stress

NPs	Plant species	Mode of application	Concentration	Effects	References
CeO_2 NPs	*Brassica napus*	Soil	0, 200, 1000 mg/kg	Improve photosynthesis, biomass, physiology, and growth	[177]
Ag NPs	*Triticum aestivum*	Seed priming	1 mg/L	Minimize the negative effects of salinity via improving photosynthetic efficiency, phytohormonal balance, and growth parameters	[7]
TiO_2 NPs	*Dracocephalum moldavica*	Hydroponics	0, 50, 100, 200 mg/L	Enhance antioxidant response and physiochemical characteristics	[84]
Ag NPs	*Satureja hortensis*		0, 40, 60, 80, 120 ppm	Increase growth parameters and germination percentage	[155]
TiO_2 NPs	*Zea mays*	Seed priming	40, 60, 80 ppm	Reduce MDA, proline, and Na^+ contents. Increase antioxidant enzymes activity, phenols, K^+, dry and fresh weight, relative moisture contents and germination rate	[192]
Se NPs	*Brassica campestris*	Hydroponics	30 mg/L	Enhance growth parameters, chlorophyll, proteins, phenols, flavonoid, carbohydrates, antioxidant activity, and change phytochemicals	[183]

(continued)

Table 1 (continued)

NPs	Plant species	Mode of application	Concentration	Effects	References
CaP NPs	*Vicia faba*	Soil	16 mg/L	Increase growth parameters, photosynthetic pigments, and production	[152]
ZnO NPs	*Vicia faba*	Foliar	0, 50, 100 mg/L	Enhance photosynthetic pigments, antioxidants, osmolytes, and secondary metabolites	[67]
CMgO NPs (chitosan magnesium oxide NPs)	*Oryza sativa*	Hydroponics	100 mg/L	Reduce oxidative stress, MDA, and H_2O_2. Increase free amino acid and K^+/Na^+ ratio	[198]
Ca NPs	*Solanum lycopersicon*	Soil	20 mM	Improve growth parameters, anabolic, and defense mechanism	[6]
B_2O_3 NPs	*Zea mays*			Enhance growth, photosynthetic efficiency, mineral homeostasis, and reduce oxidative stress	[63]
Si NPs	*Cymbopogon flexuosus*	Foliar	150 mg/L	Improve growth, photosynthesis, stomatal conductance, CO_2 assimilation rate, proline, and reduce oxidative stress	[149]
Cs NPs (chitosan NPs)	*Oryza sativa*	Seed priming	50 μg/mL	Increase sprouting, seedling vigor, antioxidant activity, and biochemical traits	[200]

(continued)

Table 1 (continued)

NPs	Plant species	Mode of application	Concentration	Effects	References
Biochar NPs	*Solanum lycopersicum*	Soil	2000 mg/kg	Increase chlorophyll contents and biomass. Reduce transfer of Na from root to shoot	[206]
Cs NPs (chitosan NPs)	*Phaseolus vulgaris*	Foliar	0.05%	Improve antioxidant system, carbohydrate, proline, and yield. Reduce H_2O_2, discharge of electrolytes and lipid peroxidation	[18]
Ag NPs	*Calendula officinalis*	Foliar		Increase chlorophyll, protein contents, antioxidants activities (CAT, SOD, POD, APX, PPL, PPO), secondary metabolites, and growth. Reduce MDA and H_2O_2	[240]
Fe NPs	*Zea mays*	Soil	15, 25 mg/kg	Increase photosynthetic pigments, antioxidant (CAT, SOD, POD, APX), nutrient uptake, water contents, membrane stability, growth, and yield	[25]

(continued)

Table 1 (continued)

NPs	Plant species	Mode of application	Concentration	Effects	References
ZnO NPs	*Gossypium hirsutum*	Foliar		Increase antioxidant (CAT, SOD, POD) activities, growth parameters, gene expression (CNGC, AHA3, NHX2, HAK17). Reduce MDA, H_2O_2, O_2^-, and expression of SKOR and CBL-CIPK pathway. Stabilize Na + / K + ratio	[167]
SiO_2 NPs	*Glycine max*	Foliar	12 mg	Improve plant microbe interactions and growth	[224]
Cu NPs	*Trigonella foenum-graecum*			Increase chlorophyll and other pigments, phenol, proline, anthocyanin, shikimic acid, and growth parameters. Reduce MDA and H_2O_2	[76]

in optimizing plant nutrition by enhancing the processes of nutrient uptake, translocation, and distribution [121]. For instance, SiO_2 NPs improved seedling growth by increasing leaves K^+ contents under salt stress in soybean plants [72]. Similarly, Cu NPs applied through leaves improved growth and Na^+/K^+ ratio under salinity in tomato [163]. Fe_2O_3 NPs mitigate salinity induced sensitivity in *Trachyspermum ammi*, while improved growth during saline conditions in *Capsicum annuum* [232].

4.2 Mitigation of Drought Stress

Drought is defined as "a period of drier than normal conditions due to an insufficiency of precipitation over a prolonged time period, such as a season or longer, leading to water scarcity" by the National Oceanic and Atmospheric Administration [156].

Drought can result from shifts in weather systems, including changes in the monsoon, weather patterns influenced by global warming, rainfall disturbances, or random and seasonal variations [186]. To alleviate the consequences of drought, various nanomaterials have been employed, including nanofertilizers, nanopesticides, nano-based biostimulants, nanoencapsulated plant growth hormones, and nanosensors [58]. The primary mechanisms for stress relief involve the regulated delivery of active compounds in nanofertilizers/nanopesticides, which can induce pH fluctuations, alter enzyme activities, and modify light, temperature, or oxidation–reduction potential in the agro-ecosystem [194]. The active ingredients in these nanofertilizers and nanopesticides are protected by the use of nanocarriers. The application of nanomaterials to mitigate stresses such as drought has been studied across various farming systems [58] and within the framework of the "farm-to-fork" concept [1]. The general mechanism involves enhancing drought resistance by reducing the generation of ROS and improving plant enzyme activities.

NPs alleviate the effects of drought stress mainly by inducing physiological and biochemical mitigation and regulating the expression of genes associated with drought response and tolerance (Table 2). They improve the photosynthetic rate of drought stressed plants by improving root growth, modifying intracellular water metabolism, maintaining ionic homeostasis, upregulating aquaporins, and accumulating compatible solutes. These mechanisms alleviate the osmotic stress caused by drought. Nanotechnology enhances photosynthetic efficacy by modifying activities of enzymes that are involved in Calvin cycle and maintaining photosynthetic pigments essential for growth of plants [137]. Many researchers have revealed that NPs like CuO and ZnO NPs produce drought stress resistance in various species of plants [229]. This may be associated with the greater production of lateral root hairs that improve absorption of water. Notable results have been observed in several investigations, i.e., CNTs treated plants showed improved germination and antioxidant quantities in maize, soybean, and barley [135]. Fullerenol (FNPs) having molecular formula $C_{60}(OH)_{24}$, can reduce excess drought effects by providing intercellular spaces with extra water supply. Specifically, FNPs can penetrate the plants through root and leaf tissues and attach water molecules in many parts of cells, suggesting that FNPs may be useful under drought stress in plants [43, 218]. NPs regulate antioxidant enzyme activities by targeting various cell organelles and discharge their contents [47, 195]. This modulation was evidenced by increased activity of SOD enzyme with the use of TiO_2 NPs application in plants [82]. Under drought stress, CuO NPs exposure leads to increased anthocyanin levels and proline accumulation in plants. CuO NPs treated wheat roots exhibited higher accumulation of free radicals, aligning with the plant's response to ROS bursts triggered by NPs. Increased ROS level further indicating that water deficiency causes elevated amounts of ABA which may leads to transcriptional alterations and tolerance [195].

Table 2 Role of NPs in mitigation of drought stress

NPs	Plant species	Mode of application	Concentration	Effects	References
SiO_2 NPs	*Crataegus laevigata*	Soil	0, 10, 50, 100 mg/L	Increase photosynthesis and transpiration rate, decrease malondialdehyde, and water potential	[31]
FNPs (Fullerenol NPs)	*Beta vulgaris*	Foliar	70 µmol/L, 700 µmol/L	Decrease antioxidant enzyme activities, increase glutathione, and malondialdehyde (MDA)	[43]
Y-Fe_2O_3 NPs (Maghemite NPs)	*Brassica napus*	Hydroponics	0.5, 0.8, 1, or 2 mg/mL	Scavenge ROS, decrease injurious fenton reactions (iron salts and H_2O_2 in acidic condition)	[160]
CeO NPs	*Sorghum bicolor*	Foliar	10 mg/L	Increase antioxidant enzymes (CAT, SOD, APX) activities, decrease H_2O_2, ROS, MDA, stomatal conductance and photosynthesis	[51]
TiO_2 NPs	*Verbascum sinuatum*	Hydroponics	20 ppm	Increase photosynthetic pigments, total phenols, flavonoids, and nitrogen assimilation	[111]
ZnO, Fe, Co, Cu NPs	*Glycine max*	Seed	0.05 mg/L of ZnO, Co NPs, 50 mg/L of Fe, Cu, ZnO NPs	Increase physiological traits, biomass reduction rate, and tolerance, upregulate drought tolerant genes and abcissic acid biosynthesis	[133]

(continued)

Table 2 (continued)

NPs	Plant species	Mode of application	Concentration	Effects	References
ZnO NPs	*Solanum melongea*	Foliar	0, 50, 100 ppm	Improve macro and micronutrients uptake, increase moisture contents and biomass, relieve harms to membranes	[190]
SiNPs	*Calendula officinalis*	Seed	0, 100, 200, 500 mg/L	Increase germination index and rate of germination	[168]
ZnO NPs	*Solanum lycopersicum*	Foliar	25, 50, 100 mg/L	Boost activities of antioxidant enzymes, increase phenols ascorbic acid, and biomass. Reduce MDA and H_2O_2	[68]
Magnetite NPs	*Trigonella foenumgraecum*	Foliar	20, 50, 100 and 200 mg/L	Higher chlorophyll contents, photosynthetic rate and development	[42]
SiO_2 NPs	*Pisum sativum*	Foliar	12.5, 25, 50 ppm	Increase antioxidant activities (SOD, CAT, APX, GSH), relative water content, phenolic contents, and growth parameters. Reduce lipid peroxidation and H_2O_2	[204]
Se NPs	*Chenopodium quinoa*	Seed	0.5, 1.5, 3, 4.5, and 6 mg/L	Increase antioxidant activities (SOD, CAT, APX), proline, protein contents, and germination parameters	[83]

(continued)

Table 2 (continued)

NPs	Plant species	Mode of application	Concentration	Effects	References
ZnO NPs	*Arachis hypogea*	Foliar	10, 20, 40, 50, 100, 300, 1000 ppm	Enhance antioxidant enzyme activities, biomass, and yield	[127]
Cu NPs	*Zea mays*	Seed	52, 69.4, 86.8 μM	Enhance chlorophyll and carotenoid contents, leaf water potential, anthocyanin, biomass and decrease ROS	[214]
Fe_2O_3 NPs	*Vitis vinifera*		10, 30, 40 μM	Increase chlorophyll contents and fluorescence, concentration of non-enzymatic antioxidants (AsA, GSH/GSSG), enzymatic activities (GPX, CAT, GR, MDHAR, DHAR), reduce MDA, and H_2O_2	[41]
ZnO NPs	*Triticum aestivum*	Foliar	50, 100, 150 ppm	Improve antioxidant defense system, growth, and physiology	[86]
Zn, Si, zeolite NPs	*Coriandrum sativum*	Foliar	2.5 (Zn), 2.5 (Si), 1.3 (zeolite) L. ha^{-1}	Improve growth parameters, physiological, and biochemical responses	[142]

(continued)

Table 2 (continued)

NPs	Plant species	Mode of application	Concentration	Effects	References
Ca NPs	*Brassica napus*	Foliar	100 ppm	Increase chlorophyll, carotenoids, xanthophylls, nutrients, photosynthetic and gas exchange parameters, ROS homeostasis genes and photosynthetic protein expressions	[34]
SiO_2 NPs	*Ehretia macrophylla*		50, 100, 200, 500, 1000 mg/L	Enhance antioxidants activity, reduce MDA and H_2O_2. Accumulate secondary metabolites and regulate differentially expressed genes	[45]
Ce NPs	*Sorghum bicolor*	Foliar	25 mg/L	Increase photosynthetic rate, reproductive traits, and grain yield	[52]

4.3 Mitigation of Nutrient Stress

Macronutrient and micronutrient deficiencies exacerbate plant sensitivity to abiotic stresses [196]. Mitigating hunger and enhancing food safety is a multifaceted challenge, aggravated by a growing population that drives additional food requirement, leading to climate change and food scarcity. Modern agriculture system currently sustains approximately 6 billion individuals, but the world population is expected to attain 9.8 billion till 2050, about 70% expansion in agriculture productivity is essential to fulfill this requirement. To achieve food production of 3 billion tons annually, the use of commercial fertilizers is projected to exceed 200 million tons. However, this dependence on chemical fertilizers is not enough to fulfill food production necessities in future [213].

Nanotechnology has the potential to enhance plant growth and yield in nutrient limited environments by promoting absorption and use efficiency of nutrients.

Nanotechnology-based solutions contain beneficial microorganisms capable of solubilizing and mobilizing soil nutrients, enhancing their availability for plants even under stress conditions. This can significantly boost plant growth, ultimately resulting in higher crop production. The latest developments in nanotechnology have directed to the establishment of nano-assisted materials designed to address nutrient deficiencies, improve crop yields, and diminish loss of nutrients in agricultural systems [49, 226]. Several effective strategies, i.e., the application of nanofertilizers, are being employed to diminish nutrient loss and mitigate soil and ground water pollution. Nanofertilizers are encapsulated with nanomaterials that regulate nutrient discharge based on the plant's requirements, thus improving nutrient use efficiency [189].

Nanotechnology is extensively utilized in agricultural applications, with nanocapsules or NPs being applied by either slow or controlled release fertilizers (SRFs, CRFs, respectively). SRFs discharge nutrients more slowly as compared to conventional fertilizers, while CRFs are designed to regulate the release rate through their preparation [181]. CRFs are encapsulated with NPs to ensure a controlled diffusion rate, SRFs are soluble in water to a little extent and degrade through microbial actions. More nutrient assimilation efficiency is indicated by enhanced plant nutrient uptake and decrease nutrient loss [169]. The applications of different types of NPs have proved efficacious under nutrient limited condition (Table 3). ZnO NPs have been demonstrated to enhance leaf surface area, root length, and sprouting in *Solanum melongena* at various doses [209]. Seeds priming with NPs have been shown to benefit germination via facilitating effective uptake of water and nutrients. For example, *Citrullus lanatus* seeds priming with Ag NPs enhanced growth, quality, germination, and productivity, even after extended storage periods [8]. Ag NPs diminish phosphorus, sulfur, and magnesium concentrations while enhancing potassium concentrations in eastern lilies [180]. Chitosan NPs increase the supply of vital elements including nitrogen, potassium, and phosphorus in the soil, promoting plant roots to uptake them [77].

4.4 Mitigation of Cold Stress

Plants are often subjected to two types of cold stress, i.e., chilling and freezing. Chilling temperature varies from 0–15 °C in range, depend on the plant species and their tolerance level, with ambient temperature and wind velocity also influencing the impact. Contrastingly, plants respond differently to freezing temperatures, which fall below 0 °C [32]. Both chilling and freezing temperatures cause injuries or kill plant species, significantly affecting their survival, environmental distribution and yield [10]. At lower temperatures, protein and enzyme activities are slowed down consequently reducing plants growth [237]. The NPs application has been proved suitable for their usefulness in enhancing cold stress endurance in plants (Table 4). Photosynthetic activity, electrolyte leakage, and membrane stability have been proven to be increased in chickpea plants by TiO_2 NPs under cold stress by modulating transcription [28]. In *Glycyrrhiza glabra*, TiO_2 NPs relieve the detrimental concerns

Table 3 Role of NPs in mitigation of nutrient stress

NPs	Plant species	Mode of application	Concentration	Effects	References
Zn NPs	*Pennisetum americanum*	Foliar	10 mg/L	Enhance growth parameters, activity of antioxidant enzyme, chlorophyll, proteins and rate of photosynthesis	[207]
Zn NPs	*Coffea arabica*	Foliar	10 mg/L	Improve photosynthetic efficacy, growth and biomass	[176]
ZnO NPs	*Cyamopsis tetragonoloba*	Foliar	10 ppm	Increase chlorophyll contents, nutrients, soluble proteins, antioxidant activity, phytase, phosphatases, growth, and biomass	[172]
SiO_2 NPs	*Carthamus tinctorius*	Foliar		Enhance growth parameters and number of achenes	[105]
Zn NPs	*Lupinus termis*	Seed priming	20, 40, 60 mg/L	Enhance antioxidants activities, photosynthesis, and growth	[2]
Fe_2O_3 NPs	*Glycine max*	Foliar and soil	15, 30, 60 mg/pot	Increase chlorophyll contents, root development, and biomass	[230]
ZnO NPs	*Solanum melongena*	Foliar	0, 50, 100 ppm	Enhance plant pigments, relative water contents, and growth parameters	[190]
TiO_2 NPs	*Triticum aestivum*	Soil	0, 100, 200, 400, and 600 mg/kg	Improve growth, chlorophyll contents, nutrients (P, Fe, K, Mn), antioxidant activities, and productivity	[19]

(continued)

Table 3 (continued)

NPs	Plant species	Mode of application	Concentration	Effects	References
Si NPs	*Oryza sativa*	Foliar		Increase pigments, proteins and activities of antioxidants	[98]
Ca NPs	*Brassica napus*	Foliar	100 ppm	Increase photosynthetic pigments and nutritional profile (K, P, Ca, Mg, Mn, B)	[34]

of low temperature stress and enhance glycyrrhizin constituents [78]. Chitosan NPs efficiently reduce ROS accumulation and promote osmoprotectant accumulation in banana plants under cold stress [220]. Moreover, the foliar spraying of ZnO NPs on *Oryza sativa* has been found to reduce cold stress via enhancing the antioxidants activity and regulating transcriptional elements accompanied with stress response [199].

4.5 Mitigation of Heat Stress

Temperatures above a certain threshold are regarded as temperature or heat stress. Extreme temperatures induce both heat and oxidative stress, prompting to overproduction of ROS and subsequent physiological and biochemical alterations involving the formulation of heat shock proteins (HSP) and osmolytes which secure cell structures and proteins, and boost antioxidant mechanisms to regulate homeostasis and oxidation–reduction potential [88]. NPs application in varying concentrations has been proven to increase plants growth and hydration, thereby alleviating the consequences of temperature extremes [20]. At low concentrations, NPs induce antioxidative properties, however, enormous amounts lead to the oxidative stress. Plants synthesize HSP and molecular chaperones that impart significant roles in maintaining protein stability and conferring heat stress resistance. Notably, HSP including HSP90 are stimulated as stress response. Moreover, the application of multiwalled CNTs has found to promote the biosynthesis of HSP and gene expression, further increasing the capability of plants to tolerate elevated temperature stress [191].

SeNPs have proven effective in mitigating temperature extremes. [51] demonstrated that SeNPs improved the antioxidant activity in sorghum, helping to scavenge ROS produced under heat stress, thus alleviating its harms. AgNPs improved the photosynthetic machinery of wheat plants, severely exaggerated by temperate stress thereby inducing tolerance. Their application increased morphological features, e.g., number of leaves and roots, root and shoot length, dry and fresh weight, area [100]. Additionally, ZnNPs improved wheat crop resilience by ameliorating the oxidative

Table 4 NPs in mitigation of cold stress

NPs	Plant species	Mode of application	Concentration	Effects	References
TiO_2 NPs	*Cicer arietinum*		0, 5 mg/L	Increase chlorophyll binding proteins, gene encoding Rubisco's expression, and activity of PEP carboxylase. Reduction in H_2O_2 level	[87]
Ca NPs-AA	*Mangifera indica*		0, 6, 9 mM	Reduce MDA and internal browning. Accumulate phenolic compounds and DPPH	[136]
SiO_2, ZnO, Se, GNPs	*Saccharum officinarum*	Foliar	300 ppm (SiO_2), 50 ppm (ZnO), 15 ppm (Se), 50 ppm (GNPs)	Increase photosynthetic pigments and non-photochemical quenching	[65]
TiO_2 NPs	*Glycyrrhiza glabra*		0, 2, 5 ppm	Increase proteins, phenols, and osmolytes. Reduce H_2O_2 level and lipid peroxidation	[112]
CTS Arg NPs	*Prunus domestica*	Fruit coating	0.5 and 1%	Increase proline, phenols, flavonoids, anthocyanin, and antioxidant activity. Reduce MDA, H_2O_2, and electrolyte leakage. Delay weight loss and maintain firmness of fruits	[143]
Au NPs	*Triticum aestivum*	Seed priming	5, 10, 20, 50 μg/ml	Increase photosynthesis, chlorophyll and unsaturated fatty acids	[217]

(continued)

Table 4 (continued)

NPs	Plant species	Mode of application	Concentration	Effects	References
ZnO NPs	*Oryza sativa*	Foliar	25, 50, 100, 200 mg/L	Enhance growth parameters, antioxidant activity, and chlorophyll	[144]
SiO_2 NPs	*Gossypium hirsutum*		100 mg/L	Enhance gas exchange, photosynthetic and transpiration rate and leaf water concentration. Reduce MDA, H_2O_2, and ABA	[131]
ZnO NPs, Se NPs	*Capsicum annuum*	Foliar	10 mg/L	Enhance antioxidants, growth and productivity. Decrease MDA and H_2O_2	[184]
CH NPs (chitosan NPs)	Banana	Foliar	0, 200, 400, 800 mg/L	Increase fresh and dry weight, reduce MDA and ROS	[153]

stress reduction and lipid peroxidation [90]. NPs treated tomato leaves through foliar spraying is triggered when temperatures exceed certain limits, protecting the plants from heat stress. SiNPs have also been reported to help cope with extreme temperature stress [120]. Various types of NPs have been shown to alleviate temperature stress in several species of plants (Table 5).

5 Conclusion

During recent years, plants face various environmental challenges due to climatic changes causing detrimental effects on global agriculture and food production. Numerous research findings have underscored the potential of NPs in ameliorating plant resistance to diverse range of abiotic stresses. The use of several kinds of NPs has been proven as a promising approach for mitigating various stresses by improving nutrients and water uptake, growth parameters, physiological and biochemical responses, stress signaling response and reducing oxidative stress. Further investigations should be directed to explore the molecular mechanisms to elucidate the underlying interactions of NPs with plants, their fate in plants and environment and toxicity mechanisms unlocking the full potential of NPs while safeguarding ecological and environmental health. Moreover, future investigations

Table 5 NPs in mitigation of heat stress

NPs	Plant species	Mode of application	Concentration	Effects	Reference
AgNPs	*Triticium aestivum*	Soil	25, 50, 75, 100 mg/L	Enhance morphological and growth parameters	[100]
TiO_2 NPs	*Solanum lycopesrsicum*	Foliar	0.05, 0.1, 0.2 g/L	Enhance photosynthesis, stomatal conductance	[166]
Si NPs	*Triticun aestivum*	Soaking	1.66 mM	Increase photosynthetic rate, pigments and relative water contents	[234]
ZnO, TiO_2 NPs	*Triticum aestivum*	Soaking	1.5 (ZnO), 10 ppm (TiO_2)	Increase antioxidant enzymes activity. Decrease MDA and H_2O_2	[208]
BioSe NPs	*Triticum aestivum*		50, 75, 100, 125, 150 μg/mL	Improve leaf water contents, gas exchange and pigments	[61]
ZnO_2 NPs	*Vigna radiate*	Foliar	0, 15, 30, 45 60 mg/L	Enhance gas exchange and chlorophyll	[113]
ZnO NPs	*Triticum aestivum*		10 ppm	Enhance growth, chlorophyll, nutrients uptake, proteins, antioxidant enzymes, indole acetic acid, soluble sugars, and biomass	[37]

(continued)

Table 5 (continued)

NPs	Plant species	Mode of application	Concentration	Effects	Reference
Se NPs	*Triticum aestivum*	Foliar	10 mg/L	Increase transpiration and photosynthetic rate, antioxidant enzymes and stress responsive genes	[158]
Ag NPs	*Triticum aestivum*		50, 100 mg/L	Increase carbohydrate metabolism, heat shock protein, crude fibers and minerals	[101]
TiO_2 NPs	*Carthamus tinctorious*	Seed	100, 200, 500 ppm	Increase germination and physiological parameters	[50]

must explore more beneficial aspects and practical challenges in real-world scenarios to produce stress resilient crops, enhance food security and sustainable agriculture.

References

1. Abdalla Z, El-Ramady H, Omara AE, Elsakhawy T, Bayoumi Y, Shalaby T, Prokisch J (2022) From farm-to-fork: a pictorial mini review on nano-farming of vegetables. Environ Biodiv Soil Secur 6:149–163
2. Abdel Latef AAH, Alhmad MFA, Abdelfattah KE (2017) The possible roles of priming with ZnO nanoparticles in mitigation of salinity stress in lupine (*Lupinus termis*) plants. J Plant Growth Regul 36:60–70
3. Abdeldym EA, El-Mogy MM, Abdellateaf HR, Atia MA (2020) Genetic characterization, agro-morphological and physiological evaluation of grafted tomato under salinity stress conditions. Agron 10:1948
4. Abdelgawad KF, El-Mogy MM, Mohamed MIA, Garchery C, Stevens RG (2019) Increasing ascorbic acid content and salinity tolerance of cherry tomato plants by suppressed expression of the ascorbate oxidase gene. Agron 9:51
5. Abdelsalam IM, Ghosh S, AlKafaas SS, Bedair H, Malloum A, ElKafas SS, Saad-Allah KM (2023) Nanotechnology as a tool for abiotic stress mitigation in horticultural crops. Biol 78(1):163–178
6. Abeed AH, Al-Huqail AA, Albalawi S, Alghamdi SA, Ali B, Alghanem SM, Al-Haithloul HAS, Amro A, Tammam SA, El-Mahdy MT (2023) Calcium nanoparticles mitigate severe salt stress in *Solanum lycopersicon* by instigating the antioxidant defense system and renovating the protein profile. S Afr J Bot 161:36–52
7. Abou-Zeid H, Ismail G (2018) The role of priming with biosynthesized silver nanoparticles in the response of *Triticum aestivum* L to salt stress. Egypt J Bot 58:73–85

8. Acharya P, Jayaprakasha GK, Crosby KM, Jifon JL, Patil BS (2020) Nanoparticle-mediated seed priming improves germination, growth, yield, and quality of watermelons (*Citrullus lanatus*) at multi-locations in texas. Sci Rep 10:5037
9. Acosta-Motos JR, Ortuño MF, Bernal-Vicente A, Diaz-Vivancos P, Sanchez-Blanco MJ, Hernandez JA (2017) Plant responses to salt stress: adaptive mechanisms. Agron 7:18
10. Adhikari L, Baral R, Paudel DR, Min D, Makaju SO, Poudel HP, Acharya JP, Missaoui AM (2022) Cold stress in plants: strategies to improve cold tolerance in forage species. Plant Stress 4:100081
11. Ahanger MA, Qi M, Huang Z, Xu X, Begum N, Qin C, Zhang C, Ahmad N, Mustafa NS, Ashraf M, Zhang L (2021) Improving growth and photosynthetic performance of drought stressed tomato by application of nano-organic fertilizer involves up-regulation of nitrogen, antioxidant and osmolyte metabolism. Ecotoxicol Environ Saf 216:112195
12. Ahluwalia O, Singh PC, Bhatia R (2021) A review on drought stress in plants: implications, mitigation and the role of plant growth promoting rhizobacteria. Resour Environ Sustain 5:100032
13. Ahmad HM, Fiaz S, Hafeez S, Zahra S, Shah AN, Gul B, Aziz O, Mahmood-Ur-Rahman FA, Rafique M, Chen Y, Yang SH, Wang X (2022) Plant growth-promoting Rhizobacteria eliminate the effect of drought stress in plants: a review. Front Plant Sci 13:875774
14. Ahmad F, Kamal A, Singh A, Ashfaque F, Alamri S, Siddiqui MH (2020) Salicylic acid modulates antioxidant system, defense metabolites, and expression of salt transporter genes in *Pisum sativum* under salinity stress. J Plant Growth Reg 1–14
15. Ahmadov IS, Ramazanov MA, Gasimov EK, Rzayev FH, Veliyeva SB (2020) The migration study of nanoparticles from soil to the leaves of plants. Biointerface Res Appl Chem 10:6101–6111
16. Ahmed T, Noman M. Manzoor N, Shahid M, Abdullah M, Ali L, Wang G, Hashem A, AL-Arjani B, AL-qarawi AA, Abd-Allah EF (2021) Nanoparticle-based amelioration of drought stress and cadmium toxicity in rice via triggering the stress responsive genetic mechanisms and nutrient acquisition. Ecotoxicol Environ Saf 209:111829
17. Airaki M, Leterrier M, Mateos RM, Valderrama R, Chaki M, Barroso JB, del Río LA, Palma JM, Corpas FJ (2012) Metabolism of reactive oxygen species and reactive nitrogen species in pepper (*Capsicum annuum* L.) plants under low-temperature stress. Plant Cell Environ 35:281–295
18. Alenazi MM, El-Ebidy AM, El-Shehaby OA, Seleiman MF, Aldhuwaib KJ, Abdel-Aziz HM (2024) Chitosan and chitosan nanoparticles differentially alleviate salinity stress in *Phaseolus vulgaris* L. plants. Plants 13(3):398
19. Alharby HF, Rizwan M, Iftikhar A, Hussaini KM, Rehman MZ, Bamagoos AA, Alharbi BM, Asrar M, Yasmeen T, Ali S (2021) Effect of gibberellic acid and titanium dioxide nanoparticles on growth, antioxidant defense system and mineral nutrient uptake in wheat. Ecotoxicol Environ Saf 221:112436
20. Ali E, El-Shehawi AM, Ibrahim O, Abdul-Hafeez E, Moussa M, Hassan F (2021) A vital role of chitosan nanoparticles in improvisation the drought stress tolerance in Catharanthus roseus (L.) through biochemical and gene expression modulation. Plant Physiol Biochem 161:166–175
21. Al-Khayri JM, Rashmi R, Surya Ulhas R, Sudheer WN, Banadka A, Nagella P, Aldaej MI, Rezk AAS, Shehata WF, Almaghasla MI (2023) The role of nanoparticles in response of plants to abiotic stress at physiological, biochemical, and molecular levels. Plants 12(2):292
22. Almeida DM, Oliveira MM, Saibo NJ (2017) Regulation of Na+ and K+ homeostasis in plants: towards improved salt stress tolerance in crop plants. Genet Mol Biol 40:326–345
23. Almutairi ZM (2016) Effect of nano-silicon application on the expression of salt tolerance genes in germinating tomato (*Solanum lycopersicum* L.) seedlings under salt stress. Plant Omics 9:106–114
24. Alsaeedi A, Show Y (2019) Synthesis of nano-carbon by in-liquid plasma method and its application to a support material of Pt catalyst for fuel cell. Nanomater Nanotechnol 9:1847980419853159

25. Alsamadany H, Anayatullah S, Zia-ur-Rehman M, Usman M, Ameen T, Alharby HF, Alharbi BM, Abdulmajeed AM, Yong JWH, Rizwan M (2024) Residual efficiency of iron-nanoparticles and different iron sources on growth, and antioxidants in maize plants under salts stress: life cycle study. Heliyon 10(7)
26. Alzahrani O, Abouseadaa H, Abdelmoneim TK, Alshehri MA, Mohamed EM, El-Beltagi HS, Atia MAM (2021) Agronomical, physiological and molecular evaluation reveals superior salt-tolerance in bread wheat through salt-induced priming approach. Not Bot Horti Agrobot Cluj-Napoca 49:12310
27. Amin S, Anwar S, Ali B, Ahmad P, Shakir A (2019) Chitosan nanoparticles alleviate salinity-induced oxidative stress in *Brassica juncea* by enhancing antioxidant defense and photosynthetic efficiency. Environ Sci Pollut Res 26(14):14604–14614
28. Amini S, Maali-Amiri R, Mohammadi R, Shahandashti SSK (2017) CDNA-AFLP analysis of transcripts induced in chickpea plants by TiO_2 nanoparticles during cold stress. Plant Physiol Biochem 111:39–49
29. Angon PB, Mondal S, Akter S, Abdul Jalil MA (2023) Roles of CRISPR to mitigate drought and salinity stresses on plants. Plant Stress 8:100169
30. Arif Y, Singh P, Siddiqui H, Bajguz A, Hayat S (2020) Salinity induced physiological and biochemical changes in plants: an omic approach towards salt stress tolerance. Plant Physiol Biochem 156:64–77
31. Ashkavand P, Tabari M, Zarafshar M, Tomášková I, Struve D (2015) Effect of SiO2 nanoparticles on drought resistance in hawthorn seedlings. Lesne Pr Badaw 76(4):350–359
32. Aslam M, Fakher B, Ashraf MA, Cheng Y, Wang B, Qin Y (2022) Plant low-temperature stress: signaling and response. Agron 12:702
33. Avellan A, Schwab F, Masion A, Chaurand P, Borschneck D, Vidal V, Rose J, Santaella C, Levard C (2017) Nanoparticle uptake in plants: gold nanomaterial localized in roots of *Arabidopsis thaliana* by X-ray computed nanotomography and hyperspectral imaging. Environ Sci Technol 51:8682–8691
34. Ayyaz A, Zhou Y, Batool I, Hannan F, Huang Q, Zhang K, Shahzad K, Sun Y, Farooq MA, Zhou W (2024) Calcium nanoparticles and abscisic acid improve drought tolerance, mineral nutrients uptake and inhibitor-mediated photosystem II performance in *Brassica napus*. J Plant Growth Regul 43(2):516–537
35. Azim Z, Singh N, Singh A, Amist N, Niharika KS, Yadav RK, Bano C, Yadav V (2023) A review summarizing uptake, translocation and accumulation of nanoparticles within the plants: current status and future prospectus. J Plant Biochem Biotechnol 32:211–224
36. Aziz Y, Shah GA, Rashid MI (2019) ZnO nanoparticles and zeolite influence soil nutrient availability but do not affect herbage nitrogen uptake from biogas slurry. Chemosphere 216:564–575
37. Azmat A, Tanveer Y, Yasmin H, Hassan MN, Shahzad A, Reddy M, Ahmad A (2022) Coactive role of zinc oxide nanoparticles and plant growth promoting rhizobacteria for mitigation of synchronized effects of heat and drought stress in wheat plants. Chemosphere 297:133982
38. Badawy SA, Zayed BA, Bassiouni SMA, Mahdi AHA, Majrashi A, Ali EF, Seleiman MF (2021) Influence of nano silicon and nano selenium on root characters, growth, ion selectivity, yield, and yield components of rice (*Oryza sativa* L.) under salinity conditions. Plants 10:1657
39. Bagheri Faradonbeh HR, Bistgani ZE, Barker AV (2022) Tuber yield and physiological characteristics of potato under irrigation and fertilizer application. Commun Soil Sci Plant Anal 53(11):1432–1443
40. Bano H, Athar HUR, Zafar ZU, Ogbaga CC, Ashraf M (2021) Peroxidase activity and operation of photo-protective component of NPQ play key roles in drought tolerance of mung bean [*Vigna radiata* (L.) Wilcziek]. Physiologia Plant 172(2):603–614
41. Bidabadi SS, Sabbatini P, VanderWeide J (2023) Iron oxide (Fe2O3) nanoparticles alleviate PEG-simulated drought stress in grape (*Vitis vinifera* L.) plants by regulating leaf antioxidants. Sci Hortic 312:111847
42. Bisht S, Sharma V, Kumari N (2022) Biosynthesized magnetite nanoparticles from *Polyalthia longifolia* leaves improve photosynthetic performance and yield of *Trigonella foenum-graecum* under drought stress. Plant Stress 5:100090

43. Borisev M, Borisev I, Zupunski M, Arsenov D, Pajevic S, Curcic Z, Vasin J, Djordjevic A (2016) Drought impact is alleviated in sugar beets (*Beta vulgaris* L.) by foliar application of fullerenol nanoparticles. PLoS ONE 11(11):e0166248
44. Bulut F, Tetiker AT, Celikkol A, Yılmaz A, Ballica B (2023) Low antioxidant enzyme levels and oxidative stress in laryngopharyngeal reflux (LPR) patients. J Voice 37(6):924–931
45. Chen M, Jiao SQ, Xie L, Geng X, Qi S, Fan J, Cheng S, Shi J, Cao X (2024) Integrated physiological, transcriptomic, and metabolomic analyses of drought stress alleviation in *Ehretia macrophylla* Wall. seedlings by SiO_2 NPs (silica nanoparticles). Front Plant Sci 15:1260140
46. Chichiricco G, Poma A (2015) Penetration and toxicity of nanomaterials in higher plants. Nanomater 5:851–873
47. Cunningham FJ, Goh NS, Demirer GS, Matos JL, Landry MP (2018) Nanoparticle-mediated delivery towards advancing plant genetic engineering. Trends Biotechnol 36:882–897
48. Dhakate P, Kandhol N, Raturi G, Ray P, Bhardwaj A, Srivastava A, Kaushal L, Singh A, Pandey S, Chauhan DK, Dubey NK (2022) Silicon nanoforms in crop improvement and stress management. Chemosphere 305:135165
49. Dimkpa CO, White JC, Elmer WH, Gardea-Torresdey J (2017) Nanoparticle and ionic Zn promote nutrient loading of sorghum grain under low NPK fertilization. J Agri Food Chem 65(39):8552–8559
50. Dinler BS, Cetinkaya H, Koc FN, Gül V, Sefaoğlu F (2024) Effects of titanium dioxide nanoparticles against salt and heat stress in safflower cultivars. Acta Bot Bras 38:e20230136
51. Djanaguiraman M, Nair R, Giraldo JP, Prasad PVV (2018) Cerium oxide Nanoparticles decrease drought-induced oxidative damage in sorghum leading to higher photosynthesis and grain yield. ACS Omega 3(10):14406–14416
52. Djanaguiraman M, Vidhya Bharathi KS, Raghu R, Jeyakumar P (2024) Sorghum drought tolerance is enhanced by cerium oxide nanoparticles via stomatal regulation and osmolyte accumulation. Plant Physiol Biochem 212:108733
53. Dubchak S, Ogar A, Mietelski JW, Turnau K (2010) Influence of silver and titanium nanoparticles on arbuscular mycorrhiza colonization and accumulation of radiocaesium in Helianthus annuus. Span J Agric Res 8:103–108
54. Eichert T, Goldbach HE (2008) Equivalent pore radii of hydrophilic foliar uptake routes in stomatous and astomatous leaf surfaces-further evidence for a stomatal pathway. Physiol Plant 132:491–502
55. Eichert T, Kurtz A, Steiner U, Goldbach HE (2008) Size exclusion limits and lateral heterogeneity of the stomatal foliar uptake pathway for aqueous solutes and water-suspended nanoparticles. Physiol Plant 134:151–160
56. Ejaz S, Fahad S, Anjum MA, Nawaz A, Naz S, Hussain S, Ahmad S (2020) Role of osmolytes in the mechanisms of antioxidant defense of plants. Sus Agric Rev 39:95–117
57. El-Mogy MM, Garchery C, Stevens R (2018) Irrigation with salt water affects growth, yield, fruit quality, storability and marker-gene expression in cherry tomato. Acta Agric Scand Sect B Soil Plant Sci 68:727–737
58. El-Ramady H, Abdalla N, Sari D, Ferroudj A, Muthu A, Prokisch J, Fawzy ZF, Brevik EC, Solberg SO (2023) Nanofarming: promising solutions for the future of the global agricultural industry. Agron 13:1600
59. El-Saadony MT, Abd El-Hack ME, Taha AE, Fouda M, Ajarem JS, N Maodaa S, Allam AA, Elshaer N (2020) Ecofriendly synthesis and insecticidal application of copper nanoparticles against the storage pest *Tribolium castaneum*. Nanomater 10:587
60. El-Saadony MT, Sitohy MZ, Ramadan MF, Saad AM (2021) Green nanotechnology for preserving and enriching yogurt with biologically available iron (II). Innov Food Sci Emerg Technol 69:102645
61. El-Saadony MT, Saad AM, Najjar AA, Alzahrani SO, Alkhatib FM, Shafi ME, Selem E, Desoky ESM, Fouda SE, El-Tahan AM, Hassan MA (2021b) The use of biological selenium nanoparticles to suppress *Triticum aestivum* L. crown and root and rot diseases induced by Fusarium species and improve yield under drought and heat stress. Saud J Biol Sci 28(8):4461–4471

62. El-Saadony MT, Saad AM, Soliman SM, Salem HM, Desoky ESM, Babalghith AO, El-Tahan AM, Ibrahim OM, Ebrahim AA, Abd El-Mageed TA, Elrys AS (2022) Role of nanoparticles in enhancing crop tolerance to abiotic stress: a comprehensive review. Front Plant Sci 13:946717
63. El-Shafey NM, Avramova V, Beemster GT, Korany SM, AbdElgawad H (2023) B2O3 nanoparticles alleviate salt stress in maize leaf growth zones by enhancing photosynthesis and maintaining mineral and redox status. Physiol Plant 175(5):e14033
64. Elsheery NI, Helaly MN, El-Hoseiny HM, Alam-Eldein SM (2020) Zinc oxide and silicon nanoparticles to improve the resistance mechanism and annual productivity of salt-stressed mango trees. Agron 10(4):558
65. Elsheery NI, Sunoj VSJ, Wen Y, Zhu JJ, Muralidharan G, Cao KF (2020) Foliar application of nanoparticles mitigates the chilling effect on photosynthesis and photoprotection in sugarcane. Plant Physiol Biochem 149:50–60
66. El-Sheikh MA, El-Nagar MY, Abogadallah GM (2021) Ameliorative impact of TiO2 nanoparticles on maize seedlings grown under salt stress. J Plant Growth Regul 1–18
67. El-Temsahy SA, Almaghrabi OA, Elsayed HE (2019) Zinc oxide nanoparticles enhance the growth and the uptake of potassium and calcium in salt-stressed wheat (*Triticum aestivum* L.) seedlings. J Plant Interact 14(1):442–452
68. El-Zohri M, Al-Wadaani NA, Bafeel SO (2021) Foliar sprayed green zinc oxide nanoparticles mitigate drought-induced oxidative stress in tomato. Plants 10(11):2400
69. Emamverdian A, Ding Y, Mokhberdoran F, Xie Y (2015) Heavy metal stress and some mechanisms of plant defense response. Sci World J 756120
70. Etesami H, Fatemi H, Rizwan M (2021) Interactions of nanoparticles and salinity stress at physiological, biochemical and molecular levels in plants: a review. Ecotoxicol Environ Saf 225:112769
71. Fadiji AE, Santoyo G, Yadav AN, Babalola OO (2022) Efforts towards overcoming drought stress in crops: revisiting the mechanisms employed by plant growth-promoting bacteria. Front Microbiol 13:962427
72. Farhangi-Abriz S, Torabian S (2018) Nano-silicon alters antioxidant activities of soybean seedlings under salt toxicity. Protoplasma 255:953–962
73. Fernández-Escobar R (2019) Olive nutritional status and tolerance to biotic and abiotic stresses. Front Plant Sci 10:461742
74. Fincheira P, Tortella G, Duran N, Seabra AB, Rubilar O (2020) Current applications of nanotechnology to develop plant growth inducer agents as an innovation strategy. Crit Rev Biotechnol 40(1):15–30
75. Forni C, Duca D, Glick BR (2017) Mechanisms of plant response to salt and drought stress and their alteration by rhizobacteria. Plant Soil 410:335–356
76. Fouda HM, Saied E, Abdelmouty ES, Osman MS (2024) Ameliorative role of copper nanoparticle in alleviating salt-induced oxidative stress in fenugreek (*Trigonella foenum-graecum* L.) plants. Biocatal Agric Biotechnol 57:103095
77. Gao X, Xiong H, Li X, Liu L, Wang X (2016) Chitosan nanoparticles as a delivery system for ammonium improves nitrogen uptake and use efficiency in rice. J Plant Growth Regul 35(3):695–708
78. Ghabel VK, Karamian R (2020) Effects of TiO2 nanoparticles and spermine on antioxidant responses of *Glycyrrhiza glabra* L. to cold stress. Acta Bot Croat 79:137–147
79. Ghafariyan MH, Tavakoli H, Fotovat R, Fathi-Achachlouei B, Salehi M (2020) The effect of iron oxide nanoparticles on the photosynthesis, chlorophyll content and nutrient uptake of tomato plants under salt stress. Ecotoxicol Environ Saf 194:110428
80. Ghafariyan MH, Malakouti MJ, Dadpour MR, Stroeve P, Mahmoudi M (2013) Effects of magnetite nanoparticles on soybean chlorophyll. Environ Sci Technol 47:10645–10652
81. Ghani MI, Saleem S, Rather SA, Rehmani MS, Alamri S, Rajput VD, Kalaji HM, Saleem N, Sial TA, Liu M (2022) Foliar application of zinc oxide nanoparticles: an effective strategy to mitigate drought stress in cucumber seedling by modulating antioxidant defense system and osmolytes accumulation. Chemosphere 289:133202

82. Ghasemlou F, Amiri H, Karamian R, Mirzaie-asl A (2019) Alleviation of the effects of on drought stress *Verbascum nudicuale* by methyl jasmonate and titanium dioxide nanoparticles. Plant Physiol 9:2911–2920
83. Gholami S, Dehaghi MA, Rezazadeh A, Naji AM (2022) Seed germination and physiological responses of quinoa to selenium priming under drought stress. Bragantia 81:e0722
84. Gohari G, Mohammadi A, Akbari A, Panahirad S, Dadpour MR, Fotopoulos V, Kimura S (2020) Titanium dioxide nanoparticles (TiO2 NPs) promote growth and ameliorate salinity stress effects on essential oil profile and biochemical attributes of *Dracocephalum moldavica.* Sci Rep 10(1):912
85. Gupta A, Rayeen F, Mishra R, Tripathi M, Pathak N (2023) Nanotechnology applications in sustainable agriculture: an emerging ecofriendly approach. Plant Nano Biol 4:100033
86. Haq RU, Kausar A, Hussain S, Javed T, Zafar S, Anwar S, Hussain S, Zahra N, Saqib M (2023) Zinc oxide nanoparticles as potential hallmarks for enhancing drought stress tolerance in wheat seedlings. Plant Physiol Biochem 195:341–350
87. Hasanpour H, Maali-Amir R, Zeinali H (2015) Effect of TiO2 nanoparticles on metabolic limitations to photosynthesis under cold in chickpea. Russ J Plant Physiol 62:779–787
88. Hasanuzzaman M, Nahar K, Alam MM, Roychowdhury R, Fujita M (2013) Physiological, biochemical, and molecular mechanisms of heat stress tolerance in plants. Int J Mol Sci 14:9643–9684
89. Hasanuzzaman M, Nahar K, Alam MM, Rahman A, Mahmud JA, Suzuki T, Fujita M (2018) Exogenous titanium dioxide modulates antioxidant defense system and improves drought tolerance in *Brassica napus* L. seedlings. Biomed Res Int
90. Hassan NS, El Din TAS, Hendawey MH, Borai IH, Mahdi AA (2018) Magnetite and zinc oxide nanoparticles alleviated heat stress in wheat plants. Curr Nanomater 3:32–43
91. He M, He CQ, Ding NZ (2018) Abiotic stresses: general defenses of land plants and chances for engineering multi-stress tolerance. Front Plant Sci 9:1771
92. Hnilickova H, Kraus K, Vachova P, Hnilicka F (2021) Salinity stress affects photosynthesis, malondialdehyde formation, and proline content in Portulaca oleracea L. Plants 10:845
93. Hu P, An J, Faulkner MM, Wu H, Li Z, Tian X, Giraldo JP (2020) Nanoparticle charge and size control foliar delivery efficiency to plant cells and organelles. ACS Nano 14:7970–7986
94. Huang Q, Ayyaz A, Farooq MA, Zhang K, Chen W, Hannan F, Sun Y, Shahzad K, Ali B, Zhou W (2024) Silicon dioxide nanoparticles enhance plant growth, photosynthetic performance, and antioxidants defence machinery through suppressing chromium uptake in Brassica napus L. Environ Pollut 342:123013
95. Hubbard JD, Lui A, Landry MP (2020) Multiscale and multidisciplinary approach to understanding nanoparticle transport in plants. Curr Opin Chem Eng 30:135–143
96. Hussain I, Singh A, Singh N, Singh P (2019) Plant-nanoceria interaction: toxicity, accumulation, translocation and biotransformation. S Afr J Bot 121:239–247
97. Iannone MF, Groppa MD, Zawoznik MS, Coral DF, van Raap MBF, Benavides MP (2021) Magnetite nanoparticles coated with citric acid are not phytotoxic and stimulate soybean and alfalfa growth. Ecotoxicol Environ Saf 211:111942
98. Ijaz U, Ahmed T, Rizwan M, Noman M, Shah AA, Azeem F, Hesham F, Alharby AA, Bamagoos BM, Alharbi AS (2023) Rice straw based silicon nanoparticles improve morphological and nutrient profile of rice plants under salinity stress by triggering physiological and genetic repair mechanisms. Plant Physiol Biochem 201:107788
99. Ilyas M, Nisar M, Khan N, Hazrat A, Khan AH, Hayat K, Fahad S, Khan A, Ullah A (2021) Drought tolerance strategies in plants: a mechanistic approach. J Plant Growth Reg 40:926–944
100. Iqbal M, Raja NI, Mashwani ZUR, Hussain M, Ejaz M, Yasmeen F (2019) Effect of silver nanoparticles on growth of wheat under heat stress. Iran J Sci Technol Trans A Sci 43:387–395
101. Iqbal M, Raja NI, Mashwani ZUR, Yasmeen F, Hussain M, Ejaz M, Abasi F, Ehsan M, Ikram M, Proćków J (2024) Insight into carbohydrate metabolism, protein quantification and mineral regulation in wheat (*Triticum aestivum* L.) by the action of green synthesized silver nanoparticles (AgNPs) against heat stress. J Biomol Struct Dyn 1–15

102. Verma N, Sao P, Srivastava A, Singh S (2021) Physiological and molecular responses to drought, submergence and excessive watering in plants. In harsh environment and plant resilience, pp 305–321
103. Jalil SU, Ansari MI (2019) Nanoparticles and abiotic stress tolerance in plants: synthesis, action, and signaling mechanisms. In: Plant signaling molecules. Woodhead Publishing, Sawston UK, pp 549–561
104. Jan N, Majeed U, Andrabi KI, John R (2018) Cold stress modulates osmolytes and antioxidant system in *Calendula officinalis*. Acta Physiol Plant 40:73
105. Janmohammadi M, Amanzadeh T, Sabaghnia N, Ion V (2016) Effect of nano-silicon foliar application on safflower growth under organic and inorganic fertilizer regimes. Bot Lith 22:53–64
106. Javed T, Li Y, Rashid S, Li F, Hu Q, Feng H, Chen X, Ahmad S, Liu F, Pulatov B (2021) Performance and relationship of four different agricultural drought indices for drought monitoring in China's mainland using remote sensing data. Sci Total Environ 759:143530
107. Javed Z, Tripathi GD, Mishra M, Gattupalli M, Dashora K (2022) Cow dung extract mediated green synthesis of zinc oxide nanoparticles for agricultural applications. Sci Rep 12(1):20371
108. Kadyampakeni DM, Chinyukwi T (2021) Are macronutrients and micronutrients therapeutic for restoring performance of trees affected by citrus greening? A discussion of current practices and future research opportunities. J Plant Nutr 44(19):2949–2969
109. Kandhol N, Jain M, Tripathi DK (2022) Nanoparticles as potential hallmarks of drought stress tolerance in plants. Physiol Plant 174(2):e13665
110. Karami MS, Maali AR, Kazemi SS (2018) Effect of cold stress on oxidative damage and mitochondrial respiratory properties in chickpea. Plant Physiol Biochem 122:31–39
111. Karamian R, Ghasemlou F, Amiri H (2020) Physiological evaluation of drought stress tolerance and recovery in *Verbascum sinuatum* plants treated with methyl jasmonate, salicylic acid and titanium dioxide nanoparticles. Plant Biosyst 154(3):277–287
112. Kardavan Ghabel V, Karamian R (2020) Effects of TiO2 nanoparticles and spermine on antioxidant responses of *Glycyrrhiza glabra* L. to cold stress. Acta Bot Croat 79:137–147
113. Kareem HA, Saleem MF, Saleem S, Rather SA, Wani SH, Siddiqui MH, Alamri S, Kumar R, Gaikwad NB, Guo Z, Niu J (2022) Zinc oxide nanoparticles interplay with physiological and biochemical attributes in terminal heat stress alleviation in mungbean (*Vigna radiata* L.). Front Plant Sci 13:842349
114. Khalid MF, Iqbal Khan R, Jawaid MZ, Shafqat W, Hussain S, Ahmed T, Rizwan M, Ercisli S, Pop OL, Alina Marc R (2022) Nanoparticles: the plant saviour under abiotic stresses. Nanomater 12(21):S3915
115. Khan I, Awan SA, Rizwan M, Akram MA, Zia-ur-Rehman M, Wang X, Zhang X, Huang L (2023) Physiological and transcriptome analyses demonstrate the silver nanoparticles mediated alleviation of salt stress in pearl millet (*Pennisetum glaucum* L). Environ Pollut 318:120863
116. Khan I, Saeed K, Khan I (2019) Nanoparticles: properties, applications and toxicities. Arab J Chem 12:908–931
117. Khan I, Raza MA, Awan SA, Shah GA, Rizwan M, Ali B, Tariq R, Hassan MJ, Alyemeni MN, Brestic M, Zhang X (2020) Amelioration of salt induced toxicity in pearl millet by seed priming with silver nanoparticles (AgNPs): the oxidative damage, antioxidant enzymes and ions uptake are major determinants of salt tolerant capacity. Plant Physiol Biochem 156:221–232
118. Khan MI, Rizwan M, Imran Q, Faisal M, Saqib M, Ahmad M (2021) Emerging role of omics approaches in understanding abiotic stress tolerance mechanisms in crops. Plants 10(2):271
119. Khan Z, Upadhyaya H (2019) Impact of nanoparticles on abiotic stress responses in plants: an overview. In: Tripathi DK, Ahmad P, Sharma S, Chauhan DK, Dubey NK (eds) Nanomaterials in plants, algae and microorganisms. Academic Press, Cambridge, MA, pp 305–322
120. Kim YH, Khan AL, Waqas M, Lee IJ (2017) Silicon regulates antioxidant activities of crop plants under abiotic-induced oxidative stress: a review. Front Plant Sci 8:510

121. Kopittke PM, Lombi E, Wang P, Schjoerring JK, Husted S (2019) Nanomaterials as fertilizers for improving plant mineral nutrition and environmental outcomes. Environ Sci Nano 6:3513–3524
122. Kumar A, Kaushik P (2021) Heat stress and its impact on plant function: an update
123. Kumar R, Sagar V, Verma VC, Kumari M, Gujjar RS, Goswami SK, Kumar Jha S, Pandey H, Dubey AK, Srivastava S, Singh SP, Mall AK, Pathak AD, Singh H, Jha PK, Prasad PVV (2023) Drought and salinity stresses induced physio-biochemical changes in sugarcane: an overview of tolerance mechanism and mitigating approaches. Front Plant Sci 14:1225234
124. Kumar V, Yadav SK, Yadav SK (2021) Silver nanoparticles alleviate salt stress in tomato plants by regulating growth, photosynthesis, and antioxidant metabolism. J Plant Growth Regul 40(3):1233–1248
125. Lal MK, Tiwari RK, Gahlaut V et al (2022) Physiological and molecular insights on wheat responses to heat stress. Plant Cell Rep 41:501–518
126. Larue C, Castillo-Michel H, Sobanska S, Cécillon L, Bureau S, Barthès V et al (2014) Foliar exposure of the crop Lactuca sativa to silver nanoparticles: evidence for internalization and changes in Ag speciation. J Hazard Mater 264:98–106
127. Latha P, Sudhakar P, Prasad TN (2022) Enhanced activity of antioxidant enzymes by foliar spray of nanoscale zinc oxide under drought stress conditions in peanut (*Arachis hypogaea* L.). Legume Res 1:6
128. Lhissoui R, El Harti A, Chokmani K (2014) Mapping soil salinity in irrigated land using optical remote sensing data. Eurasian J Soil Sci 3:82–88
129. Li Y, Liang L, Li W, Ashraf U, Ma L, Tang X, Pan S, Tian H, Mo Z (2021) ZnO nanoparticle-based seed priming modulates early growth and enhances physio-biochemical and metabolic profiles of fragrant rice against cadmium toxicity. J Nanobiotechnol 19:1–19
130. Liang Q, Dong W, Wang F, Wang W, Zhang J, Liu X (2021) *Ficus hirta* Vahl. promotes antioxidant enzyme activity under ammonia stress by inhibiting miR-2765 expression in *Penaeus vannamei*. Ecotoxicol Environm Saf 228:112989
131. Liang Y, Liu H, Fu Y, Li P, Li S, Gao Y (2023) Regulatory effects of silicon nanoparticles on the growth and photosynthesis of cotton seedlings under salt and low-temperature dual stress. BMC Plant Biol 23(1):504
132. Lin D, Xing B (2007) Phytotoxicity of nanoparticles: inhibition of seed germination and root growth. Environ Pollut 150:243–250
133. Linh TM, Mai NC, Hoe PT, Lein LQ, Ban NK, Hien LTT, Chau NH, Van NT (2020) Metal-based nanoparticles enhance drought tolerance in soybean. J Nanomater 1:4056563
134. Lipiec J, Doussan C, Nosalewicz A, Kondracka K (2013) Effect of drought and heat stresses on plant growth and yield: a review. Int Agrophy 27:463–477
135. Liu FY, Xiong FX, Fan YK, Li J, Wang HZ, Xing GM, He R (2016) Facile and scalable fabrication engineering of fullerenol nanoparticles by improved alkaline-oxidation approach and its antioxidant potential in maize. J Nanoparticle Res 18:338
136. Lo'Ay AA, Ameer NM (2019) Performance of calcium nanoparticles blending with ascorbic acid and alleviation internal browning of 'Hindi Be-Sennara'mango fruit at a low temperature. Scient Hortic 254:199–207
137. Lowry GV, Avellan A, Gilbertson LM (2019) Opportunities and challenges for nanotechnology in the agritech revolution. Nat Nanotechnol 14:517–522
138. Lv J, Christie P, Zhang S (2019) Uptake, translocation, and transformation of metal-based nanoparticles in plants: recent advances and methodological challenges. Environ Sci Nano 6:41–59
139. Ma Y, He X, Zhang P, Zhang Z, Ding Y, Zhang J, Wang G, Xie C, Lu W, Zhang J, Zheng L (2017) Xylem and phloem based transport of CeO2 nanoparticles in hydroponic cucumber plants. Environ Sci Technol 51:5215–5221
140. Machado RMA, Serralheiro RP (2017) Soil salinity: Effect on vegetable crop growth. Management practices to prevent and mitigate soil salinization. Hortic 3:30
141. Mahawar L, Živčák M, Barboricova M, Kovár M, Filaček A, Ferencova J, Vysoká DM, Brestič M (2024) Effect of copper oxide and zinc oxide nanoparticles on photosynthesis and physiology of *Raphanus sativus* L. under salinity stress. Plant Physiol Biochem 206:108281

142. Mahmoud AWM, Rashad HM, Esmail SE, Alsamadany H, Abdeldaym EA (2023) Application of silicon, zinc, and zeolite nanoparticles—a tool to enhance drought stress tolerance in coriander plants for better growth performance and productivity. Plants 12(15):2838
143. Mahmoudi R, Razavi F, Rabiei V, Palou L, Gohari G (2022) Postharvest chitosan-arginine nanoparticles application ameliorates chilling injury in plum fruit during cold storage by enhancing ROS scavenging system activity. BMC Plant Biol 22(1):555
144. Mai Y, Ren Y, Deng S, Ashraf U, Tang X, Duan M, Mo Z (2024) Influence of ZnO nanoparticles on early growth stage of fragrant rice at low temperature (LT) stress. J Soil Sci Plant Nutr 1–17
145. Manzoor N, Ali L, Ahmed T, Noman M, Adrees M, Shahid MS, Ogunyemi SO, Radwan KS, Wang G, Zaki HE (2022) Recent advancements and development in nano-enabled agriculture for improving abiotic stress tolerance in plants. Front Plant Sci 13:951752
146. Medina-Puche L, Lozano-Durán R (2022) Plasma membrane-to-organelle communication in plant stress signaling. Curr Opin Plant Biol 69:102269
147. Megha S, Basu UNK (2017) Regulation of low-temperature stress in plants by microRNAs. Plant Cell Environ 41:12956
148. Mogazy AM, Hanafy RS (2022) Foliar spray of biosynthesized zinc oxide nanoparticles alleviate salinity stress effect on *Vicia faba* plants. J Soil Sci Plant Nutr 22(2):2647–2662
149. Mukarram M, Khan MMA, Kurjak D, Lux A, Corpas FJ (2023) Silicon nanoparticles (SiNPs) restore photosynthesis and essential oil content by upgrading enzymatic antioxidant metabolism in lemongrass (*Cymbopogon flexuosus*) under salt stress. Front Plant Sci 14:1116769
150. Mustafa H, Ilyas N, Akhtar N, Raja NI, Zainab T, Shah T, Ahmad A, Ahmad P (2021) Biosynthesis and characterization of titanium dioxide nanoparticles and its effects along with calcium phosphate on physicochemical attributes of wheat under drought stress. Ecotoxicol Environ Saf 223:112519
151. Nagajyoti PC, Lee KD, Sreekanth TVM (2010) Heavy metals, occurrence and toxicity for plants: a review. Environ Chem Lett 8:199–216
152. Nasrallah AK, Kheder AA, Kord MA, Fouad AS, El-Mogy MM, Atia MAM (2022) Mitigation of salinity stress effects on broad bean productivity using calcium phosphate nanoparticles application. Hortic 8:75
153. Nayab N, Alam MA (2024) Chitosan nanoparticles: a promising stimulant for augmenting cold stress resistance in safed velchi cultivars of banana plants. J Himalayan Eco Sustain Dev 18:0973–7502
154. Neal CA (2000) Fertilizing trees and shrubs. University of New Hampshire, Cooperative Extension
155. Nejatzadeh F (2021) Effect of silver nanoparticles on salt tolerance of *Satureja hortensis* l. during in vitro and in vivo germination tests. Heliyon 7
156. NOAA (2023) National integrated drought information system, Drought Basics. https://www.drought.gov/what-is-drought/droughtbasics.
157. Noohpisheh Z, Amiri H, Mohammadi A, Farhadi S (2021) Effect of the foliar application of zinc oxide nanoparticles on some biochemical and physiological parameters of *Trigonella foenum-graecum* under salinity stress. Plant Biosyst Int J Deal All Asp Plant Biol 155:267–280
158. Omar AA, Heikal YM, Zayed EM, Shamseldin SA, Salama YE, Amer KE, Basuoni MM, Abd Ellatif S, Mohamed AH (2023) Conferring of drought and heat stress tolerance in wheat (*Triticum aestivum* L.) genotypes and their response to selenium nanoparticles application. Nanomater 13(6):998
159. Ou LJ, Wei G, Zhang ZQ, Dai XZ, Zou XX (2015) Effects of low temperature and low irradiance on the physiological characteristics and related gene expression of different pepper species. Photosynthetica 53:85–94
160. Palmqvist NGM, Seisenbaeva GA, Svedlindh P, Kessler VG (2017) Maghemite nanoparticles acts as nanozymes, improving growth and abiotic stress tolerance in *Brassica napus*. Nanoscale Res Lett 12(631):1–9

161. Paramo LA, Feregrino-Pérez AA, Guevara R, Mendoza S, Esquivel K (2020) Nanoparticles in agroindustry: applications, toxicity, challenges, and trends. Nanomater 10:1654
162. Pereira ADES, Oliveira HC, Fraceto LF, Santaella C (2021) Nanotechnology potential in seed priming for sustainable agriculture. Nanomater 11:267
163. Perez-Labrada F, López-Vargas ER, Ortega-Ortiz H, Cadenas-Pliego G, Benavides-Mendoza A, Juárez-Maldonado A (2019) Responses of tomato plants under saline stress to foliar application of copper nanoparticles. Plants 8:151
164. Pimentel D, Berger B, Filiberto D, Newton M (2004) Water resources: agricultural and environmental issues. Biosci 54:909–918
165. Qi J, Sun S, Yang L, Li M, Ma F, Zou Y (2019) Potassium uptake and transport in apple roots under drought stress. Hortic Plant J 5(1):10–16
166. Qi M, Liu Y, Li T (2013) Nano-TiO2 improve the photosynthesis of tomato leaves under mild heat stress. Biol Trace Element Res 156:323–328
167. Qian J, Shan R, Shi Y, Li H, Xue L, Song Y, Zhao T, Zhu S, Chen J, Jiang M (2024) Zinc oxide nanoparticles alleviate salt stress in cotton (*Gossypium hirsutum* l.) by adjusting Na+/K+ ratio and antioxidative ability. Life 14(5):595
168. Rahimi M, Kordrostami M, Mohamadhasani F, Chaeikar SS (2021) Antioxidant gene expression analysis and evaluation of total phenol content and oxygen-scavenging system in tea accessions under normal and drought stress conditions. BMC Plant Biol 21(1):1–2
169. Rahman H, Haque KS, Khan ZH (2021) A review on application of controlled released fertilizers influencing the sustainable agricultural production: a Cleaner production process. Environ Tech Innov 23:101697
170. Rajanna GA, Suman A, Venkatesh P (2023) Mitigating drought stress effects in arid and semi-arid agroecosystems through bio-irrigation strategies—a review. Sustain 15(4):3542
171. Rajput VD, Minkina T, Kumari A, Singh VK, Verma KK, Mandzhieva S, Sushkova S, Srivastava S, Keswani C (2021) Coping with the challenges of abiotic stress in plants: new dimensions in the field application of nanoparticles. Plants 10:1221
172. Raliya R, Tarafdar JC (2013) ZnO nanoparticle biosynthesis and its effect on phosphorous-mobilizing enzyme secretion and gum contents in clusterbean (*Cyamopsis tetragonoloba* L.). Agric Res 2:48–57
173. Ranjan A, Rajput VD, Minkina T, Bauer T, Chauhan A, Jindal T (2021) Nanoparticles induced stress and toxicity in plants. Environ Nanotechnol Monit Manage 15:100457
174. Rasheed A, Li H, Tahir MM, Mahmood A, Nawaz M, Shah AN, Aslam MT, Negm S, Moustafa M, Hassan MU, Wu Z (2022) The role of nanoparticles in plant biochemical, physiological, and molecular responses under drought stress: a review. Front Plant Sci 13:976179
175. Rastogi A, Tripathi DK, Yadav S, Chauhan DK, Zivcak M, Ghorbanpour M, El-Sheey NI, Brestic M (2019) Application of silicon nanoparticles in agriculture. Biotechnol 9:90
176. Rossi L, Fedenia LN, Sharifan H, Ma X, Lombardini L (2019) Effects of foliar application of zinc sulfate and zinc nanoparticles in coffee (*Coffea arabica* L.) plants. Plant Physiol Biochemi 135:160–166
177. Rossi L, Zhang W, Lombardini L, Ma X (2016) The impact of cerium oxide nanoparticles on the salt stress responses of Brassica napus L. Environ Pollut 219:28–36
178. Rossi L, Zhang W, Ma X (2017) Cerium oxide nanoparticles alter the salt stress tolerance of Brassica napus L. by modifying the formation of root apoplastic barriers. Environ Pollut 229:132–138
179. Safari S, Gholami M, Fard JR, Ghorbanpour M (2020) Chitosan nanoparticles improve growth and physiological responses of rice (*Oryza sativa* L.) under salt stress. Ecotoxicol Environ Saf 190:110091
180. Salachna P, Byczyńska A, Zawadzińska A, Piechocki R, Mizielińska M (2019) Stimulatory effect of silver nanoparticles on the growth and flowering of potted oriental lilies. Agron 9:610
181. Salama DM, El-Aziz MA, Rizk FA, Elwahed MA (2021) Applications of nanotechnology on vegetable crops. Chemosphere 266:129026
182. Sarkar D, Singh S, Parihar M, Rakshit A (2021) Seed bio-priming with microbial inoculants: a tailored approach towards improved crop performance, nutritional security, and agricultural sustainability for smallholder farmers. Curr Res Environ Sustain 3:100093

183. Sarkar RD, Kalita MC (2022) Green synthesized Se nanoparticle-mediated alleviation of salt stress in field mustard, TS-36 variety. J Biotechnol 359:95–107
184. Sayed EG, Desoukey SF, Desouky AF, Farag MF, EL-kholy R I, Azoz SN (2024) Synergistic influence of arbuscular mycorrhizal fungi inoculation with nanoparticle foliar application enhances chili (*Capsicum annuum* L.) antioxidant enzymes, anatomical characteristics, and productivity under cold-stress conditions. Plants 13(4):517
185. Scudiero E, Skaggs TH, Corwin DL (2016) Comparative regional-scale soil salinity assessment with near-ground apparent electrical conductivity and remote sensing canopy reflectance. Ecol Indic 70:276–284
186. Seleiman MF, Al-Suhaibani N, Ali N, Akmal M, Alotaibi M, Refay Y, Dindaroglu T, Abdul-Wajid HH, Battaglia ML (2021) Drought stress impacts on plants and different approaches to alleviate its adverse effects. Plants (Basel) 10(2):259
187. Seleiman MF, Almutairi KF, Alotaibi M, Shami A, Alhammad BA, Battaglia ML (2021) Nano-fertilization as an emerging fertilization technique: why can modern agriculture benefit from its use? Plants 10:2
188. Seleiman MF, Alotaibi MA, Alhammad BA, Alharbi BM, Refay Y, Badawy SA (2020) Effects of ZnO nanoparticles and biochar of rice straw and cow manure on characteristics of contaminated soil and sunflower productivity, oil quality, and heavy metals uptake. Agron 10:790
189. Seleiman MF, Almutairi KF, Alotaibi M, Shami A, Alhammad BA, Battaglia ML (2020) Nano-fertilization as an emerging fertilization technique: why can modern agriculture benefit from its use? Plants 10:2
190. Semida WM, Abdelkhalik A, Mohamed GF, Abd El-Mageed TA, Abd El-Mageed SA, Rady MM, Ali EF (2021) Foliar application of zinc oxide nanoparticles promotes drought stress tolerance in eggplant (*Solanum melongena* L.). Plants 10(2):421
191. Shafqat W, Jaskani MJ, Maqbool R, Chattha WS, Ali Z, Naqvi SA, Haider MS, Khan IA Vincent CI (2021) Heat shock protein and aquaporin expression enhance water conserving behavior of citrus under water deficits and high temperature conditions. Environ Exp Bot 181:104270
192. Shah T, Latif S, Saeed F, Ali I, Ullah S, Alsahli AA, Jan S, Ahmad P (2021) Seed priming with titanium dioxide nanoparticles enhances seed vigor, leaf water status, and antioxidant enzyme activities in maize (*Zea mays* L.) under salinity stress. J King Saud Univ Sci 33(1):101207
193. Sharma P, Lakra N, Goyal A, Ahlawat YK, Zaid A, Siddique KHM (2023) Drought and heat stress mediated activation of lipid signaling in plants: a critical review. Front Plant Sci 14:1216835
194. Shen M, Liu S, Jiang C, Zhang T, Chen W (2023) Recent advances in stimuli-response mechanisms of nano-enabled controlled-release fertilizers and pesticides. Eco-Environ Health 2(3):161–175
195. Siddiqi KS, Husen A (2016) Engineered gold nanoparticles and plant adaptation potential. Nanoscale Res Lett 11:400
196. Soares JC, Santos CS, Carvalho SMP, Pintado MM, Vasconcelos MW (2019) Preserving the nutritional quality of crop plants under a changing climate: importance and strategies. Plant Soil 443:1–26
197. Solomon PS, Tan KC, Oliver RP (2003) The nutrient supply of pathogenic fungi; a fertile field for study. Mol Plant Pathol 4(3):203–210
198. Song Y, Zheng C, Li S, Chen J, Jiang M (2023) Chitosan-magnesium oxide nanoparticles improve salinity tolerance in rice (*Oryza sativa* L.). ACS Appl Mater Interfaces 15(17):20649–20660
199. Song Y, Jiang M, Zhang H, Li R (2021) Zinc oxide nanoparticles alleviate chilling stress in rice (*Oryza Sativa* L.) by regulating antioxidative system and chilling response transcription factors. Mol 26:2196
200. Soni AT, Rookes JE, Arya SS (2023) Chitosan nanoparticles as seed priming agents to alleviate salinity stress in rice (*Oryza sativa* L.) seedlings. Polysaccharides 4(2):129–141

201. Spielman-Sun E, Lombi E, Donner E, Howard D, Unrine JM, Lowry GV (2017) Impact of surface charge on cerium oxide nanoparticle uptake and translocation by wheat (*Triticum aestivum*). Environ Sci Technol 51:7361–7368
202. Sreelakshmi B, Induja S, Adarsh P, Rahul H, Arya S, Aswana S, Haripriya R, Aswathy B, Manoj P, Vishnudasan D (2020) Drought stress amelioration in plants using green synthesized iron oxide nanoparticles. Mater Today Proc 41:723–727
203. Sun D, Hussain HI, Yi Z, Rookes JE, Kong L, Cahill DM (2016) Mesoporous silica nanoparticles enhance seedling growth and photosynthesis in wheat and lupin. Chemosphere 152:81–91
204. Sutulienė R, Ragelienė L, Samuolienė G, Brazaitytė A, Urbutis M, Miliauskienė J (2022) The response of antioxidant system of drought-stressed green pea (*Pisum sativum* L.) affected by watering and foliar spray with silica nanoparticles. Hortic 8(1):35
205. Takahashi F, Kuromori T, Urano K, Yamaguchi-Shinozaki K, Shinozaki K (2020) Drought stress responses and resistance in plants: from cellular responses to long-distance intercellular communication. Front Plant Sci 11:556972
206. Tao R, Zhang Y, Yang J, Yang T, White JC, Shen Y (2023) Biochar nanoparticles alleviate salt stress in tomato (*Solanum lycopersicum*) seedlings. Environ Sci Nano 10(7):1800–1811
207. Tarafdar JC, Raliya R, Mahawar H, Rathore I (2014) Development of zinc nanofertilizer to enhance crop production in pearl millet (*Pennisetum americanum*). Agric Res 3:257–262
208. Thakur S, Asthir B, Kaur G, Kalia A, Sharma A (2021) Zinc oxide and titanium dioxide nanoparticles influence heat stress tolerance mediated by antioxidant defense system in wheat. Cereal Res Commun 1–12
209. Thunugunta T, Reddy AC, Seetharamaiah SK, Hunashikatti LR, Chandrappa SG, Kalathil NC, Reddy LRDC (2018) Impact of Zinc oxide nanoparticles on eggplant (*S. melongena*): studies on growth and the accumulation of nanoparticles. IET Nanobiotechnol 12:706–713
210. Tombuloglu H, Slimani Y, Tombuloglu G, Almessiere M, Baykal A, Ercan I, Sozeri H (2019) Tracking of NiFe2O4 nanoparticles in barley (Hordeum vulgare L.) and their impact on plant growth, biomass, pigmentation, catalase activity, and mineral uptake. Environ Nanotechnol, Monit Manag 11:100223
211. Tripathi DK, Shweta SS, Singh S, Pandey R, Singh VP, Sharma NC, Prasad SM, Dubey NK, Chauhan DK (2017) An overview on manufactured nanoparticles in plants: uptake, translocation, accumulation and phytotoxicity. Plant Physiol Biochem 110:2–12
212. Tripathi R, Tewari R, Singh KP, Keswani C, Minkina T, Srivastava AK, De Corato U, Sansinenea E (2022) Plant mineral nutrition and disease resistance: a significant linkage for sustainable crop protection. Front Plant Sci 13:883970
213. Usman M, Farooq M, Wakeel A, Nawaz A, Alam Cheema SA, Rehman HU, Ashraf I, Sanaullah M (2020) Nanotechnology in agriculture: current status, challenges and future opportunities. Sci Total Environ 721:137778
214. Van Nguyen D, Nguyen HM, Le NT, Nguyen KH, Nguyen HT, Le HM, Nguyen AT, Dinh NTT, Hoang SA, Van Ha C (2022) Copper nanoparticle application enhances plant growth and grain yield in maize under drought stress conditions. J Plant Growth Reg 41(1):364–375
215. Van Wallendael A, Soltani A, Emery NC, Peixoto MM, Olsen J, Lowry DBA (2019) Molecular view of plant local adaptation: incorporating stress-response networks. Annu Rev Plant Biol 70:559–583
216. Venkatchalam P, Jayaraj M, Manikandan R, Geetha N, Rene ER, Sharma NC, Sahi SV (2017) Zinc oxide nanoparticles (ZnONPs) alleviate heavy metal-induced toxicity in Leucaenaleucocephala seedlings: a physiochemical analysis. Plant Physiol Biochem 110:59–69
217. Venzhik Y, Deryabin A, Popov V, Dykman L, Moshkov I (2022) Gold nanoparticles as adaptogens increazing the freezing tolerance of wheat seedlings. Environ Sci Pollut Res 29(36):55235–55249

218. Verma SK, Das AK, Gantait S, Kumar V, Gurel E (2019) Applications of carbon nanomaterials in the plant system: a perspective view on the pros and cons. Sci Total Environ 667:485–499
219. Wahid I, Kumari S, Ahmad R, Hussain SJ, Alamri S, Siddiqui MH, Khan MIR (2020) Silver nanoparticle regulates salt tolerance in wheat through changes in ABA concentration, ion homeostasis, and defense systems. Biomol 10(10):1506
220. Wang A, Li J, Al-Huqail AA, Al-Harbi MS, Ali EF, Wang J, Ding Z, Rekaby SA, Ghoneim AM, Eissa MA (2021) Mechanisms of chitosan nanoparticles in the regulation of cold stress resistance in banana plants. Nanomater 11:2670
221. Wang J, Wang Z, Mao H, Zhao H, Huang D (2013) Increasing Se concentration in maize grain with soil-or foliar-applied selenite on the Loess Plateau in China. Field Crops Res 150:83–90
222. Wang L, Dong S, Liu L, Ma Y, Li S, Zu W (2018) Transcriptome profiling reveals PEG-simulated drought, heat and combined stress response mechanisms in soybean. Comput Biol Chem 77:413–419
223. Wang L, Ning C, Pan T, Cai K (2022) Role of silica nanoparticles in abiotic and biotic stress tolerance in plants: a review. Int J Mol Sci 23:1947
224. Wang P, Zhang H, Hu X, Xu L, An X, Jin T, Ma R, Li Z, Chen S, Du S, Wei G (2024) Comparing the potential of silicon nanoparticles and conventional silicon for salinity stress alleviation in soybean (*Glycine max* L.): growth and physiological traits and rhizosphere/endophytic bacterial communities. J Agric Food Chem 71(19):10781–10793
225. Wang X, Mao Z, Zhang J, Hemat M, Huang M, Cai J, Zhou Q, Dai T, Jiang D (2019) Osmolyte accumulation plays important roles in the drought priming induced tolerance to post-anthesis drought stress in winter wheat (*Triticum aestivum* L.). Environ Exp Bot 166:103804
226. White JC, Gardea-Torresdey J (2018) Achieving food security through the very small. Nat Nanotechnol 13:627–629
227. Yadav S, Modi P, Dave A, Vijapura A, Patel D, Patel M (2020) Effect of Abiotic Stress on Crops. In: Hasanuzzaman M, Fujita M, Teixeira Filho MCM, Nogueira TAR, Galindo FS (eds) Sustainable crop production. London, UK
228. Yang Y, Yu J, Qian Q, Shang L (2022) Enhancement of heat and drought stress tolerance in rice by genetic manipulation: a systematic review. Rice (NY) 15(1):67
229. Yang KY, Doxey S, McLean JE, Britt D, Watson A, Al Qassy D, Anderson AJ (2017) Remodeling of root morphology by CuO and ZnO nanoparticles: effects on drought tolerance for plants colonized by a beneficial pseudomonad. Bot 96:175–186
230. Yang X, Alidoust D, Wang C (2020) Effects of iron oxide nanoparticles on the mineral composition and growth of soybean (*Glycine max* L.) plants. Acta Physiol Plant 42:1–11
231. Yang X, Lu M, Wang Y, Wang Y, Liu Z, Chen S (2021) Response mechanism of plants to drought stress. Hortic 7(3):50
232. Ye Y, Cota-Ruiz K, Hernández-Viezcas JA, Valdés C, Medina-Velo IA, Turley RS, Peralta-Videa JR, Gardea-Torresdey JL (2020) Manganese nanoparticles control salinity-modulated molecular responses in *Capsicum annuum* L. through priming: a sustainable approach for agriculture. ACS Sustain Chem Eng 8:1427–1436
233. Yeats TH, Rose JKC (2013) The formation and function of plant cuticles. Plant Physiol 163:5–20
234. Younis AA, Khattab H, Emam MM (2020) Impacts of silicon and silicon nanoparticles on leaf ultrastructure and TaPIP1 and TaNIP2 gene expressions in heat stressed wheat seedlings. Biol Plant 64(1)
235. Zahedi SM, Karimi M, Teixeira da Silva JA (2020) The use of nanotechnology to increase quality and yield of fruit crops. J Sci Food Agric 100(1):25–31
236. Zandalinas SI, Fritschi FB, Mittler R (2021) Global warming, climate change, and environmental pollution: recipe for a multifactorial stress combination disaster. Trends Plant Sci 26:588–599
237. Zhang H, Zhao Y, Zhu JK (2020) Thriving under stress: how plants balance growth and the stress response. Dev Cell 55:529–543
238. Zhang H, Zhu J, Gong Z, Zhu JK (2022) Abiotic stress responses in plants. Nat Rev Genet 23:104–119

239. Zhao J, Lu Z, Wang L, Jin B (2020) Plant responses to heat stress: physiology, transcription, noncoding RNAs, and epigenetics. Int J Mol Sci 22:117
240. Zulfiqar F, Moosa A, Chen J, Şimşek Ö, Jabeen R, Tütüncü M, İzgü T, Mohammadi M, Masood HA, Ali HM, Alsakkaf WA (2024) Application of biogenic silver nanoparticles ameliorates salt stress and improves growth and flower quality of *Calendula officinalis* L. Sci Hortic 334:113339
241. Zulfiqar F, Ashraf M (2021) Nanoparticles potentially mediate salt stress tolerance in plants. Plant Physiol Biochem 160:257–268

Plant-Mediated Synthesis of Nanoparticles for Sustainable Management of Plant Diseases

Mahabuba Mostafa, Humayra Ferdus, Suvankar Kumar Biswas, Md. Abdullah Al Sabbir, and Md. Motaher Hossain

Abstract Plant diseases are the most pressing threat to global efforts to sustain and improve agricultural outputs. Nanotechnology offers potential solutions for enhancing agricultural output amidst rising global populations and escalating challenges from pests and diseases. While conventional pesticides are effective, they pose significant environmental and human health risks. Researchers are thus seeking sustainable alternatives through nanotechnology. Green nanoparticles (NPs) produced using plant extracts show promise in enhancing pesticide delivery, effectiveness, and environmental safety for crop protection. Plant-based NPs can be synthesized through intracellular and extracellular methods, where plant metabolites reduce and stabilize the nanoparticles. These NPs possess intricate structures and diverse physicochemical properties, essential for managing phytopathogens. This plant-mediated manufacturing approach yields substantial quantities of high-quality NPs with precise dimensions, ensuring biocompatibility and non-toxicity. NPs exhibit antimicrobial properties through multiple mechanisms, including surface charge attraction, reactive oxygen species (ROS) production, and membrane damage, which combat phytopathogens and improve crop health. Despite their potential, the deployment of NPs raises a few concerns about environmental dispersal, bioaccumulation, and potential ecological impacts, necessitating comprehensive toxicological studies and standardized protocols for safe utilization. Thus, successful use of plant-mediated green synthesis of NPs and mitigating associated hazards require multidisciplinary collaboration and leveraging technological advancements. This chapter presents an in-depth study of nanoparticle production using plant-based techniques and their application as antimicrobial agents against plant diseases.

Keywords Physicochemical properties · Green synthesis · Sources · Plant pathogens · Potential risks

M. Mostafa · H. Ferdus · S. K. Biswas · Md. A. Al Sabbir · Md. M. Hossain (✉)
Department of Plant Pathology, Gazipur Agricultural University, Gazipur 1706, Bangladesh
e-mail: hossainmm@gau.edu.bd

L. F. Fraceto et al. (eds.), *Emerging Nanotechnologies for Agroecosystem Management*, Smart Nanomaterials Technology, https://doi.org/10.1007/978-981-95-0187-8_6

1 Introduction

The world population reached 8,019,876,189 on January 1, 2024 [282], a 0.95% increase from 2023. This growth is projected to continue, with the population exceeding 9.6 billion by 2050. To meet the food demands of this expanding population, global agricultural and food production must increase by 70–100% annually [101, 184]. Despite 19% of the world's population relying on agriculture as their principal income source, crop production faces significant threats from various stresses. Biotic stressors, including pests such as weeds, pathogens, insects, and nematodes, remain among the most significant obstacles to agricultural productivity and food security [102], causing a substantial 20–40% reduction in agricultural yield [64]. The impact is concerning, as global farmers harvest only about 50% of their potential crop output due to considerable crop losses and a significant yield gap caused by these stressors [63]. Among the biotic stresses, plant diseases are particularly worrisome for crop quality and productivity, causing annual economic losses exceeding $220 billion [65]. Throughout history, food crises induced by phytopathogens have led to starvation and forced migration for millions. This financial burden underscores the urgent need for effective plant disease management solutions.

Farmers often depend on chemical-based pesticides to control phytopathogens, but these pesticides pose severe threats to crop productivity, nutritional value, and soil fertility [100, 257, 269]. Additionally, pesticide application inefficiency is a major concern, with more than 50% of pesticides sprayed not reaching the intended target crops, thus posing risks to the environment, non-target species, and public health [208, 243]. Therefore, addressing the challenge of increasing food production while mitigating environmental concerns is crucial in modern agriculture [99, 98].

Increased public awareness of the negative effects of pesticides on health and the environment and a growing demand for safe foods have shifted research focus to ecologically friendly farming tools and practices [181]. Significant technological advancements have been made in agriculture in recent years, leading to the adaptation of ground-breaking approaches for the sustainable management of phytopathogens. One such promising approach is nanotechnology, which shows the use of nanoparticles (NPs) for plant disease management [139, 151, 223]. NPs are among the top ten chemical innovations that are expected to change the world [81]. NPs are discrete nano-objects with dimensions of 1 to 100 nm (nm) [47]. Due to their large surface area and small size, NPs have drawn considerable attention across various fields [186]. NPs can occur naturally in various inorganic and organic sources or be synthesized artificially. NPs can be synthesized from inorganic and organic sources, including microorganisms and plants of NPs from biological sources, known as green synthesis, is particularly successful for the safe and eco-friendly production of NPs [294]. Although the use of inorganic and organic NPs in agriculture is relatively new, their application has shown enormous potential to reduce yield losses from various plant pests and diseases (Fig. 1).

Different nanoparticles have shown the potential to control weeds, insects, and phytopathogens. Compared to inorganic NPs, biological NPs are thought to be safer

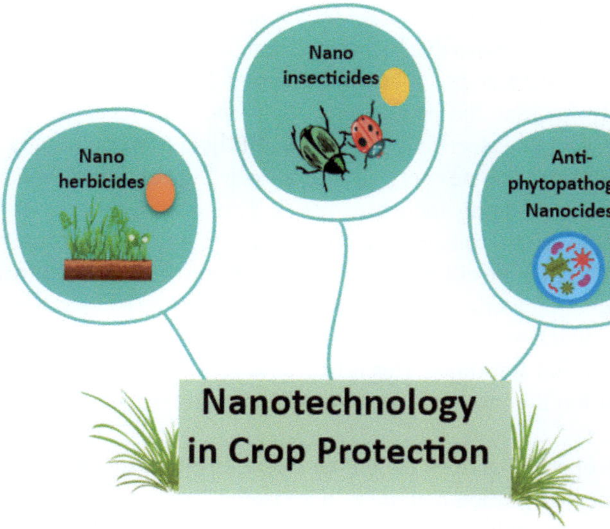

Fig. 1 Application of nanoparticles in crop protection against diverse plant pests and insects

for the environment because of their lower possibility of carrying toxic substances. Biological or green synthesis of NPs primarily utilizes microorganisms (fungi, bacteria, and algae) or plant parts [27]. Production of green NPs from plants is cost-effective because plant extracts contain abundant phytochemicals that stabilize and reduce metal ions into metal NPs [291]. Botanical NPs hold enormous promise for developing novel plant diseases. This chapter reviews the plant-mediated synthesis of NPs and discusses recent advances in plant disease management using NPs.

2 Nature and Properties of Nanoparticles

NPs show special characteristics that distinguish them from their bulk counterparts, making them a subject of extensive research and application across various scientific fields [47]. The primary feature that makes NPs intriguing is their high surface area-to-volume ratio. As particles shrink to the nanoscale, a significant proportion of their atoms reside on the surface rather than within the bulk of the material [50]. This high surface area enhances their reactivity, making NPs excellent catalysts for chemical reactions and bioagents for biological actions [186].

The optical properties of NPs are equally fascinating. Metallic NPs, such as those made from silver and gold, display surface plasmon resonance (SPR) [148]. This phenomenon occurs when conduction electrons on the nanoparticle's surface oscillate in resonance with incident light, resulting in strong absorption and scattering of light. This gives rise to vibrant colors and is exploited in applications like biological imaging, where gold NPs can provide contrast in cellular imaging techniques

[114]. Similarly, quantum dots, which are semiconductor NPs, have size-dependent electronic and optical properties due to quantum confinement effects. These NPs can emit light of specific wavelengths when excited, which is invaluable in developing high-resolution displays and biomedical imaging tools.

NPs also possess unique electronic properties. The quantum confinement in semiconductor NPs results in discrete energy levels, allowing the tuning of their electronic properties by simply changing their size [212]. This has significant implications for developing photovoltaic cells, where NPs can be engineered to absorb different wavelengths of sunlight more efficiently [124]. In addition to electronic attributes, the mechanical qualities of materials can be significantly enhanced by incorporating NPs. Nanocomposites, which are materials reinforced with NPs, often exhibit improved strength, toughness, and durability. This is because the NPs can hinder the movement of dislocations within the material, thereby enhancing its mechanical performance [9].

The thermal properties of NPs are another area of interest. Carbon-based NPs, including graphene and, carbon nanotubes have exceptionally high thermal conductivity and are ideal for applications in thermal management systems in electronics. Conversely, some NPs can exhibit reduced thermal conductivity, which can be beneficial in creating thermal insulation materials [38]. In the biomedical field, NPs are revolutionizing drug delivery systems. For example, liposomal NPs can encapsulate drugs, allowing for controlled delivery and release, minimizing side effects, and improving therapeutic efficacy [205]. Magnetic NPs are employed in magnetic resonance imaging (MRI) to enhance contrast, making it easier to visualize and diagnose medical conditions [199].

Chemically, NPs are categorized into organic, inorganic, and semi-organic groups. They can also be categorized based on their nanoscale dimensions (zero: quantum dots, one: nanorods and nanowires, two: nanolayers and nanoplates, and three: nanoflowers and nanocoils) and shapes (such as spherical, cubic, rod-shaped, triangular, octahedral, star-shaped) [85]. The biogenic reduction of elements such as Ag, Au, Al, Cu, Cd, Ce, Fe, Se, Ti, and Zn is a widely recognized method for nanoparticle production in recent years [94]. Among green biological approaches, plant-based NP synthesis is considered the gold standard for its versatility and ease of use [87]. These NPs can be administered through the methods of spraying or drenching/soaking seeds, foliage, or roots. Plant-based green synthesized NPs enhance the ability to interact with and combat harmful microorganisms, proving more effective than bulk materials. Because of their nanoscale size, NPs display unique chemical, physical, and biological properties, making them valuable for gene transfer [215] and in controlling plant diseases and pests [171, 170, 214].

Despite their potential applications, NPs pose challenges that need to be addressed. Their small size allows them to penetrate biological membranes and accumulate in organisms, raising concerns about their toxicity and long-term environmental impact. Understanding and minimizing the potential health concerns linked with NPs is an important area of ongoing research [189]. Moreover, the large-scale production of NPs requires careful consideration of sustainability and environmental impact.

With uses in medicine, electronics, environmental research, and beyond, NPs represent a significant advancement in material science and technology. Their unique properties, driven by their nanoscale dimensions, open up possibilities for innovation and improved performance in various applications. However, responsible development, thorough research on safety, and environmental considerations are essential to fully utilize their usefulness while minimizing possible risks. As research continues, the role of NPs in defining the upcoming technology and medicine becomes increasingly prominent, providing solutions to contemporary challenges.

3 Sources of Nanoparticles

Nanoparticles can originate from a variety of natural and anthropogenic sources, each contributing to their presence in the environment and their utilization in various applications. Naturally occurring nanoparticles are generated through numerous physical, chemical, and biological processes. For instance, volcanic eruptions and forest fires release nanoparticles into the atmosphere, while weathering of rocks and soil erosion also contribute to their presence in terrestrial and aquatic environments [180]. Biological processes, such as the activities of microorganisms, plants, and animals, can result in the formation of nanoparticles. For instance, certain bacteria produce magnetite nanoparticles as part of their metabolic processes, and plants can uptake and translocate nanoparticles from the soil, subsequently releasing them into the atmosphere through transpiration [33].

Anthropogenic sources of nanoparticles are primarily associated with industrial processes, combustion, and engineered applications. Combustion processes, including those occurring in vehicles, power plants, and industrial facilities, release substantial amounts of nanoparticles into the air. These particles, often composed of carbon, metal oxides, and other compounds, are byproducts of fuel burning and can have adverse effects on air quality and human health [211]. Industrial processes, such as welding, grinding, and other high-temperature operations, also generate nanoparticles. Additionally, nanoparticles are deliberately engineered and synthesized for use in a broad array of products and applications, including cosmetics, medicine, electronics, and environmental remediation. Engineered nanoparticles, such as zinc oxide, titanium dioxide, and silver nanoparticles, are incorporated into products for their unique properties, such as UV protection, antibacterial activity, and enhanced conductivity [137].

The synthesis of engineered NPs involves both top-down and bottom-up approaches. Top-down methods, such as milling and lithography, involve breaking down bulk materials into nanoscale particles. These techniques are widely used in the production of nanomaterials for electronic and structural applications [83]. Bottom-up methods, on the other hand, involve assembling nanoparticles from atomic or molecular precursors through chemical reactions or self-assembly processes. Sol–gel processes, chemical vapor deposition, and biological synthesis are examples of bottom–up approaches [231]. Biological synthesis, or green synthesis, leverages natural biological processes, using plant extracts, bacteria, and fungi to produce nanoparticles in an environmentally friendly manner. This method is receiving admiration due to its sustainability and reduced environmental impact compared to traditional chemical synthesis methods [261].

The increasing production and use of engineered nanoparticles have raised concerns about their potential environmental and health impacts. Once released into the environment, nanoparticles can undergo various transformations, including aggregation, dissolution, and chemical modification, which can affect their mobility, bioavailability, and toxicity [188]. Understanding the sources, behaviors, and impacts of nanoparticles is key for formulating tactics to manage their use and mitigate potential risks. Regulatory frameworks and guidelines are being established to guarantee the safe production, use, and disposal of nanoparticles [236]. Ongoing research is attentive to understanding the life cycle of nanoparticles, from their synthesis and application to their eventual release and interaction with the environment. This knowledge is essential for balancing the benefits of nanotechnology with the need to protect public health and the environment [137].

4 Synthesis Routes of Nanoparticles

A diverse array of chemical, biological, and physical means of nanoparticle synthesis is available [278]. Chemical routes for nanoparticle synthesis offer an array of options, including the colloidal method, chemical reduction method, sonochemical method, sol–gel method, chemical solution deposition, chemical vapor deposition, catalytic route, hydrolysis, Langmuir–Blodgett method, electrodeposition, soft chemical method, wet chemical method, and coprecipitation method. Physical methods include the electrochemical method, microwave method, solvothermal decomposition, laser ablation, plasma arcing, ball milling, lithographic method, sputter deposition, pulsed laser desorption, layer-by-layer growth, spray pyrolysis, ultra-thin film, molecular beam epistaxis, and diffusion flame-based synthesis. Biological methods include plant-based synthesis, bacterial system-based synthesis, fungal system-based synthesis, insect-based synthesis, and microalgae systembased synthesis [62, 88]) (Fig. 2). However, the chemical and physical techniques used for nanoparticle synthesis could be more economically efficient and require hazardous and toxic substances. It is widely recognized that these procedures could affect the environment and public health negatively due to the release of dangerous radiation and the utilization of condensed synthetic and stabilizing chemicals [71, 192]. Green synthesis methods are gradually replacing physical and chemical processes because of concerns over energy consumption, the emission of poisonous and dangerous substances, and the requirement of sophisticated synthesis conditions and equipment [17].

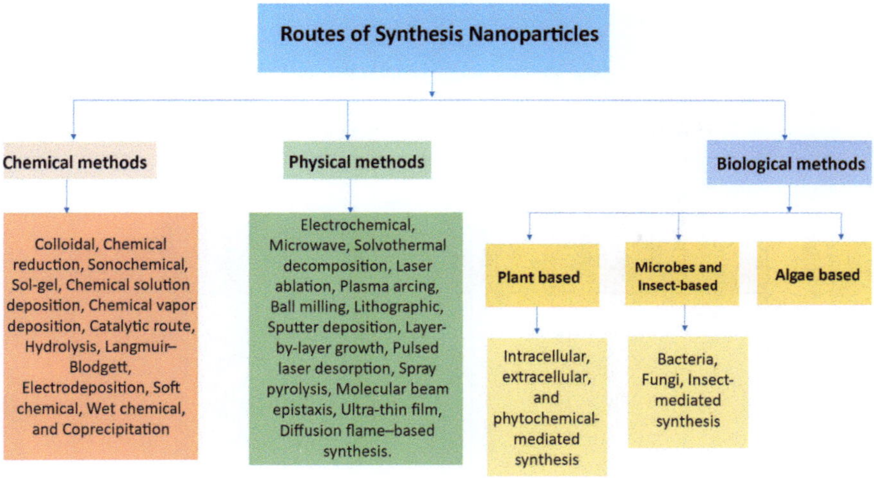

Fig. 2 Different routes of chemical, physical, and biological synthesis of nanoparticles

5 Plant-Mediated Synthesis of Nanoparticles

Plants are one of the most effective sources for biosynthetically producing NPs [266]. The process of reducing precious metals using inactive plant materials has been well-documented for almost a century. [173] and [110] conducted research on the process of silver reduction by plant chlorophyll and gold reduction by fresh leaf extracts, respectively. Approximately 66 years ago, [183] documented the process of $AgNO_3$ reduction by plant cells. In a study, [25] showed that powdered wheat and oat biomasses can be used to bio reduce $KAuCl_4$, resulting in the production of gold NPs.

Various in vitro methods are devised to use plant extracts as reducing agents for NP synthesis (Table 1). The green NP synthesis employs extracts from a variety of plant species of medicinal plants, culinary herbs, spices, weeds, and agroforest plants (Table 1). Diverse plant parts (leaves, seeds, flowers, bark, etc.) are used as precursors for the synthesis of different metal nanoparticles such as Ag, CuO, ZnO, and Au [219, 226, 234]. The resulting nanoparticles exhibit a range of sizes, shapes, and structural properties, from spherical and cubic forms to irregular and hexagonal shapes, with particle sizes varying from as small as 2 nm to over 400 nm (Table 1). The essential components in nanoparticle synthesis include reducing agents, capping agents, metal salts, and solvents [159]. Commonly used potent reducing agents include citric acid, amino acids, aldehydes, flavonoids, tartaric acid, NADP reductase, and secondary metabolites [133]. The reduction potential of each metal varies and significantly influences the reduction process. A higher positive reduction potential accelerates the reduction rate of the metal precursor. Equilibrium in the nucleation and growth stages is achieved when the reduction rate becomes gradual [304].

Table 1 Green synthesis of NPs from different plant species (medicinal, culinary herbs and spices, weeds, flower, fruit, agroforestry plants) and their physical characteristics

Plant Type	Plant species	Plant parts	Precursor	Metal NPs	Size (nm)	Shape	References
Medicinal plants	*Tridax procumbens*	Leaf	$AgNO_3$	Ag	65–100	Spherical and truncated	[67]
		Leaf	–	Cr_2O_3	–	Strips with rough surfaces	[229]
		Leaf	ZnSO4	ZnO	75.8 ± 36.6	Spherical	[8]
		Leaf	$C_2H_4CuO_2$	Cu	85–118	Spherical	[122]
		Leaf	$FeCl_3$	Fe	72	Spherical	[202]
	Justicia adhatoda	Leaf	$ZnSO_4$, $Zn(NO_3)_2$, $Zn(CH_3CO_2)_2$	ZnO	15–20	Orthogonal/ nanorod	[194]
		Leaf	$AgNO_3$	Ag	9 -24	Spherical or oval	[44]
		Leaf	$HAuCl_4$	Au	13–57	Spherical	[145]
		Leaf	$[Y(NO_3)_3.6H_2O] + [AgNO_3] + [HAuCl_4. 3H_2O] +$ [BMIM] PF6	Ag-Au/ Y_2O_3	30	Cubic	[197]
		Leaf	CuCl2	CuO	45.07	Spherical	[109]
		Leaf	$Pb(CH_3COO)_2$	PbO	30.35	Spherical, irregular sponge	[150]
		Leaf	–	MgO	22.96	Spherical	[106]
	Andrographis paniculata	Leaf	$C_4H_6CdO_4$	CdO	22	Spherical	[125]
		Leaf	$Ni(CH_3COO)0.24H_2O$	NiO	24	Cubic	[126]

(continued)

Table 1 (continued)

Plant Type	Plant species	Plant parts	Precursor	Metal NPs	Size (nm)	Shape	References
		Leaf	$Zn(NO_3)_2.6H_2O$	ZnO	57 ± 0.23	Spherical and hexagonal	[217]
		Leaf	$NdCl_3.6H_2O$	Nd_2O_3	20–80	Irregular	[273]
		Leaf	$AgNO_3$	Ag	40–60	Cubic	[263]
		Leaf	$HAuCl_4$	Au	10–15	Spherical	[51]
		Leaf	TiO_2	TiO_2	50	Monodisperse and spherical	[221]
	Rauvolfia serpentina	Leaf	$Cu(NO_3)_20.3\ H_2O$	CuO	10–20	Sponge-like	[149]
		Leaf	$AgNO_3$	Ag	7–10	Spherical	[198]
		Leaf	$HAuCl_4$	Au	68	Spherical	[18]
	Calotropis gigantea	Leaf	$Ni\ (NO_3)_2 \cdot 6H_2O$	NiO and Ni	<60	–	[57]
		Leaf	–	AgCl	10–50	Spherical	[153]
		Leaf	$AgNO_3$	Ag	20–35	Clustered and Irregular shapes	[210]
		Leaf	$Cu(NO_3)_2$	CuO	20	Spherical	[252]
		Leaf	$C_4H_6CdO_4 + Na_2S$	CdS	20	–	[29]
		Leaf	$Zn(NO_3)_2$	ZnO	30–35	Hexagonal	[293]
		Leaf	$Mg(NO_3)_2.6H_2O$	MgO	50.75	Spherical	[95]
		Leaf	$FeCl_3 + FeCl_2$	Fe_3O_4	11.1	Irregular	[123]

(continued)

Table 1 (continued)

Plant Type	Plant species	Plant parts	Precursor	Metal NPs	Size (nm)	Shape	References
		Leaf, seed, flower	$TiCl_4$	TiO_2	–	Tetragonal	[204]
		Latex	$Fe_2(SO_4)_3$	Fe_2O_3	3–5	Spherical	[140]
	Sphagneticola trilobata	Leaf, stem, root	$Zn(CH_3CO_2)_2$	ZnO	65–80	Irregular and complex	[249]
		Leaf	$FeCl_3$	Fe_2O_3	42.2–62.54	Spherical	[55]
		Leaf	$Zr(NO_3)_4$	Zr	20–100	Oval-shaped	[130]
		Leaf, flower	$HauCl_4.3H_2O$	Au	7.49–57.42	–	[157]
Culinary herbs and spices	*Allium sativum*	Cloves	$Cu(NO3)_2$	CuO	20 – 40	Spherical and oval	[289]
		Cloves	$Na_2S_2O_3 \cdot 5H_2O$	S	55.613	Spherical	[132]
		Cloves	$HAuCl_4$ $0.4H_2O$	Au	9–15	Spherical	[48]
		Cloves	$Ag(NH_3)_2$	Ag	7.3 ± 4.4	Spherical	[233]
		Cloves	Na_2SeSO_3	Se	40–100	Spherical	[22]
		Bulb	$Zn(CH_3COO)_2 \cdot 2H_2O$	ZnO	19.8–23.86	Wurtzite	[286]
		Cloves	$ZnCl_2$, Na_2S	ZnS	6.5	–	[246]
		Leaf	$CuSO_4.5H_2O$	Cu	100	Spherical	[120]
	Coriandrum sativum	Leaf, seed	$AgNO_3$	Ag	6.45, 13.09	Spherical	[134, 185]
		Leaf	$Zn(CH_3COO)_2 \cdot 2H_2O$	ZnO	78–84	–	[283]

(continued)

Table 1 (continued)

Plant Type	Plant species	Plant parts	Precursor	Metal NPs	Size (nm)	Shape	References
		Leaf	$Ce(NO_3)_3.6H_2O$	CeO_2	13.56	Cubic	[241]
		Leaf	$NaAuCl_4$	Au	12 ± 4	–	[155]
		Leaf	$Ni(NO_3)_2 \cdot 6H_2O$	NiO	15–16	Spherical	[216]
	Trigonella foenum-graecum	Leaf	$TiOSO_4$	TiO_2	20–90	Spherical	[268]
		Seed	$HAuCl_4$	Au	15–25	Spherical	[26]
		Leaf	$Mg(NO_3)_2$	MgO	13	Spherical	[292]
		Leaf, seed	$AgNO_3$	Ag	9–57	Spherical, sphere	[235, 288]
		Seed	$FeCl_3 \cdot 6H_2O$	Fe	11	–	[213]
		Seed	$Zn(NO_3)_2 \cdot 6H_2O$	ZnO	24–44	Hexagonal	[131]
		Leaf	$AgNO_3$	AgO	20–35	Spherical	[19]
		Leaf	$CuSO_4$	CuO	47	Cubic	[142]
		Seed	$SnCl_2.2H_2O$	SnO_2	10	Spherical	[82]
		Seed	Na_2SeO_3	Se	97	Spherical	[265]
	Piper nigrum	Seed, leaf, stem	$AgNO_3$	Ag	9–50	Nanorod, spherical and irregular	[172, 203]
		Leaf	$Mg(NO_3)_2$	MgO	20	Rod	[206]
		Fruit	$C_2H_4CuO_2$	Cu	30–32	Spindle	[136]
		Seed	$HAuCl_4 \cdot 3H_2O$	Au	40–60	Irregular	[36]
		Leaf	$AgNO_3$	AgO	20–35	Spherical	[19]

(continued)

Table 1 (continued)

Plant Type	Plant species	Plant parts	Precursor	Metal NPs	Size (nm)	Shape	References
		Leaf	$CuSO_4$	CuO	47	Cubic	[143]
	Cinnamomum tamala	Leaf	$AgNO_3$	Ag	25–30	Spherical	[11]
		Leaf	ZnO	ZnO	26.57	Spherical and hexagonal	[3]
		Leaf	$HAuCl_4.3H_2O$	Au	–	Prism	[264]
		Leaf	CH_6ClN_3S & $AgNO_3$	Ag_2S	50	Spherical	[285]
Weed	*Solanum nigrum*	Leaf	$Zn(NO_3)_2$	ZnO	29.79	Quasi-spherical	[228]
		Leaf	$AgNO_3$	Ag	10–50	Spherical	[118]
		Leaf	$HAuCl_4$	Au	50	Cubic	[182]
		Fruit	Na_2SeO_3	Se	87	Spherical	[295]
		Leaf	$FeCl_3$	Fe_2O_3	24.1	Hexagonal	[270]
		Leaf	$Cu(NO_3)_2$	CuO	–	Featureless	[290]
	Cynodon dactylon	Leaf	$AgNO_3$	Ag	30–50	–	[274]
		Grass	$Cu(NO_3)_2{\cdot}3H_2O$	CuO	1–500	Rice spikelet-like	[275]
		Leaf	$Zn(CH_3COO)_2{\cdot}2H_2O$	ZnO	Variable	Hexagonal rod	[164]
		Grass	$HAuCl_4$	Au	36.01	Crumbled	[296]
		Leaf	–	Si	7–80	Spherical	[30]
		Leaf	$Ti\{OCH(CH_3)_2\}_4$	TiO_2	16	Irregular	[89]
	Chenopodium album	Leaf	$AgNO_3$	Ag	10–30	Spherical	[20]
		Leaf	$Co(NO_3)_2.6H_2O,FeCl_3.6H_2O$	$CoFe_2O_4$	13	Spinel	[175]

(continued)

Table 1 (continued)

Plant Type	Plant species	Plant parts	Precursor	Metal NPs	Size (nm)	Shape	References
		Leaf	–	CuO	17 ± 3	Spherical	[24]
		Leaf	$Pd(OAC)_2$, $FeCl_3$ and $FeCl_2$	Pd/Fe_3O_4	70–90	Irregular	[195]
	Eichhornia crassipes	Leaf	$AgNO_3$	Ag	16–65	Sphere-like	[103]
		Leaf	$CuSO_4$	CuO	20 ± 2	Nearly spherical	[93]
		Leaf	$FeSO_4$	Fe	–	Rod	[112]
	Oxalis stricta	Leaf	$AgNO_3$	Ag	104.52	Spherical	[262]
Flower	*Allamanda cathartica*	Flower	$AgNO_3$	Ag	39	Spherical	[127]
		Leaf	$Zn(NO_3)_2$	ZnO	500	Spherical	[256]
		Latex	MgO	MgO	32.3	Spherical, cylindrical, rectangular, hexagonal and round	[196]
		Flower	SeO_2	Se	60.31	Spherical	[240]
		Flower	$HAuCl_4$	Au	5–70	Spherical and triangular	[73]
	Plumeria rubra	Leaf	$C_{12}H_{28}O_4Ti$	TiO_2	77.6	–	[40]
		Flower	$HAuCl_4$	Au	15.6 ± 3.4–28 ± 5.6	Spherical	[160]
		Flower	$AgNO_3$	Ag	36.19	Spherical	[161]
		Leaf	$Zn(CH_3COO)_2{\cdot}2H_2O$	ZnO	19.4	Agglomerated	[162]
	Nerium oleander	Leaf	$HAuCl_4$	Au	2–10	Spherical	[277]

(continued)

Table 1 (continued)

Plant Type	Plant species	Plant parts	Precursor	Metal NPs	Size (nm)	Shape	References
		Leaf	$AgNO_3$	Ag	37.5–75	Spherical	[84]
		Leaf	$CuSO_4$	CuO	21	Monodispersed and agglomerated	[244]
		Flower	$FeCl_3.6H_2O$	Fe_2O_3	47.2 ± 20	Agglomerated	[253]
		Leaf	Na_2SiO_3	SiO_2	–	Spherical	[166]
		Leaf	$Zn(NO_3)_2{\cdot}6H_2O$	ZnO	–	Hexagonal wurtzite	[144]
	Vinca rosea	Leaf	$AgNO_3$	Ag	32	Spherical	[138]
		Leaf	$Zn(CH_3COO)_2{\cdot}2H_2O$	ZnO	16	Hexagonal	[45]
		Leaf	$CuSO_4$ and $MnCl_2$	Mn + Cu	412	Spherical	[6]
	Bombax malabaricum/ Bombax ceiba	Gum extract	–	Fe_3O_4	09 ± 02	Round	[28]
		Thorn	$AgNO_3$	Ag	20–30	Cylindrical	[279]
Fruit	*Spondias pinnata*	Bark	$AgNO_3$	Ag	34.52	Spherical	[121]
		Leaf	$Zn(NO_3)_2{\cdot}6H_2O$	ZnO	30–48.5	Hexagonal	[5]
	Anacardium occidentalis	Leaf	$PdCl_2$	Pd	2.5–4.5	Spherical	[255]
		Leaf	$HAuCl_4$	Au	40	Spherical	[272]
		Pseudocarp	$AgNO_3$	Ag	39–56	Spherical	[177]
		Leaf	$[Zn(CH_3COO)_2{\bullet}2H_2O]$ and $[ZnCl_2]$	ZnO	$107.03 \pm 1.54–206.58 \pm 1.86$	Spherical and flake-like	[58]
	Mangifera indica	Leaf	$HAuCl_4{\cdot}3H_2O$	Au	17–20	Spherical	[207]

(continued)

Table 1 (continued)

Plant Type	Plant species	Plant parts	Precursor	Metal NPs	Size (nm)	Shape	References
		Leaf	$AgNO_3$	Ag	32.4	Cubic	[271]
		Leaf	$FeCl_3$	ZVI, Fe	50–100	–	[56]
		Seed	$Zn(NO_3)_2{\cdot}6H_2O$	ZnO	40–60	Spherical, cylindrical, rectangular, and triangular	[220]
	Annona reticulata	Leaf	$AgNO_3$	Ag	6.48–8.13	Cubic	[201]
		Fresh fruit Peel	$HAuCl_4$	Au	5 ± 2	Spherical	[72]
		Stem bark	$CuSO_4$	Cu	13.7 ± 3.3	Spherical	[163]
		Seed	$Zn(NO_3)_2{\cdot}6H_2O$	ZnO	70	Spherical	[298]
	Carissa carandas	Fruit	$AgNO_3$	Ag	23 ± 2	Spherical	[23]
Agroforest plant	*Shorea robusta*	Leaf	$AgNO_3$	Ag	12–37	Spherical	[250]
		Leaf	$FeSO_4.7H_2O$	Fe	54–80	Spherical	[119]
	Leucaena leucocephala	Leaf	$[Cd(NO_3)_2]0.4H_2O$	CdO	36–57	Spherical	[242]
		Leaf	$AgNO_3$	Ag	26	Spherical to quasi-spherical	[224]
		Leaf	$Cu(CH_3COO)_2.H_2O$	CuO	10–25	Spherical	[4]
	Acacia nilotica	Pod	$FeSO_4{\cdot}7H_2O$	Fe	229	Irregular	[52]
		Leaf	$Ni(NO_3)_2.6H_2O$	NiO	16	Spherical	[104]
		Gum	$AgNO_3$	Ag	10–40	Spherical	[78]

(continued)

Table 1 (continued)

Plant Type	Plant species	Plant parts	Precursor	Metal NPs	Size (nm)	Shape	References
		Leaf	$HAuCl_4$	Au	6–12	Spherical	[152]
		Fruit	$Zn(CH_3COO)_2 \cdot 2H_2O$	ZnO	16–90	Spherical	[232]
		Plant	TiO_2	TiO_2	60	Round/spherical	[129]
		Pod	$Cu(NO_3)_2 0.3H_2O$	CuO	–	Irregular	[230]
	Artocarpus heterophyllus	Seed	$AgNO_3$	Ag	10.78	Irregular	[113]
		Peel	$FeCl_3.6H_2O$	Fe	33	Spherical	[115]
		Fruit	$HAuCl_4$	Au	20–25	Polyhedron	[34]
		Leaf	$CuSO_4$	Cu	132	Crystals	[254]
	Tectona grandis	Seed	$AgNO_3$	Ag	10–30	Oval, spherical	[234]
		Leaf	$Zn(NO_3)_2 \cdot 6H_2O$	ZnO	5	Spherical	[247]
		Leaf	$MgCl_2$	MgO	90	Rod	[2]

The benefits of using plant-based synthesis for NPs are numerous. This method utilizes readily available aqueous solvents and abundant plant materials. Extracts derived from plant species are biocompatible and safe for various applications. The synthesis process is straightforward and energy-savvy as it utilizes an aqueous medium and room temperature conditions [46]. Additionally, this approach is cost-effective, and environmentally friendly, making it well-suited for large-scale production [21].

6 Methods of Plant-Derived Synthesis of Nanoparticles

Plant materials play a crucial role in NP production because they can synthesize NPs through the uptake, accumulation, utilization, and recycling of various mineral species [200]. There are three main techniques for producing plant-derived NPs: intracellular synthesis, extracellular synthesis, and phytochemical-mediated synthesis [21].

6.1 Plant-Mediated Intracellular Synthesis of Nanoparticles

Intracellular NP synthesis involves producing NPs within the cells of living organisms. This technique leverages the cellular biological machinery, utilizing the cells' inherent metabolic activities. When mediated by plants, this process is commonly referred to as 'Living Plant-Mediated Synthesis of NPs'. Several plant species demonstrate the capacity for detoxification and accumulation of heavy metals [169, 303]. These plants have shown the ability to produce various metallic NPs. Studies have confirmed that live plants can be employed in NP synthesis, with the ability to produce mixed metal NPs suggesting the potential for creating catalysts with specific compositions that are difficult to achieve using conventional approaches. *Brassica juncea* and *Medicago sativa* have demonstrated significant uptake, reduction, and dispersion of silver NPs within their cellular structures [31, 37, 92, 156, 158, 91]. Further research has shown that *Festuca rubra, Medicago sativa, Zea mays*, and *Arabidopsis thaliana* also produce silver NPs within their cells [141, 156, 302]. Alfalfa roots can uptake silver from an agar medium and transport it to the aboveground parts of the plant without changing its oxidation state. Inside the plant, silver atoms arrange themselves into NPs, which then merge to form larger structures, illustrating multiple levels of organization [74].

Gold NPs are produced within the cells of peanut seedlings, resulting in oval-shaped NPs varying in size from 5 to 50 nm [225]. Plants used for the production of gold NPs include *Brassica juncea, Chilopsis linearis, Medicago sativa, Cucumis sativus, Trifolium pratense, Sesbania bispinosa, Lolium perenne, Ulva armoricana, Helianthus annuus, Arabidopsis thaliana,* and *Origanum vulgare* [37, 75, 92, 116, 158, 176, 237, 267]. *Medicago sativa, Phaseolus vulgaris,* and *Raphanus sativus* can synthesize both gold (Au) and silver (Ag) NPs within their cells [37]. Metallic NPs of Cu, Co, Zn, and Ni can be synthesized from plants such as *Helianthus annuus, Brassica juncea,* and *Medicago sativa* [31, 92]. While intracellular synthesis is simpler than other techniques, it complicates the separation and purification of NPs, contributing to their morphological diversity. Recovering NPs created using this method is laborious and time-consuming [21].

6.2 Plant-Mediated Extracellular Synthesis of Nanoparticles

The extracellular production of NPs involves generating NPs outside the plant cells. This method leverages the inherent chemical components present in plants, which act as reducing and stabilizing agents, thus enabling the formation of NPs in an ecologically safe manner. For example, the conversion of silver ions to silver NPs outside of cells has been demonstrated using living *Arachis hypogaea* and *Tephrosia apollinea* plants [15, 227]. Other living plants, such as *Arachis hypogaea, Brassica juncea, Medicago sativa,* and *Vigna unguiculata,* can synthesize gold NPs extracellularly [90, 225, 248].

Plant extracts contain raw phyto-constituents responsible for nanoparticle production. Plant extracts are obtained using either hot or cold extraction techniques, making this approach highly appropriate for synthesis because of its convenient downstream processing and scalability. Using plant extracts, Ag NPs, selenium NPs, ZnO NPs, CuO NPs, MnO_2 NPs, and NiO NPs have been manufactured from *Lallemantia royleana, Moringa oleifera, Myrtus communis, Macroptilium lathyroides,* and *Psidium guajava* [32, 79, 167, 209, 245, 251]. An example of plant-mediated silver NP synthesis from neem leaves is shown in Fig. 3.

Fig. 3 Hypothetical model showing plant-based silver nanoparticle synthesis from neem leaves

6.3 Phytochemical-Mediated Synthesis of Nanoparticles

Synthesizing NPs using phytochemicals is an essential aspect of extracellular nanoparticle production. This method, primarily conducted in a water-based medium, is considered a green approach to nanoparticle synthesis. Identifying the correct reducing and stabilizing agents required for NP synthesis is challenging due to the diverse range of phytochemicals in plant extracts [21]. For instance, Au NPs can be synthesized using guavanoic acid and *Solanum torvum*/β-glucosidase [35]. Ag NPs can be produced using β-sitosterol-D-glucopyranoside from *Desmostachya bipinnata*, sodium parahydroxybenzoate tetrahydrate from *Vitex negundo*, and geraniol from *Pelargonium graveolens* [7, 59, 239]. Gold nanoparticles with enhanced catalytic activity can be synthesized using polyphenols derived from *Salvia officinalis* [193]. Numerous studies have demonstrated that flavonoids can bind to metal nanoparticles, particularly those of Ag, Au, and oxides of Fe, Zn, and Ti. This binding has led to synthesizing materials with intriguing physical, chemical, and biological properties [276].

7 Physicochemical Properties of Plant-Based Nanoparticles

NPs are tiny particles with a complex composition, consisting of three distinct layers: a surface layer, a shell layer, and a core region [259]. The central area is the nanoparticle itself [284]. The key physicochemical qualities of NPs include shape, size, surface area and charge, chemical composition, solubility, dispersibility, optical properties, and magnetic properties. These unique characteristics give NPs their individuality and make them ideal for a broad range of applications. The stability, toxicity, and efficacy of NPs in different conditions are influenced by their various forms, such as quasi-spherical, lumpy, and agglomerated morphologies [68, 97, 187]. Even if particles are made of the same substance, their levels of cytotoxicity can vary significantly based on their size and surface reactivity [259].

NPs can have four shapes based on the dimensionality: 0D, 1D, 2D, and 3D. 0D shapes include spherical, pseudo-spherical, dodecahedral, tetrahedral, octahedral, and cubic shapes. 1D shapes cover nanotubes, nanoneedles, nanorods, nanowires, nano-shuttles, nano-capsules, and hollow structures. 2D shapes embrace round disks, hexagonal, triangular, or quadrangular plates or sheets, belts, mesoporous-hollow nanospheres, and hollow rings. 3D shapes include nano-urchins, nanoflowers, nanostars, polygonal nano-frames, multiple hollow-shelled NPs, and hollow bunches [300]. Surface charge can ensure colloidal stability, preventing NPs from adhering to each other. The charge of NPs also controls how they interact with biological entities. For example, serum proteins may attach to NPs, triggering their uptake by endocytic pathways or even altering the shape of the proteins [53, 301].

A basic approach to checking nanoparticle synthesis involves visually inspecting the color change of the solution and observing precipitation. Ag NPs typically range in color from yellow to brown, while gold NPs exhibit a light yellow to ruby-red color. For zinc NPs, the reaction mixture changes from pale yellow to pale white [13, 145, 283]. Various analytical methods are employed to characterize the properties of NPs, including UV–vis spectroscopy, Fourier Transform Infrared Spectroscopy (FTIR), X-ray diffraction (XRD), transmission electron microscopy (TEM), scanning electron microscopy (SEM), infrared spectroscopy (IR), zeta sizing, X-ray photoelectron spectroscopy (XPS), and energy-dispersive X-ray (EDX) spectroscopy (Fig. 4).

UV–vis absorption spectrophotometry is used to confirm the presence of green-synthesized NPs in the resulting reaction mixtures [251]. The UV–vis absorption spectra of nanoparticle mixtures are typically obtained within the 200–800 nm range at various time intervals. The characteristic absorption peaks for silver, TiO_2, and Zn NPs are observed in the 350–450 nm and 350–420 nm ranges, respectively [178, 190, 268].

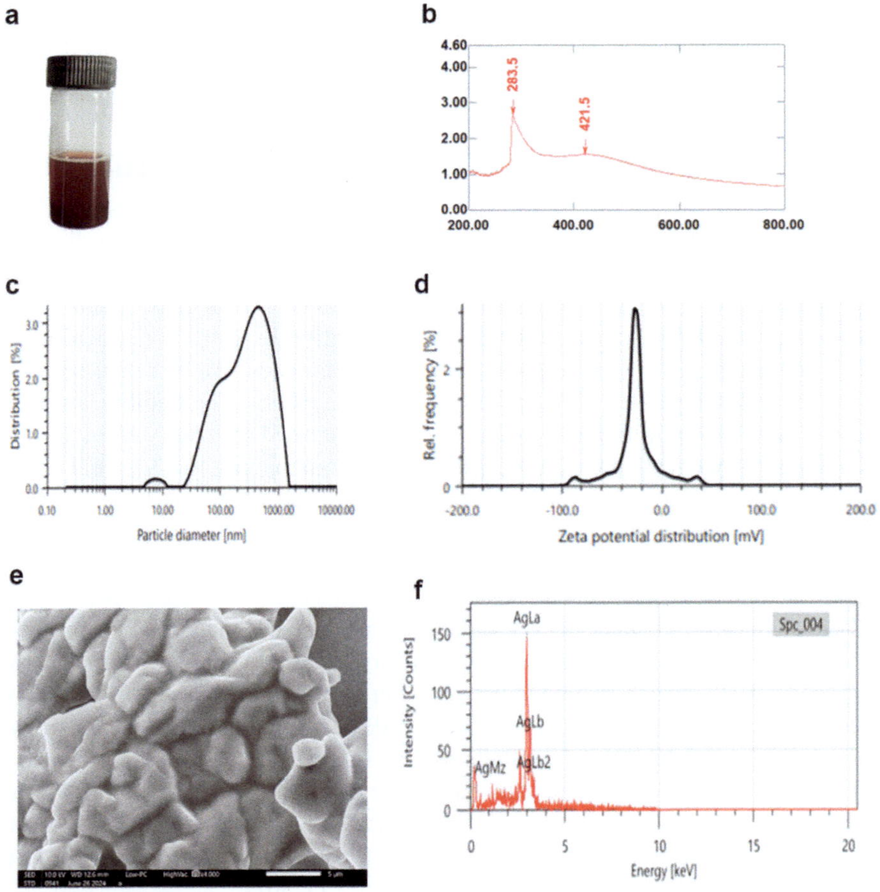

Fig. 4 Physiochemical analysis of silver NPs through different studies. **a** Visual observation, **b** UV–vis spectroscopy, **c** Particle Diameter Analysis, **d** Zeta Potential **e** SEM Images, **f** EDX Analysis

TEM and selected area electron diffraction (SAED) patterns are used to observe the size, shape, and size distribution of NPs in TEM images [60]. The hydrodynamic radius of NPs is verified through dynamic light scattering (DLS) analysis [77]. Although DLS is capable of measuring particles in the range from 1 to 500 nm, estimating the agglomerated particle size is challenging [10]. SEM establishes the surface morphology, including the size, shape, and size distribution of NPs [238].

Zeta potential is an important factor affecting NP dispersion, stability, and the electric charge on their surfaces. XRD is an effective method for qualitative and quantitative characterization of NPs. XRD tests substantiate the NPs formation and define their crystal crystallinity [54]. XPS is considered the most accurate method for determining the elemental composition and the types of bonds between elements in NPs [154].

8 Application of Plant-Derived Nanoparticles Against Plant Pathogens

Notorious plant pathogens attack and cause various forms of damage to plants, leading to significant economic losses. Recent investigations have shown that the application of plant-derived NPs holds promising potential in combating notorious plant pathogens. These nanoparticles demonstrate significant inhibitory activity against phytopathogens, which can translate into reduced disease incidence, increased crop yields, and minimized environmental pollution (Fig. 5). By harnessing the natural defenses of plants through these nanoparticles, researchers are exploring sustainable and effective strategies to enhance agricultural productivity while mitigating the adverse effects of pathogens on crops. This approach not only addresses economic losses but also aligns with efforts towards eco-friendly agricultural practices.

Table 2 provides a detailed account of various plant-derived nanoparticles used for controlling different plant pathogens. Derived from diverse plants such as *Justicia adhatoda*, *Ipomoea carnea*, and *Calotropis procera*, these nanoparticles come in various compositions, sizes, and shapes, demonstrating significant antimicrobial properties. For instance, zinc NPs from *Justicia adhatoda* (~15–20 nm, orthogonal/nanorod) are effective against *Aspergillus niger* and *A. flavus* [194]. Ag NPs from *Ipomoea carnea* (66.2 nm, spherical) inhibit *Rhizoctonia solani* [108], while iron oxide NPs from *Calotropis procera* (11.1 nm, irregular) target *A. niger* and *Fusarium oxysporum* [123]. Similarly, zinc NPs from *Calotropis gigantea* (80–100 nm, irregularly spherical) are effective against *F. solani, A. niger*, and *Rhizopus nigricans* [66]. Additionally, NPs from *Citrus limon* (ZnO and TiO_2, 20–200 nm, polymorphic) inhibit *Dickeya dadantii* [96], and magnesium oxide NPs from *Rosmarinus officinalis* (<20 nm, flower-shaped) target *Xanthomonas oryzae* pv. *oryzae* [1].

There are also studies on NPs from *Matricaria chamomilla*, where MgO and MnO2 nanoparticles (9–112 nm, disk-shaped spherical) inhibit *Acidovorax oryzae*

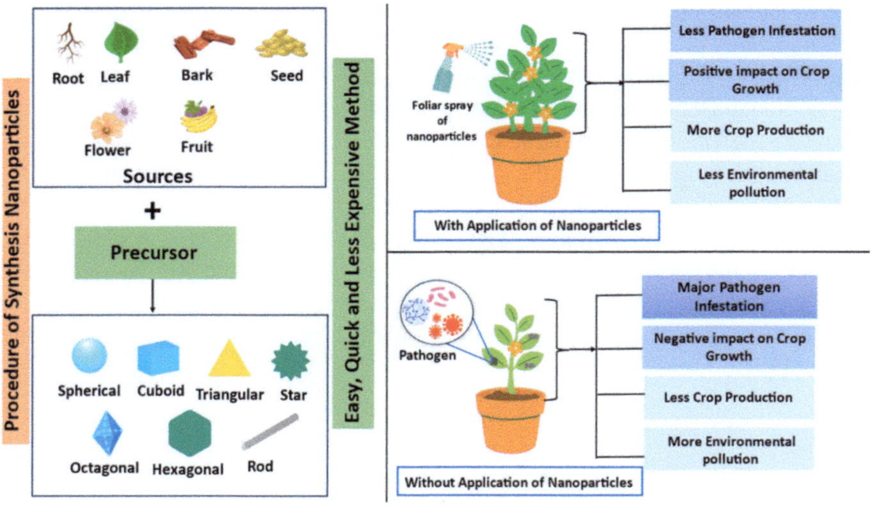

Fig. 5 Plant-based synthesis of NPs, their applications in plant disease management, and potential implications

[190], and ZnO NPs (50–192 nm, crystalline) target *X. oryzae* pv. oryzae [191]. *Olea europaea-derived* ZnO NPs (41–124 nm, crystalline) also show effectiveness against *X. oryzae* pv. *oryzae* [191]. Additionally, *Lycopersicon esculentum* ZnO NPs (66–133 nm, crystalline) are used against *X. oryzae* pv. *oryzae* [191], while *Piper nigrum* Ag NPs (9–30 nm, crystalline) target *Erwinia cacticida* [203].

Other significant examples include *Citrus maxima*-derived Ag NPs (11–13 nm, spherical) effective against *A. oryzae* [16] and *Artemisia absinthium* Ag NPs (5–100 nm, spherical) inhibiting *Phytophthora parasitica* [14]. *Trachyspermum ammi* Ni NPs (68 nm) target *Colletotrichum musae* [111], and *Abelmoschus esculentus* AuNPs (45–75 nm, spherical) are effective against *Puccinia graminis* pv. *tritici* [117]. Furthermore, *Parthenium hysterophorus* ZnO NPs (28–84 nm, spherical and hexagonal) inhibit *F. culmorum* [222], and *Syzygium aromaticum* Cu NPs (15 nm, spherical) are used against *A. niger*, *A. flavus*, and *Penicillium* spp. [218]. These studies underscore the versatility and effectiveness of plant-derived nanoparticles in combating plant pathogens, offering a promising direction for sustainable agriculture and plant disease management.

9 Mechanisms of Nanoparticles Against Pathogens

Despite multiple studies demonstrating their antibacterial effects, the exact mechanisms underlying the antimicrobial activity of metallic NPs against pathogenic microbes remain unclear [80]. Various mechanisms have been documented, including attraction to pathogen cell walls due to opposite surface charges, protein dysfunction

Table 2 Application of plant-derived nanoparticles for controlling different plant pathogens

Source plant	Nanoparticle	Size (nm)	Shape	Inhibited plant pathogen	References
Justicia adhatoda	Zn	~15–20	Orthogonal/ nanorod	*Aspergillus niger, Aspergillus flavus*	[194]
Ipomoea carnea	Ag	66.2	Spherical	*Rhizoctonia solani*	[108]
Calotropis procera	Fe_3O_4	11.1	Irregular	*A. niger, Fusarium oxysporum*	[123]
Calotropis gigantea	Zn	80–100	Irregular spherical	*F. solani, A. niger, R. nigricans*	[66]
Citrus limon	ZnO and TiO_2	20–200 nm	Polymorphic	*Dickeya dadantii*	[96]
Phyllanthus emblica	Ag	20–93	Spherical	*Acidovorax oryzae*	[159]
Rosmarinus officinalis	MgO	<20	Flower	*Xanthomonas oryzae* pv *oryzae*	[1]
Matricaria chamomilla	MgO and MnO_2	9–112	Disk-shaped spherical	*Ac. oryzae*	[190]
Matricaria chamomilla	ZnO	50–192	Crystalline	*X. oryzae pv. oryzae*	[191]
Olea europaea	ZnO	41–124	Crystalline	*X. oryzae pv. oryzae*	[191]
Lycopersicon esculentum	ZnO	66–133	Crystalline	*X. oryzae pv. oryzae*	[191]
Piper nigrum	Ag NPs	9–30	Crystalline	*Erwinia cacticida*	[203]
Citrus maxima	Ag	11–13	Spherical	*Ac. oryzae*	[16]
Artemisia absinthium	Ag	5–100	Spherical	*Phytophthora parasitica*	[14]
Trachyspermum ammi	Ni	68	–	*Colletotrichum musae*	[111]
Abelmoschus esculentus	Au	45–75	Spherical	*Puccinia graminis* pv.*tritci*	[117]
Parthenium hysterophorus	ZnO	28–84	Spherical and hexagonal	*F. culmorum*	[222]
Syzygium aromaticum	Cu	15	Spherical	*A. niger, A. flavus, and Penicillium spp.*	[218]

(cysteine oxidation in the Fe-binding site, Fe-S cluster destruction, and catalytic and structural metal exchange), reactive oxygen species (ROS) production, antioxidant depletion, disrupted membrane function, nutrient uptake interference (Fe (III) transporter gene expression inhibition), and chromosomal abnormalities (chromosome clumping, acentric fragments, stickiness, translocations, breaks, and micronucleus formation) leading to genotoxicity in microorganisms. The release of metal ions and modification of the signaling pathway are also observed [49, 146, 165]. These mechanisms likely work together to combat various phytopathogens [12] (Fig. 6).

Exposure of cells to NPs causes direct damage to the cell membrane [297]. When cells absorb NPs, it triggers ROS generation, resulting in oxidative stress and nanotoxicity. This process includes DNA damage, altered cell signaling, changes in cell movement, cytotoxicity, and programmed cell death [168]. The chemical configuration of nanomaterials affects the extent of ROS generation in specific cells [43]. Reducing molecular oxygen to water within the mitochondria generates ATP through interconnected proton and electron transport processes. In the process, radicals containing oxygen, including superoxide anion radicals, are produced by incompletely reduced oxygen. ROS, as cellular oxidative metabolism byproducts, primarily targets DNA within cells. They cause DNA–protein crosslinks, base and sugar lesions, strand breaks, and the formation of basic sites [69]. ROS are relatively

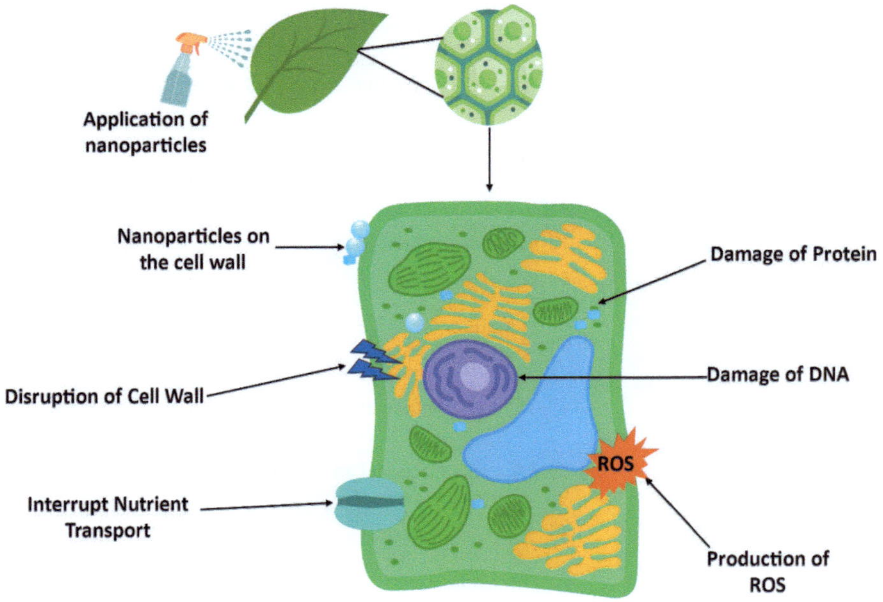

Fig. 6 Mode of action of nanoparticles against invading plant pathogens in a plant cell. Once applied to plants, nanoparticles adhere to the cell wall. When pathogens come in contact, nanoparticles disrupt the cell wall and interrupt nutrient transport within the pathogen cell. The interaction with the plant cell leads to the production of ROS, which can induce significant damage to the DNA and proteins within the pathogen cell, compromising the cell's structural integrity and function

less reactive and can interact with DNA from a distance. However, more potent radicals like hydroxyl radicals rapidly inflict damage on nearby DNA. Peroxidases, catalases, and superoxide dismutase (SODs) are prominent antioxidant enzymes protecting against harmful biotic activities. For example, SOD transforms superoxide into hydrogen peroxide, which exhibits limited reactivity but can be transformed into highly reactive radicals, particularly the hydroxyl radical, responsible for numerous detrimental effects [287].

TEM and SEM studies observe damages in the membrane and pit formation on cell surfaces induced by NPs [70]. Application of green-synthesized Ag NPs showed extensively ruptured cell walls, cytoplasmic and nucleic content leakage, and swollen structures causing bacterial death [61]. Similarly, the application of bio-synthesized Ag NPs to *F. graminearum* strain PH-1 demonstrated antimicrobial activity as a result of hyphal distortion and cell wall injury [105]. Silver NPs induced hyphal shrinkage, deformation, and vacuole formation, demonstrating substantial inhibitory activity against fungal pathogens and resulting in the degradation of organelles, structures, and cells [147]. Zinc nitrate-derived ZnO NPs altered the fungal cell wall structure and induced cell death [179]. The sporangial membrane of *Cercospora beticola* was damaged by TiO_2 NPs, which thereby prevented the spread of the fungus in sugar beetroot [86]. TiO_2 NPs effectively inhibited the mycelial growth of *Ustilago tritici* mycelium, hence preventing wheat rust [107]. Ag NPs caused chromosomes to remain bound together and break into smaller pieces, while ZnO NPs treatment caused chromosomes to break apart in viruses [165]. When exposed for an extended period of time, TiO_2 NPs disrupt mitosis by causing spindle misalignment and improper chromosome separation. Ag NPs also reacted with the -SH group of tubulin protein, altering chromosome arrangement during the cell cycle [260]. Additionally, ZnO and Fe_3O_4 NPs inhibited plant pathogens by interacting with the glycoprotein viral envelope, directly damaging the *Tobacco mosaic virus* (TMV) shell protein, changing its aggregation, and preventing TMV from functioning [41, 42].

10 Potential Risks and Environmental Impact of Nanotechnology in Agriculture

Despite huge potential, the use of NPs in agriculture also presents significant threats and challenges to human health and the environment. Preliminary investigations into NPs have revealed significant health risks and demonstrated toxicological consequences, including tissue damage and effects on essential organs when introduced into the human body. The impact of NPs on plants can be either beneficial or harmful, depending on factors such as dosage, size, exposure duration, level of exposure, and whether the NPs are synthetic or biological. The toxicity of metal NPs may link to oxidative stress events, the generation of ROS within cells, and the stimulation of pro-inflammatory mediators. This can lead to various cellular dysfunctions, including

DNA and protein damage, lysosomal hydrolase disruption, mitochondrial dysfunction, apoptosis, cell membrane damage, cytoplasmic disorders, ATP level alterations, and changes in cell membrane permeability [128]. NPs can disperse from croplands into soil, water, and the atmosphere through leaching, surface runoff caused by rain, and transportation via air circulation or trophic transfer [174]. Research indicates that microorganisms in soil, sediments, and plant roots can absorb NPs, which are then transported from the roots to other parts of the plants, leading to their accumulation [76]. Particles can also enter into the food chain as various organisms and animals consume plant products, or their wastes [281, 280]. This can result in the negative effects being passed on to their offspring [258]. This phenomenon is common among marine organisms [39] as well as in food webs comprising plants, herbivores, and carnivores [135, 299]. The use of plant-mediated nanoparticles can address these potential risks and the environmental impact of nanotechnology in agriculture.

11 Future Perspective and Challenge

Research on NPs and their potential use in agriculture has advanced significantly. Many studies have already described the green synthesis of metallic NPs from various plant species and their application for plant disease control. However, there are significant challenges to green synthesis and the widespread application of plant-mediated NPs, which include addressing issues related to scalability, reproducibility, and environmental impact. While green NPs offer a sustainable alternative to conventional synthesis techniques, achieving uniform particle size and consistent yields remains challenging due to the variability in plant metabolites. Scaling up the process from laboratory to industrial levels requires careful optimization to maintain green principles and cost-effectiveness. Additionally, the purification of NPs from the reaction mixture is a crucial part that needs further exploration. Research is desirable to comprehend the mechanistic pathways of nanoparticle formation and to develop standardized protocols for reproducibility. Ensuring the safety and biocompatibility of these NPs for various applications, including medical and environmental, is crucial. Comprehensive toxicological studies of the NPs on plants and animals are mandatory to expand their use in various fields. Therefore, multidisciplinary collaboration and technological advancements are essential to overcome these challenges and realize the full potential of plant-mediated green synthesis of NPs.

12 Conclusion

Nanotechnology can revolutionize agriculture through enhanced crop protection and sustainable practices. Plant-based synthesis of NPs exemplifies a shift towards eco-friendly solutions, offering advantages such as scalability, energy efficiency, and biocompatibility. However, challenges remain in ensuring reproducibility, managing environmental impacts, and comprehensively assessing the safety of nanomaterials across diverse ecosystems. The benefits of NPs in agriculture are multifaceted, including improved pesticide efficacy, reduced environmental contamination, and targeted delivery systems that minimize ecological harm. Metallic NPs synthesized from plant extracts exhibit significant antimicrobial properties, which are crucial for combating plant diseases caused by bacteria, fungi, and viruses. Their ability to disrupt pathogen structures and functions underscores their potential in sustainable crop management. Yet, concerns persist regarding the potential adverse effects of NPs on public health and the environment. Issues such as nanoparticle dispersion, accumulation in ecosystems, and long-term impacts necessitate thorough toxicological evaluations and the development of regulatory frameworks to mitigate risks effectively.

Acknowledgements This research was financially supported by the Ministry of Education, Bangladesh.

References

1. Abdallah Y, Ogunyemi SO, Abdelazez A, Zhang M, Hong X, Ibrahim E, Hossain A, Fouad H, Li B, Chen J (2019) The green synthesis of MgO nano-flowers using rosmarinus officinalis L. (Rosemary) and the antibacterial activities against *Xanthomonas oryzae* pv. *oryzae*. BioMed Res Int (1):5620989
2. Abubakar MH, Kuta FA, Oladejo TJ, Usman AN, David BJ (2019) Biosynthesis and antibacterial potential of *Tectona grandis* mediated magnesium oxide nanorods. J Bio-Sci 27:109–120
3. Agarwal H, Nakara A, Menon S, Shanmugam V (2019) Eco-friendly synthesis of zinc oxide nanoparticles using *Cinnamomum Tamala* leaf extract and its promising effect towards the antibacterial activity. J Drug Deliv Sci Technol 53:101212
4. Aher YB, Jain GH, Patil GE, Savale AR, Ghotekar SK, Pore DM, Pansambal SS, Deshmukh KK (2017) Biosynthesis of copper oxide nanoparticles using leaves extract of *Leucaena leucocephala L.* and their promising upshot against diverse pathogens. Int J Mol Clin Microbiol 7(1):776–786
5. Ahlam AA, Shaniba VS, Jayasree PR, Kumar PRM (2021) *Spondias pinnata(L.f.)* Kurz leaf extract derived zinc oxide nanoparticles induce dual modes of apoptotic-necrotic death in HCT 116 and K562 cells. Biol Trace Elem Res 199:1778–1801
6. Ahmad M, Kotb H, Mushtaq S, Waheed-Ur-Rehman M, Maghanga C, Alam M (2022) Green synthesis of Mn + Cu bimetallic nanoparticles using *Vinca rosea* extract and their antioxidant, antibacterial, and catalytic activities. Crystals 12:72
7. Ahmed KBA, Subramaniam S, Veerappan G, Hari N, Sivasubramanian A, Veerappan A (2014) β-Sitosterol-d-glucopyranoside isolated from *Desmostachya bipinnata* mediates photoinduced rapid green synthesis of silver nanoparticles. RSC Adv 4:59130–59136
8. Ahmed SS, Alqahtani AM, Alqahtani T, Alamri AH, Menaa F, Mani RK, R BD, Kavitha K (2022) Green synthesis, characterizations of zinc oxide nanoparticles from aqueous leaf extract of *Tridax procumbens linn.* and assessment of their anti-hyperglycemic activity in streptozoticin-induced diabetic rats. Materials 15(22):8202
9. Ajayan PM, Schadler LS, Braun PV (2003) Nanocomposite science and technology. John Wiley & Sons

10. Akbari B, Tavandashti MP, Zandrahimi M (2011) Particle size characterization of nanoparticles–a practical approach. Iran J Mater Sci Eng 8:48–56
11. Al Mashud MA, Moinuzzaman M, Hossain MS, Ahmed S, Ahsan G, Reza A, Anwar Ratul RB, Uddin MH, Momin MA, Hena Mostofa, Jamal MA (2022) Green synthesis of silver nanoparticles using *Cinnamomum tamala* (Tejpata) leaf and their potential application to control multidrug resistant Pseudomonas aeruginosa isolated from hospital drainage water. Heliyon 8(7):09920
12. Alghuthaymi MA, Almoammar H, Rai M, Said-Galiev E, Abd-Elsalam KA (2015) Myconanoparticles: synthesis and their role in phytopathogens management. Biotechnol Biotechnol Equip 29:221–236
13. Alharbi NS, Alsubhi NS, Felimban AI (2022) Green synthesis of silver nanoparticles using medicinal plants: characterization and application. J Radiat Res Appl Sci 15:109–124
14. Ali M, Kim B, Belfield KD, Norman D, Brennan M, Ali GS (2015) Inhibition of *Phytophthora parasitica* and *P. capsici* by silver nanoparticles synthesized using aqueous extract of *Artemisia absinthium*. Phytopathology 105:1183–1190
15. Ali M, Mosa K, El-Keblawy A et al (2019) Exogenous production of silver nanoparticles by *Tephrosia apollinea* living plants under drought stress and their antimicrobial activities. Nanomaterials 9:1716
16. Ali KA, Yao R, Wu W, Masum MMI, Luo J, Wang Y, Zhang Y, An Q, Sun G, Li B (2020) Biosynthesis of silver nanoparticle from pomelo (*Citrus maxima*) and their antibacterial activity against *Acidovorax oryzae* RS-2. Mater Res Express 7:015097
17. Alsammarraie FK, Wang W, Zhou P, Mustapha A, Lin M (2018) Green synthesis of silver nanoparticles using turmeric extracts and investigation of their antibacterial activities. Colloids Surf B 171:398–405
18. Alshahrani MY, Rafi Z, Alabdallah NM, Shoaib A, Ahmad I, Asiri M, Zaman GS, Wahab S, Saeed M, Khan S (2021) A comparative antibacterial, antioxidant, and antineoplastic potential of *Rauwolfia serpentina* (L.) leaf extract with its biologically synthesized gold nanoparticles (R-AuNPs). Plants 10:2278
19. Amrin MI, Roshan MM, SaiGowri R, Durai SCV (2024) Green synthesis of silver oxide nanoparticles using *Trigonella foenum-graecum* leaf extract and their characterization. Semicond Phys Quant Electron Optoelectron 27:162–168
20. Anandalakshmi K (2021) Green synthesis, characterization and antibacterial activity of silver nanoparticles using *Chenopodium album* leaf extract. Indian J Pure Appl Phys 59:456–461
21. Ankamwar B, Kirtiwar S, Shukla AC (2020) Plant-mediated green synthesis of nanoparticles. In: Springer eBooks, pp 221–234
22. Anu K, Singaravelu G, Murugan K, Benelli G (2017) Green-synthesis of selenium nanoparticles using garlic cloves (*Allium sativum*): biophysical characterization and cytotoxicity on vero cells. J Cluster Sci 28:551–563
23. Anupama N, Madhumitha G (2017) Green synthesis and catalytic application of silver nanoparticles using Carissa carandas fruits. Inorg Nano-metal Chem 47:116–120
24. Arif H, Yasir M, Ali F, Nazir A, Ali A, Huwayz MA, Alwadai N, Iqbal M (2023) Photocatalytic degradation of atrazine and abamectin using *Chenopodium album* leaves extract mediated copper oxide nanoparticles. Z Phys Chem 237:689–705
25. Armendariz V, Herrera I, Peralta-Videa JR, Jose-Yacaman M, Troiani H, Santiago P, Gardea-Torresdey JL (2004) Size controlled gold nanoparticle formation by *Avena sativa* biomass: use of plants in nanobiotechnology. J Nanopart Res 6:377–382
26. Aromal SA, Philip D (2012) Green synthesis of gold nanoparticles using *Trigonella foenum-graecum* and its size-dependent catalytic activity. Spectrochim Acta Part A Mol Biomol Spectrosc 97:1–5
27. Arsiya F, Sayadi MH, Sobhani S (2017) Green synthesis of palladium nanoparticles using *Chlorella vulgaris*. Mater Lett 186:113–115
28. Awais S, Munir H, Najeeb J, Anjum F, Naseem K, Kausar N, Shahid M, Irfan M, Najeeb N (2023) Green synthesis of iron oxide nanoparticles using *Bombax malabaricum* for antioxidant, antimicrobial and photocatalytic applications. J Clean Prod 406:136916

29. Ayodhya D, Veerabhadram G (2017) One-pot green synthesis, characterization, photocatalytic, sensing and antimicrobial studies of *Calotropis gigantea* leaf extract capped CdS NPs. Mater Sci Eng B, Solid-state Mater Adv Technol 225:33–44
30. Babu RH, Yugandhar P, Savithramma N (2018) Synthesis, characterization and antimicrobial studies of bio silica nanoparticles prepared from *Cynodon dactylon*L.: a green approach. Bull Mater Sci 41
31. Bali R, Razak N, Lumb A, Harris AT (2006) The synthesis of metallic nanoparticles inside live plants. IEEE 2150–3591
32. Banerjee M, Rajeswari VD (2024) Green synthesis of selenium nanoparticles using leaf extract of *Moringa oleifera*, their biological applications, and effects on the growth of *Phaseolus vulgaris*: agricultural synthetic biotechnology for sustainable nutrition. Biocatal Agric Biotechnol 55:102978
33. Banfield JF, Zhang H (2001) Nanoparticles in the environment. Rev Mineral Geochem 44(1):1–58
34. Basavegowda N, Kumar GD, Tyliszczak B, Wzorek Z, Sobczak-Kupiec A (2015) One-step synthesis of highly-biocompatible spherical gold nanoparticles using *Artocarpus heterophyllus Lam.* (jackfruit) fruit extract and its effect on pathogens. Ann Agric Environ Med 22:84–89
35. Basha SK, Govindaraju K, Manikandan R, Ahn JS, Bae EY, Singaravelu G (2010) Phytochemical mediated gold nanoparticles and their PTP 1B inhibitory activity. Colloids Surf, B 75:405–409
36. Bawazeer S, Khan I, Rauf A, Aljohani ASM, Alhumaydhi FA, Khalil AA, Qureshi MN, Ahmad L, Khan SA (2022) Black pepper (*Piper nigrum*) fruit-based gold nanoparticles (BP-Au NPs): synthesis, characterization, biological activities, and catalytic applications-a green approach. Green Process Synth 11:11–28
37. Beattie IR, Haverkamp RG (2011) Silver and gold nanoparticles in plants: sites for the reduction to metal. Metallomics 3:628
38. Berber S, Kwon YK, Tománek D (2000) Unusually high thermal conductivity of carbon nanotubes. Phys Rev Lett 84(20):4613–4616
39. Bielmyer-Fraser GK, Jarvis TA, Lenihan HS, Miller RJ (2014) Cellular partitioning of nanoparticulate versus dissolved metals in marine phytoplankton. Environ Sci Technol 48:13443–13450
40. Bundele M, Rane N, Lande V, Dani A, Shende S (2023) Green synthesis of TiO_2 nanoparticle from *Plumeria rubra* L. leaves for anticorrosive application. Mater Today Proceed 72:1685–1691
41. Cai L, Cai L, Jia H, Liu C, Wang D, Sun X (2020) Foliar exposure of Fe3O4 nanoparticles on *Nicotiana benthamiana*: evidence for nanoparticles uptake, plant growth promoter and defense response elicitor against plant virus. J Hazard Mater 393:122415
42. Cai L, Liu C, Fan G, Liu C, Sun X (2019) Preventing viral disease by ZnO NPs through directly deactivating TMV and activating plant immunity in *Nicotiana benthamiana*. Environ Sci Nano 6:3653–3669
43. Carlson C, Hussain SM, Schrand AM, Braydich-Stolle LK, Hess KL, Jones RL, Schlager JJ (2008) Unique cellular interaction of silver nanoparticles: size-dependent generation of reactive oxygen species. J Phys Chem B 112:13608–13619
44. Chandraker SK, Lal M, Kumar A, Shukla R (2022) *Justicia adhatoda* L. mediated green synthesis of silver nanoparticles and assessment of their antioxidant, hydrogen peroxide sensing and optical properties. Mater Technol 37:1355–1365
45. Chandrasekaran S, Anbazhagan V (2024) Green synthesis of ZnO and V-Doped ZnO nanoparticles using *Vinca rosea* plant leaf for biomedical applications. Appl Biochem Biotechnol 196:50–67
46. Chen JC, Lin ZH, Ma XX (2003) Evidence of the production of silver nanoparticles via pretreatment of *Phoma* sp.3.2883 with silver nitrate. Lett Appl Microbiol 37:105–108
47. Chowdappa P, Gowda S (2013) Nanotechnology in crop protection: status and scope. Pest Manag Hortic Ecosyst 19(2):131–151

48. Coman C, Leopold LF, Rugină OD, Barbu-Tudoran L, Leopold N, Tofană M, Socaciu C (2014) Green synthesis of gold nanoparticles by *Allium sativum* extract and their assessment as SERS substrate. J Nanopart Res 16:1–9
49. Dakal TC, Kumar A, Majumdar RS, Yadav V (2016) Mechanistic basis of antimicrobial actions of silver nanoparticles. Front Microbiol 7:1831
50. Daniel MC, Astruc D (2004) Gold nanoparticles: assembly, supramolecular chemistry, quantum-size-related properties, and applications toward biology, catalysis, and nanotechnology. Chem Rev 104(1):293–346
51. Dat TD, Cong CQ, Nhi TLH, Khang PT, Nam NTH, Tinh NT, Hue DT, Hieu NH (2023) Green synthesis of gold nanoparticles using *Andrographis paniculata* leaves extract for lead ion detection, degradation of dyes, and bioactivities. Biochem Eng J 200:109103
52. Da'na E, Taha A, Afkar E (2018) Green synthesis of iron nanoparticles by *Acacia nilotica* pods extract and its catalytic, adsorption, and antibacterial activities. Appl Sci 8(10):1922
53. Deng ZJ, Liang M, Monteiro M, Toth I, Minchin RF (2011) Nanoparticle-induced unfolding of fibrinogen promotes Mac-1 receptor activation and inflammation. Nat Nanotechnol 6:39–44
54. Dey SK, Mandal B, Bhowmik M, Ghosh LK (2009) Development and in vitro evaluation of Letrozole loaded biodegradable nanoparticles for breast cancer therapy. Braz J Pharm Sci 45:585–591
55. Dhavale RP, Patil SS, Jadhav SU, Dhavale RP, Agawane SB (2022) Green synthesis-mediated iron oxide nanoparticles using Sphagneticola trilobata *(L.)* for antibacterial and anticancer assessment. Pharmacogn Mag 18:953–961
56. Dhuper S, Panda D, Nayak PL (2012) Green synthesis and characterization of zero valent iron nanoparticles from the leaf extract of *Mangifera indica*. Nano Trends J Nanotech App 13(2):16–22
57. Din MI, Nabi AG, Rani A, Aihetasham A, Mukhtar M (2018) Single step green synthesis of stable nickel and nickel oxide nanoparticles from *Calotropis gigantea*: catalytic and antimicrobial potentials. Environ Nanotechnol Monit Manag 9:29–36
58. Droepenu EK, Asare EA, Wee BS, Wahi RB, Ayertey F, Kyene MO (2021) Biosynthesis, characterization, and antibacterial activity of ZnO nanoaggregates using aqueous extract from *Anacardium occidentale* leaf: comparative study of different precursors. Beni-Seuf Univ J Basic Appl Sci 10:1–10
59. Durai P, Chinnasamy A, Gajendran B, Ramar M, Pappu S, Kasivelu G, Thirunavukkarasu A (2014) Synthesis and characterization of silver nanoparticles using crystal compound of sodium para-hydroxybenzoate tetrahydrate isolated from *Vitex negundo*. L leaves and its apoptotic effect on human colon cancer cell lines. Eur J Med Chem 84:90–99
60. Edison TJI, Sethuraman MG (2012) Instant green synthesis of silver nanoparticles using *Terminalia chebula* fruit extract and evaluation of their catalytic activity on reduction of methylene blue. Process Biochem 47:1351–1357
61. Elbeshehy EKF, Elazzazy AM, Aggelis G (2015) Silver nanoparticles synthesis mediated by new isolates of *Bacillus* spp., nanoparticle characterization and their activity against Bean Yellow Mosaic Virus and human pathogens. Front Microbiol 6:453
62. Ewunkem AJ, Johnson N, Beard AF, Tshimanga I, Justice B, Meixner J (2024) Synthesis of silver nanoparticles from honeybees and its antibacterial potential. Open J Med Microbiol 14:77–92
63. FAO (2017) Food and agriculture organization of the United Nations Rome. https://www.fao.org/3/i6583e/i6583e.pdf.
64. FAO (2021) News article: new standards to curb the global spread of plant pests and diseases. https://www.fao.org/newsroom/detail/Climate-change-fans-spread-of-pests-and-threatens-plants-and-crops-new-FAO-study/en
65. FAO (2022) FAOSTAT (Food and agriculture organization of the united nations statistical database). https://www.fao.org/3/cc2211en/cc2211en.pdf
66. Farooq A, Khan UA, Ali H, Sathish M, Naqvi SAH, Iqbal S, Ali H, Mubeen I, Amir MB, Mosa WFA, Baazeem A, Moustafa M, Alrumman S, Shati A, Negm S (2022) Green chemistry based synthesis of zinc oxide nanoparticles using plant derivatives of *Calotropis gigantea* (Giant

Milkweed) and its biological applications against various bacterial and fungal pathogens. Microorganisms 10:2195
67. Fatima F, Aldawsari MF, Ahmed MM, Anwer MdK, Naz M, Ansari MJ, Hamad AM, Zafar A, Jafar M (2021) Green synthesized silver nanoparticles using *Tridax procumbens* for topical application: excision wound model and histopathological studies. Pharmaceutics 13:1754
68. Fouad H, Yang G, El-Sayed AA, Mao G, Khalafallah D, Saad M, Ga'al H, Ibrahim E, Mo J (2021) Green synthesis of Ag NP–ligand complexes and their toxicological effects on *Nilaparvata lugens*. J Nanobiotechnology 19:1–17
69. Fu PP, Xia Q, Hwang H-M, Ray PC, Yu H (2014) Mechanisms of nanotoxicity: generation of reactive oxygen species. Yàowù Shípǐn Fēnxī 22:64–75
70. Gahlawat G, Choudhury AR (2019) A review on the biosynthesis of metal and metal salt nanoparticles by microbes. RSC Adv 9:12944–12967
71. Gan PP, Ng SH, Huang Y, Li SFY (2012) Green synthesis of gold nanoparticles using palm oil mill effluent (POME): a low-cost and eco-friendly viable approach. Biores Technol 113:132–135
72. Gangapuram BR, Bandi R, Alle M, Dadigala R, Kotu GM, Guttena V (2018) Microwave assisted rapid green synthesis of gold nanoparticles using *Annona squamosa* L peel extract for the efficient catalytic reduction of organic pollutants. J Mol Struct 1167:305–315
73. Gangwar RK, Dhumale VA, Gosavi SW, Sharma RB, Datar SS (2013) Catalytic activity of Allamanda mediated phytosynthesized anisotropic gold nanoparticles. Adv Nat Sci Nanosci Nanotechnol 4:045005
74. Gardea-Torresdey JL, Gomez E, Peralta-Videa JR, Parsons JG, Troiani H, Jose-Yacaman M (2003) Alfalfa sprouts: a natural source for the synthesis of silver nanoparticles. Langmuir 19:1357–1361
75. Gardea-Torresdey JL, Parsons JG, Gomez E, Peralta-Videa J, Troiani HE, Santiago P, Yacaman MJ (2002) Formation and growth of Au nanoparticles inside live alfalfa plants. Nano Lett 2(4):397–401
76. Gardea-Torresdey JL, Rico CM, White JC (2014) Trophic transfer, transformation, and impact of engineered nanomaterials in terrestrial environments. Environ Sci Technol 48:2526–2540
77. Gavade NL, Kadam AN, Suwarnkar MB, Ghodake VP, Garadkar KM (2015) Biogenic synthesis of multi-applicative silver nanoparticles by using *Ziziphus Jujuba* leaf extract. Spectrochim Acta Part A Mol Biomol Spectrosc 136:953–960
78. Gavade SJM, Nikam GH, Sabale SR, Tamhankar BV (2016) Green synthesis of fluorescent silver nanoparticles using *Acacia nilotica* gum extract for kinetic studies of 4-nitrophenol reduction. Mater Today Proceed 3:4109–4114
79. Gemachu LY, Birhanu AL (2024) Green synthesis of ZnO, CuO and NiO nanoparticles using neem leaf extract and comparing their photocatalytic activity under solar irradiation. Green Chem Lett Rev 17(1):2293841
80. Girma A (2023) Alternative mechanisms of action of metallic nanoparticles to mitigate the global spread of antibiotic-resistant bacteria. Cell Surf 10:100112
81. Gomollón-Bel F (2019) Ten chemical innovations that will change our world: IUPAC identifies emerging technologies in chemistry with potential to make our planet more sustainable. Chem Int 41:12–17
82. Goyal V, Singh A, Singh J, Kaur H, Kumar S, Rawat M (2022) Biogenically structural and morphological engineering of *Trigonella foenum-graecum* mediated SnO_2 nanoparticles with enhanced photocatalytic and antimicrobial activities. Mater Chem Phys 282:125946
83. Guzmán KA, Taylor MR, Banfield JF (2009) Environmental risks of nanotechnology: national nanotechnology initiative funding, 2000–2004. Environ Sci Technol 40(5):1401–1407
84. Hadi RG, Samaka HM, Abdulridha WAM (2020) Green synthesis of silver nanoparticles from *Nerium oleander* and evaluate its effects as an antifungal agent. Biochem Cell Arch 20(2):1–7
85. Hamida RS, Ali MA, Abdelmeguid NE, Al-Zaban MI, Baz L, Bin-Meferij MM (2021) Lichens-a potential source for nanoparticles fabrication: a review on nanoparticles biosynthesis and their prospective applications. J Fungi 7:291

86. Hamza A, El-Mogazy S, Derbalah A (2016) Fenton reagent and titanium dioxide nanoparticles as antifungal agents to control leaf spot of sugar beet under field conditions. J Plant Prot Res 56:270–278
87. Hano C, Abbasi BH (2021) Plant-based green synthesis of nanoparticles: production, characterization and applications. Biomolecules 12:31
88. Haq IU, Ijaz S (2019) Use of metallic nanoparticles and nanoformulations as nanofungicides for sustainable disease management in plants. In: Nanotechnology in the life sciences, pp 289–316
89. Hariharan D, Srinivasan K, Lc N (2017) Synthesis and characterization of TiO2 nanoparticles using *Cynodon dactylon* leaf extract for antibacterial and anticancer (A549 Cell Lines) activiity. J Nanomed Res Synth 5:1–5
90. Harris AT, Bali R (2010) On the formation and extent of uptake of silver nanoparticles by live plants. J Nanopart Res 10:691–695
91. Harris AT, Bali R (2008) On the formation and extent of uptake of silver nanoparticles by live plants. J Nanopart Res 10(4):691–695
92. Haverkamp RG, Marshall AT, Van Agterveld D (2007) Pick your carats: nanoparticles of gold–silver–copper alloy produced in vivo. J Nanopart Res 9:697–700
93. Hemalatha S, Makeswari M (2017) Green synthesis, characterisation, and antibacterial studies of CuO nanoparticles from *Eichhornia crassipes*. Rasayan J Chem 10:838–843
94. Hernández-Díaz JA, Garza-García JJ, Zamudio-Ojeda A, León-Morales JM, López-Velázquez JC, García-Morales S (2020) Plant-mediated synthesis of nanoparticles and their antimicrobial activity against phytopathogens. J Sci Food Agric 101:1270–1287
95. Hii YS, Jeevanandam J, Chan YS (2018) Plant mediated green synthesis and nanoencapsulation of MgO nanoparticle from *Calotropis gigantea*: characterisation and kinetic release studies. Inorg Nano-Met Chem 48:620–631
96. Hossain A, Abdallah Y, Ali MdA, Masum MdMI, Li B, Sun G, Meng Y, Wang Y, An Q (2019) Lemon-fruit-based green synthesis of zinc oxide nanoparticles and titanium dioxide nanoparticles against soft rot bacterial pathogen Dickeya dadantii. Biomolecules 9:863
97. Hossain MM, Robinson NA, Mok YS, Wu S (2023) Investigation of silver nanoparticle synthesis with various nonthermal plasma reactor configurations. Arab J Chem 16:105174
98. Hossain MM, Sultana F (2020) Application and mechanisms of plant growth promoting fungi (PGPF) for phytostimulation. In IntechOpen eBooks. https://doi.org/10.5772/intechopen.92338
99. Hossain MM, Sultana F, Islam S (2017) Plant growth-promoting fungi (PGPF): Phytostimulation and induced systemic resistance. In: Singh D, Singh H, Prabha R (eds) Plant-microbe interactions in agro-ecological perspectives. Springer, Singapore. https://doi.org/10.1007/978-981-10-6593-4_6
100. Hossain MM, Sultana F, Kubota M et al (2008) Systemic resistance to bacterial leaf speck pathogen in *Arabidopsis thaliana* induced by the culture filtrate of a plant growth-promoting fungus (PGPF) *Phoma* sp. GS8–1. J Gen Plant Pathol 74:213–221. https://doi.org/10.1007/s10327-008-0093-5
101. Hossain MM (2024a) Upscaling plant defense system through the application of plant growth-promoting fungi (PGPF). In Microbial technology for agro-ecosystems: crop productivity, sustainability, and biofortification, pp 61–95. Elsevier. https://doi.org/10.1016/B978-0-443-18446-8.00018-8
102. Hossain MM (2024b) Diseases of Lablab. In: Elmer WH, McGrath M, McGovern RJ (eds) Handbook of vegetable and herb diseases. Handbook of plant disease management. Springer, Cham. https://doi.org/10.1007/978-3-030-35512-8_20-1
103. Hublikar LV, Ganachari SV, Raghavendra N, Patil VB, Banapurmath NR (2021) Green synthesis silver nanoparticles via *Eichhornia crassipes* leaves extract and their applications. Curr Res Green Sustain Chem 4:100212
104. Hussain S, Muazzam MA, Ahmed M, Ahmad M, Mustafa Z, Murtaza S, Ali J, Ibrar M, Shahid M, Imran M (2023) Green synthesis of nickel oxide nanoparticles using *Acacia nilotica* leaf extracts and investigation of their electrochemical and biological properties. J Taibah Univ Sci 17(1):2170162

105. Ibrahim E, Zhang M, Zhang Y, Hossain A, Qiu W, Chen Y, Wang Y, Wu W, Sun G, Li B (2020) Green-synthesization of silver nanoparticles using endophytic bacteria isolated from garlic and its antifungal activity against wheat fusarium head blight pathogen *Fusarium graminearum*. Nanomaterials 10:219
106. Ilavenil KK, Senthilkumar V, Kasthuri A, Sivaraman S (2024) Nanoparticles (MgO) utilising *Justicia adhatoda* and an evaluation of their antimicrobial properties. Materials today: proceedings
107. Irshad MA, Nawaz R, Rehman MZU, Imran M, Ahmad J, Ahmad S, Inam A, Razzaq A, Rizwan M, Ali S (2020) Synthesis and characterization of titanium dioxide nanoparticles by chemical and green methods and their antifungal activities against wheat rust. Chemosphere 258:127352
108. Islam AKMS, Bhuiyan R, Nihad SAI, Akter R, Khan MAI, Akter S, Islam MR, Khokon MAR, Latif MA (2024) Green synthesis and characterization of silver nanoparticles and its efficacy against *Rhizoctonia solani*, a fungus causing sheath blight disease in rice. PLoS ONE 19:e0304817
109. Islam MJ, Khatun MT, Rahman MR, Alam MM (2021) Green synthesis of copper oxide nanoparticles using *Justicia adhatoda* leaf extract and its application in cotton fibers as antibacterial coatings. AIP Adv 11
110. Iwase E (1928) The preparation of red-gold sols by using as reducing agents the extracts of fresh leaves and plants. Kolloid Z 44:42–43
111. Jagana D, Hegde YR, Lella R (2017) Green nanoparticles-a novel approach for the management of banana anthracnose caused by *Colletotrichum musae*. Int J Curr Microbiol App Sci 6:1749–1756
112. Jagathesan G, Rajiv P (2018) Biosynthesis and characterization of iron oxide nanoparticles using *Eichhornia crassipes* leaf extract and assessing their antibacterial activity. Biocatal Agric Biotechnol 13:90–94
113. Jagtap UB, Bapat VA (2013) Green synthesis of silver nanoparticles using *Artocarpus heterophyllus* Lam. seed extract and its antibacterial activity. Ind Crops Prod 46:132–137
114. Jain PK, Huang X, El-Sayed IH, El-Sayed MA (2008) Noble metals on the nanoscale: optical and photothermal properties and some applications in imaging, sensing, biology, and medicine. Acc Chem Res 41(12):1578–1586
115. Jain R, Mendiratta S, Kumar L, Srivastava A (2021) Green synthesis of iron nanoparticles using *Artocarpus heterophyllus* peel extract and their application as a heterogeneous Fenton-like catalyst for the degradation of Fuchsin Basic dye. Curr Res Green Sustain Chem 4:100086
116. Jain A, Sinilal B, Starnes DL, Sanagala R, Krishnamurthy S, Sahi SV (2014) Role of Fe-responsive genes in bioreduction and transport of ionic gold to roots of Arabidopsis thaliana during synthesis of gold nanoparticles. Plant Physiol Biochem 84:189–196
117. Jayaseelan C, Ramkumar R, Rahuman AA, Perumal P (2013) Green synthesis of gold nanoparticles using seed aqueous extract of *Abelmoschus esculentus* and its antifungal activity. Ind Crops Prod 45:423–429
118. Jenifer AA, Malaikozhundan B, Arsiya kumar S, Anjugam M, Iswarya A, Vaseeharan B (2020) Green synthesis and characterization of silver nanoparticles (Ag NPs) using leaf extract of *Solanum nigrum* and assessment of toxicity in vertebrate and invertebrate aquatic animals. J Clust Sci 31:989–1002
119. Jha AK, Chakraborty S (2020) Photocatalytic degradation of Congo red under UV irradiation by zero valent iron nano particles (nZVI) synthesized using *Shorea robusta* (Sal) leaf extract. Water Sci Technol 82(11):2491–2502
120. Joseph AT, Prakash P, Narvi S (2016) Phytofabrication and characterization of copper nanoparticles using *Allium sativum* and its antibacterial activity. Int J Sci Eng Technol 4:463–472
121. Jyothi D, Priya S, James JP (2022) Microwave-assisted green synthesis of silver nanoparticles using extract of *Spondias pinnata*bark. J Microbiol Biotechnol Food Sci 12:1951
122. Kalpana V, Chakraborthy P, Palanichamy V, Rajeswari VD (2016) Synthesis and characterization of copper nanoparticles using *Tridax procumbens* and its application in degradation of bismarck brown. Int J ChemTech Res 9:498–507

123. Kalu AO, Egwim EC, Jigam AA, Muhammed HL (2022) Green synthesis of magnetite nanoparticles using *Calotropis procera* leaf extract and evaluation of its antimicrobial activity. Nano Express 3:045004
124. Kamat PV (2002) Photophysical, photochemical and photocatalytic aspects of metal nanoparticles. J Phys Chem B 106(32):7729–7744
125. Karthik K, Dhanuskodi S, Gobinath C, Prabukumar S, Sivaramakrishnan S (2017) *Andrographis paniculata* extract mediated green synthesis of CdO nanoparticles and its electrochemical and antibacterial studies. J Mater Sci Mater Electron 28:7991–8001
126. Karthik K, Shashank M, Revathi V, Tatarchuk T (2018) Facile microwave-assisted green synthesis of NiO nanoparticles from *Andrographis paniculata* leaf extract and evaluation of their photocatalytic and anticancer activities. Mol Cryst Liq Cryst 673:70–80
127. Karunakaran G, Jagathambal M, Gusev A, Kolesnikov E, Mandal AR, Kuznetsov D (2016) *Allamanda cathartica* flower's aqueous extract-mediated green synthesis of silver nanoparticles with excellent antioxidant and antibacterial potential for biomedical application. MRS Communications 6:41–46
128. Kausar H, Mehmood A, Khan RT, Ahmad KS, Hussain S, Nawaz F, Ullah TS (2022) Green synthesis and characterization of copper nanoparticles for investigating their effect on germination and growth of wheat. PLoS One 17(6):e0269987
129. Kazi MA (2019) Exploration of dye degradation potential of eco-friendly green synthesized TiO2 nanoparticles using extract of *Acacia nilotica*. Pak J Anal Environ Chem 20:135–140
130. Kazi S, Nirwan S, Kunde S, Jadhav S, Rai M, Kamble D, Sayyed S, Chavan P (2022) Green synthesis, characterization and bio-evaluation of zirconium nanoparticles using the dried biomass of *Sphagneticola trilobata* Plant Leaf. BioNanoScience 12:731–740
131. Kermani M, Mostafapour A, Sabouri Z, Gheibihayat SM, Darroudi M (2023) The photocatalytic, cytotoxicity, and antibacterial properties of zinc oxide nanoparticles synthesized using *Trigonella foenum-graecum* L extract. Environ Sci Pollut Res Int 30:19313–19325
132. Khairan K, Zahraturriaz N, Jalil Z (2019) Green synthesis of sulphur nanoparticles using aqueous garlic extract (*Allium sativum*). Rasayan J Chem 12:50–57
133. Khan F, Shariq M, Asif M, Siddiqui MA, Malan P, Ahmad F (2022) Green Nanotechnology: plant-Mediated nanoparticle synthesis and application. Nanomaterials 12:673
134. Khan MZH, Tareq FK, Hossen MA, Roki MNAM (2018) Green synthesis and characterization of silver nanoparticles using *Coriandrum savitum*leaf extract. J Eng Sci Technol 13:158–166
135. Kim JI, Park H-G, Chang K-H, Nam DH, Yeo M-K (2016) Trophic transfer of nano-TiO_2 in a paddy microcosm: a comparison of single-dose versus sequential multi-dose exposures. Environ Pollut 212:316–324
136. Kiranmayee M, Rajesh N, Vani MV, Khadri H, Mohammed A, Chinni SV, Ramachawolran G, Riazunnisa K, Moussa AY (2023) Green synthesis of *Piper nigrum* copper-based nanoparticles: in silico study and ADMET analysis to assess their antioxidant, antibacterial, and cytotoxic effects. Front Chem 11
137. Klaine SJ, Alvarez PJ, Batley GE, Fernandes TF, Handy RD, Lyon DY, Mahendra S, McLaughlin MJ, Lead JR (2008) Nanomaterials in the environment: behavior, fate, bioavailability, and effects. Environ Toxicol Chem 27(9):1825–1851
138. Kotakadi VS, Rao YS, Gaddam SA, Prasad TNVKV, Reddy AV, Gopal DVRS (2013) Simple and rapid biosynthesis of stable silver nanoparticles using dried leaves of *Catharanthus roseus*. Linn. G. Donn and its antimicrobial activity. Colloids Surf B 105:194–198
139. Kumar A, Choudhary A, Kaur H, Guha S, Mehta S, Husen A (2022) Potential Applications of engineered nanoparticles in plant disease management: a critical update. Chemosphere 295:133798
140. Kumar S, Kumari S, Kumari A, Kondal N, Choudhary P, Dhiman V, Sharma R (2024) Green synthesis of iron oxide nanoparticles from *Calotropis procera* latex: an eco-friendly catalyst for biodiesel production from Calotropis procera seed oil. Clean Technol Environ Policy 1–13
141. Kumari R, Singh JS, Singh DP (2017) Biogenic synthesis and spatial distribution of silver nanoparticles in the legume mungbean plant (*Vigna radiata* L.). Plant Physiol Biochem 110:158–166

142. Lakshmanan SP, Jostar ST, Arputhavalli GJ, Jebasingh S, Josephine CMR (2021) Role of green synthesized CuO Nanoparticles of Trigonella foenum-graecum L. leaves and their impact on structural, optical and antimicrobial activity. Int J Nanosci Nanotechnol 17(2):109–121
143. Lakshmanan SP, Jostar ST, Arputhavalli GJ, Jebasingh S, Josephine CMR (2021) Role of green synthesized CuO nanoparticles of *Trigonella foenum-graecum* L. leaves and their impact on structural, optical and antimicrobial activity
144. Lakshmeesha TR, Sateesh MK, Prasad BD, Sharma SC, Kavyashree D, Chandrasekhar M, Nagabhushana H (2014) Reactivity of crystalline ZnO superstructures against fungi and bacterial pathogens: synthesized using *Nerium oleander* leaf extract. Cryst Growth Des 14:4068–4079
145. Latha D, Sampurnam S, Arulvasu C, Prabu P, Govindaraju K, Narayanan V (2018) Biosynthesis and characterization of gold nanoparticle from *Justicia adhatoda* and its catalytic activity. Mater Today Proceed 5:8968–8972
146. Lemire JA, Harrison JJ, Turner RJ (2013) Antimicrobial activity of metals: mechanisms, molecular targets and applications. Nat Rev Microbiol 11:371–384
147. Li L, Pan H, Deng L, Qian G, Wang Z, Li W, Zhong C (2022) The antifungal activity and mechanism of silver nanoparticles against four pathogens causing kiwifruit post-harvest rot. Front Microbiol 13:988633
148. Liang A, Liu Q, Wen G, Jiang Z (2012) The surface-plasmon-resonance effect of nanogold/silver and its analytical applications. TrAC Trends Anal Chem 37:32–47
149. Lingaraju K, Naika HR, Manjunath K, Nagaraju G, Suresh D, Nagabhushana H (2015) Rauvolfia serpentina-mediated green synthesis of Cuo nanoparticles and its multidisciplinary studies. Acta Metall Sin Engl Lett 28:1134–1140
150. Logambal S, Chandrasekar M, Kumar RA, Inmozhi C, Aravindan S, Uthrakumar R, Naveenkumar S, Muthukumaran A, Kaviyarasu K (2024) Biosynthesis of PbO nanoparticles via Adhatoda vesica (*Justicia adhatoda*) leaves extract for antimicrobial and photocatalytic applications. J King Saud Univ Sci/Mağallaẗ Ǧāmi'aẗ Al-malik Sa'ūd al-'Ulūm 36:103169
151. Mahmud QM, Khan MAI, Akanda AM, Hossain MM, Latif MA, Akter R, Hossain MM (2024) Antifungal potential of commercial silver nanoparticles against rice blast pathogen *Magnaporthe oryzae*. Ann Bangladesh Agric 27(1):17–30. https://doi.org/10.3329/aba.v27i1.70887
152. Majumdar R, Bag BG, Maity N (2013) *Acacia nilotica* (Babool) leaf extract mediated size-controlled rapid synthesis of gold nanoparticles and study of its catalytic activity. Int Nano Lett 3:1–6
153. Mandal BK, Mandal R, Limbu D, Adhikari MD, Chauhan PS, Das R (2022) Green synthesis of AgCl nanoparticles using *Calotropis gigantea*: characterization and their enhanced antibacterial activities. Chem Phys Lett 801:139699
154. Mansha M, Qurashi A, Ullah N, Bakare FO, Khan I, Yamani ZH (2016) Synthesis of In2O3/graphene heterostructure and their hydrogen gas sensing properties. Ceram Int 42:11490–11495
155. Mansou SE, Pattanayak M, Nayak PL (2014) Green synthesis of gold nano particles vii: green synthesis and characterization of gold nano particles using the extract of dhania (Coriandrum sativum) and study of its cytotoxicity properties. Middle-East J Sci Res 21(12):2257–2263
156. Marchiol L, Mattiello A, Pošćić F, Giordano C, Musetti R (2014) In vivo synthesis of nanomaterials in plants: location of silver nanoparticles and plant metabolism. Nanoscale Res Lett 9:1–11
157. Mardina V, Fadlly TA, Harmawan T, Sufriadi E, Iqramullah M, Umar H, Ilyas S (2024) Green synthesis of gold nanoparticles from the aqueous extracts of *Sphagneticola trilobata* (L.) J.F Pruski as anti-breast cancer agents. J Adv Pharm Technol Res 15:75–80
158. Marshall AT, Haverkamp RG, Davies CE, Parsons JG, Gardea-Torresdey JL, Van Agterveld D (2007) Accumulation of gold nanoparticles in *Brassic Juncea*. Int J Phytorem 9:197–206
159. Masum MMI, Siddiqa MM, Ali KA, Zhang Y, Abdallah Y, Ibrahim E, Qiu W, Yan C, Li B (2019) Biogenic synthesis of silver nanoparticles using *Phyllanthus emblica* fruit extract and its inhibitory action against the pathogen Acidovorax oryzae strain rs-2 of rice bacterial brown stripe. Front Microbiol 10:820

160. Mata R, Bhaskaran A, Sadras SR (2016) Green-synthesized gold nanoparticles from *Plumeria alba* flower extract to augment catalytic degradation of organic dyes and inhibit bacterial growth. Particuology 24:78–86
161. Mata R, Nakkala JR, Sadras SR (2015) Catalytic and biological activities of green silver nanoparticles synthesized from *Plumeria alba* (frangipani) flower extract. Mater Sci Eng C Biomim Mater Sens Syst 51:216–225
162. Mathur S, Gautam V, Kumari S, Mishra P (2024) Photocatalytic degradation of organic dye pollutants using ZnO nanoparticles synthesized from *Plumeria alba* L. leaf extract. Int J Life Sci 10:3734–3742
163. Maulana I, Ginting B, Azizah K (2023) Green synthesis of copper nanoparticles employing *Annona squamosa L* extract as antimicrobial and anticancer agents. S Afr J Chem Eng 46:65–71
164. Meenatchi T, Palanimurugan A, Dhanalakshmi A, Maheshkumar V, Natarajan B (2020) Green synthesis of *Cynodon dactylon* capped concentrations on ZnO nanoparticles for antibacterial activity, ROS/ML-DNA treatment and compilation of best controlling microbes by mathematical comparisons. Chem Phys Lett 749:137429
165. Mehrian SK, De Lima R (2016) Nanoparticles cyto and genotoxicity in plants: mechanisms and abnormalities. Environ Nanotechnol Monit Manag 6:184–193
166. Messaoudi NE, Khomri ME, Ablouh E-H, Bouich A, Lacherai A, Jada A, Lima E, Sher F (2022) Biosynthesis of SiO_2 nanoparticles using extract of *Nerium oleander* leaves for the removal of tetracycline antibiotic. Chemosphere 287:132453
167. Messaoudi NE, Ciğeroğlu Z, Şenol ZM, Kazan-Kaya ES, Fernine Y, Gubernat S, Lopicic Z (2024) Green synthesis of $CuFe_2O_4$ nanoparticles from bioresource extracts and their applications in different areas: a review. Biomass Convers Biorefinery 1–22
168. Miao Y, Zhao S, Gao Y, Wang R, Wu Q, Wu H, Luo T (2016) Curcumin pretreatment attenuates inflammation and mitochondrial dysfunction in experimental stroke: the possible role of Sirt1 signaling. Brain Res Bull 121:9–15
169. Milner MJ, Kochian LV (2008) Investigating heavy-metal hyperaccumulation using *Thlaspi caerulescens* as a model system. Ann Bot 102:3–13
170. Mishra S, Singh HB (2014) Biosynthesized silver nanoparticles as a nanoweapon against phytopathogens: exploring their scope and potential in agriculture. Appl Microbiol Biotechnol 99:1097–1107
171. Mishra S, Singh BR, Singh A, Keswani C, Naqvi AH, Singh HB (2014) Biofabricated silver nanoparticles act as a strong fungicide against *Bipolaris sorokiniana* causing spot blotch disease in wheat. PLoS ONE 9:e97881
172. Mohapatra B, Kuriakose S, Mohapatra S (2015) Rapid green synthesis of silver nanoparticles and nanorods using *Piper nigrum* extract. J Alloy Compd 637:119–126
173. Molisch H (1921) The microchemistry of plants XVI. The reduction of silver by chlorophyll granules. Ber Botan Ges 39:136–139
174. Morales-Díaz AB, Ortega-Ortíz H, Juárez-Maldonado A, Cadenas-Pliego G, González-Morales S, Benavides-Mendoza A (2017) Application of nanoelements in plant nutrition and its impact in ecosystems. Adv Nat Sci Nanosci Nanotechnol 8:013001
175. Mostaghni F, Daneshvar A, Sakhaie M (2017) Green synthesis of nano size CoFe2O4 using *Chenopodium album* leaf extract for photo degradation of organic pollutants. Iran Chem Commun 5(17):429–435
176. Mukhoro OC, Roos WD, Jaffer M, Bolton JJ, Stillman MJ, Beukes DR, Antunes E (2017) Very Green photosynthesis of gold nanoparticles by a living aquatic plant: photoreduction of AuIII by the seaweed Ulva armoricana. Chemistry 24:1657–1666
177. Mukunthan KS, Balaji S (2012) Cashew apple juice (*Anacardium occidentale* L.) speeds up the synthesis of silver nanoparticles. Int J Green nanotechnol 4:71–79
178. Mulvaney P (1996) Surface plasmon spectroscopy of nanosized metal particles. Langmuir 12:788–800
179. Munir N, Gulzar W, Abideen Z, Hancock JT, El-Keblawy A, Radicetti E (2023) Nanotechnology improves disease resistance in plants for food security: applications and challenges. Biocatal Agric Biotechnol 51:102781

180. Murr LE, Bang JJ, Esquivel EV, Guerrero PA, Lopez DA (2004) Carbon nanotubes, nanocrystals, and other nanoparticulates in the environment: a TEM study. J Mater Sci Eng A 44(1):13–15
181. Muthuramalingam R, Barroso K, Milagres J, Tedardi V, De Oliveira FF, Takeshita V, Karmous I, El-Tanbouly R, Da Silva W (2024) Tiny but mighty: nanoscale materials in plant disease management. Plant Dis 108:241–255
182. Muthuvel A, Adavallan K, Balamurugan K, Krishnakumar N (2014) Biosynthesis of gold nanoparticles using *Solanum nigrum* leaf extract and screening their free radical scavenging and antibacterial properties. Biomed Prev Nutr 4:325–332
183. Nagai S (1951) The reduction of silver nitrate by plant cell. II. Nature and responsibility of substances which cause the reduction. J Osaka City Univ Inst Polytech Ser D Biol 2:1–8
184. Nawaz S, Maqsood I, Batool F, Sandhu ZY, Hassan S, Akram F, Rashid B (2023) Improvement of abiotic stress tolerance in plants with the application of nanoparticles. In: IntechOpen eBooks
185. Nazeruddin GM, Prasad NR, Waghmare SR, Garadkar KM, Mulla IS (2014) Extracellular biosynthesis of silver nanoparticle using Azadirachta indica leaf extract and its anti-microbial activity. J Alloys Compd 583:272–277
186. Negrescu AM, Killian MS, Raghu SNV, Schmuki P, Mazare A, Cimpean A (2022) Metal oxide nanoparticles: review of synthesis, characterization and biological effects. J Funct Biomater 13:274
187. Nguyen NPU, Dang NT, Doan L, Nguyen TTH (2023) Synthesis of silver nanoparticles: from conventional to 'modern' methods-a review. Processes 11:2617
188. Nowack B, Bucheli TD (2007) Occurrence, behavior and effects of nanoparticles in the environment. Environ Pollut 150(1):5–22
189. Oberdörster G, Oberdörster E, Oberdörster J (2005) Nanotoxicology: an emerging discipline evolving from studies of ultrafine particles. Environ Health Perspect 113(7):823–839
190. Ogunyemi SO, Abdallah Y, Zhang M, Fouad H, Hong X, Ibrahim E, Masum MMI, Hossain A, Mo J, Li B (2019) Green synthesis of zinc oxide nanoparticles using different plant extracts and their antibacterial activity against *Xanthomonas oryzae pv. oryzae*. Artif Cells Nanomed Biotechnol 47:341–352
191. Ogunyemi SO, Zhang F, Abdallah Y, Zhang M, Wang Y, Sun G, Qiu W, Li B (2019) Biosynthesis and characterization of magnesium oxide and manganese dioxide nanoparticles using *Matricaria chamomilla* L. extract and its inhibitory effect on *Acidovorax oryzae* strain RS-2. Artif Cells Nanomed Biotechnol 47:2230–2239
192. Oliveira MM, Ugarte D, Zanchet D, Zarbin AJG (2005) Influence of synthetic parameters on the size, structure, and stability of dodecanethiol-stabilized silver nanoparticles. J Colloid Interface Sci 292:429–435
193. Oueslati MH, Tahar LB, Harrath AH (2020) Synthesis of ultra-small gold nanoparticles by polyphenol extracted from *Salvia officinalis* and efficiency for catalytic reduction of p-nitrophenol and methylene blue. Green Chem Lett Rev 13:18–26
194. Pachaiappan R, Rajendran S, Ramalingam G, Vo D-VN, Priya PM, Soto-Moscoso M (2021) Green synthesis of zinc oxide nanoparticles by *Justicia adhatoda* leaves and their antimicrobial activity. Chem Eng Technol 44:551–558
195. Pai M, Ahmed E, Batakurki S, Kumar SG, Kusanur R (2023) Green synthesis of Palladium magnetic nanoparticles decorated on carbon nanospheres using Chenopodium and their application as heterogenous catalyst in the Suzuki-Miyaura coupling reaction. Appl Surf Sci Adv 16:100427
196. Pamidimukkala K, N P, HS N, AVM K (2021) *Allamanda cathartica* L. latex mediated magnesium oxide nanoparticles as antiproliferative agents. Nanomed Res J 6:257–268
197. Pandiyan N, Murugesan B, Arumugam M, Chinnaalagu D, Samayanan S, Mahalingam S (2021) Ionic liquid mediated green synthesis of Ag-Au/Y2O3 nanoparticles using leaves extracts of *Justicia adhatoda*: structural characterization and its biological applications. Adv Powder Technol 32:2213–2225

198. Panja S, Chaudhuri I, Khanra K, Bhattacharyya N (2016) Biological application of green silver nanoparticle synthesized from leaf extract of *Rauvolfia serpentina* Benth. Asian Pac J Trop Dis 6:549–556
199. Pankhurst QA, Connolly J, Jones SK, Dobson J (2003) Applications of magnetic nanoparticles in biomedicine. J Phys D Appl Phys 36(13):R167–R181
200. Parajuli D, Kawakita H, Inoue K, Ohto K, Kajiyama K (2007) Persimmon peel gel for the selective recovery of gold. Hydrometallurgy 87:133–139
201. Parthiban E, Manivannan N, Ramanibai R, Mathivanan N (2019) Green synthesis of silver-nanoparticles from *Annona reticulata* leaves aqueous extract and its mosquito larvicidal and anti-microbial activity on human pathogens. Biotechnol Rep 21:00297
202. Patil YY, Sutar VB, Tiwari AP (2020) Green synthesis of magnetic iron nanoparticles using medicinal plant *Tridax procumbens* leaf extracts and its application as an antimicrobial agent against e. Coli Inter J Appl Pharm 34–39
203. Paulkumar K, Gnanajobitha G, Vanaja M, Rajeshkumar S, Malarkodi C, Pandian K, Annadurai G (2014) *Piper nigrum* leaf and stem assisted green synthesis of silver nanoparticles and evaluation of its antibacterial activity against agricultural plant pathogens. Sci World J 2014:1–9
204. Pavithra S, Bessy TC, Bindhu VR, Parimaladevi R, Alam MM, Mayandi J, Umadevi M (2023) Photocatalytic and photovoltaic applications of green synthesized titanium oxide (TiO_2) nanoparticles by *Calotropis gigantea* extract. J Alloy Compd 960:170638
205. Peer D, Karp JM, Hong S, Farokhzad OC, Margalit R, Langer R (2007) Nanocarriers as an emerging platform for cancer therapy. Nat Nanotechnol 2(12):751–760
206. Periakaruppan R, Naveen V, Danaraj J (2022) Green synthesis of magnesium oxide nanoparticles with antioxidant potential using the leaf extract of *Piper nigrum*. JOM 74:4817–4822
207. Philip D (2010) Rapid green synthesis of spherical gold nanoparticles using *Mangifera indica* leaf. Spectrochim Acta Part A Mol Biomol Spectrosc 77:807–810
208. Pimentel D, Burgess M (2011) Small amounts of pesticides reaching target insects. Environ Dev Sustain 14:1–2
209. Prabu P, Losetty V (2024) Green synthesis of copper oxide nanoparticles using *Macroptilium Lathyroides (L)* leaf extract and their spectroscopic characterization, biological activity and photocatalytic dye degradation study. J Mol Struct 1301:137404
210. Priya S, Murugan K, Priya A, Dinesh D, Panneerselvam C, Devi GD, Ramohan DB, Ch KPM, Barnard D, Xue R, Hwang J, Nicoletti M, Rasekar RCh, Amsath A, Bhagooli R, Wei H (2014) Green synthesis of silver nanoparticles using *Calotropis gigantea* and their potential mosquito larvicidal property. Int J Pure Appl Zool 2(2):128–137
211. Pöschl U (2005) Atmospheric aerosols: composition, transformation, climate and health effects. Angew Chem Int Ed 44(46):7520–7540
212. Rabouw FT, de Mello DC (2016) Excited-state dynamics in colloidal semiconductor nanocrystals. Top Curr Chem (Z) 374:58
213. Radini IA, Hasan N, Malik MA, Khan Z (2018) Biosynthesis of iron nanoparticles using *Trigonella foenum-graecum* seed extract for photocatalytic methyl orange dye degradation and antibacterial applications. J Photochem Photobiol B 183:154–163
214. Rai M, Ingle A (2012) Role of nanotechnology in agriculture with special reference to management of insect pests. Appl Microbiol Biotechnol 94:287–293
215. Rai M, Gade A, Gaikwad S, Marcato PD, Durán N (2012) Biomedical applications of nanobiosensors: the state-of-the-art. J Braz Chem Soc
216. Raj RA, Al Salhi M, Devanesan S (2017) Microwave-assisted synthesis of nickel oxide nanoparticles using *Coriandrum sativum* leaf extract and their structural-magnetic catalytic properties. Materials 10:460
217. Rajakumar G, Thiruvengadam M, Mydhili G, Gomathi T, Chung I-M (2018) Green approach for synthesis of zinc oxide nanoparticles from *Andrographis paniculata* leaf extract and evaluation of their antioxidant, anti-diabetic, and anti-inflammatory activities. Bioprocess Biosyst Eng 41:21–30

218. Rajesh KM, Ajitha B, Reddy YAK, Suneetha Y, Reddy PS (2018) Assisted green synthesis of copper nanoparticles using *Syzygium aromaticum* bud extract: physical, optical and antimicrobial properties. Optik 154:593–600
219. Rajeshkumar S (2016) Synthesis of silver nanoparticles using fresh bark of *Pongamia pinnata* and characterization of its antibacterial activity against gram positive and gram negative pathogens. Resour Effic Technol 2:30–35
220. Rajeshkumar S, Parameswari RP, Sandhiya D, Al-Ghanim KA, Nicoletti M, Govindarajan M (2023) Green synthesis, characterization and bioactivity of *Mangifera indica* seed-wrapped zinc oxide nanoparticles. Molecules/Molecules Online/Molecules Annual 28:2818
221. Rajeshkumar S, Santhoshkumar J, Jule LT, Ramaswamy K (2021) Phytosynthesis of titanium dioxide nanoparticles using king of bitter *Andrographis paniculata* and its embryonic toxicology evaluation and biomedical potential. Bioinorg Chem Appl 2021:1–11
222. Rajiv P, Rajeshwari S, Venckatesh R (2013) Bio-Fabrication of zinc oxide nanoparticles using leaf extract of *Parthenium hysterophorus* L. and its size-dependent antifungal activity against plant fungal pathogens. Spectrochim Acta Part A Mol Biomol Spectrosc 112:384–387
223. Rajpal VR, Dhingra Y, Khungar L, Mehta S, Minkina T, Rajput VD, Husen A (2024) Exploring metal and metal-oxide nanoparticles for nanosensing and biotic stress management in plant systems. Curr Res Biotechnol 100219
224. Raju S, Ashok D, Boddu AR (2022) *Leucaena leucocephala* mediated green synthesis of silver nanoparticles and their antibacterial, dye degradation and antioxidant properties. Int J Nanosci Nanotechnol 18(1):65–78
225. Raju D, Mehta UJ, Ahmad A (2012) Phytosynthesis of intracellular and extracellular gold nanoparticles by living peanut plant (*Arachis hypogaea L.*). Biotechnol Appl Biochem 59:471–478
226. Raju D, Mehta UJ, Ahmad A (2015) Simple recovery of intracellular gold nanoparticles from peanut seedling roots. J Nanosci Nanotechnol 15:1575–1581
227. Raju D, Paneliya N, Mehta UJ (2014) Extracellular synthesis of silver nanoparticles using living peanut seedling. Appl Nanosci 4:875–879
228. Ramesh M, Anbuvannan M, Viruthagiri G (2015) Green synthesis of ZnO nanoparticles using *Solanum nigrum* leaf extract and their antibacterial activity. Spectrochim Acta Part A Mol Biomol Spectrosc 136:864–870
229. Ramesh C, Kumar KTM, Latha N, Ragunathan V (2012) Green synthesis of Cr2O3 nanoparticles using *Tridax procumbens* leaf extract and its antibacterial activity on Escherichia coli. Curr Nanosci 8:603–607
230. Ramesh S, Vinitha UG, Anthony SP, Muthuraman MS (2021) Pods of *Acacia nilotica* mediated synthesis of copper oxide nanoparticles and it's in vitro biological applications. Mater Today Proceed 47:751–756
231. Rao CNR, Müller A, Cheetham AK (2017) The chemistry of nanomaterials: synthesis, properties and applications. John Wiley & Sons
232. Rasha E, Monerah A, Manal A, Rehab A, Mohammed D, Doaa E (2021) Biosynthesis of Zinc Oxide nanoparticles from *Acacia nilotica* (L.) extract to overcome carbapenem-resistant Klebsiella Pneumoniae. Molecules/Molecules Online/Molecules Annual 26:1919
233. Rastogi L, Arunachalam J (2011) Sunlight based irradiation strategy for rapid green synthesis of highly stable silver nanoparticles using aqueous garlic (*Allium sativum*) extract and their antibacterial potential. Mater Chem Phys 129:558–563
234. Rautela A, Rani J, Debnath M (2019) Green synthesis of silver nanoparticles from *Tectona grandis* seeds extract: characterization and mechanism of antimicrobial action on different microorganisms. J Anal Sci Technol 10(1):1–10
235. Rizwana H, Alwhibi MS, Aldarsone HA, Awad MA, Soliman DA, Bhat RS (2021) Green synthesis, characterization, and antimicrobial activity of silver nanoparticles prepared using *Trigonella foenum-graecum L.* leaves grown in Saudi Arabia. Green Process Synth 10:421–429
236. Roco MC (2005) Environmentally responsible development of nanotechnology. Environ Sci Technol 39(5):106A-112A

237. Rodriguez E, Parsons JG, Peralta-Videa JR, Cruz-Jimenez G, Romero-Gonzalez J, Sanchez-Salcido BE, Saupe GB, Duarte-Gardea M, Gardea-Torresdey JL (2007) Potential of *Chilopsis Linearis* for gold phytomining: using Xas to determine gold reduction and nanoparticle formation within plant tissues. Int J Phytorem 9:133–147
238. Sadeghi B, Gholamhoseinpoor F (2015) A study on the stability and green synthesis of silver nanoparticles using *Ziziphora tenuior* (Zt) extract at room temperature. Spectrochim Acta Part A Mol Biomol Spectrosc 134:310–315
239. Safaepour M, Shahverdi AR, Shahverdi HR, Khorramizadeh MR, Gohari AR (2009) Green synthesis of small silver nanoparticles using geraniol and its cytotoxicity against Fibrosarcoma-Wehi 164. Avicenna J Med Biotechnol 1(2):111–115
240. Sarkar RD, Kalita MC (2022) Se nanoparticles stabilized with *Allamanda cathartica* L. flower extract inhibited phytopathogens and promoted mustard growth under salt stress. Heliyon 8:09076
241. Sathiyapriya R, Balaji M, Rajesh S (2020) Bio-Synthesis of cerium oxide nanoparticles from *Coriandrum sativum L.* leaf extract and their antibacterial activity. Int J Adv Sci Eng 6:1439–1444
242. Savale A, Ghotekar S, Pansambal S, Pardeshi O (2017) Green synthesis of fuorescent CdO nanoparticles using *Leucaena leucocephala* L. extract and their biological activities. J Bacteriol Mycol Open Access 5(5):00148
243. Schäfer RB, Liess M, Altenburger R, Filser J, Hollert H, Roß-Nickoll M, Schäffer A, Scheringer M (2019) Future pesticide risk assessment: narrowing the gap between intention and reality. Environ Sci Eur 31:1–5
244. Sebeia N, Jabli M, Ghith A (2019) Biological synthesis of copper nanoparticles, using *Nerium oleander* leaves extract: characterization and study of their interaction with organic dyes. Inorg Chem Commun 105:36–46
245. Sedefoglu N (2024) Green synthesis of ZnO nanoparticles by Myrtus communis plant extract with investigation of effect of precursor, calcination temperature and study of photocatalytic performance. Ceram Int 50:9884–9895
246. Senapati US, Athparia R (2022) Green synthesis of ZnS nanoparticles using *Allium sativum* l. extract and study of their structural, optical and electrical properties. Chalcogenide Lett 19:203–216
247. Senthilkumar N, Nandhakumar E, Priya P, Soni D, Vimalan M, Potheher IV (2017) Synthesis of ZnO nanoparticles using leaf extract of *Tectona grandis* (L.) and their anti-bacterial, anti-arthritic, anti-oxidant and in vitro cytotoxicity activities. New J Chem 41:10347–10356
248. Shabnam N, Pardha-Saradhi P, Sharmila P (2014) Phenolics impart Au3+-stress tolerance to cowpea by generating nanoparticles. PLoS ONE 9:e85242
249. Shaik AM, Raju MD, Reddy DRS (2020) Green synthesis of zinc oxide nanoparticles using aqueous root extract of *Sphagneticola trilobata* Lin and investigate its role in toxic metal removal, sowing germination and fostering of plant growth. Inorg Nano-Met Chem 50:569–579
250. Shaikh WA, Chakraborty S, Islam RU (2020) Photocatalytic degradation of rhodamine B under UV irradiation using *Shorea robusta* leaf extract-mediated bio-synthesized silver nanoparticles. Int J Environ Sci Technol 17:2059–2072
251. Sharifi-Rad M, Elshafie HS, Pohl P (2024) Green synthesis of silver nanoparticles (Ag NPs) by *Lallemantia royleana* leaf extract: their bio-pharmaceutical and catalytic properties. J Photochem Photobiol, A 448:115318
252. Sharma JK, Akhtar MS, Ameen S, Srivastava P, Singh G (2015) Green synthesis of CuO nanoparticles with leaf extract of *Calotropis gigantea* and its dye-sensitized solar cells applications. J Alloy Compd 632:321–325
253. Sharma V, Sharma JK, Kansay V, Dutta A, Raj M, Singh M, Kapoor A, Pahwa C, Sharma A, Kumar S, Sharma AK, Bera MK (2024) Green synthesis, characterization and drug-loaded iron oxide nanoparticles derived from *Nerium oleander* flower extract as a nanocarrier for in vitro antibacterial efficacy. Nano Express 5:015014

254. Sharon EA, Velayutham K, Ramanibai R (2019) Biosynthesis of copper nanoparticles using *Artocarpus heterophyllus* against dengue vector Aedes aegypti. Int J Life-Sci Sci Res/SSR Inst Int J Life Sci 4:1872–1879
255. Sheny DS, Philip D, Mathew J (2012) Rapid green synthesis of palladium nanoparticles using the dried leaf of *Anacardium occidentale.* Spectrochim Acta Part A Mol Biomol Spectrosc 91:35–38
256. Shilpa VP, Samuel BT, Roshni ER, Sangeetha UV, Lekshmi MS, Panicker SB, Jilsha G, Muddukrishnaiah K (2020) Green synthesis Zinc Oxide nanoparticle using *Allamanda cathartica* leaf extract and their cytotoxic and antibacterial activity. Nanomed Res J 5(3):298–305
257. Shimizu K, Hossain MM, Kato K, Kubota M, Hyakumachi M (2013) Induction of defense responses in cucumber plants by using the cell-free filtrate of the plant growth-promoting fungus *Penicillium simplicissimum* GP17-2. J Oleo Sci 62(8):613–621. https://doi.org/10.5650/jos.62.613
258. Shimizu M, Tainaka H, Oba T, Mizuo K, Umezawa M, Takeda K (2009) Maternal exposure to nanoparticulate titanium dioxide during the prenatal period alters gene expression related to brain development in the mouse. Part Fibre Toxicol 6:1–8
259. Shin WK, Cho J, Kannan AG, Lee YS, Kim DW (2016) Cross-linked composite gel polymer electrolyte using mesoporous methacrylate-functionalized SiO_2 nanoparticles for lithium-ion polymer batteries. Sci Rep 6(1):26332
260. Shukla R, Singh A, Singh KK (2024) Vincristine-based nanoformulations: a preclinical and clinical studies overview. Drug Deliv Transl Res 14:1–16
261. Singh P, Kim YJ, Zhang D, Yang DC (2016) Biological synthesis of nanoparticles from plants and microorganisms. Trends Biotechnol 34(7):588–599
262. Singla S, Jana A, Thakur R, Kumari C, Goyal S, Pradhan J (2022) Green synthesis of silver nanoparticles using *Oxalis griffithii* extract and assessing their antimicrobial activity. OpenNano 7:100047
263. Sinha SN, Paul D (2015) Phytosynthesis of silver nanoparticles using *Andrographis paniculata* leaf extract and evaluation of their antibacterial activities. Spectrosc Lett 48:600–604
264. Smitha SL, Philip D, Gopchandran KG (2009) Green synthesis of gold nanoparticles using *Cinnamomum zeylanicum* leaf broth. Spectrochim Acta Part A Mol Biomol Spectrosc 74:735–739
265. Soheyli F, Hassani H, Darroudi M (2024) Biosynthesis of selenium nanoparticles using *Trigonella foenum-graecum L* extract and examination of their photocatalytic and cytotoxic properties. Int J Environ Sci Technol 1–12
266. Srikar SK, Giri DD, Pal DB, Mishra PK, Upadhyay SN (2016) Green synthesis of silver nanoparticles: a review. Green Sustain Chem 06:34–56
267. Starnes DL, Jain A, Sahi SV (2010) In planta engineering of gold nanoparticles of desirable geometries by modulating growth conditions: an environment-friendly approach. Environ Sci Technol 44:7110–7115
268. Subhapriya S, Gomathipriya P (2018) Green synthesis of titanium dioxide (TiO_2) nanoparticles by *Trigonella foenumgraecum* extract and its antimicrobial properties. Microb Pathog 116:215–220
269. Sultana F, Hossain MM, Kubota M, Hyakumachi M (2008) Elicitation of systemic resistance against the bacterial speck pathogen in *Arabidopsis thaliana* by culture filtrates of plant growth-promoting fungi. Can J Plant Path 30(2):196–205. https://doi.org/10.1080/07060661.2008.10540535
270. Sulthana MR, Madhavan MS (2022) Green synthesis, characterization and antibacterial activity of iron oxide nanoparticles derived from *Solanum nigrum* leaf extract. Int J Adv Res Biol Sci (IJARBS) 9:131–139
271. Sundeep D, Kumar TV, Rao PSS, Ravikumar RVSSN, Krishna AG (2017) Green synthesis and characterization of Ag nanoparticles from *Mangifera indica* leaves for dental restoration and antibacterial applications. Prog Biomater 6:57–66

272. Sunderam V, Thiyagarajan D, Lawrence AV, Mohammed SSS, Selvaraj A (2019) In-vitro antimicrobial and anticancer properties of green synthesized gold nanoparticles using *Anacardium occidentale* leaves extract. Saudi J Biol Sci 26:455–459
273. Sundrarajan M, Muthulakshmi V (2021) Green synthesis of ionic liquid mediated neodymium oxide nanoparticles by *Andrographis paniculata* leaves extract for effective bio-medical applications. J Environ Chem Eng 9:104716
274. Supraja S, Ali SM, Chakravarthy N, Jaya Prakash Priya A, Sagadevan E, Kasinathan MK, Sindhu S, Arumugam P (2013) Green synthesis of silver nanoparticles from *Cynodon dactylon*leaf extract. Int J Chem Tech 5(1):271–277
275. Suresh S, Ilakiya R, Kalaiyan G, Thambidurai S, Kannan P, Prabu KM, Suresh N, Jothilakshmi R, Kumar SK, Kandasamy M (2020) Green synthesis of copper oxide nanostructures using *Cynodon dactylon* and *Cyperus rotundus* grass extracts for antibacterial applications. Ceram Int 46:12525–12537
276. Sysak S, Czarczynska-Goslinska B, Szyk P, Koczorowski T, Mlynarczyk DT, Szczolko W, Lesyk R, Goslinski T (2023) Metal nanoparticle-flavonoid connections: synthesis, physicochemical and biological properties, as well as potential applications in medicine. Nanomaterials 13:1531
277. Tahir K, Nazir S, Li B, Khan AU, Khan ZUH, Gong PY, Khan SU, Ahmad A (2015) *Nerium oleander* leaves extract mediated synthesis of gold nanoparticles and its antioxidant activity. Mater Lett 156:198–201
278. Tarrahi R, Mahjouri S, Khataee A (2021) A review on in vivo and in vitro nanotoxicological studies in plants: a headlight for future targets. Ecotoxicol Environ Saf 208:111697
279. Telrandhe R, Mahapatra DK, Kamble MA (2017) *Bombax ceiba* thorn extract mediated synthesis of silver nanoparticles: evaluation of anti-Staphylococcus aureus activity. Int J Pharm Drug Anal 5(9):376–379
280. Tiwari A, Pandey K, Singh S, Chawla S (2023) Grass and their waste products for nanoparticles synthesis and applications. In: Husen A (eds) Nanomaterials from agricultural and horticultural products. Smart nanomaterials technology. Springer, Singapore. https://doi.org/10.1007/978-981-99-3435-5_14
281. De La Torre RR, Servin A, Hawthorne J, Xing B, Newman LA, Ma X, Chen G, White JC (2015) Terrestrial trophic transfer of bulk and nanoparticle La_2O_3 does not depend on particle size. Environ Sci Technol 49:11866–11874
282. USCB (2024) United States Census Bureau, Census Bureau Projects U.S. and World Populations on New Year's Day. https://www.census.gov/en.html
283. Ukidave VV, Ingale LT (2022) Green synthesis of zinc oxide nanoparticles from *Coriandrum sativum* and their use as fertilizer on Bengal gram, Turkish gram, and green gram plant growth. Int J Agron 2022:1–14
284. Ullah H, Khan I, Yamani ZH, Qurashi A (2017) Sonochemical-driven ultrafast facile synthesis of SnO_2 nanoparticles: growth mechanism structural electrical and hydrogen gas sensing properties. Ultrason Sonochem 34:484–490
285. Upadhyay C, Kothari R (2023) Anti–oxidant and anti-inflammation activities of nanostructured assemblies of silver sulfide nanoparticles using an extract of Cinnamomum tamala leaves. J Optoelectron Biomed Mater 15:65–79
286. Urge SK, Dibaba ST, Gemta AB (2023) Green synthesis method of ZnO nanoparticles using extracts of zingiber officinale and garlic bulb (*Allium sativum*) and their synergetic effect for antibacterial activities. J Nanomater 2023:1–9
287. Valko M, Rhodes CJ, Moncol J, Izakovic M, Mazur M (2006) Free radicals, metals and antioxidants in oxidative stress-induced cancer. Chem Biol Interact 160:1–40
288. Varghese R, Almalki MA, Ilavenil S, Rebecca J, Choi KC (2019) Silver nanopaticles synthesized using the seed extract of Trigonella foenum-graecum L. and their antimicrobial mechanism and anticancer properties. Saudi J Biol Sci 26:148–154
289. Velsankar K, RM AK, R P, V M, Sudhahar S (2020) Green synthesis of CuO nanoparticles via *Allium sativum* extract and its characterizations on antimicrobial, antioxidant, antilarvicidal activities. J Environ Chem Eng 8:104123

290. Venkata ALK, Anthony SP, Muthuraman MS (2019) Synthesis of *Solanum nigrum* mediated copper oxide nanoparticles and their photocatalytic dye degradation studies. Mater Res Express 6:125402
291. Venkataraman S (2022) Plant molecular pharming and plant-derived compounds towards generation of vaccines and therapeutics against coronaviruses. Vaccines 10:1805
292. Vergheese M, Vishal SK (2018) Green synthesis of magnesium oxide nanoparticles using *Trigonella foenum-graecum* leaf extract and its antibacterial activity. www.phytojournal.com
293. Vidya C, Hiremath S, Chandraprabha M, Antonyraj ML, Gopal IV, Jain A, Bansal K (2013) Green synthesis of ZnO nanoparticles by *Calotropis gigantea.* Int J Curr Eng Technol 1:118–120
294. Vijayaram S, Razafindralambo H, Sun Y-Z, Vasantharaj S, Ghafarifarsani H, Hoseinifar SH, Raeeszadeh M (2023) Applications of green synthesized metal nanoparticles—a review. Biol Trace Elem Res 202:360–386
295. Vijilvani C, Bindhu FFC, AlSalhi MS, Sabitha S, Saravanakumar K, Devanesan S, Umadevi M, Aljaafreh MJ, Atif M (2020) Antimicrobial and catalytic activities of biosynthesized gold, silver and palladium nanoparticles from *Solanum nigurum* leaves. J Photochem Photobiol B 202:111713
296. Vinayagam R, Santhoshkumar M, Lee KE, David E, Kang SG (2021) Bioengineered gold nanoparticles using *Cynodon dactylon* extract and its cytotoxicity and antibacterial activities. Bioprocess Biosyst Eng 44:1253–1262
297. Wardman P, Candeias LP (1996) Fenton chemistry: an introduction. Radiat Res 145:523
298. Ware PS (2020) Biosynthesis and characterization of Zno nanoparticals by *Annona reticulata* seeds. Eur J Pharm Med Res 7(6):624–630
299. Werlin R, Priester JH, Mielke RE, Krämer S, Jackson S, Stoimenov PK, Stucky GD, Cherr GN, Orias E, Holden PA (2011) Biomagnification of cadmium selenide quantum dots in a simple experimental microbial food chain. Nat Nanotechnol 6:65–71
300. Wu Z, Yang S, Wu W (2016) Shape control of inorganic nanoparticles from solution. Nanoscale 8:1237–1259
301. Xia T, Kovochich M, Liong M, Zink JI, Nel AE (2007) Cationic Polystyrene Nanosphere toxicity depends on cell-specific endocytic and mitochondrial injury pathways. ACS Nano 2:85–96
302. Xu H, Yu T, Fu Y, Dang Z, Wang L, Xie S, Chang F, Shen H, Ren Q (2020) Biosynthesis of Ag nanoparticles and two-dimensional element distribution in Arabidopsis. IET Nanobiotechnol 14:325–330
303. Yang X, Feng Y, He Z, Stoffella PJ (2005) Molecular mechanisms of heavy metal hyperaccumulation and phytoremediation. J Trace Elem Med Biol 18:339–353
304. Zhang G, Liu Z, Xiao Z, Huang J, Li Q, Wang Y, Sun D (2015) Ni_2P-Graphite Nanoplatelets supported Au–Pd core–shell nanoparticles with superior electrochemical properties. J Phys Chem 119:10469–10477

Sensor-Based Approaches

Nanosensors for Detection of Plant Pathogens

Seweta Srivastava, Meenakshi Rana, Jay Prakash Singh, Shivam Singh, and Shweta Meshram

Abstract Plant diseases represent one of the key factors limiting crop yields which still pose the greatest danger to global agriculture and significantly lower crop yields globally, which has a negative impact on food supply. To stop dispersion disease and to promote efficient control techniques, crop health monitoring depends heavily on the correct and early noticing of plant pathogenic microbes. Various techniques have been utilized to diagnose crop diseases, such as visual examination of symptoms, serological tests, and detection of pathogens based on DNA. To tackle these challenges, nanotechnology, particularly nanobiosensors, has emerged as a promising tool for diagnosing plant diseases and pathogens, assessing microbial infections, managing diseases, and ultimately improving productivity. Moreover, they are labour-intensive, require expensive equipment, produce inaccurate findings from the cross contamination, and needs professional experts in the field. Nanobiosensor-based technology introduces a novel approach to plant disease diagnostics by offering non-destructive, minimally invasive, affordable, and user-friendly systems. It enhances detection limit, sensitivity, and specificity, enabling efficient on-site pathogen detection. This chapter provides a brief outline of the use of nano integrated nanobiosensors in plant pathogen diagnosis.

Keywords Plant pathogen diagnosis · Nanobiosensors · FRET nanosensors · Quantum dots · Plant diseases

S. Srivastava · M. Rana · S. Meshram (✉)
Department of Plant Pathology, School of Agriculture, Lovely Professional University, Phagwara, Punjab, India
e-mail: shweta.26662@lpu.co.in

J. P. Singh
Department of Plant Pathology, SMMTD College, Balia, Uttar Pradesh, India

S. Singh
Krishi Vigyan Kendra Baghpat, Sardar Vallabhbhai Patel University of Agriculture and Technology, Meerut, Uttar Pradesh, India

L. F. Fraceto et al. (eds.), *Emerging Nanotechnologies for Agroecosystem Management*, Smart Nanomaterials Technology, https://doi.org/10.1007/978-981-95-0187-8_7

1 Introduction

Given the significant reliance of the global economy on agriculture and food, the increasing world population has placed immense pressure on sustainable agricultural practices [93]. Over the past 300 years, agricultural ecosystems have expanded to cover approximately 40% of the planet's surface to meet the demands of a growing global population [23, 65]. It's crucial to note that a significant 20–40% of crop losses result from abiotic and biotic stresses, including pathogen invasions, insect pest attacks, herbicide damage, and insect infestations [19, 61]. Therefore, a major challenge is the management of plant diseases against the biotic factors. Plant diseases have three major causes: bacterial, fungal, and viral infections. These infections disseminate widely in plantations through the introduction of the etiological agents of the disease into the field or through the infected plant components [73]. In fundamental and applied agricultural science, it is easy to make a premature assessment of plant disease incidence and severity in field crops [25]. There is a need to make accurate and timely decisions because the decisions made are the basis for the field and plant protection measures.

Furthermore, the application of conventional molecular techniques in a new financial state is limited by its high cost and short shelf life of certain molecular biology ingredients, like as primers and catalysts. These common methods also have some negative impacts on human health, the environment, and plant ecosystems [74]. Conventional molecular diagnostic procedures are often employed to identify phytopathogenic microorganisms with an extended degree of sensitivity as well as specificity [27, 84]. Recent developments in the nanotechnology have contributed to the creation of nanoscale sensors with exceptional sensitivity and versatility [72]. Nanotechnology has turned splicing of nanoparticles into a biosensor for diagnostics and therapeutic applications in the pharmaceutical sector, agriScience, industries, and medicine among other wide industries [6].

2 Nanosensor and Their Types

The term nanosensor is defined as a selective transducer with a characteristic dimension that is in nanometre scale [78]. Nanobiosensors are made up of organic, inorganic or hybrid nanocomposite materials which are developed through unique nanotechnology methods. Nanobiosensors include a wide range of nanomaterials like metals, carbon nanotubes (CNTs), 'magnetic nanoparticles', 'nanowires', 'nano-probes', and 'quantum dots' (QDs) [9, 63]. Nanobiosensors offer advanced applications in agriculture, enabling real-time monitoring of crop conditions. They detect critical soil parameters such as nutrient levels, temperature, moisture, and pH. Piezoelectric nanosensors, in particular, are effective for sensing various abiotic factors. This technology helps optimize crop management by providing precise, timely data on soil health. Overall, nanobiosensors support sustainable agriculture by enhancing

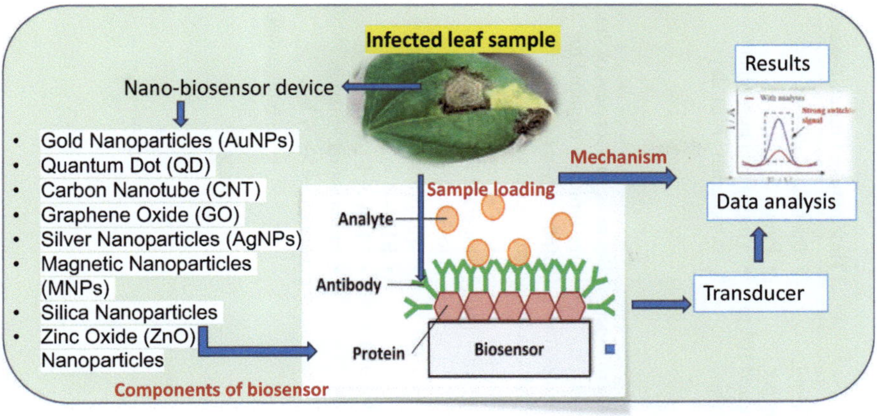

Fig. 1 Types of nanobiosensors used for rapid and sensitive detection of plant pathogens

decision-making for nutrient and resource management [16]. Wider applications of nanocomposites can instead substitute the applications of fungicides, pesticides, and synthetic chemicals in crops to arrest pests attacking crops. The nanocomposite on a small scale penetrates through all the barriers that allow nutrients to get transferred via roots or as foliar applications in the most effective manner.

A number of nano structure-based sensors have been developed which are used to diagnose and detect plant pathogens (Fig. 1), such as gold nanoparticles (AuNPs)-based nanosensors, aluminium nanoparticles (AlNPs)-based nanosensors, Quantum dots, carbon-based nanosensors, silica nanoparticles-based nanosensors, carbon nanotubes nanosensors, silver nanoparticles (AgNPs)-based nanosensors, copper nanoparticles (CuNPs)-based nanosensors, nanowire nanosensors, metallic nanoparticle-based biosensors, and antibody nanosensors [59, 65]. However, based on the signal transduction mechanisms, and bio-recognition principles nanosensors are categorized into four types [43] viz., fluorescence resonance energy transfer (FRET) nanosensors, Surface-Enhanced Raman Scattering (SERS) nanosensors, electrochemical-based nanosensors (ECN), and piezoelectric-based nanosensors (PZN). Nanosensors have enabled the study of cellular functions [41] and metabolic flux [8], the monitoring of spatiotemporal dynamics of analytes [14, 86], and the detection of viral and fungal pathogens [18, 45, 47]. The summary of mechanism, advantages, and disadvantages of each type of nanosensor is presented in Table 1.

3 Förster Resonance Energy Transfer-Based Nanosensors

Optical nanosensors are based on Förster resonance energy transfer have been extensively used to study protein interactions, cell contents, and biophysical parameters [37, 56]. FRET nanosensor operates on the principle of transfer of fluorescence

Table 1 Comparison among detection methods used in plant pathogen diagnosis [79]

Detection methods	Principle	Advantages	Disadvantages	Detection limit	Time required for an Assay
Conventional methods	Traditional approaches include culturing the pathogen on selective media, observing symptoms, and microscopic examination	Direct observation and isolation of pathogens as well as low assay cost	Needed an experienced worker, insensitive, and time-consuming	About 10^3 to 10^4 CFU/mL (colony-forming units per millilitre) for bacteria and 10^3 to 10^4 spores/mL for fungi	• Bacteria: 2–7 days • Fungi: 5–14 days • Viruses (via host plant symptoms): 7–21 days
Immunological/ serological methods	Based on the detection of specific antigens or antibodies	Inexpensive, portable, able to quantify and differentiate infections according to taxonomy	Inadequate sensitivity and precision, incongruous responses	The detection limit of serological methods refers to the lowest concentration of antibodies (or antigens) in a sample that can be reliably detected by the assay • ELISA and Western Blot—1 to 10 ng/mL • Lateral Flow Assays—10 to 100 ng/mL • RIA, CLIA and FIA—0.01 to 1 ng/mL	Below is an overview of the time typically required for different types of serological assays: • ELISA—2 to 5 h • Lateral Flow Assay—10 to 30 min • Western blot—1 to 2 days • RIA—4 to 8 h • CLIA & FIA—1 to 3 h • Immunoprecipitation assay—4 to 24 h

(continued)

Table 1 (continued)

Detection methods	Principle	Advantages	Disadvantages	Detection limit	Time required for an Assay
Nucleic acid-based methods	Detection of specific DNA or RNA sequences of pathogens using techniques like PCR, qPCR, RT-PCR, LAMP, next-generation sequencing, etc	Widely used technology that is simple to use, highly sensitive, and accurate	Assay efficiency is influenced by the purity of the DNA, the existence of inhibitors, and the substantial investment required	The detection limit of a nucleic acid-based assay refers to the smallest quantity of nucleic acid that can be reliably detected and quantified by the assay. The detection limits of some frequently used assays are— • PCR—10^3–10^4 $cfuml^{-1}$ • RT-PCR—10^2–10^3 $cfuml^{-1}$ • LAMP—10–10^2 $cfuml^{-1}$	The time required for nucleic acid-based assays varies depending on the specific method, the difficulty of the assay, and the various steps involved. Some are mentioned below— • PCR—2 to 4 h • qPCR—1.5 to 2 h • LAMP—30 min to 1 h • NGS—24 h to several days • CRISPR-based assays—1 to 2 h • Southern blotting, Northern blotting, and In Situ Hybridization—1 to 3 days

(continued)

Table 1 (continued)

Detection methods	Principle	Advantages	Disadvantages	Detection limit	Time required for an Assay
Nanotechnology-based methods	Utilizes nanoparticles, nanosensors, and other nanomaterials for pathogen detection	Highly efficient analysis that is quick, manageable, economical, has good on-field portability, high accuracy, and outstanding sensitivity	The toxicity of nanomaterials and their effects on the environment. Still emerging with limited commercial availability High cost of development and implementation. Requires technical expertise. Stability and reproducibility issues in some cases	The detection limit of nanotechnological assays depends on the type of nanomaterial used, the target molecule, and the specific detection method. Here's an overview: • Gold Nanoparticles (AuNPs): 10^{-18} to 10^{-15} M (femtomolar to attomolar) • Quantum Dots (QDs): 10^{-15} to 10^{-12} M (femtomolar to picomolar) • Magnetic Nanoparticles (MNPs): 10^{-12} to 10^{-9} M (picomolar to nanomolar) • Carbon Nanotubes (CNTs) and Graphene: 10^{-15} to 10^{-12} M (femtomolar to picomolar) • Nanopore Technologies: 10^{-21} M (zeptomolar) • Plasmonic Nanosensors: 10^{-18} to 10^{-15} M (femtomolar to attomolar)	Below is an overview of the time required for different types of nanotechnological assays: • AuNPs—30 min to 2 h • QDs—1 to 3 h • MNPs—1 to 4 h • CNTs and Graphene—1 to 2 h • Nanopore Technologies—Minutes to several hours (depending on the complexity) • Plasmonic Nanosensors—30 min to 2 h • Surface-Enhanced Raman Scattering (SERS)—1 to 2 h

resonance energy and its measurement [40]. The reason for its name, FRET, is that German physicist Theodor Förster discovered this phenomenon [91]. In biochemistry and biophysics, it is a commonly used auxiliary technique for measuring molecular dynamics for biomolecular interactions such as protein–DNA and protein–protein interactions, as well as protein confirmational changes. FRET enables precise and sensitive detection of biomolecules without altering them or adding labels. This technique leverages the proximity-dependent energy transfer between a donor and an acceptor fluorophore. When these molecules are close, the donor's energy excites the acceptor, causing it to fluoresce. Any environmental change that affects the distance between the fluorophores alters the fluorescence intensity, allowing for the detection of subtle changes, like the presence of specific biomolecules [52, 63]. FRET is also a non-radiative process, eliminating the risk of ionizing radiation associated with other biosensor types, making FRET-based sensors safer to use and handle [76]. Moreover, FRET biosensors are highly specific, targeting a particular biomolecule or environmental change without interference from other factors. This specificity arises from the biosensor's high affinity for the target molecule and the close proximity of the donor and acceptor fluorophores [106]. The high sensitivity and the specificity of FRET-based assays make them ideal for the rapid and accurate for plant pathogens detection [92]. Due to their adaptability, FRET-based biosensors are useful for research, diagnosis, and monitoring in a wide range of fields, such as biochemistry, cell biology, and drug discovery. In the last few years, the biosensor has been applied to nearly every area of study, from pathology to physiology, from diagnosis to treatment, and from environment to health [4, 92].

Nanosensors FRET-based technology can be introduced exogenously as externally synthesized compounds, or they can be genetically encoded within the plant itself. Applications for genetically encoded FRET-based nanosensors include detecting reactive oxygen species, monitoring levels of ions or metabolites, monitoring protein interactions with nucleic acids, plant viruses, and transgenes in plants [5, 29].

4 Quantum Dots-Based Biosensor

Fluorescent semiconductor (nano-scale) crystals with a brilliant and adjustable emission spectrum are known as quantum dots, or QDs. Based on the quantum dot size, the emission range is selected from the daylight to the infrared region. FRET, or the transmission of energy between two reactive molecules, is a special optical feature that quantum dots provide [65]. QT-FRET immunoassays use this characteristic to visually identify infection caused by pathogens. To identify plant pathogens like *Aspergillus amstelodami*, *Candidatus Phytoplasma aurantifolia*, and *Polymyxa beta*, QD-FRET has been employed as an effective detection method [70]. QD-based biosensors, tailored to specific phytopathogens, leverage the physicochemical properties of interface components and bioconjugates. A notable example is the nanobiosensor for Citrus Tristeza Virus (CTV) detection. It employs cadmium telluride quantum dots (CdTe-QDs) conjugated to CTV coat protein (CTV-CP) antibodies. This biosensor effectively immobilizes CTV-CP on carbon nanoparticles (CNPs) using CdTe-QDs and antibodies. The limit of detection (LOD) for this biosensor is approximately 220 ng/mL of CTV [65].

5 Surface-Enhanced Raman Scattering Nanosensors

Surface-Enhanced Raman Scattering (SERS) is a sensitive non-destructive spectroscopic method that can identify analytes at the single-molecule level [82]. The inelastic scattering of the photons by lighted substance is known as Raman scattering. Both an energy exchange and a direction shift of the light occur during this phase. Photons interact with the vibrational states of materials to produce the energy difference between the incident and dispersed light. A molecule gains vibrational energy as a result of incident photons being shifted to a lower energy state in this effect. A. Smekal proposed the effect theoretically in 1923 [80], and C.V. Raman experimentally proved it in 1928 [67].

The inability to create nanostructured materials that could consistently boost signal was a hindrance to initial attempts to use the SERS technology as a quantitative analytical tool for diagnosing bacterial pathogens [10, 89]. Because of more thorough research into numerous basic fabrication factors, recent studies have begun to alter this scenario to the reproducible creation of nanoparticle structures with well-defined sizes and shapes for SERS applications [13, 62]. For the implementation of SERS, continuous innovation in spectroscopic instruments is also essential. In addition to guaranteeing superior instrumental performance, advancements have drastically decreased equipment prices [89]. Using a single laser excitation, SERS is a technique that may be employed to the surfaces of metal nanoparticles to produce enhanced Raman scattering patterns typical of the adsorbed molecules [75]. According to [77] and [46], SERS has been suggested as a very effective readout technique for quick diagnostic tests and is a very strong molecular spectroscopic detection tool [60]. SERS may be more appropriate for highly multiplexed applications than conventional fluorescent-based techniques due to its narrow distinctive spectrum peaks [44].

Recombinase polymerase amplification (RPA) is an isothermal method that has been increasingly applied in innovative diagnostic technologies in recent years [55, 98]. In contrast to PCR, RPA uses enzymes at a single low temperature for DNA separation, primer recognition, and extension. This enables efficient amplification and stable low-temperature operation. A multiplex diagnostic platform combining RPA and SERS has been developed for detecting plant pathogens (see Table 2) [30]. This technique was utilized to detect three economically significant plant pathogens affecting Arabidopsis thaliana and tomato (*Solanum lycopersicum L.*) viz., '*Fusarium oxysporum* f.sp. *conglutinans*', '*F. oxysporum* f.sp. *lycopersici*', '*Pseudomonas syringae*', and '*Botrytis cinerea*' [88].

Table 2 Diverse applications of nanobiosensors in detecting various plant pathogens, leveraging the unique properties of nanomaterials for enhanced sensitivity and specificity in agricultural diagnostics

Nanobiosensor type	Pathogen detected	Sensitivity	Detection mechanism	References
Virus /viroid pathogen(s)				
Gold NP-coated QCM Crystals	*Maize Chlorotic Mottle Virus* (MCMV)	250 ng mL^{-1}	Piezoelectric nanosensor (PZN)	[31]
QCM crystals	*Cymbidium Mosiac Potexvirus* (CymMV), *Odontoglossum Ringspot Tobamovirus* (ORSV)	1 ng	Piezoelectric nanosensor (PZN)	[21, 40]
Chitosan/ muti-walled carbon nanotubes nanocomposite	*Cauliflower Mosaic Virus* (CMV)	High with 8.5 × 10–14 M	Electrochemical response	[95]
Gold nanoparticles (GNPs) Based Biosensor	*Tobacco Mosaic Virus* (TMV)	The sensors can detect TMV concentrations ranges from 100 ng/mL—mg/ mL within a minute	Electrochemical response	[15, 49]
TwistAmp basic/ AuNP probe	*Tomato yellow leaf curl virus* (TYLCV)	10–50 nM	Colorimetric	[96]
TwistAmp basic/ AuNP-modified SPCE/EIS/ electronic nose (E-Nose)	*Citrus Tristeza Virus* (CTV)	1 pg/uL	Electrochemical response	[28]
CdTe QD-CD surface immunological functionalization	*Citrus Tristeza Virus* (CTV)	High sensitivity of 0.9 ng mL^{-1}	FRET	[50]
CdTe QD-Rd surface immunological functionalization	*Citrus Tristeza Virus* (CTV)	High sensitivity of 0.9 ng mL^{-1}	FRET	[24]
TwistAmp basic/ Cas12a/AuNP probe	*Apple RNA viroid*	250–2500 copies/ reaction	Colorimetric	[35]
Bacterial/phytoplasma pathogen(s)				

(continued)

Table 2 (continued)

Nanobiosensor type	Pathogen detected	Sensitivity	Detection mechanism	References
Microfluidic electrochemical immunosensor	*Xanthomonas arboricola*	Covalently immobilization of monoclonal anti-XA antibody were used	Electrochemical response	[68]
Silver nanoparticles (AgNPs)-based Biosensor	*Pseudomonas syringae*	Surface plasmon with metal micro-nano structure and material were used	Electrochemical response	[51]
TwistAmp basic/ AuNP probestreptavidin beads/DPV	*Pseudomonas syringae*	214 pM	Electrochemical Response	[47]
Magnetic nanoparticles (MNPs)-based biosensor	*Ralstonia solanacearum*	Provide a substantial adsorption capacity for pathogenic DNA	Optical, electrochemical, or magnetic signals	[2]
Gold nanoparticles (AuNPs)	*Acidovorax avenae subsp. citrulli*	Qualitative and semi-quantitative detections of target DNA by lateral flow assay	Colorimetric	[104]
Quartz crystal	*Campylobacter jejuni*	1.30 log CFU ml^{-1}	Piezoelectric nanosensor (PZN)	[57]
Quartz crystal	*Staphylococcus aureus*	7.41 log CFU ml^{-1}	Piezoelectric nanosensor (PZN)	[64]
Quantum dots-based biosensor	*Candidatus Phytoplasma aurantifolia*,	High sensitivity of 0.9 ng mL^{-1}	QD-FRET	[70]
Fungal pathogen(s)				
Quantum dots-based biosensor	*Aspergillus amstelodami* and *Polymyxa beta*	high sensitivity of 0.9 ng mL^{-1}	QD-FRET	[70]
polydopamine (PDA) nanoparticles	*Fusarium oxysporum* f. sp. *cucumerinum*	Evaluated bioactivity of the prepared hymexazol-loaded nanocomposite	Electrochemical response	[87]

(continued)

Table 2 (continued)

Nanobiosensor type	Pathogen detected	Sensitivity	Detection mechanism	References
TwistAmp basic/ SERS nanotagstreptavidin beads/Raman microscope	*Fusarium oxysporum, Pseudomonas syringae, Botrytis cinerea*	2.32 genomic copies of pathogen DNA	Spectroscopy	[47]
Graphene Oxide (GO) Based Biosensor	*Phytophthora infestans*	The linear range of detection was 10 fM–50 nM and the detection limit was 10 fM	Electrochemical response	[103]
SnO_2 and TiO_2 NPs	*Phytophthora cactorum*	High, 35–62 nM	Electrochemical response	[22]
Silica nanoparticles-based biosensor	*Botrytis cinerea*	As low as 2.103 CFU/mL	Electrochemical and immune biosensors	[12]
Zinc oxide (ZnO) nanoparticles-based biosensor	*Alternaria alternata*	Photo-catalytic, fluorescent, and electron generating aspects associated with zinc nanomaterials were explored	Electrochemical response	[39]

6 Electrochemical Nanosensors (ECN)

Due to its built-in benefits of simplicity, relatively low cost, high sensitivity, ease of miniaturization, and the ability for direct data analysis, electrochemical detection is an appealing alternative method for the detection of heavy metals and other redox active species in plants [99]. A reference electrode, counter electrode, and working electrode are the standard components of electrochemical nanosensors. It has been demonstrated that voltammetric and amperometric methods are helpful in the qualitative investigation of plant sensing [100]. These methods can also be applied in other domains (air and water, for example) where industrial environmental sensors are employed. Because nanomaterial-designed electrodes have a comparatively large active surface area that increases sensitivity, ECN can be used to detect pathogens in horticulture crops and the field [81]. Due to their distinct electronic properties, metallic nanoparticles like gold [83, 102] and nanocarbon materials like carbon nanotubes and graphene-based materials have been the subject of recent advances in electrochemical nanosensor research [38, 43]. Because of its great sensitivity and ability to analyse data directly, electrochemical detection is a desirable technique for identifying biological compounds in plants [99]. Table 2 lists a few newly created electrochemical nanosensors (ECNs) for the spotting of plant pathogens.

7 Piezoelectric Nanosensors

The coupling of mechanical and electrical properties of materials is known as piezoelectricity. It is a reversible process whereby, in certain solid materials, an electric charge builds up in response to mechanical stress and, conversely, whereby a solid material contracts or elongates when placed in an electric field [17, 97]. A 'piezoelectric sensor' converts mechanical stress into an electrical charge, allowing for the measurement of pressure, acceleration, temperature, strain, or force [78]. It can facilitate real-time monitoring by converting the energy of biological mechanical processes, acoustic/ultrasonic vibrations, and biofluid hydraulic processes into electric signals. Quartz crystal microbalance is the material used in PZN the most frequently. Quartz crystal is coated with a thin layer of NPs, allowing biomolecules to create a covalent or hydrophobic interaction with them [43]. Numerous PZN-based nanosensors have been created by taking advantage of this phenomena (Table 2).

In PZN-based nanosensors, AuNPs are carefully layered onto Quartz Crystal Microbalance (QCM) surfaces to act as electrodes. This greatly increases the formation of hydrophobic or covalent bonds with biomolecules such as particular DNA probes or other selective receptors [40]. Any contact between an analyte, such as viruses or pathogen DNA molecules, and the receptor causes a decrease in frequency because of the inverse connection between frequency and mass. This drop results from an increase in mass that happens when the selective receptor binds to the analyte (pathogens), proving beyond a reasonable doubt that a pathogen is present [17].

8 Role of Nanosensors in Plant Disease Management

Nanosensors play a pivotal role in plant disease management by providing advanced tools for early detection, monitoring, and analysis of plant health [78]. These sensors are engineered at the nanoscale to detect specific biochemical markers, pathogens, or environmental conditions (Fig. 2). Their unique properties, such as high sensitivity and specificity, enable them to detect pathogens and disease markers at very low concentrations, which is crucial for timely intervention in agricultural practices. This capability is particularly important in the context of precision agriculture, where the health of individual plants can significantly impact overall crop yield and quality [54]. These advanced technologies leverage the unique properties of nanomaterials to enhance agricultural productivity and sustainability. The following sections outline the key aspects of nanosensors in this context—

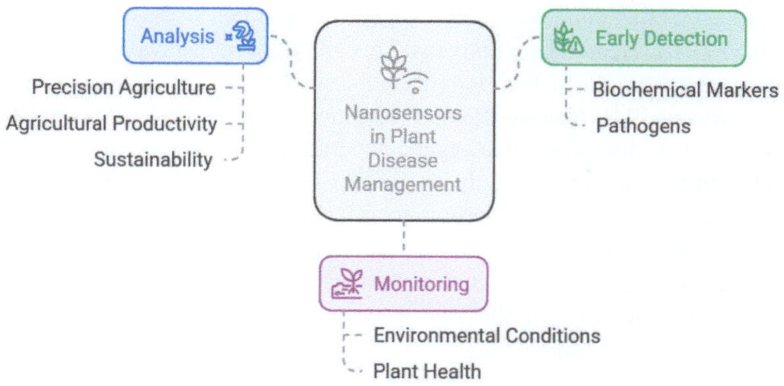

Fig. 2 Sustainable applications of nanosensors in plant disease management

8.1 Early Detection of Pathogens

Nanosensors enable the rapid identification of plant pathogens, including bacteria, fungi, and viruses, facilitating early disease management [93]. Techniques such as quantum dots and plasmonic nanoparticles are utilized for their sensitivity and specificity in detecting pathogens [66]. The application of nanosensors extends beyond mere detection, they also contribute to the understanding of plant-pathogen interactions. By employing various nanosensor types, such as those based on surface-enhanced Raman spectroscopy or fluorescence resonance energy transfer, researchers can gain insights into the biochemical processes occurring within plants under stress [78]. This knowledge is essential for developing targeted disease management strategies that can mitigate the effects of pathogens on crops.

Furthermore, the use of nanomaterials, such as gold nanoparticles and carbon nanotubes, in the design of biosensors has shown promising results in enhancing the sensitivity and specificity of disease detection [42, 71]. These materials can facilitate the development of portable and cost-effective sensing devices that are crucial for smallholder farmers who may lack access to sophisticated laboratory equipment [53]. The ability to conduct rapid tests in the field empowers farmers to make informed decisions about crop management and disease control.

8.2 Real-Time Monitoring

One of the primary advantages of nanosensors is their ability to monitor plant health in real-time. Nanosensors are now being used to detect and monitor physical characteristics such as the temperature at nanoscale, with chemical species and nanoparticles [34]. Continuous monitoring of plant health parameters like stress hormones, nutrient levels, or oxidative stress markers. For instance, nanosensors can be designed to detect specific biochemical markers associated with plant stress or disease, such as extracellular DNA released by pathogens. This allows for non-invasive monitoring of plant tissues, enabling farmers to identify potential disease outbreaks before visible symptoms appear [94]. Moreover, the integration of machine learning techniques with nanosensor technology enhances the accuracy and efficiency of disease detection, facilitating rapid on-site analysis [7, 48]. In addition to pathogen detection, nanosensors can also monitor environmental factors that contribute to plant health, such as soil contamination and nutrient levels [58]. By providing comprehensive data on both biotic and abiotic stressors, nanosensors enable a holistic approach to plant disease management, ensuring that interventions are timely and effective.

8.3 Disease-Specific Identification

Nanosensors play a transformative role in disease-specific identification in plants (Table 3) by enabling precise, rapid, and sensitive detection of pathogens [90]. They are designed to target specific biochemical markers, such as pathogen DNA, RNA, proteins, or toxins, which allows them to differentiate between diseases caused by various pathogens. Nanosensors are equipped with probes (e.g., DNA/RNA sequences, antibodies, or aptamers) that specifically bind to unique markers of a pathogen. This specificity ensures accurate identification of the causal agent, even in the presence of multiple pathogens. They can detect and measure pathogen-secreted metabolites, enzymes, or toxins that are unique to specific diseases. This helps in distinguishing between diseases with overlapping symptoms. Certain nanosensors detect volatile organic compounds (VOCs) released by infected plants, which vary depending on the pathogen. This non-invasive approach enables disease-specific diagnosis without harming the plant [11]. Advanced nanosensors can distinguish between strains or races of a pathogen, aiding in targeted disease resistance strategies. Some diseases involve multiple pathogens, and nanosensors can pinpoint specific contributors [79]. Nanosensors revolutionize plant disease-specific identification by providing farmers and researchers with an efficient tool to tackle plant health challenges, enhancing both productivity and sustainability [85, 93].

9 Conclusion and Further Outlook

The function of nanosensors in the diagnosis and detection of plant diseases has been covered in this chapter. Nanosensors provide insights to enhance plant growth, addressing major challenges like energy and food security. They enable in vivo study of analyte distribution, plant signalling, and responses to environmental factors. Nanosensors also facilitate infection detection for containment and offer nutrient analysis in the field, assessing supplementation needs for optimal growth.

Table 3 Specific plant diseases detected by specific nanosensors

Plant disease	Causal agent	Type of nanosensor	Detection target	References
Late Blight	*Phytophthora infestans*	DNA-based nanosensor	Pathogen DNA	[103]
Early Blight	*Alternaria solani*	Electrochemical nanosensor	Fungal metabolites	[26]
Fusarium Wilt	*Fusarium oxysporum*	Carbon nanotube nanosensor	Mycotoxins	[3]
Rice Blast	*Magnaporthe oryzae*	Optical nanosensor	Fungal proteins	[36]
Powdery Mildew	*Erysiphe* spp.	Surface plasmon resonance nanosensor	Spore surface proteins	[103]
Stem Rust of Wheat	*Puccinia graminis*	Piezoelectric nanosensor	Rust spores	[26]
Downy Mildew	*Plasmopara viticola*	Surface plasmon resonance nanosensor	Spore-related biomarkers	[3]
Gray Mold	*Botrytis cinerea*	VOC-sensitive carbon nanotube nanosensor	Fungal VOCs, mycotoxins	[26]
Bacterial Wilt	*Ralstonia solanacearum*	Electrochemical nanosensor	Bacterial DNA, VOCs	[3]
Citrus Canker	*Xanthomonas axonopodis*	Quantum dot-based nanosensor	Toxins, bacterial DNA	[36]
Fire Blight	*Erwinia amylovora*	Carbon-based nanosensor	VOCs, bacterial biomarkers	[26]
Bacterial Leaf Blight of Rice	*Xanthomonas oryzae pv. oryzae*	DNA-based nanosensor	Pathogen DNA	[103]
Tomato Yellow Leaf Curl Virus	Tomato Yellow Leaf Curl Virus (TYLCV)	RNA-based nanosensor	Viral RNA, coat proteins	[3]
Tobacco Mosaic Virus	Tobacco Mosaic Virus (TMV)	Electrochemical nanosensor	Viral proteins	[36]
Banana Bunchy Top Virus	Banana Bunchy Top Virus (BBTV)	Gold nanoparticle-based nanosensor	Viral nucleic acids	[26]
Papaya Ringspot Virus	Papaya Ringspot Virus (PRSV)	Fluorescent nanosensor	Viral RNA	[103]

To sum up, plant disease management could undergo a revolutionary shift with the introduction of nanosensors for the detection of plant pathogens. Target species or strains must bind to a nanosensor in order for it to eventually change into detectable signals. By enabling the rapid, accurate, and early identification of pathogens, nanosensors contribute to the production of high-quality food. It has the advantage of having excellent sensitivity and specificity for identifying plant pathogens. The key benefit of a nanosensor for pathogen detection in outdoor conditions is its mobility and ease of usage. Technology is expected to become more and more important in preserving sustainable agriculture and global food security as it develops and becomes more affordable. It would be advantageous to have an integrated disease management system with a nanosensors system. To detect infections and pesticide residues in foods and feeds more effectively, current nanodevices should be improved, and new ones should be developed quickly and reliably.

References

1. Ahmad M, Mohsin M, Iqrar S, Manzoor O, Siddiqi TO, Ahmad A (2018) Live cell imaging of vitamin B12 dynamics by genetically encoded fluorescent nanosensor. Sens Actuators B Chem 257:866–874
2. Alghuthaymi MA, Ali AA, Hashim AF, Abd-Elsalam KA (2016) A rapid method for the detection of *Ralstonia solanacearum* by isolation DNA from infested potato tubers based on magnetic nanotools. Philipp Agric Sci 99(1):113–118
3. Alharbi M, Smith RJ, Wilson ME (2023) Carbon-based nanomaterials for detection and control of plant pathogens. Nanotechnol Agric 45(3):211–225
4. Andrews DL, Bradshaw DS (2004) Virtual photons, dipole fields and energy transfer: a quantum electrodynamical approach. Eur J Phys 25:845–858
5. Anjum NA, Amreen TAY, Khan NA, Ahmad A (2020) Reactive oxygen species detection-approaches in plants: insights into genetically encoded FRET-based sensors. J Biotechnol 308:108–117
6. Ansari MA (2023) Nanotechnology in food and plant science: challenges and future prospects. Plants 12(13):2565
7. Arellano Vidal CL, Govan JE (2024) Machine learning techniques for improving nanosensors in agroenvironmental applications. Agronomy 14:341
8. Bagal-Kestwal D, Kestwal RM, Chiang BH (2015) Invertase-nanogold clusters decorated plant membranes for fluorescence-based sucrose sensor. J Nanobiotechnol 13:30
9. Beeguma S, Dasb S (2022) Nanosensors in agriculture 17. In: Agricultural nanobiotechnology: biogenic nanoparticles, nanofertilizers and nanoscale biocontrol agents, p 465
10. Bell SE, Sirimuthu NM (2008) Quantitative surface-enhanced Raman spectroscopy. Chem Soc Rev 37:1012–1024
11. Bharti A, Jain U, Chauhan N (2024) From lab to field: nano-biosensors for real-time plant nutrient tracking. Plant Nano Biology 9:100079
12. Bilkiss M, Shiddiky MJ, Ford R (2019) Advanced diagnostic approaches for necrotrophic fungal pathogens of temperate legumes with a focus on *Botrytis* spp. Front Microbiol 10:428798
13. Burda C, Chen X, Narayanan R, El-Sayed MA (2005) Chemistry and properties of nanocrystals of different shapes. Chem Rev 105:1025–1102
14. Chandra S, Chakraborty N, Dasgupta A, Sarkar J, Panda K, Acharya K (2015) Chitosan nanoparticles: a positive modulator of innate immune responses in plants. Sci Rep 5:15195
15. Dickert FL, Hayden O, Bindeus R, Mann KJ, Blaas D, Waigmann E (2004) Bioimprinted QCM sensors for virus detection—screening of plant sap. Anal Bioanal Chem 378:1929–1934
16. Dimri A, Pathak N, Sharma S (2020) Nanosensors for root zone parameters influencing plant growth. In: Nanomaterials for agriculture and forestry applications. Elsevier, pp 387–406
17. Dorjee L, Haniya K (2023) Electrochemical and piezoelectric-based nanosensor: a cuttingedge technology for plant disease detection. Agri Articles 03(05):594–600

18. Duhan JS, Kumar R, Kumar N, Kaur P, Nehra K, Duhan S (2017) Nanotechnology: the new perspective in precision agriculture. Biotechnol Rep 15:11–23
19. Dutta P, Sharma V, Bhardwaj H, Agrawal VV, Rajesh SG (2022) Fabrication of electrochemical biosensor using zinc oxide nanoflowers for the detection of uric acid. Mapan 37(3):585–595
20. Escouboué M, Camborde L, Jauneau A, Gaulin E, Deslandes L (2019) Preparation of plant material for analysis of protein-nucleic acid interactions by FRET-FLIM. Methods Mol Biol 1991:69–77
21. Eun AJC, Huang L, Chew FT, Li SFY, Wong SM (2002) Detection of two orchid viruses using quartz crystal microbalance (QCM) immunosensors. J Virological Methods 99(1–2):71–79
22. Fang Y, Umasankar Y, Ramasamy RP (2014) Electrochemical detection of p-ethylguaiacol, a fungi infected fruit volatile using metal oxide nanoparticles. Analyst 139(15):3804–3810
23. Foley J, Ramankutty N, Brauman K, Cassidy E, Gerber J, Johnston M et al (2011) Solutions for a cultivated planet. Nature 478(7369):337–342
24. Gao W, Wang X, He Y, Yu H, Zheng Y, Yin R et al (2024) Sub-ppm NO_2 gas sensing in CdTe quantum dots functionalized hollow-core anti-resonant fiber. Sens Actuators B Chem 405:135350
25. Gautam A, Kumar S (2020) Techniques for the detection, identification, and diagnosis of agricultural pathogens and diseases. Natural remedies for pest, disease and weed control, pp 135–142
26. Gupta S, Khare P (2020) Emerging nanotechnology-based tools for the rapid detection and management of plant diseases. J Agric Food Chem 68(12):3434–3445
27. Hariharan G, Prasannath K (2021) Recent advances in molecular diagnostics of fungal plant pathogens: a mini review. Front Cell Infect Microbiol 10:600234
28. Hazarika S, Choudhury R, Montazer B, Medhi S, Goswami MP, Sarma U (2020) Detection of citrus tristeza virus in mandarin orange using a custom-developed electronic nose system. IEEE Trans Instrum Meas 69(11):9010–9018
29. Hong S, Lee C (2018) The current status and future outlook of quantum dot-based biosensors for plant virus detection. Plant Pathol J 34:85–92
30. Hongdan G, Yao D, Qiang C, Meng H, Xiaorong L, Zhihao X, Dongli M (2024) A multiplex recombinase polymerase amplification assay combined with CRISPR/Cas12a for the detection of respiratory syncytial virus and respiratory adenovirus. J Int Med Res 52(1):3000605231223083
31. Huang X, Xu J, Ji HF, Li G, Chen H (2014) Quartz crystal microbalance based biosensor for rapid and sensitive detection of maize chlorotic mottle virus. Anal Methods 6(13):4530–4536
32. Ijaz U, Ahmed T, Rizwan M, Noman M, Shah AA, Azeem F et al (2023) Rice straw based silicon nanoparticles improve morphological and nutrient profile of rice plants under salinity stress by triggering physiological and genetic repair mechanisms. Plant Physiology Biochem 201:107788
33. Jamieson LE, Asiala SM, Gracie K, Faulds K, Graham D (2017) Bioanalytical measurements enabled by surface-enhanced Raman scattering (SERS) probes. Annu Rev Anal Chem 10:415–437
34. Javaid M, Haleem A, Singh RP, Rab S, Suman R (2021) Exploring the potential of nanosensors: a brief overview. Sens Int 2:100130
35. Jiao J, Kong K, Han J, Song S, Bai T, Song C, Wang M, Yan Z, Zhang H, Zhang R, Feng J (2021) Field detection of multiple RNA viruses/viroids in apple using a CRISPR/Cas12a-based visual assay. Plant Biotechnol J 19(2):394–405
36. Jones AM, Brown KA, Green RD (2021) Optical nanosensors for rapid pathogen detection in agriculture. Sens Actuators B Chem 328:129031
37. Jones AM, Grossmann G, Danielson JÅ, Sosso D, Chen LQ, Ho CH, Frommer WB (2013) *In vivo* biochemistry: applications for small molecule biosensors in plant biology. Curr Opin Plant Biol 16:389–395
38. Justino CIL, Gomes AR, Freitas AC, Duarte AC, Rocha-Santos TAP (2017) Graphene based sensors and biosensors. TrACTrends Anal Chem 91:53–66

39. Kalia A, Abd-Elsalam KA, Kuca K (2020) Zinc-based nanomaterials for diagnosis and management of plant diseases: ecological safety and future prospects. J Fungi 6(4):222
40. Kashyap PL, Kumar S, Jasrotia P, Singh DP, Singh GP (2019) Nanosensors for plant disease diagnosis: current understanding and future perspective. In: Nanoscience for sustainable agriculture. Springer Nature Switzerland, pp 189–205
41. Kosaka PM, Pini V, Ruz JJ, Da Silva RA, González MU, Ramos D, Calleja M, Tamayo J (2014) Detection of cancer biomarkers in serum using a hybrid mechanical and optoplasmonic nanosensor. Nat Nanotechnol 9:1047–1053
42. Kulabhusan PK, Tripathi A, Kant K (2022) Gold nanoparticles and plant pathogens: an overview and prospective for biosensing in forestry. Sensors 22(3):1259
43. Kwak SY, Wong MH, Lew TTS, Bisker G, Lee MA, Kaplan A, Strano MS (2017) Nanosensor technology applied to living plant systems. Annu Rev Anal Chem 10:113–140
44. Laing S, Gracie K, Faulds K (2016) Multiplex in vitro detection using SERS. Chem Soc Rev 45:1901–1918
45. Lattanzio VMT, Nivarlet N (2017) Multiplex dipstick immunoassay for semiquantitative determination of *Fusarium* mycotoxins in oat. In: Methods in molecular biology. Elsevier: Amsterdam, The Netherlands, pp 137–142
46. Lau HY, Wang Y, Wee EJH, Botella JR, Trau M (2016) Field demonstration of a multiplexed point-of-care diagnostic platform for plant pathogens. Anal Chem 88:8074–8081
47. Lau HY, Wu H, Wee EJH, Trau M, Wang Y, Botella JR (2017) Specific and sensitive isothermal electrochemical biosensor for plant pathogen DNA detection with colloidal gold nanoparticles as probes. Sci Rep 7:38896
48. Leong YX, Tan EX, Leong SX, Lin Koh CS, Thanh Nguyen LB, Ting Chen JR, Xia K, Ling XY (2022) Where nanosensors meet machine learning: prospects and challenges in detecting disease X. ACS Nano 16:13279–13293
49. Li G, Fei X, Liu H, Gao J, Nie J, Wang Y, et al (2020b) Fluorescence and optical activity of chiral CdTe quantum dots in their interaction with amino acids. ACS Nano 14(4):4196–4205
50. Li L, Zhang M, Chen W (2020) Gold nanoparticle-based colorimetric and electrochemical sensors for the detection of illegal food additives. J Food Drug Anal 28(4):641–653
51. Li Y, Liao Q, Hou W, Qin L (2023) Silver-based surface plasmon sensors: fabrication and applications. Int J Molecular Sci 24(4):4142
52. Liu L, He F, Yu Y, Wang Y (2020) Application of FRET biosensors in mechanobiology and mechanopharmacological screening. Front Bioeng Biotechnol 8:595497
53. Liu W, Mirzoeva S, Yuan Y, Deng J, Chen S, Lai B, Vogt S, Shah K, Shroff R, Bleher R, Jin Q, Vo N, Bazak R, Ritner C, Gutionov S, Raha S, Sedlmair J, Hirschmugl C, Jacobsen C, Paunesku T, Kalapurkal J, Woloschak GE (2021) Development of Fe_3O_4 Core-TiO_2 shell nanocomposites and nanoconjugates as a foundation for neuroblastoma radiosensitization. Cancer Nano 12(1):12
54. Liyakat KSS, Liyakat KKS (2024) Nanosensors in agriculture field: a study. Int J Appl Nanotechnol 10(2):12–22
55. Lobato IM, O'Sullivan CK (2018) Recombinase polymerase amplification: basics, applications and recent advances. Trends Analyt Chem 98:19–35
56. Long Y, Stahl Y, Weidtkamp-Peters S, Postma M, Zhou W, Goedhart J, Sánchez-Pérez MI, Gadella TWJ, Simon R, Scheres B, Blilou I (2017) *In vivo* FRET–FLIM reveals cell-type-specific protein interactions in *Arabidopsis* Roots. Nature 548:97–102
57. Masdor NA (2021) Determination of the detection limit of the detection of GMO in food using the isothermal solid-phase recombinase polymerase amplification on microfluidic DVDs. Asian J Plant Biol 3(2):17–19
58. Miguel-Rojas C, Pérez-de-Luque A (2023) Nanobiosensors and nanoformulations in agriculture: new advances and challenges for sustainable agriculture. Emerg Top Life Sci 7(2):229–238
59. Mokhtarzadeh A, Eivazzadeh-Keihan R, Pashazadeh P, Hejazi M, Gharaatifar N, Hasanzadeh M, de la Guardia M (2017) Nanomaterial-based biosensors for detection of pathogenic virus. TrAC, Trends Anal Chem 97:445–457

60. Moskovits M (2010) Spectroscopy: expanding versatility. Nature 464:357
61. Munir N, Gulzar W, Abideen Z, Hancook JT, El-Keblawy A, Radicetti E (2023) Nanotechnology improves disease resistance in plants for food security: applications and challenges. Biocatal Agric Biotechnol 51:102781
62. Murphy CJ, Sau TK, Gole AM, Orendorff CJ, Gao J, Gou L, Hunyadi SE, Li T (2005) Anisotropic metal nanoparticles: synthesis, assembly, and optical applications. J Phys Chem B 109:13857–13870
63. Naresh V, Lee N A review on biosensors and recent development of nanostructured materials-enabled biosensors. Sensors 21:1109
64. Noi K, Iijima M, Kuroda SI, Ogi H (2019) Ultrahigh-sensitive wireless QCM with bio-nanocapsules. Sens Actuators B Chem 293:59–62
65. Patel R, Mitra B, Vinchurkar M, Adami A, Patkar R, Giacomozzi F, Lorenzelli L, Baghini MS (2022) A review of recent advances in plant-pathogen detection systems. Heliyon 8(12):e11855
66. Patil J, Bhattacharya S (2024) Exploring the potential of quantum dots and plasmonic nanoparticles for imaging and phototherapy in colorectal neoplasia. Results Chem 10:101689
67. Raman CV, Krishnan KS (1928) A new type of secondary radiation. Nature 121:501–502
68. Regiart M, Rinaldi-Tosi M, Aranda PR, Bertolino FA, Villarroel-Rocha J, Sapag K, Messina GA, Raba J, Fernández-Baldo MA (2017) Development of a nanostructured immunosensor for early and in situ detection of *Xanthomonas arboricola* in agricultural food production. Talanta 175:535–541
69. Romanovski V, Periakaruppan R (2023) Why metal oxide nanoparticles are superior to other nanomaterials for agricultural application? In: Nanometal oxides in horticulture and agronomy. Academic Press, pp 7–18
70. Safarpour H, Salimi A, Hallaj R (2012) Quantum dots fluorescence resonance energy transfer (QD-FRET) nanosensor for detection of fungal pathogens and plant viruses. Biosens Bioelectron 38(1):15–21
71. Safdar M, Kim W, Park S, Gwon Y, Kim YO, Kim J (2022) Engineering plants with carbon nanotubes: a sustainable agriculture approach. J Nanobiotechnol 20(1):1–30
72. Sagadevan S, Periasamy M (2014) Recent trends in nanobiosensors and their applications—a review. Rev Adv Mater Sci 36:62–69
73. Sankaran S, Mishra A, Ehsani R, Davis C (2010) A review of advanced techniques for detecting plant diseases. Comput Electron Agric 72(1):1–13
74. Sarkar R, Farghaly AA, Arachchige IU (2022) Oxidative self-assembly of Au/Ag/Pt alloy nanoparticles into high-surface area, mesoporous, and conductive aerogels for methanol electro-oxidation. Chem Mater 34(13):5874–5887
75. Schlücker S (2014) Surface-enhanced Raman spectroscopy: concepts and chemical applications. Angew Chem Int Ed Engl 53(19):4756–4795
76. Schneider R, Lamien-Meda A, Auer H, Wiedermann-Schmidt U, Chiodini PL, Walochnik J (2021) Validation of a novel FRET real-time PCR assay for simultaneous quantitative detection and discrimination of human *Plasmodium* parasites. PLoS ONE 16:e0252887
77. Sharma B, Frontiera RR, Henry AI, Ringe E, Van Duyne RP (2012) SERS: materials, applications, and the future. Mater Today 15(1–2):16–25
78. Shaw DS, Honeychurch KC (2022) Nanosensor applications in plant science. Biosensors 12:675
79. Shivashakarappa K, Reddy V, Tupakula VK, Farnian A, Vuppula A, Gunnaiah R (2022) Nanotechnology for the detection of plant pathogens. Plant Nano Biology 2:100018
80. Smekal A (1923) Zur Quantentheorie Der Dispersion. Naturwissenschaften 11:873–875
81. Srivastava AK, Dev A, Karmakar S (2018) Nanosensors and nanobiosensors in food and agriculture. Environ Chem Lett 16(1):161–182
82. Stiles PL, Dieringer JA, Shah NC, Van Duyne RP (2008) Surface-enhanced raman spectroscopy. Annu Rev Anal Chem 1:601–626
83. Tan X, He S, Liu X, Zhao G, Huang T, Yang L (2019) Ultrasensitive electrochemical sensing of dopamine by using Dihydroxylatopillar [5]Arene-Modified Gold Nanoparticles and Anionic Pillar [5]Arene-Functionalized Graphitic Carbon Nitride. Microchim Acta 186:703

84. Tanwar A, Sushil S (2019) Role and effects of nanotechnology used in pesticides and agriculture field. In: AIP conference proceedings (AIP Publishing), 2142
85. Thakur M, Wang B, Verma ML (2022) Development and applications of nanobiosensors for sustainable agricultural and food industries: recent developments, challenges and perspectives. Environ Technol Innov 26:102371
86. Tiwari M, Krishnamurthy S, Shukla D, Kiiskila J, Jain A, Datta R, Sharma N, Sahi SV (2016) Comparative transcriptome and proteome analysis to reveal the biosynthesis of gold nanoparticles in *Arabidopsis*. Sci Rep 6:21733
87. Tong Y, Shao L, Li X, Lu J, Sun H, Xiang S, Zhang Z, Wu Y, Wu X (2018) Adhesive and stimulus-responsive polydopamine-coated graphene oxide system for pesticide-loss control. J Agric Food Chem 66(11):2616–2622
88. Trusov Y, Chakravorty D, Botella JR (2013) *Fusarium oxysporum* infection assays in *Arabidopsis*. Methods Mol Biol 1043:67–72
89. Usman M, Tang JW, Li F, Lai JX, Liu QH, Liu W, Wang L (2023) Recent advances in surface enhanced Raman spectroscopy for bacterial pathogen identifications. J Adv Res 51:91–107
90. Valenzuela-Amaro HM, Aguayo-Acosta A, Meléndez-Sánchez ER, de la Rosa O, Vázquez-Ortega PG, Oyervides-Muñoz MA, Sosa-Hernández JE, Parra-Saldívar R (2023) Emerging applications of nanobiosensors in pathogen detection in water and food. Biosensors (Basel) 13(10):922
91. Verma AK, Ansari ZA (2020) Fluorescent ZnO quantum dot probe to study glucose–glucose oxidase interaction via fluorescence resonance energy transfer. Sens Lett 18:351–365
92. Verma AK, Noumani A, Yadav AK, Solanki PR (2023) FRET based biosensor: principle applications recent advances and challenges. Diagnostics 13:1375
93. Virk V, Deepak H, Taneja K, Srivastava R, Giri S (2024) Amelioration in nanobiosensors for the control of plant diseases: current status and future challenges. Front Nanotechnol 6:1310165
94. Voke E, Pinals RL, Goh NS, Landry MP (2021) In planta nanosensors: understanding biocorona formation for functional design. ACS Sens 6(8):2802–2814
95. Wang Q, Zhang B, Lin X, Weng W (2011) Hybridization biosensor based on the covalent immobilization of probe DNA on chitosan–mutiwalled carbon nanotubes nanocomposite by using glutaraldehyde as an arm linker. Sens Actuators B Chem 156(2):599–605
96. Wang TM, Yang JT (2019) Visual DNA diagnosis of tomato yellow leaf curl virus with integrated recombinase polymerase amplification and a gold-nanoparticle probe. Sci Rep 9(1):1–8
97. Wang ZL (2007) Nanopiezotronics. Adv Mater 19:889–892
98. Wee EJH, Trau M (2016) Simple Isothermal strategy for multiplexed, rapid, sensitive, and accurate MiRNA detection. ACS Sens 1:670–675
99. Wujcik EK, Wei H, Zhang X, Guo J, Yan X, Sutrave N, Wei S, Guo Z (2014) Antibody nanosensors: a detailed review. RSC Adv 4:43725–43745
100. Yotova L, Yaneva S, Marinkova D (2013) Biomimetic nanosensors for determination of toxic compounds in food and agricultural products (Review). J Univ Chem Technol Metall 48:215–227
101. Yu M, Zhao K, Zhu X, Tang S, Nie Z, Huang Y, Zhao P, Yao S (2017) Development of near-infrared ratiometric fluorescent probe based on cationic conjugated polymer and CdTe/CdS QDs for label-free determination of glucose in human body fluids. Biosens Bioelectron 95:41–47
102. Yusoff N, Pandikumar A, Ramaraj R, Lim HN, Huang NM (2015) Gold nanoparticle based optical and electrochemical sensing of dopamine. Microchim Acta 182:2091–2114
103. Zhang W, Li X, Chen Y, Zhao H (2022) Nanosensors for pathogen detection in agricultural systems: advances and applications. Agric Nanotechnol 45(3):123–140
104. Zhao W, Lu J, Ma W, Xu C, Kuang H, Zhu S (2011) Rapid on-site detection of *Acidovorax avenae* subsp. *citrulli* by gold-labeled DNA strip sensor. Biosens Bioelectron 26(10):4241–4244

105. Zhou W, Hu Z, Wei J, Dai H, Chen Y, Liu S, Duan Z, Xie F, Zhang W, Guo R (2022) Quantum dots-hydrogel composites for biomedical applications. Chin Chem Lett 33:1245–1253
106. Zhu H, Zhang H, Xu Y, Laššáková S, Korabečná M, Neužil P (2020) PCR past, present and future. Biotechniques 69:317–325

Advanced Nanosensors for Pesticide Detection in Water: Ensuring Environmental Safety and Public Health

Diego Maroso da Silva, Clarice Steffens, and Juliana Steffens

Abstract The detection and monitoring of pesticides in water are critical for environmental management and public health protection, as pesticides used in agriculture can contaminate water bodies through runoff, leaching, and direct application, posing risks to aquatic ecosystems and human health. Various sensor technologies, including electrochemical, optical, and biorecognition sensors, have been developed for efficient pesticide detection, offering advantages like cost-effectiveness, portability, and rapid response times. Among these, various biosensors (enzymatic, whole cell, immunosensors, DNA) show high sensitivity and specificity for detecting pesticides. Recent advancements in nanotechnology have further enhanced these sensors. Nanomaterials are used to amplify signals, increasing sensitivity and allowing for more accurate detection at lower concentrations. Additionally, nanoscale materials improve the performance by increasing their surface area and enhancing electron transfer rates, which boosts their sensitivity and response time. Integrating these advanced nanomaterials with traditional sensor technologies enables real-time monitoring of water contamination with greater accuracy, aiding timely interventions to mitigate risks. Incorporating nanotechnology into sensor development not only improves detection limits but also supports the development of portable and cost-effective solutions for on-site analysis. Combining these sensors with comprehensive monitoring programs and sustainable pesticide management practices is essential for preserving water quality and protecting ecosystems. Continued research in nanotechnology and sensor development is crucial for enhancing our ability to detect and manage pesticide pollution, ultimately safeguarding human and environmental health.

D. M. da Silva · C. Steffens · Juliana Steffens (✉)
Food Engineering Pos-Graduate Program, URI, Erechim, RS, Brazil
e-mail: julisteffens@uricer.edu.br

D. M. da Silva
e-mail: diegomaroso852@hotmail.com

C. Steffens
e-mail: julisteffens@uricer.edu.br

L. F. Fraceto et al. (eds.), *Emerging Nanotechnologies for Agroecosystem Management*, Smart Nanomaterials Technology, https://doi.org/10.1007/978-981-95-0187-8_8

Keywords Monitoring · Environmental · Agricultural fields · Contaminants · Nanotechnology

1 Introduction

Pesticides play a crucial role in modern agriculture by controlling pests and increasing crop yields, directly contributing to food security and agricultural productivity. However, their widespread use has raised concerns about their impact on the environment and human health, particularly when contaminating water sources. Monitoring pesticides in water is essential not only for assessing environmental quality and protecting aquatic ecosystems but also for safeguarding public health.

Water is a vital resource for various purposes, including drinking, irrigation, and industrial processes. Contamination of water bodies with pesticides can have detrimental effects on aquatic organisms, disrupt ecosystems, and pose risks to human health through the consumption of contaminated water or food products. Pesticides can enter water bodies through various pathways, such as runoff from agricultural fields, leaching from soil, and improper disposal of pesticide containers [27].

Effective monitoring of pesticides in water is necessary to identify contamination sources, track trends over time, and implement appropriate mitigation measures. This is particularly beneficial for agriculture, where advanced sensor technologies can play a transformative role. By enabling real-time monitoring and early detection of pesticide contamination, these sensors help ensure that pesticide application is managed more effectively. This not only protects water quality but also aids in optimizing pesticide use, reducing waste, and minimizing environmental impact.

Recent advancements in sensor technology have revolutionized pesticide monitoring. Sensors offer rapid, sensitive, and cost-effective methods for detecting pesticides in water, providing valuable data for assessing the efficacy of pesticide management practices. This allows farmers and agricultural managers to make informed decisions about pesticide application and management, improving both the sustainability of agricultural practices and the health of agroecosystems.

The role of sensors in detecting pesticides in water is paramount, as they provide an efficient and reliable means of monitoring pesticide levels in various environmental settings. Sensors can detect a wide range of pesticide compounds with high sensitivity and specificity, allowing for accurate quantification and identification of contamination hotspots. Additionally, sensor-based monitoring systems can be deployed in remote or hard-to-reach areas, providing comprehensive coverage of water bodies and enhancing monitoring capabilities [65].

Sensor technologies also contribute to evaluating the impact of regulatory measures, ensuring compliance with environmental standards, and enhancing overall agricultural sustainability. By offering detailed insights into contamination levels and sources, sensors help in refining agricultural practices to reduce runoff and leaching, thereby protecting water resources and maintaining ecosystem health.

This chapter aims to provide an overview of the importance of monitoring pesticides in water and the role of sensor technology in addressing this critical environmental and public health issue, for advanced agricultural sustainability. Through a comprehensive examination of the challenges, opportunities, and advancements in pesticide monitoring, this chapter seeks to contribute to the understanding of effective strategies for ensuring water quality and sustainability in agricultural and industrial landscapes.

2 Micropollutants in Water

Organic contaminants, also known as micropollutants, are compounds such as pesticides, pharmaceuticals, personal care products, plasticizers, waterproofing agents, and insulating foams. These micropollutants are deposited in sediments or transported by hydrological impact to other locations, and their byproducts are released into the water surface due to chemical and biological processes.

Micropollutants can originate from various sources, such as industrial wastewater, runoff from agriculture, livestock and aquaculture, and leaching from landfills, as well as domestic and hospital effluents. In aquatic systems, they can be found in a concentration range from ng/L to μg/L, and it is precisely this trace level that brings complications in their identification and analysis methods [13].

The sources of pollution can be classified as point and non-point sources. A typical example of a point source is a wastewater treatment plant. Point and non-point sources also differ in how pollutants are released into the receiving body. Point sources usually result in direct discharges into watercourses, whereas the pathway for non-point sources may involve a route that results in partial deposition before reaching the watercourse. Additionally, contaminant concentrations from non-point sources can vary significantly with time and space, reflecting both seasonal usage patterns and the relative contribution of destination processes [17].

Among point sources, we can mention industrial effluent treatment plants, domestic sewage networks, accidental spills, and mining. Emissions from point sources are more easily detected and controlled and usually result in direct discharge of pollutants into water bodies. In contrast, emissions from non-point sources, which are difficult to control, vary with time and space and may involve routes that result in partial deposition of pollutants before reaching water bodies. A typical example of a non-point source is the use of pesticides and fertilizers in soil [105].

The occurrence of micropollutants in aquatic systems has been associated with numerous adverse impacts, such as reproductive deficiencies, endocrine disruption, physical abnormalities, short and long-term toxicity, and feminization of fish species, among others. Some studies have observed that groundwater is less contaminated with micropollutants than surface water [103]. The contamination of groundwater by micropollutants mainly results from leachate from landfills, groundwater-surface water interaction, infiltration of contaminated water from agricultural lands, or septic tanks and sewage system infiltration [55]. Soil is the main pathway for groundwater pollution by certain micropollutants, for example, pesticides [23].

Among the most important contaminants found in groundwater and surface waters are heavy metals (such as lead, arsenic, cadmium, and mercury); pesticides (nitrates, organochlorine compounds, organophosphates, and carbamates); and volatile organic compounds (such as fuel products and halogenated solvents) [89].

The reason for concern is that these small molecules are resistant to conventional water treatments. The complete biodegradation of these substances in every ecosystem is not guaranteed, as their breakdown products also generate transformation products that in turn interact with organic matter under specific conditions.

They can even become increasingly toxic and recalcitrant, polluting our most valuable natural resources: water and food. Therefore, even treated water may contain residual contaminants that, upon reaching their concentration limits, become toxic. Studies show that many of these substances are resistant to the processes of sewage treatment plants (STPs), persisting in the environment for long periods [82]. The wastewater treatment plants use combinations of physical, biological, and/or chemical treatment methods to removal pesticides. The effluent may still contain a combined concentration of pesticides that fails to meet regulatory standards, posing dual risks to human populations and the environment [26].

Modern agriculture uses pesticides to increase productivity to an economically viable level [43]. The increase in pesticide concentration in surface waters is related to the use of these compounds in crops and landscaping areas [64], and washing of tools and machinery used in preparation and spraying, as well as traces in water used for washing contaminated fruits and vegetables. Additionally, there are wastewater streams originating from pesticide production by the industry [82].

Pesticides can pollute the aquatic environment through various processes, such as adsorption by soil particles, leaching through soil by rain or irrigation water, contaminating groundwater, or through surface runoff, appearing in surface waters [67].

3 Types of Pesticides Found in Water

Pesticides are chemicals used to manage or eliminate pests such as insects, weeds, and fungi across various environments, including agriculture, homes, and industrial areas. These substances are extensively applied to boost crop productivity, safeguard public health, and maintain food security. Nevertheless, their widespread application can lead to environmental issues, such as the contamination of water bodies. Identifying the types of pesticides present in water is crucial for evaluating their possible effects on aquatic life and human health [70].

Among the pesticides, the herbicides are specifically formulated to control or eliminate unwanted vegetation, such as weeds and grasses. Common herbicides used in agricultural practices include glyphosate, atrazine, paraquat, diuron, metribuzin, s-metolachlor, clopyralid, hexazinone, simazine, 2,4-D, and others. Water bodies can be contaminated with herbicides, especially through surface runoff from agricultural fields, urban areas, and residential properties [101].

Another category of pesticides is insecticides, which are chemicals designed to target and control insect pests that harm crops, spread diseases, or infest buildings and structures. Examples of insecticides include carbanates, organochlorine, organophosphates, pyrethrin, pyrethroid, and neonicotinoids. Insecticides can enter water bodies through runoff from agricultural fields, aerial spraying, and application to urban or residential areas. In water, insecticides can affect aquatic insects, fish, and other aquatic organisms, disrupting food chains and ecological balance [99].

The fungicides are pesticides designed to prevent or control fungal diseases in crops, turf, and ornamental plants. Common fungicides include azoxystrobin, chlorothalonil, thiophanate-methyl, benomyl, carbendazim, thiabendazole, and thiophanate. Fungicides may reach water bodies through surface runoff, leaching from soil, and direct application to water for disease control in aquaculture. In aquatic environments, fungicides can impact aquatic fungi, algae, and other microorganisms [70].

Rodenticides are also a type of pesticide used to manage rodent populations in agricultural, urban, and residential environments. Examples include anticoagulant compounds like warfarin and bromadiolone. It may enter water bodies through runoff from treated areas or direct application to water bodies for rodent control. In aquatic environments, rodenticides can pose risks to non-target species, including fish, birds, and mammals, through secondary poisoning and direct toxicity [1].

There are some pesticides frequently detected in water sources and their sources, impacts, and implications for water management are discussed [31]. Glyphosate which has been found in drinking water sources has raised concerns about potential human exposure to this herbicide through consumption [94]. In a study conducted by the U.S. Geological Survey National Water Quality Program, it was found that the glyphosate was detected in 66 out of 70 streams and rivers across various land-use settings in the United States. It was detected in at least 80% of samples collected [61].

Atrazine can persist in water for extended periods and has been associated with adverse effects on aquatic organisms [88]. Throughout the monitoring period, atrazine concentrations showed significant fluctuations. It was present in 91% of surface water samples taken before crop planting, and in 98% of samples collected after planting. After the growing season, atrazine was detected in 76% of the water samples collected [4].

S-metolachlor is one of the top five commonly found compounds in European freshwaters. Its concentrations vary across countries: 0.33 to 3.00 μg/L in Spain, 0.16 to 3.68 μg/L in Italy, and 0.9 to 1.5 μg/L in France [22].

In different countries, allowable levels have been frequently surpassed by chlorpyrifos in various water sources such as seawater, rivers, groundwater, and rainwater. Concentrations of chlorpyrifos have been detected as high as 303.8 μg/L, greatly exceeding environmental regulations [37].

Neonicotinoid concentrations were found to be highest in agricultural watersheds (12 ng/L) in various water sources such as rivers, streams, lakes, groundwater, and treated wastewater monitored in Minnesota. This was followed by urban areas (2.9 ng/L) and undeveloped watersheds (1.9 ng/L) [12].

By assessing pesticide contamination levels and identifying sources of pollution, stakeholders can work toward sustainable pesticide management practices and minimize the adverse effects of pesticides on water resources and biodiversity [47].

3.1 Potential Risks and Water Contamination

Pesticide contamination in water poses risks to human health and the environment. It can harm aquatic life even at low levels, disrupting reproduction and growth, leading to population declines and ecosystem imbalances [96]. Humans may be exposed through drinking water, swimming, or consuming contaminated seafood, which is linked to health issues such as cancer, reproductive problems, and neurological disorders [69]. Chronic exposure can also disrupt hormone balance and cognitive development, particularly in children [108].

Pesticides primarily enter water bodies through runoff from fields and roads. The process starts with the application of pesticides to crops, where some chemicals are absorbed by plants while others degrade due to environmental factors like UV light. Pesticides can also evaporate into the atmosphere, traveling long distances before being redeposited by rainfall. This redeposition results in pesticide residues returning to the ground and being carried by surface runoff into lakes and rivers, causing water pollution [111].

Simultaneously, pesticides can also seep into the soil, where they may adhere to soil particles. This interaction with soil can result in a slower release of pesticides into groundwater, a process known as leaching. Once in the groundwater, pesticides can eventually discharge into streams, contributing to groundwater pollution. In addition to these processes, pesticides undergo biodegradation through bacterial oxidation or chemical hydrolysis, which helps break down these substances in the soil.

Pesticides can also enter water bodies through direct spraying, air drifting, intentional dumping, improper mixing, and contaminated groundwater. Their movement into groundwater is influenced by the volume of water and interactions with organisms and solid particles. This interaction involves a balance between absorption, where pesticides are taken up by organisms, and adsorption, where they bind to solid particles [56].

Pesticides not absorbed by plants may either stay in the soil or decompose. Water-soluble pesticides can leach into groundwater during rainfall, while insoluble pesticides accumulate in topsoil and can be washed into surface waters through runoff and erosion [6]. Runoff is most likely right after pesticide application, within about 0.25–0.85 cm from the soil surface. Pesticides can also enter water bodies through volatilization and subsequent deposition with rain, although this is less significant.

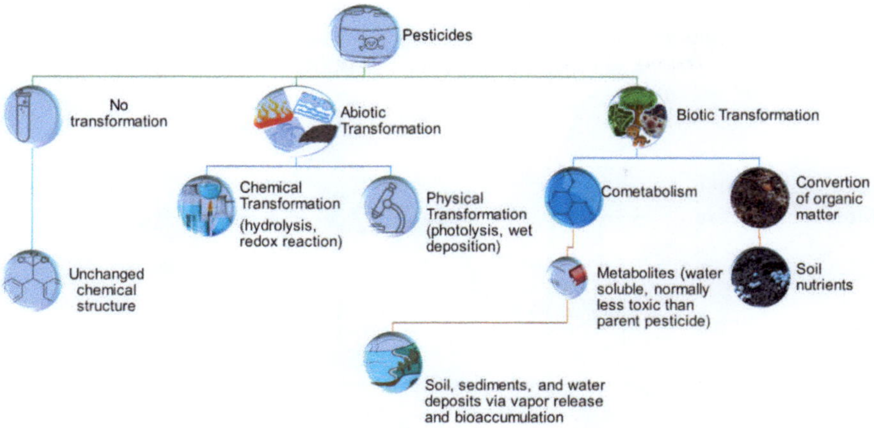

Fig. 1 Diagram depicting the movement and transformation of pesticides in the environment, including their circulation, retention, and potential for contaminating water sources

Overall, the movement of pesticides into water is influenced by the pesticide's properties, soil characteristics, environmental conditions, and how the pesticide is used [91].

Understanding how pesticides behave and move through the environment is crucial for managing their impact on ecosystems. Pesticides can follow various paths once applied, as illustrated in Fig. 1.

The movement of pesticides into water sources through runoff and leaching depends on their characteristics, such as half-life, solubility, and adsorption capacity. Pesticides degrade over time via microbial, photochemical, or chemical processes. Microbial degradation breaks them down into simpler forms, while photochemical degradation occurs with UV light. Chemical degradation involves reactions with soil substances. Pesticides with low biodegradation rates persist longer and can contaminate water. Their mobility is influenced by how well they bind to soil; those that bind strongly are less likely to leach but can be transported via soil erosion into surface water [92].

Accurate detection and monitoring of pesticide contamination in water are essential for managing and reducing associated risks. Numerous techniques have been developed, each providing unique benefits in sensitivity, specificity, and practicality.

4 Sensor Technology

Sensors are now widespread and essential in various industrial and consumer applications. A sensor is a device that detects a signal or stimulus and produces an electrical signal in response. This output, which can be in the form of current or voltage, is straightforward to measure. Sensors monitor changes in their environment and relay

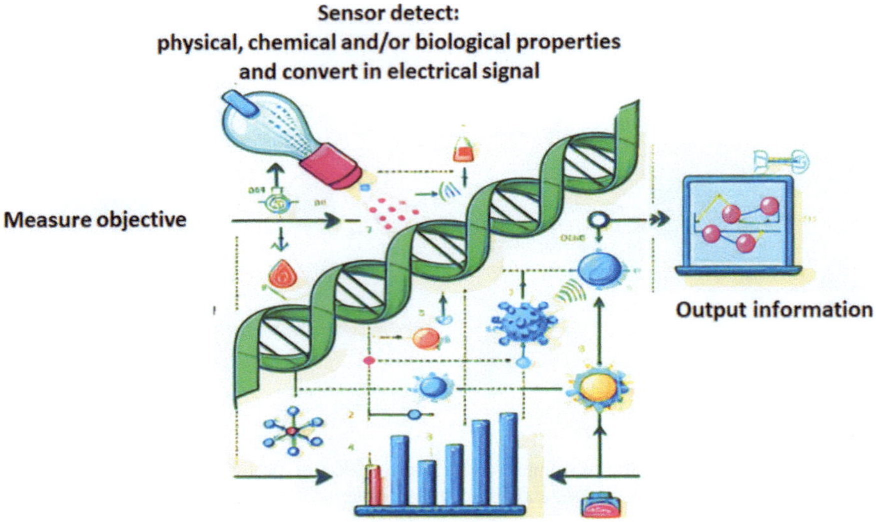

Fig. 2 Illustration of sensor operation

this information to other electronic systems, typically a computer processor. They convert physical phenomena into digital signals that can be displayed, interpreted, or further processed (Fig. 2).

Sensors are categorized into active and passive types. Active sensors, like GPS and RADAR, require an external power source to operate. In contrast, passive sensors, such as thermal sensors and metal detectors, generate their own electrical signals without external power [28]. Another classification is based on the conversion processes they use, including thermoelectric, photoelectric, and electrochemical methods [30].

In the context of sensor operation, the process begins with a recognition element within the sensor that selectively binds to the target pesticide molecule. This element may consist of enzymes, antibodies, aptamers, or specific receptors. Upon binding, the pesticide molecule induces a change in the sensor's physical, chemical, or electrical properties, thereby generating a response related to the concentration. This signal is transduced into a readable output using various mechanisms such as optical, electrochemical, and piezoelectric transduction. In some cases, the signal may be too weak to detect directly. Amplification techniques, such as enzymatic reactions or signal enhancement strategies, can be employed to increase the signal strength [87].

4.1 Sensors Used for Pesticide Detection

Sensors offer several advantages over traditional lab-based methods, such as lower costs, reduced reagent use, portability, rapid response, and the ability to continuously monitor complex wastewater. They are particularly effective at detecting low levels of pollutants in wastewater. Various sensor types included electrochemical, optical, and biorecognition [44].

4.1.1 Electrochemical Sensors

These sensors detect pesticides based on their electrochemical properties, once they measure electrical changes resulting from chemical interactions. Amperometric sensors are particularly useful as they gauge the current generated during electrochemical reactions involving pesticides, with the current level correlating to pesticide concentration. Potentiometric sensors measure variations in electrical potential in response to pesticide presence, while conductometric sensors detect changes in electrical conductivity caused by pesticide interactions with the sensor surface.

These sensors offer high sensitivity, allowing them to detect even trace amounts of pesticides with precision. Their cost-effectiveness is another significant benefit, making them an economical choice for both routine and large-scale monitoring. Despite these advantages, one major drawback is their sensitivity to temperature and humidity. Temperature variations influence the reaction kinetics, while high humidity can introduce interference or alter the sensor readings [3]. Additionally, electrochemical sensors may face difficulties with selectivity in complex water matrices where other substances could interfere with the sensing process. This interference can result in false positives or inaccuracies for quantification, necessitating careful calibration and signal processing. Another concern is the potential for sensor degradation over time. Exposure to water and its various components can cause fouling or deterioration of the sensor, which may require regular maintenance to ensure consistent performance [79]. Furthermore, the need for frequent calibration to maintain accuracy can be a drawback, as sensor responses may drift due to environmental changes or chemical interactions.

4.1.2 Optical Sensors

Optical sensors detect pesticides in water by measuring light absorption, often using UV–Visible spectroscopy. Infrared spectroscopy, another optical method, uses infrared light to detect molecular vibrations, providing detailed information about pesticide types based on their absorption patterns. Fluorescence sensors, on the other hand, detect the light emitted by pesticides when excited by certain wavelengths, with the intensity of this emission indicating pesticide concentration. Raman spectroscopy also falls under optical methods, measuring the inelastic scattering of monochromatic light to provide molecular information and detect pesticide residues [34].

Optical sensors, while offering advanced detection capabilities, are generally more expensive compared to other types of sensors. This is particularly useful in environmental monitoring where preserving the sample's integrity is important. They can deliver rapid and real-time results, which is essential for timely management and intervention in cases of pesticide contamination. Their ability to provide instant feedback helps in making prompt decisions to address pollution issues and prevent further environmental damage. On the other hand, a drawback is their high cost for maintenance and calibration. Environmental factors can also impact the performance, such as water turbidity, which is common in many natural water bodies, and can interfere with optical measurements by scattering light and affecting the accuracy of the results. Additionally, water temperature and other substances can affect readings, requiring complex calibration to maintain accuracy. Also, the sensor can be sensitive to fouling and degradation due to exposure to water and its components, necessitating regular upkeep to prevent performance issues [45].

4.1.3 Biorecognition Sensors

Specific pesticides are identified and bound by biorecognition sensors, which use biomolecules like antibodies or enzymes. They are highly selective and sensitive, but may be affected by environmental conditions and require careful storage and handling [84]. Enzyme-based sensors use specific enzymes that react with pesticides, producing measurable signals such as color changes or fluorescence. Immunosensors, utilizing antibodies that bind specifically to pesticides, generate signals that can be detected and quantified, providing high specificity and sensitivity. DNA-based sensors use nucleic acid sequences to bind with complementary pesticide molecules, and the resulting hybridization is detected through various methods, including fluorescence or electrochemical changes. Aptasensors employ specific aptamers for receptor functions, whereas biological sensors use microorganisms as their receptors (Fig. 3).

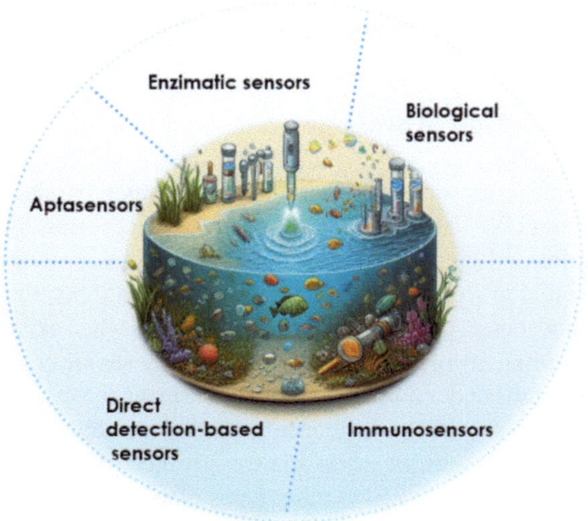

Fig. 3 Main biosensors for detection of pesticides in water

For the electrode to work it can be modified with different materials and recognition elements such as nanoparticles (Ag, Au, Cu, Ni, and Pt), graphene, nanotubes, enzymes, quantum dots, antibodies, magnetic beads, and biological agents [71].

4.2 Biosensors for Pesticide Detection

Biosensors deliver sensitive, specific, cost-effective, and rapid real-time detection. They are reusable and ideal for on-site monitoring of trace pesticide levels. They come in various types, such as enzymatic, whole cell, immunosensors, and DNA-based [102].

Enzymatic biosensors detect pesticides by measuring their impact on enzyme activity. They identify pesticide effects either through inhibition of enzyme activity or by detecting changes in products formed, such as proton release or chromophoric/ electro-active substances [88, 102]. Enzymes such as cholinesterase, tyrosinase, peroxidase, alkaline phosphatase, organophosphorus hydrolase, and glutathione-s-transferase have been used to detect organophosphates, carbamates, atrazine, thiodicarb, and paraoxon (Table 1).

Whole cell biosensors utilize cells as immobilization elements during the detection process. These biosensors are categorized based on the type of cell used, including microbial and plant cells as bioreporters through coupling with physiochemical transducers to produce signals for specific analyte(s) (Table 2). They operate using amperometric, potentiometric, and conductometric methods [41].

Immunosensors are based on antigen–antibody interactions, where antibodies are fixed on a substrate and bind with antigens to form a complex. Detecting changes in the optical and electrochemical properties of this complex is key to the detection process. An effective immunosensor should detect and quantify antibodies in real samples (Table 3) [14]. However, the use of immunosensors for pesticide detection is limited due to the high cost and complexity of producing monoclonal antibodies and the associated animal care requirements [102].

DNA biosensors use immobilized DNA on electrodes to detect pesticides by monitoring changes in DNA's redox properties [68]. They have been used to detect various pesticides (Table 4), including atrazine, 2,4-D, and glufosinate ammonium. However, their effectiveness is often limited by issues such as selectivity, cost, and reusability.

4.3 Nanotechnology-Based Sensors

In this, sensors leverage the unique properties of nanomaterials to enhance detection capabilities. Nanoparticle sensors use materials like gold or silver nanoparticles

Table 1 Applications and types of enzymatic biosensors for pesticide detection in water

Pesticide	Enzyme	Type of sensor	Application	References
Organophosphates	Cholinesterase	Amperometric	Water	[106]
Organophosphate paraoxon	Acetylcholinesterase	Amperometric	Real time	[73]
Organophosphate	Esterase-2	Optical (fluorescent)	Water	[10]
Atrazine	Tyrosinase	Cantilever	River water	[60]
Glyphosate	Peroxidase	Cantilever	Water sources	[63]
Dichlorofenthion	Hydrolase and horseradish peroxidase	Amperometric	Water	[80]
Chlortoluron	Tyrosinase	Electrochemically	Tap and river water	[29]
Paraoxon Methyl	Acetylcholinesterase	Resistive	River and well water	[57]
Glyphosphate	Peroxidase	Colorimetric	River and ground water	[49]
Malathion	Peroxidase and Oxidase	Colorimetric	Tap water and river water	[51]
Malathion	Oxidase	Colorimetric	Lake water	[21]
Methyl parathion	Oxidase	Colorimetric	East lake water and tap water	[15]
Malathion	Peroxidase	Fluorescence	Tap water	[16]
Dichlovos	Peroxidase	Fluorescence	Tap and field water	[81]
Dichlovos	Peroxidase	Fluorescent	Lake water	[110]
Ethoprophos	Peroxidase	Chemiluminescence	Tap and river water	[33]
Triazophos	Peroxidase	Colorimetric and SERS	Water	[109]
Chlorpyrifos	Acetylcholinesterase	Electrochemical	Drinking water	[90]
Isocarbophos, profenofos, and omethoate	Oxidase	Colorimetric	Lake and river water	[93]
Paraoxon, 2,4-D, and Atrazine	Butyrylcholinesterase, alkaline phosphatase, and tyrosinase	Chronoamperometry	River water	[2]
Chlorpyrifos	Laccases and catalases	Voltametric	Wastewater	[76]

(continued)

Table 1 (continued)

Pesticide	Enzyme	Type of sensor	Application	References
Paraoxon	Acetylcholinesterase	Colorimetric	Tap and irrigation waters	[5]
Paraoxon methyl	Acetylcholinesterase	Colorimetric	Ground water	[40]

Table 2 Applications and types of cell biosensors for pesticide detection in water

Pesticide	Biosensor Type	Detection Method	Application	References
Fenitrothion	Recombinant *Pseudomonas putida* JS444	Amperometric	Lake water	[46]
Atrazine	*Spinacia oleracea, Senecio vulgaris*	Screen-printed electrodes	River water	[97]
Atrazine and diuron	*Anabaena variabilis*	Amperometric	Freshwater	[98]
Atrazine	*Scenedesmus obliquus*	Self-Powered Photosynthetic	Water quality	[25]
2,4-D and chlorpyrifos	*Anabaena torulosa*	Fluorescence	Deionized and distilled water	[107]

Table 3 Applications and types of immunosensors for pesticide detection in water

Pesticide	Immunosensor	Type of Sensor	Application	References
Triazophos	Bio-barcode	Fluorescence	Tap water	[113]
Paraoxon	Single wall nanotubes paraoxon antibody	Amperometric	Tap and purified water	[53]
Endosulfan and paraoxon	Single wall nanotubes	Electrochemical	Tap and purified water	[52]
Carbofuran	Polyclonal anti-carbofuran antibody	Impedimetric	Drinking water	[54]
Atrazine	Anti-atrazine antibodies	Carbon nanotubes field effect transistors	Seawater and River water	[11]
Atrazine	Phage anti-immunocomplex with an anti-atrazine monoclonal antibody	Electrochemical	River water	[24]
Imidacloprid	Monoclonal antibodies	Electrochemical	Tap water	[72]
Cypermethrin and esfenvalerate	Antibody conjugated gold nanoparticles	Surface-enhanced Raman scattering	Tap water and river water	[50]
Atrazine and acetochlor	Immuno-chromatographic strips	Electrochemical	River water	[78]

Table 4 Applications and types of DNA biosensors for pesticide detection in water

Pesticide	DNA	Type of Sensor	Application	References
Acetamiprids	DNA	Electrochemical	Tap water	[19]
Acetamiprid and atrazine	DNA	Impedimetric	Tap and mineral water	[58]
Malathion	DNA	Colorimetric	Lake water	[8]
Methamidophos, isocarbophos, phosalone, methamidophos, acephate, trichlorfon, and dursban	DNA	Colorimetric	River water	[7]

to amplify optical signals, improving sensitivity. Nanosensors incorporate nanomaterials that provide distinct electrical, optical, or chemical properties, enhancing performance for detecting pesticides due to surface area and reactivity [89].

Studies have demonstrated the effectiveness of nanosensors in detecting pesticides in water through the integration of carbon nanomaterials, graphene, and metals. This combination enhancing detection capabilities such as surface area, electrical conductivity, and mechanical properties serves as an excellent substrate for sensor development. When combined with metals, the properties of graphene are further enhanced, leading to improved sensitivity.

For instance, the integration of silver with graphene creates a synergistic effect that enhances the sensor's electrical conductivity and catalytic activity. Silver-graphene modified boron-doped diamond electrodes have been utilized to detect pesticides such as paraquat and carbaryl [74]. This combination not only improves the electron transfer rate but also provides a stable and reproducible sensing platform.

Similarly, reduced graphene oxide (rGO)-linked copper and silver form a highly sensitive nanocomposite. [32] developed an rGO–Cu/CuO–Ag nanocomposite sensor that effectively detected fenamiphos and carbaryl. The incorporation of metal oxides with graphene enhances the redox properties and increases the active sites for pesticide interaction, leading to improved detection performance.

Moreover, the detection of glyphosate, a common herbicide, has been significantly improved using graphene-based electrochemical sensors combined with metals such as gold and platinum. [95] developed a sensitive platform on gold electrodes, enabling the detection of glyphosate in river water samples (LOD of around 0.08 ppb).

Additionally, [39] utilized platinum-decorated laser-induced graphene along with glycine oxidase for selective glyphosate detection. This approach offers a scalable method suitable for complex matrices, combining the high catalytic activity of platinum with the robust structure of laser-induced graphene.

Furthermore, [83] detecting glyphosate at nanomolar levels in drinking water by using gold nanoelectrode arrays on graphene paper. This innovative approach leverages the high surface area and conductivity of graphene paper, coupled with the superior electrochemical properties of gold.

[112] employed the spiking method to assess the practical use of sensor in detecting simazine with high recovery rate for tap, river, and pond water samples. [75]

synthesized ZnO@Cu nanoparticles using the cetyltrimethylammonium bromide surfactant for electrochemical detection of 2,4-dichlorophenol. The real-time analysis was evaluated demonstrating remarkable stability, and reproducibility.

[35] developed a biosensor for detecting dichlorvos using acetylcholinesterase based on a titanium dioxide (TiO_2) sol–gel carrier and observed a linear response (1.13 nM to 22.6 μM) with a LOD of approximately 0.23 nM.

[62] developed an amperometric biosensor using a biocomposite electrode composed of graphite, epoxy, and acetylcholinesterase. The biosensor incorporated 16% graphite into the composite, optimizing the electrode's conductivity and sensitivity. This design showed LOD of 0.25 ppb of carbofuran. The biosensor's performance was evaluated using spiked tap water containing carbofuran and paraoxon pesticides and demonstrated recovery percentages close to 100%, indicating its high accuracy and reliability. Furthermore, the study ensured that no pesticides were present in the ground prior to the spiking process, thereby validating the biosensor's effectiveness in detecting externally introduced pesticides. The integration of graphene with metals (boron, silver, gold, and platinum) significantly enhances the performance of sensors for detecting pesticides in water.

5 Overview of Regulations and Guidelines Governing Pesticide Levels

The establishment of guideline levels for pesticides in drinking water underscores the importance of safeguarding public health and ensuring the quality of water resources. Regulatory agencies and organizations have developed guidelines and regulation levels for drinking water.

The World Health Organization (WHO) presented permitted levels for herbicides commonly detected in drinking water sources. These levels were meticulously established through the evaluation of each herbicide's toxicity data, taking into account various factors such as acute and chronic health effects. The comprehensive assessment aimed to ensure the protection of public health from potential risks associated with pesticide exposure through drinking water consumption. The calculation involved the consideration of several critical parameters. It was assumed that an average adult weighing 70 kg consumes 2 L of water per day, reflecting typical daily water intake patterns. Additionally, a precautionary approach was adopted, where 10% of the Acceptable Daily Intake (ADI) for each herbicide was allocated to drinking water. The recommended levels for herbicides in drinking water are 25 μg/L bentazone; 0.5 μg/L 2-methyl-4-chlorophenoxyacetic acid; 5 μg/L metolachlor; 17 μg/L pendimethalin; 175 μg/L propanil; 60 μg/L pyridate; 17 μg/L simazine:; and 170 μg/L trifluralin [38].

At the European level, the Water Framework Directive (WFD) mandates that Member States adopt specific measures to combat water pollution. Recognizing the importance of safeguarding water quality, one of the fundamental regulations is the

European directive on drinking water [18]. This directive establishes stringent standards, including a maximum concentration of 0.1 μg/L for each individual pesticide and 0.5 μg/L for the total amount present in the water sample.

The US Environmental Protection Agency (EPA) establishes minimum contaminant levels (MCLs) based on an ADI, the ingested amount of pesticides allowed for humans extrapolated from toxicological data. The values of MCLs in drinking water were 2 μg/L alachlor; 3 μg/L aldicarb; 3 μg/L aldicarb sulfone; 3 μg/L atrazine; 403 μg/L carbofuran, 53 μg/L carbon tetrachloride; 70 μg/L 2,4-D; 20 μg/L diquat; 0.4 μg/L heptachlor; 0.2 μg/L heptachlor epoxide; 40 μg/L methoxychlor; 200 μg/L oxamyl; 1 μg/L pentachlorphenol; 500 μg/L picloram; 4 μg/L simazine; and 3 μg/L toxaphene [100].

Sensor technology enables the early detection of contaminants in water sources, allowing regulatory authorities to promptly identify and address potential threats to water safety. By continuously monitoring parameters and specific contaminants, sensors can detect deviations from acceptable levels, triggering immediate corrective actions [9].

Unlike traditional sampling and laboratory methods, which are time-consuming and may provide delayed results, sensor technology offers real-time monitoring capabilities. Sensors installed at strategic points within water distribution networks or at intake points can continuously collect data on water quality parameters. Rapid response to changes in water quality is facilitated by this real-time data, ensuring compliance with safety guidelines and minimizing the risk of exposure to contaminants.

The utilization of real-time data is instrumental in facilitating swift responses to fluctuations in water quality, particularly in the context of detecting contaminants such as organophosphate pesticides. A study presents a fluorescence-based biosensing device for detecting organophosphate pesticides in water and consumable food. The biosensor utilizes a mutant form of thermostable esterase-2 from *Alicyclobacillus acidocaldarius* (EST2-S35C) as the bioreceptor. When applied in surface water, samples from the Tiber River observed a decrease in fluorescence, suggesting contamination by domestic and industrial activities [10].

A sensitive photonic biosensor based on a bimodal waveguide interferometer was applied directly for quantification of fenitrothion. The biosensor demonstrated outstanding recovery percentages, achieving accurate results with a remarkably short time (20 min), without sample pre-treatment [77].

The integration of sensor systems with Internet of Things (IoT) platforms and big data analytics enables real-time data transmission, storage, and analysis. By leveraging IoT connectivity, sensor networks can communicate data to centralized platforms, allowing for rapid decision-making and predictive modeling based on real-time monitoring data. A biosensor based on acetylcholinesterase and integrated with the IoT was utilized for the detection of chlorpyrifos. The detection process was facilitated using an electrochemical workstation. The study yielded promising results, with a LOD of 2 μg/L achieved. Furthermore, the integration with IoT enabled various functionalities, including wireless transmission of detection information, management of detection data, remote dynamic supervision of detection devices, and sharing of detection information. This comprehensive approach not only enhances the efficiency and accuracy of chlorpyrifos detection but also facilitates real-time monitoring and management of pesticide contamination [20].

Automation of sensor systems and the development of autonomous monitoring platforms reduce manual intervention and enable continuous, unmanned operation. Autonomous sensor platforms equipped with artificial intelligence algorithms can detect anomalies, identify trends, and trigger alerts or interventions automatically. As an example, a chemiluminescence system with data processed chemometrically using artificial neural network trained by the algorithm was applied to detect dipterex, dichlorvos, and omethoate [48].

Sensor technology allows for the simultaneous monitoring of multiple parameters, and ensures that regulatory requirements regarding various contaminants and water quality indicators are met. Additionally, sensors can be deployed in remote or hard-to-access areas, providing coverage across different geographical locations and enhancing overall monitoring capabilities.

Automating water quality monitoring with sensor technology can lead to increased efficiency for regulatory authorities. Moreover, early detection of contaminants can prevent costly water treatment interventions or regulatory penalties associated with non-compliance.

Sensor data provides valuable insights for risk management and decision-making processes. Regulatory authorities can use real-time data to assess potential risks to water safety, prioritize interventions, and allocate resources effectively. Additionally, historical sensor data can facilitate trend analysis and predictive modeling, allowing for proactive measures to prevent water quality issues before they escalate.

6 Challenges and Future Perspectives

The perspectives for expanding the use of sensor systems for real-time analysis of pesticides in drinking water are promising, driven by advancements in technology, ongoing research initiatives, interdisciplinary collaboration, and policy support. By harnessing the potential of sensor technology, stakeholders can enhance water quality monitoring, mitigate risks associated with pesticide contamination, and safeguard public health and environmental sustainability.

The utilization of sensors for pesticide detection presents several challenges, ranging from sensitivity and selectivity to cost-effectiveness. Sensors must possess appropriate characteristics, including low cost, portability, and flexibility, to be practical for widespread use. Achieving high sensitivity and selectivity while maintaining affordability can be particularly challenging. Nanozyme-based sensors targeting pesticides face operational challenges due to the complexity of their underlying mechanisms, including inhibition kinetics and the influence of sample composition [104]. Moreover, the presence of non-specific analytes in samples can affect sensor performance. However, nanozyme-based sensors offer the advantage of enhanced selectivity, as they can be designed to specifically target the analyte of interest while minimizing interference from other compounds. To address operational challenges, a deeper understanding of relevant factors is necessary to optimize inhibition-based sensors. Additionally, the use of haptens, analogues of the target compounds, has shown promise in improving the sensitivity of immunoassays for pesticide detection [66]. Overall, addressing these challenges is crucial for advancing sensor technologies and improving pesticide detection methodologies.

Portable biosensors have emerged as promising tools for on-site pesticide detection. Ensuring that biosensors can detect pesticides accurately and reliably amid complex environmental matrices is crucial for their effectiveness in real-world applications [86]. However, these techniques still face limitations. For example, colorimetric analysis, while widely used, can be hindered by the natural color of test materials, leading to poor discriminating colorimetric effects. An example of an innovative approach to portable pesticide detection involves a colorimetric paper sensor utilizing nanowires as a chromogenic indicator. This portable assembly of the acetylcholinesterase-nanozyme tandem reaction on test paper has demonstrated LODs of 10 ng/mL for omethoate, and 3 ng/mL for dichlorvos. However, the catalytic activity nanozymes is influenced by various reaction parameters, including reaction time, concentration, pH, and temperature [36].

A portable acetylcholinesterase biosensor, utilizing electrode strips connected to a radio shack digital multimeter with computer interface, was employed for methomyl, prothiofos, and chlorpyrifos detection in agricultural drainage water samples. The biosensor demonstrated acceptable sensitivity and accuracy with recoveries ranging from $94.73 \pm 2.74\%$ to $108.3 \pm 0.80\%$. However, it is worth noting that methomyl S, S,-dioxide, a degradation product of methomyl, exhibited significantly higher toxicity [42]. These findings underscore the importance of monitoring pesticide degradation products in addition to parent compounds for comprehensive pesticide residue analysis.

Nanoparticle-based colorimetric probes, integrated into nano-based chips, have been employed with smartphone-based portable devices, enabling the acquisition of measurable digital signals. This innovative approach facilitates rapid visual quantitative detection of glyphosate residues in real samples using portable smartphones. The sensing platform based on smartphones underscores its potential for sensitive and efficient pesticide residue analysis [114].

While innovative technologies for real-time and remote monitoring sensing array with polymeric layers and platinum nanoparticles hold promise for pesticide detection, their practical application in the field is not without challenges. For example, in the case of detecting insecticide (chlorpyrifos), the sensors demonstrated 73.95 ppb of LOD [85]. However, several challenges must be addressed for effective field deployment of such systems. These challenges include ensuring robustness and durability of the sensors in diverse environmental conditions, optimizing the sensing array for specificity and selectivity to minimize false positives, and integrating the sensor platform with user-friendly interfaces for seamless operation by field personnel. Additionally, factors such as calibration and maintenance of the sensors over time, data transmission and interpretation, and cost-effectiveness are crucial considerations in overcoming the challenges associated with deploying sensor technologies for pesticide detection in real-world agricultural settings.

One significant challenge is the presence of interfering compounds, which can affect the selectivity of the sensor. Developing sensors that can discriminate between target pesticides and other compounds commonly found in environmental samples is essential for reliable detection in diverse sample types, such as water, soil, and food.

Cost-effectiveness is another critical factor in sensor development, as widespread adoption of pesticide detection technologies depends on their affordability. While advanced sensor technologies may offer high sensitivity and selectivity, they may also come with higher production costs. Furthermore, ensuring the portability and flexibility of pesticide detection sensors is essential for enabling on-site monitoring in various settings. Portable sensors that can be easily deployed in the field or at points of pesticide application offer advantages in terms of real-time monitoring and early detection of contamination incidents. Flexibility in sensor design allows for adaptation to different sample types and environmental conditions, enhancing the versatility of pesticide detection technologies.

7 Conclusions

Monitoring and detecting pesticides in water are crucial for managing environmental and public health risks. Pesticides, including herbicides, insecticides, fungicides, and rodenticides, are essential for pest control and agricultural productivity but can contaminate water bodies through runoff, leaching, and direct application.

Understanding the types of pesticides found in water and their potential risks is important for evaluating their impact on aquatic ecosystems and human health. Pesticides can accumulate in water sources, affecting aquatic life and posing health risks through drinking water, swimming, or consumption of contaminated aquatic organisms.

To address these issues, various sensor technologies have been developed for efficient pesticide detection and quantification in water. Electrochemical, optical, and biorecognition sensors provide advantages such as cost-effectiveness, portability, and rapid response. Electrochemical sensors, especially those with screen-printed electrodes, are notable for their reproducibility, sensitivity, and affordability.

Biosensors, including enzymatic, whole cell, immunosensors, and DNA-based, offer sensitive and specific pesticide detection. These biosensors use recognition elements like enzymes, antibodies, or DNA to selectively bind to pesticides, producing measurable signals indicative of their presence and concentration.

Recent advancements have led to the creation of highly sensitive and selective sensors capable of detecting a broad spectrum of pesticides in water. These technologies enable real-time monitoring, facilitating timely action to reduce risks to aquatic ecosystems and human health. Integrating sensor technologies with comprehensive monitoring programs and sustainable pesticide management practices is crucial for maintaining water quality and protecting aquatic environments. Ongoing research and innovation in sensor technology promise improved detection and mitigation of pesticide pollution, ultimately benefiting both human and environmental health.

Acknowledgements The authors would like to thank the National Council for Scientific and Technological Development—Brazil (CNPq), Coordination for the Improvement of Higher Education Personnel—Brazil (CAPES)—Finance Code 001, and Research Support Foundation of the State of Rio Grande of Sul—Brazil (FAPERGS). This study also was supported in part by the INCT Nanotechnology for Sustainable Agriculture (CNPq) #405924/2022-4; CAPES and São Paulo Research Foundation (FAPESP).

References

1. Akashe MM, Pawade UV, Nikam AV (2018) Classification of pesticides: a review. Int J Res Ayurveda Pharm 9(4):144–150. https://doi.org/10.7897/2277-4343.094131
2. Arduini F, Cinti S, Caratelli V, Amendola L, Palleschi G, Moscone D (2019) Origami multiple paper-based electrochemical biosensors for pesticide detection. Biosens Bioelectron 126:346–354. https://doi.org/10.1016/j.bios.2018.10.014
3. Ariyaratne R, Elangasinghe MA, Zamora ML, Karunaratne DGGP, Manipura A, Jinadasa KBSN, Abayalath KHN (2023) Understanding the effect of temperature and relative humidity on sensor sensitivities in field environments and improving the calibration models of multiple electrochemical carbon monoxide (CO) sensors in a tropical environment. Sens Actuators B Chem 390:133935. https://doi.org/10.1016/j.snb.2023.133935
4. ATSDR (2003) Agency for Toxic Substances and Disease Registry (ATSDR). Toxicological profile for Atrazine. U.S. Department of Health and Human Services, Public Health Service, Atlanta, GA
5. Ayaz S, Uluçay S, Üzer A, Dilgin Y, Apak R (2024) A novel acetylcholinesterase inhibition based colorimetric biosensor for the detection of paraoxon ethyl using CUPRAC reagent as chromogenic oxidant. Talanta 266:124962. https://doi.org/10.1016/j.talanta.2023.124962
6. Aydinalp C, Porca MM (2004) The effects of pesticides in water resources. J Central Eur Agric 5(1):5–12
7. Bai W, Zhu C, Liu J, Yan M, Yang S, Chen A (2015) Gold nanoparticle–based colorimetric aptasensor for rapid detection of six organophosphorous pesticides. Environ Toxicol Chem 34(10):2244–2249. https://doi.org/10.1002/etc.3088
8. Bala R, Kumar M, Bansal K, Sharma RK, Wangoo N (2016) Ultrasensitive aptamer biosensor for malathion detection based on cationic polymer and gold nanoparticles. Biosens Bioelectron 85:445–449. https://doi.org/10.1016/j.bios.2016.05.042
9. Ballen SC, Ostrowski GM, Steffens J, Steffens C (2021) Graphene oxide/urease nanobiosensor applied for cadmium detection in river water. IEEE Sens J 21(8):9626–9633. https://doi.org/10.1109/JSEN.2021.3056042
10. Barbieri MV, Rodrigues AC, Febbraio F (2022) Monitoring of pesticide amount in water and drinkable food by a fluorescence-based biosensor. EFSA J 20. https://doi.org/10.2903/j.efsa.2022.e200403
11. Belkhamssa N, Justino CIL, Santos PSM, Cardoso S, Lopes I, Duarte AC, Rocha-Santos T, Ksibi M (2016) Label-free disposable immunosensor for detection of atrazine. Talanta 146:430–434. https://doi.org/10.1016/j.talanta.2015.09.015
12. Berens MJ, Capel PD, Arnold WA (2021) Neonicotinoid insecticides in surface water, groundwater, and wastewater across land-use gradients and potential effects. Environ Toxicol Chem 40(4):1017–1033. https://doi.org/10.1002/etc.4959

13. Bhatt P, Bhandari G, Bilal M (2022) Occurrence, toxicity impacts and mitigation of emerging micropollutants in the aquatic environments: recent tendencies and perspectives. J Environ Chem Eng 10(3):107598. https://doi.org/10.1016/j.jece.2022.107598
14. Cezaro AM, Ballen SC, Hoehne L, Steffens J, Steffens C (2021) Cantilever nanobiosensors applied for endocrine disruptor detection in water: a review. Water Air Soil Pollut 232(6):225. https://doi.org/10.1007/s11270-021-05179-z
15. Chen T, Zhou D, Hou S, Li Y, Liu Y, Zhang M, Zhang G, Xu H (2022) Designing hierarchically porous single atoms of Fe–N 5 catalytic sites with high oxidase-like activity for sensitive detection of organophosphorus pesticides. Anal Chem 94(44):15270–15279. https://doi.org/10.1021/acs.analchem.2c02540
16. Chen Y, Zhu Y, Zhao Y, Wang J (2021) Fluorescent and colorimetric dual-response sensor based on copper (II)-decorated graphitic carbon nitride nanosheets for detection of toxic organophosphorus. Food Chem 345:128560. https://doi.org/10.1016/j.foodchem.2020.128560
17. Dressing S, Meals D, Harcum J, Spooner J, Stribling J, Richards RP, Millard CJ, Lanberg S, Odonnell J (2016) Monitoring and evaluating nonpoint source watershed projects
18. EUR-LEX (1998) Direttiva 98/83/CE del Consiglio del 3 novembre 1998 concernente la qualità delle acque destinate al consumo umano. https://eur-lex.europa.eu/legal-content/IT/ALL/?uri=CELEX:31998L0083
19. Fei A, Liu Q, Huan J, Qian J, Dong X, Qiu B, Mao H, Wang K (2015) Label-free impedimetric aptasensor for detection of femtomole level acetamiprid using gold nanoparticles decorated multiwalled carbon nanotube-reduced graphene oxide nanoribbon composites. Biosens Bioelectron 70:122–129. https://doi.org/10.1016/j.bios.2015.03.028
20. Garrido-Momparler V, Peris M (2022) Smart sensors in environmental/water quality monitoring using IoT and cloud services. Trends Environ Anal Chem 35:e00173. https://doi.org/10.1016/j.teac.2022.e00173
21. Ge J, Yang L, Li Z, Wan Y, Mao D, Deng R, Zhou Q, Yang Y, Tan W (2022) A colorimetric smartphone-based platform for pesticides detection using Fe–N/C single-atom nanozyme as oxidase mimetics. J Hazard Mater 436:129199. https://doi.org/10.1016/j.jhazmat.2022.129199
22. Gikas GD, Vryzas Z, Tsihrintzis VA (2018) S-metolachlor herbicide removal in pilot-scale horizontal subsurface flow constructed wetlands. Chem Eng J 339:108–116. https://doi.org/10.1016/j.cej.2018.01.056
23. González-Rodríguez RM, Rial-Otero R, Cancho-Grande B, Gonzalez-Barreiro C, Simal-Gándara J (2011) A Review on the fate of pesticides during the processes within the food-production chain. Crit Rev Food Sci Nutr 51(2):99–114. https://doi.org/10.1080/10408390903432625
24. González-Techera A, Zon MA, Molina PG, Fernández H, González-Sapienza G, Arévalo FJ (2015) Development of a highly sensitive noncompetitive electrochemical immunosensor for the detection of atrazine by phage anti-immunocomplex assay. Biosens Bioelectron 64:650–656. https://doi.org/10.1016/j.bios.2014.09.046
25. Gonzalez D, Di Lorenzo M (2019) Self-powered photosynthetic biosensor for pesticide detection in water. ECS Meeting Abstracts MA2019-04(8):402–402. https://doi.org/10.1149/MA2019-04/8/402
26. Goodwin L, Carra I, Campo P, Soares A (2017) Treatment options for reclaiming wastewater produced by the pesticide industry. Int J Water Wastewater Treat 4:1–15
27. Graboski AM, Martinazzo J, Ballen SC, Steffens J, Steffens C (2020) Nanosensors for water quality control. In: Amrane A, Rajendran S, Nguyen TA, Assadi AA, Sharoba AM (eds) Nanotechnology in the beverage industry fundamentals and applications, 1st edn. Elsevier, p 744. https://doi.org/10.1016/B978-0-12-819941-1.00004-3
28. Ha S, Lee D, Kim H, Kwon S, Kim E, Yang J, Lee S (2021) Neural network for metal detection based on magnetic impedance sensor. Sensors 21(13):4456. https://doi.org/10.3390/s21134456

29. Haddaoui M, Raouafi N (2015) Chlortoluron-induced enzymatic activity inhibition in tyrosinase/ZnO NPs/SPCE biosensor for the detection of ppb levels of herbicide. Sens Actuators B: Chem 219:171–178. https://doi.org/10.1016/j.snb.2015.05.023
30. Hallil H, Dejous C, Hage-Ali S, Elmazria O, Rossignol J, Stuerga D, Talbi A, Mazzamurro A, Joubert P-Y, Lefeuvre E (2021) Passive resonant sensors: trends and future prospects. IEEE Sens J 21(11):12618–12632. https://doi.org/10.1109/JSEN.2021.3065734
31. Hangay G, Gayubo SF, Hoy MA, Goula M, Sanborn A, Morrill WL, GÄde G, Marco HG, Kabissa JCB, Ellis J, Ellis A, Lord CC, Schabel HG, Heppner JB, Schabel HG, Heppner JB, Heppner JB, Heppner JB, Nadel H et al (2008) Acaricides or miticides. In: Encyclopedia of entomology. Springer, The Netherlands, pp 9–21. https://doi.org/10.1007/978-1-4020-6359-6_18
32. Hashemi P, Karimian N, Khoshsafar H, Arduini F, Mesri M, Afkhami A, Bagheri H (2019) Reduced graphene oxide decorated on Cu/CuO–Ag nanocomposite as a high-performance material for the construction of a non-enzymatic sensor: application to the determination of carbaryl and fenamiphos pesticides. Mater Sci Eng, C 102:764–772. https://doi.org/10.1016/j.msec.2019.05.010
33. He L, Jiang ZW, Li W, Li CM, Huang CZ, Li YF (2018) In Situ synthesis of gold nanoparticles/metal–organic gels hybrids with excellent peroxidase-Like activity for sensitive chemiluminescence detection of organophosphorus Pesticides. ACS Appl Mater Interfaces 10(34):28868–28876. https://doi.org/10.1021/acsami.8b08768
34. Herrera-Domínguez M, Morales-Luna G, Mahlknecht J, Cheng Q, Aguilar-Hernández I, Ornelas-Soto N (2023) Optical biosensors and their applications for the detection of water pollutants. Biosensors 13(3):370. https://doi.org/10.3390/bios13030370
35. Hu H, Wang B, Li Y, Wang P, Yang L (2020) Acetylcholinesterase Sensor with patterned structure for detecting organophosphorus pesticides based on titanium dioxide sol-gel carrier. Electroanalysis 32(8):1834–1842. https://doi.org/10.1002/elan.202060027
36. Huang L, Sun D-W, Pu H, Wei Q, Luo L, Wang J (2019) A colorimetric paper sensor based on the domino reaction of acetylcholinesterase and degradable γ-MnOOH nanozyme for sensitive detection of organophosphorus pesticides. Sens Actuators B: Chem 290:573–580. https://doi.org/10.1016/j.snb.2019.04.020
37. Huang X, Cui H, Duan W (2020) Ecotoxicity of chlorpyrifos to aquatic organisms: a review. Ecotoxicol Environ Saf 200:110731. https://doi.org/10.1016/j.ecoenv.2020.110731
38. IUPAC (2003) Regulatory limits for pesticide residues in water. Pure Appl Chem 75:1123–1155
39. Johnson ZT, Jared N, Peterson JK, Li J, Smith EA, Walper SA, Hooe SL, Breger JC, Medintz IL, Gomes C, Claussen JC (2022). Enzymatic laser-induced graphene biosensor for electrochemical sensing of the herbicide glyphosate. Global Challenges 6(9). https://doi.org/10.1002/gch2.202200057
40. Kaur J, Bandyopadhyay D, Singh PK (2022) A simple and convenient choline oxidase inhibition based colorimetric biosensor for detection of organophosphorus class of pesticides. J Mol Liq 347:118258. https://doi.org/10.1016/j.molliq.2021.118258
41. Kaur N, Prabhakar N (2017) Current scenario in organophosphates detection using electrochemical biosensors. TrAC, Trends Anal Chem 92:62–85. https://doi.org/10.1016/j.trac.2017.04.012
42. Khaled E, Kamel MS, Hassan HNA, Malhat FM, Abdel-Gawad H (2015) Rapid detection of methomyl and organophosphorous pesticides with portable potentiometric biosensor. Anal Chem Lett 5(3):117–126. https://doi.org/10.1080/22297928.2015.1090332
43. Kılıç O, Boz İ, Eryılmaz GA (2020) Comparison of conventional and good agricultural practices farms: a socio-economic and technical perspective. J Clean Prod 258:120666. https://doi.org/10.1016/j.jclepro.2020.120666
44. Korotcenkov G, Cho BK, Sevilla F (2022) Chemical sensors: comprehensive sensor technologies. In: Korotcenkov G (ed) Chemical sensors: comprehensive sensor technologies, 1st. Momentum Press, pp 311–476

45. Kumar M, Khamis K, Stevens R, Hannah DM, Bradley C (2024) In-situ optical water quality monitoring sensors—applications, challenges, and future opportunities. Front Water 6. https://doi.org/10.3389/frwa.2024.1380133
46. Lei Y, Mulchandani P, Chen W, Mulchandani A (2007) Biosensor for direct determination of fenitrothion and EPN using recombinant Pseudomonas putida JS444 with surface-expressed organophosphorous hydrolase. 2. Modified carbon paste electrode. Appl Biochem Biotechnol 136(3):243–250. https://doi.org/10.1007/s12010-007-9023-9
47. Leskovac A, Petrović S (2023) Pesticide use and degradation strategies: food safety, challenges and perspectives. Foods 12(14):2709. https://doi.org/10.3390/foods12142709
48. Li B, He Y, Xu C (2007) Simultaneous determination of three organophosphorus pesticides residues in vegetables using continuous-flow chemiluminescence with artificial neural network calibration. Talanta 72(1):223–230. https://doi.org/10.1016/j.talanta.2006.10.023
49. Li H, Zhao S, Wang Z, Li F (2023) Controllable preparation of 2D V2O5 peroxidase-mimetic nanozyme to develop portable paper-based analytical device for intelligent pesticide assay. Small 19(14):2206465. https://doi.org/10.1002/smll.202206465
50. Li X, Yang T, Song Y, Zhu J, Wang D, Li W (2019) Surface-enhanced Raman spectroscopy (SERS)-based immunochromatographic assay (ICA) for the simultaneous detection of two pyrethroid pesticides. Sens Actuators B: Chem 283:230–238. https://doi.org/10.1016/j.snb.2018.11.112
51. Liang X, Wang X, Zhang Y, Huang B, Han L (2022) Selective inhibition toward dual enzyme-like activities of iridium nanozymes for a specific colorimetric assay of malathion without enzymes. J Agric Food Chem 70(12):3898–3906. https://doi.org/10.1021/acs.jafc.1c06954
52. Liu G, Guo W, Song D (2014) A multianalyte electrochemical immunosensor based on patterned carbon nanotubes modified substrates for detection of pesticides. Biosens Bioelectron 52:360–366. https://doi.org/10.1016/j.bios.2013.09.009
53. Liu G, Song D, Chen F (2013) Towards the fabrication of a label-free amperometric immunosensor using SWNTs for direct detection of paraoxon. Talanta 104:103–108. https://doi.org/10.1016/j.talanta.2012.11.039
54. Liu L, Xu D, Hu Y, Liu S, Wei H, Zheng J, Wang G, Hu X, Wang C (2015) Construction of an impedimetric immunosensor for label-free detecting carbofuran residual in agricultural and environmental samples. Food Control 53:72–80. https://doi.org/10.1016/j.foodcont.2015.01.009
55. Luo Y, Guo W, Ngo HH, Nghiem LD, Hai FI, Zhang J, Liang S, Wang XC (2014) A review on the occurrence of micropollutants in the aquatic environment and their fate and removal during wastewater treatment. Sci Total Environ 473–474:619–641. https://doi.org/10.1016/j.scitotenv.2013.12.065
56. Lushchak V, Matviishyn T, Husak V, Storey J, Storey K (2018) Pesticide toxicity: a mechanistic approach. EXCLI J 8(17):1101–1136. https://doi.org/10.17179/excli2018-1710
57. Maanaki H, Xu T, Chen G, Du X, Wang J (2023) Development of integrated smartphone/resistive biosensor for on-site rapid environmental monitoring of organophosphate pesticides in food and water. Biosens Bioelectron X 15:100402. https://doi.org/10.1016/j.biosx.2023.100402
58. Madianos L, Tsekenis G, Skotadis E, Patsiouras L, Tsoukalas D (2018) A highly sensitive impedimetric aptasensor for the selective detection of acetamiprid and atrazine based on microwires formed by platinum nanoparticles. Biosens Bioelectron 101:268–274. https://doi.org/10.1016/j.bios.2017.10.034
59. Manasa RL, Mehta A (2020) Wastewater: sources of pollutants and its remediation. In: Environmental biotechnology, 2nd edn. Springer International Publishing, pp 197–219. https://doi.org/10.1007/978-3-030-38196-7_9
60. Martinazzo J, Muenchen DK, Brezolin AN, Cezaro AM, Rigo AA, Manzoli A, Hoehne L, Leite FL, Steffens J, Steffens C (2018) Cantilever nanobiosensor using tyrosinase to detect atrazine in liquid medium. J Environ Sci Health B 53(4):229–236. https://doi.org/10.1080/03601234.2017.1421833

61. Medalie L, Baker NT, Shoda ME, Stone WW, Meyer MT, Stets EG, Wilson M (2020) Influence of land use and region on glyphosate and aminomethylphosphonic acid in streams in the USA. Sci Total Environ 707:136008. https://doi.org/10.1016/j.scitotenv.2019.136008
62. Montes R, Céspedes F, Gabriel D, Baeza M (2018) Electrochemical biosensor based on optimized biocomposite for organophosphorus and carbamates pesticides detection. J Nanomater 2018:1–13. https://doi.org/10.1155/2018/7093606
63. Muenchen DK, Martinazzo J, Brezolin AN, de Cezaro AM, Rigo AA, Mezarroba MN, Manzoli A, de Lima Leite F, Steffens J, Steffens C (2018) Cantilever functionalization using peroxidase extract of low cost for glyphosate detection. Appl Biochem Biotechnol 186(4):1061–1073. https://doi.org/10.1007/s12010-018-2799-y
64. Mushtaq N, Singh DV, Bhat RA, Dervash MA, Bin HO (2020) Freshwater contamination: sources and hazards to aquatic biota. In: Fresh water pollution dynamics and remediation. Springer, Singapore, pp 27–50. https://doi.org/10.1007/978-981-13-8277-2_3
65. Mutunga T, Sinanovic S, Harrison CS (2024) Integrating wireless remote sensing and sensors for monitoring pesticide pollution in surface and groundwater. Sensors 24(10). https://doi.org/10.3390/s24103191
66. Naveen Prasad S, Bansal V, Ramanathan R (2021) Detection of pesticides using nanozymes: Trends, challenges and outlook. TrAC Trends Anal Chem 144:116429. https://doi.org/10.1016/j.trac.2021.116429
67. Nieto-Sandoval J, Munoz M, de Pedro ZM, Casas JA (2022) Application of catalytic hydrodehalogenation in drinking water treatment for organohalogenated micropollutants removal: a review. J Hazard Mater Adv 5:100047. https://doi.org/10.1016/j.hazadv.2022.100047
68. Nowicka AM, Kowalczyk A, Stojek Z, Hepel M (2010) Nanogravimetric and voltammetric DNA-hybridization biosensors for studies of DNA damage by common toxicants and pollutants. Biophys Chem 146(1):42–53. https://doi.org/10.1016/j.bpc.2009.10.003
69. Panis C, Kawassaki ACB, Crestani APJ, Pascotto CR, Bortoloti DS, Vicentini GE, Lucio LC, Ferreira MO, Prates RTC, Vieira VK, Gaboardi SC, Candiotto LZP (2022) Evidence on human exposure to pesticides and the occurrence of health hazards in the brazilian population: a systematic review. Front Public Health 9. https://doi.org/10.3389/fpubh.2021.787438
70. Pathak VM, Verma VK, Rawat BS, Kaur B, Babu N, Sharma A, Dewali S, Yadav M, Kumari R, Singh S, Mohapatra A, Pandey V, Rana N, Cunill JM (2022) Current status of pesticide effects on environment, human health and it's eco-friendly management as bioremediation: a comprehensive review. Front Microbiol 13. https://doi.org/10.3389/fmicb.2022.962619
71. Pérez-Fernández B (2020) Electrochemical (Bio)sensors for pesticides detection using screen-printed electrodes. Biosensors 10(4):32. https://doi.org/10.3390/bios10040032
72. Pérez-Fernández B, Mercader JV, Abad-Fuentes A, Checa-Orrego BI, Costa-García A, de la Escosura-Muñiz A (2020) Direct competitive immunosensor for Imidacloprid pesticide detection on gold nanoparticle-modified electrodes. Talanta 209:120465. https://doi.org/10.1016/j.talanta.2019.120465
73. Pohanka M, Jun D, Kuca K (2008) Amperometric biosensors for real time assays of organophosphates. Sensors 8(9):5303–5312. https://doi.org/10.3390/s8095303
74. Pop A, Lung S, Orha C, Manea F (2018) Silver/graphene-modified boron doped diamond electrode for selective detection of carbaryl and paraquat from water. Int J Electrochem Sci 13(3):2651–2660. https://doi.org/10.20964/2018.03.02
75. Prabhu K, Malode SJ, Kulkarni RM, Shetti NP (2022) Electro-sensing base for hazardous pesticide 2, 4-DCP and its quantification in real samples at ZnO@Cu core-shell nanoparticles in the presence of cationic surfactant. Mater Chem Phys 278:125705. https://doi.org/10.1016/j.matchemphys.2022.125705
76. Rafaqat S, Raqba Ali N, Hussain A (2022) Validating role of different enzymes (laccases and catalases) based voltammetric biosensors in detection of pesticide and dye. Mater Chem Phys 290:126545. https://doi.org/10.1016/j.matchemphys.2022.126545
77. Ramirez-Priego P, Estévez M-C, Díaz-Luisravelo HJ, Manclús JJ, Montoya Á, Lechuga LM (2021) Real-time monitoring of fenitrothion in water samples using a silicon nanophotonic biosensor. Analytica Chimica Acta 1152:338276. https://doi.org/10.1016/j.aca.2021.338276

78. Ruan X, Wang Y, Kwon EY, Wang L, Cheng N, Niu X, Ding S, Van Wie BJ, Lin Y, Du D (2021) Nanomaterial-enhanced 3D-printed sensor platform for simultaneous detection of atrazine and acetochlor. Biosens Bioelectron 184:113238. https://doi.org/10.1016/j.bios.2021.113238
79. Saha A, Sedaghat S, Gopalakrishnan S, Waimin J, Yermembetova A, Glassmaker N, Mousoulis C, Shakouri A, Wei A, Rahimi R, Alam MA (2023) A new paradigm of reliable sensing with field-deployed electrochemical sensors integrating data redundancy and source credibility. Sci Rep 13(1):3101. https://doi.org/10.1038/s41598-022-25920-w
80. Sahin A, Dooley K, Cropek DM, West AC, Banta S (2011) A dual enzyme electrochemical assay for the detection of organophosphorus compounds using organophosphorus hydrolase and horseradish peroxidase. Sens Actuators B: Chem 158(1):353–360. https://doi.org/10.1016/j.snb.2011.06.034
81. Sahub C, Tuntulani T, Nhujak T, Tomapatanaget B (2018) Effective biosensor based on graphene quantum dots via enzymatic reaction for directly photoluminescence detection of organophosphate pesticide. Sens Actuators, B Chem 258:88–97. https://doi.org/10.1016/j.snb.2017.11.072
82. Saleh IA, Zouari N, Al-Ghouti MA (2020) Removal of pesticides from water and wastewater: chemical, physical and biological treatment approaches. Environ Technol Innov 19:101026. https://doi.org/10.1016/j.eti.2020.101026
83. Scandurra A, Censabella M, Gulino A, Grimaldi MG, Ruffino F (2022) Gold nanoelectrode arrays dewetted onto graphene paper for selective and direct electrochemical determination of glyphosate in drinking water. Sens Bio-Sens Res 36:100496. https://doi.org/10.1016/j.sbsr.2022.100496
84. Shanbhag MM, Manasa G, Mascarenhas RJ, Mondal K, Shetti NP (2023) Fundamentals of bio-electrochemical sensing. Chem Eng J Adv 16:100516. https://doi.org/10.1016/j.ceja.2023.100516
85. Skotadis E, Kanaris A, Aslanidis E, Michalis P, Kalatzis N, Chatzipapadopoulos F, Marianos N, Tsoukalas D (2020) A sensing approach for automated and real-time pesticide detection in the scope of smart-farming. Comput Electron Agric 178:105759. https://doi.org/10.1016/j.compag.2020.105759
86. Sohrabi H, Hemmati A, Majidi MR, Eyvazi S, Jahanban-Esfahlan A, Baradaran B, Adlpour-Azar R, Mokhtarzadeh A, de la Guardia M (2021) Recent advances on portable sensing and biosensing assays applied for detection of main chemical and biological pollutant agents in water samples: a critical review. TrAC Trends Anal Chem 143:116344. https://doi.org/10.1016/j.trac.2021.116344
87. Song J, Chen Y, Li L, Tan M, Su W (2024) Recent progress in photoelectrochemical sensing of pesticides in food and environmental samples: photoactive materials and signaling mechanisms. Molecules 29(3):560. https://doi.org/10.3390/molecules29030560
88. Steffens C, Ballen SC, Scapin E, da Silva DM, Steffens J, Jacques RA (2022) Advances of nanobiosensors and its application in atrazine detection in water: a review. Sens Actuators Rep 4:100096. https://doi.org/10.1016/j.snr.2022.100096
89. Steffens C, Steffens J, Graboski AM, Manzoli A, Leite FL (2017) Nanosensors for detection of pesticides in water. In: New pesticides and soil sensors, pp 595–635. Elsevier. https://doi.org/10.1016/B978-0-12-804299-1.00017-5
90. Suwannachat J, Saenchoopa A, Tun WST, Patramanon R, Daduang S, Daduang J, Kulchat S (2024) An electrochemical AChE-based biosensor for organophosphate pesticides using a modified CuNWs/rGO nanocomposite on a screen-printed carbon electrode. Food Chem 434:137431. https://doi.org/10.1016/j.foodchem.2023.137431
91. Syafrudin M, Kristanti RA, Yuniarto A, Hadibarata T, Rhee J, Al-onazi WA, Algarni TS, Almarri AH, Al-Mohaimeed AM (2021) Pesticides in drinking water—a review. Int J Environ Res Public Health 18(2):468. https://doi.org/10.3390/ijerph18020468
92. Székács A, Mörtl M, Darvas B (2015) Monitoring pesticide residues in surface and ground water in Hungary: surveys in 1990–2015. J Chem 2015:1–15. https://doi.org/10.1155/2015/717948

93. Tang Y, Zhan X, Zheng J, Xie Z, Zhu S, Wu Y (2023) Facile colorimetric smartphone-based biosensor for rapid detection of organophosphorus pesticides residues in environment using the aptamer-enhanced oxidase activity of octahedral Ag_2O particles. Anal Chim Acta 1264:341325. https://doi.org/10.1016/j.aca.2023.341325
94. Tarazona JV, Court-Marques D, Tiramani M, Reich H, Pfeil R, Istace F, Crivellente F (2017) Glyphosate toxicity and carcinogenicity: a review of the scientific basis of the European Union assessment and its differences with IARC. Arch Toxicol 91(8):2723–2743. https://doi.org/10.1007/s00204-017-1962-5
95. Thanh CT, Binh NH, Duoc PND, Thu VT, Van Trinh P, Anh NN, Van Tu N, Tuyen NV, Van Quynh N, Tu VC, Thao BTP, Thang PD, Abe H, Van Chuc N (2021) Electrochemical sensor based on reduced graphene oxide/double-walled carbon nanotubes/octahedral Fe_3O_4/chitosan composite for glyphosate detection. Bull Environ Contam Toxicol 106(6):1017–1023. https://doi.org/10.1007/s00128-021-03179-7
96. Topaz T, Egozi R, Suari Y, Ben-Ari J, Sade T, Chefetz B, Yahel G (2020) Environmental risk dynamics of pesticides toxicity in a Mediterranean micro-estuary. Environ Pollut 265:114941. https://doi.org/10.1016/j.envpol.2020.114941
97. Touloupakis E, Giannoudi L, Piletsky SA, Guzzella L, Pozzoni F, Giardi MT (2005) A multi-biosensor based on immobilized Photosystem II on screen-printed electrodes for the detection of herbicides in river water. Biosens Bioelectron 20(10):1984–1992. https://doi.org/10.1016/j.bios.2004.08.035
98. Tucci M, Grattieri M, Schievano A, Cristiani P, Minteer SD (2019) Microbial amperometric biosensor for online herbicide detection: Photocurrent inhibition of *Anabaena variabilis*. Electrochim Acta 302:102–108. https://doi.org/10.1016/j.electacta.2019.02.007
99. Tudi M, Daniel Ruan H, Wang L, Lyu J, Sadler R, Connell D, Chu C, Phung DT (2021) Agriculture development, pesticide application and its impact on the environment. Int J Environ Res Public Health 18(3):1112. https://doi.org/10.3390/ijerph18031112
100. USEPA (2002) Edn. of the drinking water standards and health advisories. EPA 822-R-02-038
101. USEPA (2023) Ingredients used in pesticide products
102. Vohra T, Hardik G, Shagun S, Kumar VD, Rashmi R (2020) A review on nanoparticles based biosensors for pesticide detection in water. In Importance and applications of nanotechnology, pp 1–10
103. Vulliet E, Cren-Olivé C (2011) Screening of pharmaceuticals and hormones at the regional scale, in surface and groundwaters intended to human consumption. Environ Pollut 159(10):2929–2934. https://doi.org/10.1016/j.envpol.2011.04.033
104. Wang K, Meng X, Yan X, Fan K (2024) Nanozyme-based point-of-care testing: Revolutionizing environmental pollutant detection with high efficiency and low cost. Nano Today 54:102145. https://doi.org/10.1016/j.nantod.2023.102145
105. Wang Y, Xiong H, Chen C (2022) Agricultural non-point source pollution and health of the elderly in rural China. PLoS ONE 17(10):e0274027. https://doi.org/10.1371/journal.pone.0274027
106. Wei M, Feng S (2017) Amperometric determination of organophosphate pesticides using a acetylcholinesterase based biosensor made from nitrogen-doped porous carbon deposited on a boron-doped diamond electrode. Microchim Acta 184(9):3461–3468. https://doi.org/10.1007/s00604-017-2380-3
107. Wong LS, Lee YH, Surif S (2013) Performance of a cyanobacteria whole cell-based fluorescence biosensor for heavy metal and pesticide detection. Sensors 13(5):6394–6404. https://doi.org/10.3390/s130506394
108. Yadav IC, Devi NL, Syed JH, Cheng Z, Li J, Zhang G, Jones KC (2015) Current status of persistent organic pesticides residues in air, water, and soil, and their possible effect on neighboring countries: a comprehensive review of India. Sci Total Environ 511:123–137. https://doi.org/10.1016/j.scitotenv.2014.12.041
109. Yan M, Chen G, She Y, Ma J, Hong S, Shao Y, Abd El-Aty AM, Wang M, Wang S, Wang J (2019) Sensitive and Simple competitive biomimetic nanozyme-linked immunosorbent assay for colorimetric and surface-enhanced raman scattering sensing of triazophos. J Agric Food Chem 67(34):9658–9666. https://doi.org/10.1021/acs.jafc.9b03401

110. Yao T, Liu A, Liu Y, Wei M, Wei W, Liu S (2019) Ratiometric fluorescence sensor for organophosphorus pesticide detection based on opposite responses of two fluorescence reagents to MnO_2 nanosheets. Biosens Bioelectron 145:111705. https://doi.org/10.1016/j.bios.2019.111705
111. Zaller JG, Kruse-Plaß M, Schlechtriemen U, Gruber E, Peer M, Nadeem I, Formayer H, Hutter H-P, Landler L (2022) Pesticides in ambient air, influenced by surrounding land use and weather, pose a potential threat to biodiversity and humans. Sci Total Environ 838:156012. https://doi.org/10.1016/j.scitotenv.2022.156012
112. Zeng Z, Wang X, Yang T, Li Y, Liu X, Zhang P, Feng B, Qing T (2023) Transition metal-doped germanium oxide nanozyme with enhanced enzyme-like activity for rapid detection of pesticide residues in water samples. Anal Chim Acta 1245:340861. https://doi.org/10.1016/j.aca.2023.340861
113. Zhang C, Du P, Jiang Z, Jin M, Chen G, Cao X, Cui X, Zhang Y, Li R, Abd El-Aty AM, Wang J (2018) A simple and sensitive competitive bio-barcode immunoassay for triazophos based on multi-modified gold nanoparticles and fluorescent signal amplification. Analytica Chimica Acta 999:123–131. https://doi.org/10.1016/j.aca.2017.10.032
114. Zhang Q, Zhang Z, Xu S, Da L, Lin D, Jiang C (2022) Enzyme-free and rapid visual quantitative detection for pesticide residues utilizing portable smartphone integrated paper sensor. J Hazard Mater 436:129320. https://doi.org/10.1016/j.jhazmat.2022.129320

Environmental Cleanup

Nanomaterials in Water Remediation for Irrigation Management

Jyoti Rani, Sushila Singh, Simran Kakkar, Anuradha Beniwal, Monika Moond, Seema Sangwan, and Rajni Kant Sharma

Abstract The scarcity of clean water is indeed a global challenge that impacts both human consumption and agricultural irrigation. Consequently, there are wastages of water resources and reductions in crop productions. Groundwater irrigation indeed has seen significant growth over the past few decades, becoming a vital source of water for agriculture worldwide. Various pathogens, nutrients, salts, and micropollutants reduced crop yields and, ultimately, food production. In semi-arid and arid regions, where water scarcity is more pronounced, efficient water management is essential for sustaining agricultural productivity and ensuring food security. However, even in humid climates where water availability may not be as limited, effective water management remains critical. Innovations in irrigation management and practices are indeed crucial, especially in regions where agriculture demands a significant amount of water. Over the past few years, nanotechnology has grown to be a crucial tool in many facets of environmental and water treatment. Examining various nanomaterials to sustainable irrigation water management involves assessing their effectiveness, scalability, and potential impact on productivity, equity, and sustainability. The objective of this chapter is to present nanotechnology-based methods for remediating water, which are essential to understanding and appraising the severity of the current state of affairs.

Keywords Nanomaterials · Irrigation water · Remediation · Industrial pollutants · Pesticides and pathogens

J. Rani · S. Singh (✉) · S. Kakkar · A. Beniwal · M. Moond · R. K. Sharma
Department of Chemistry, CCS Haryana Agricultural University, Hisar, Haryana, India
e-mail: singhsushila999@gmail.com

S. Sangwan
Department of Microbiology, CCS Haryana Agricultural University, Hisar, Haryana, India

L. F. Fraceto et al. (eds.), *Emerging Nanotechnologies for Agroecosystem Management*, Smart Nanomaterials Technology, https://doi.org/10.1007/978-981-95-0187-8_9

1 Introduction

Although water is one of the most abundant resources on Earth, less than 1% of it is suitable for agricultural and human use [40]. Water for irrigation can be obtained from surface water, groundwater, or even alternative sources like desalinated or treated wastewater. Due to rising demand for water and the depletion of freshwater supplies, the use of marginal quality water for agriculture has created a new problem for environmental management. In agriculture, using water from irrigation channels is a standard procedure. According to surveys, more than 50 countries use contaminated or partially purified water for irrigation, irrigating an area of 20 million hectares [80, 139]. But there's more cause for alarm because a variety of organic and inorganic pollutants (industrial pollutants, agrochemicals, etc.) frequently contaminate the water [10]. The increased usage of pesticides and the untreated discharge of sewage and industrial sludge into water bodies are to blame for the increasingly frequent pollution of water sources, especially irrigation canals [106]. In agricultural and industrial areas, where canal water is contaminated with a range of substances, this issue is particularly severe. It frequently contains low dissolved oxygen, high levels of nutrients, heavy metals and high concentrations of soluble salts, phthalic acid esters (PAEs), high concentrations of polycyclic aromatic hydrocarbons (PAHs), and pesticide residues from fields' drainage [82]. Even though farmers benefit from lower costs, using such water for irrigation generally has negative effects on the ecosystem, can have phytotoxic effects, and inhibit plant physiological processes, which manifest as yellowing and chlorosis, reduced growth parameters, and can affect crop productivity. Farmers frequently use fertilizers to increase harvest yields, while pesticides are utilized to suppress pest growth. A portion of the excess pesticides were carried downstream by irrigation or rainfall into irrigation canals, but some remained in the soil and contaminated the aquifers. Furthermore, pesticides that do not adsorb or evaporate quickly enough to volatilize are likely to be accessible for leaching, which contaminates groundwater [52]. Despite a notable increase in the awareness of contaminated water channels in developed nations, the whole picture of pesticide contamination is still relatively incomplete. There are several possible reasons for this, including (a difficulties locating low concentrations of residual pesticides (in the lower microgram range,(b a lack of historical pesticide use records, (c manpower and budgetary constraints, and (d the lack of industry participation in research initiatives in developing nations [150]. Additionally, the existence of dangerous microorganisms in irrigation water is a global concern. Poor quality water can spread localized contamination in the field and act as a direct source of contamination. The European Food Safety Agency (EFSA 2014) has identified agriculture water as a significant risk factor for contaminating leafy vegetables that are consumed fresh in salads [19]. The textile dyeing industry is considered one of the world's worst polluters due to the massive amounts of water needed in its production processes. Before being designated as wastewater, it's extremely toxic wastewaters are dumped into drains, sewers, and agricultural land without any kind of treatment [43], [64]. Continuous discharge of textile dyeing wastewater (TDW) into the environment is changing the chemical, physical, and biological characteristics of ecological resources [74], which is seen as a worldwide danger to sustainable biodiversity. However, industrialization is unavoidable in the struggle against globalization. Thus, a number of studies have been carried out to ascertain the physiochemical properties of wastewater from industrial sources (loom dye, tannery, textile, etc.), as well as their impact on crop development, yield, nutritional values, buildup of heavy metals in soil, and evaluation of health risks, utilizing just the blended effluents as the subjects of analysis [109].

For many years, we have been treating wastewater with coagulation, sedimentation, flocculation, filtration, and chlorine disinfection to eliminate pollutants. However, these conventional methods are either unable to remove pollutants efficiently or cannot remove enough to meet the ever-tougher water quality standards [107]. Conventional decontamination techniques such as ozonation and chlorination need large amounts of chemical reagents and produce potentially harmful byproducts. Surface water contamination poses a concern to water reservoirs also since contaminants possess the capacity to infiltrate into aquifers and then find routes to drinking water sources. These challenges emphasize the need to develop new technology or add to existing ones in the domains of pollution removal and water treatment [94]. Concerns about the availability and tactics required to produce contamination-free water are expanding as the demand for clean water sources rises. Thus, the application of the nanoscale approach to irrigation water treatment has not only led to the development of new procedures but also to a new degree of improvement in traditional methods such as adsorption and membrane filtration. Nanotechnology has the potential to significantly lower contamination levels to "nearly zero" and provides quick, affordable, and ecologically friendly solutions. The method of removing toxins from irrigation water by using appropriate nanomaterials (NMs) is known as nano remediation of irrigation water. Due to their high adsorption capacity (caused by their large specific surface area and numerous adsorption sites), excellent selectivity, and ease of regeneration after utilization, NMs are primarily used for treating water and soil [14]. Metallic NMs, carbonaceous NMs, and zeolites are the utmost popularly used NMs in water purification methods that act via redox reactions, catalysis, and adsorption. In order to comprehend and assess the gravity of the current situation, the primary objective of this chapter is to present nanotechnology-based techniques for water purification. A special effort has been paid to hazardous heavy metals, dyes, pathogenic microorganisms, and pesticides as these are most frequently found in wastewater, polluted irrigation water, industrial systems, and agricultural systems.

2 Challenges and Remediation Technique for Polluted Irrigation Water

Water for irrigation can be obtained from surface water, groundwater, or even alternative sources like treated wastewater. However, in order to reduce pollutants or pathogens in the water stored, treated, or extracted for irrigation, multiple treatments

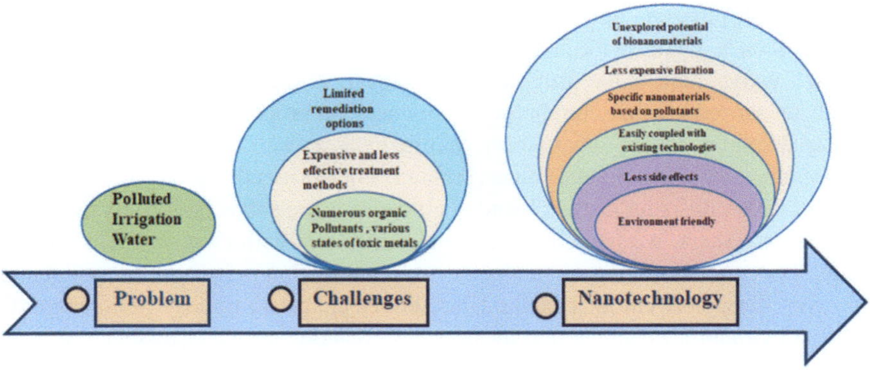

Fig. 1 Challenges and nanotechnology for polluted irrigation water

are often necessary. As a result, nanotechnologies have been used to synthesize novel materials with unique properties that have never been seen before, which may be utilized to raise the standard of irrigation water. Figure 1 shows the irrigation water challenges with remediation method, i.e. nanotechnology.

3 Pollutants in Irrigation Water and Their Remediation Using NMs

3.1 Metal Ion

As there is less good quality water available for agriculture, people are using other different sources of water for the irrigation of a variety of crops. Numerous carcinogenic substances, particularly heavy metals, are found in irrigation water [72]. The release and accumulation of metals, particularly heavy metals, is the most dangerous component of industrial effluents and other wastes [119]. Metallic elements having a relatively high density that are dangerous even in low concentrations are known as toxic heavy metals [140]. Water bodies can also get contaminated by runoff from mining waste erosion, dust from moving crude ores, metal corrosion, and heavy metal leaching into soil and ground water. It has been found that vegetables grown by irrigating with untreated water have high levels of heavy metals. Heavy metal pollution adversely affects human health both directly and indirectly. These substances are present in more than permissible limit, and negatively affect the productivity of plants, animals, and soils, as well as the environment's overall health. Heavy metals are absorbed by plants and eventually make their way into the food chain because they stay in the soil for a longer amount of time, even in low concentrations in the soil–water system [135]. These heavy metals are more concentrated in the area close to the root zone of irrigation water. When combined with nutrients, heavy metals are taken up by the plant system and accumulate in different parts of the plant, with the roots and leaves frequently accumulating the greatest amount of residue [129, 135].

Plants can absorb heavy metals, which can then build up in animals and people as they move up the food chain. One of the primary channels via which heavy metals enter the food chain is through plant roots [98]. The primary causes of environmental contamination with heavy metals are human activities. However, some of the man-made sources of heavy metals in irrigation water are mining, smelting, vehicle exhaust, sewage sludge, and the use of pesticides and fertilizers [135].

Certain heavy metals, such as Pb, Cd, Hg, As, and Se, are not necessary for the growth of plants because they are not recognized to have any physiological functions in plants. Others, such as Co, Ni, Mn, Zn, Cu, Fe, and, Mo, are necessary for plants to grow and function normally, but when concentrations of these elements exceed recommended levels, they can quickly cause toxicity [98].

In order to reduce the negative environmental consequences of these substances, it is better to treat these heavy metals and transform them into less hazardous forms before disposing them into water. However, before the water may be used for irrigation, the right set of procedures for treating the contamination must be followed. We need to find a technique that can serve as a magic bullet for eliminating heavy metal contamination from the environment in order to solve the issue and prevent its intake.

Nanotechnology gives NMs having diverse geometries and unique properties that are not present in bulk materials. This can be attributed to their exceptional characteristics, such as large surface area, their porosity, functionality, specific surface charge, and ion-binding capacity. NMs are appealing and cost-effective because they can be used repeatedly [83].

3.1.1 Mechanism Involved

The various methods that NMs use to extract heavy metals from water can be described as shown in the following Fig. 2.

Transformation

One efficient way to eliminate inorganic and organic contaminants from contaminated environment is by transformation via oxidation or reduction. These processes could result in metal speciation and lower metal toxicity. The state and/or site of toxicity are altered as a result of the electron-transfer process to stable oxidation states. Decontaminating heavy metal contaminants from contaminated locations can be accomplished successfully through transformation by oxidation or reduction [25] (Fig. 3).

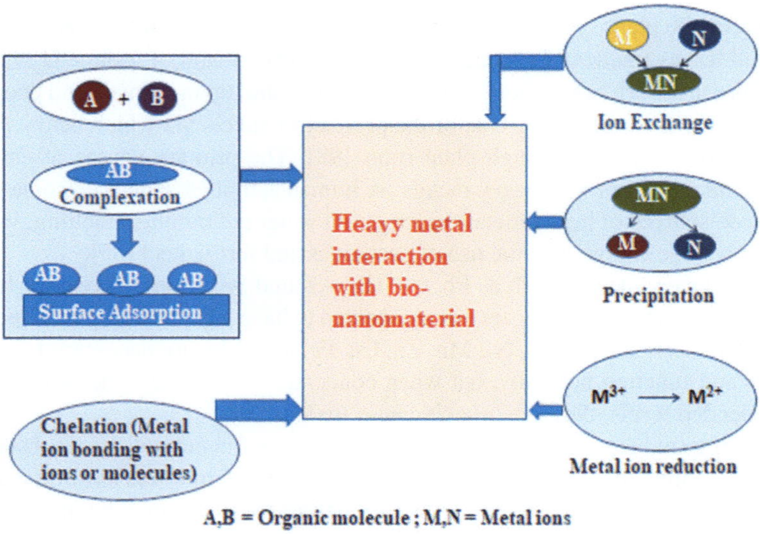

Fig. 2 Possible mechanism for the elimination of metal ions

Fig. 3 Metal ion removal through transformation

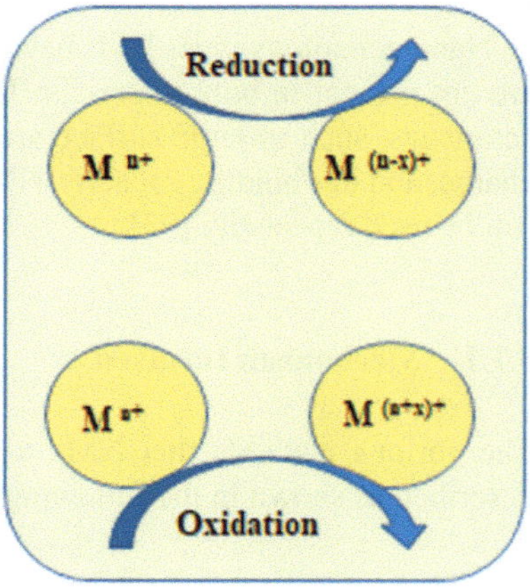

Catalysis

The energy, chemical, and environmental industries of the present and the future depend heavily on catalytic technologies. In order to enhance its surface area, an active catalyst needs more sites, and it also needs to decrease the size of the catalytic particles. As a result, the NMs offer both the requirement of active catalyst as well as being affordable and environmentally friendly [25].

Photocatalysis

In catalysis, photocatalysis is the term used to describe the acceleration of a photoreaction when a catalyst is present. It can be used for both inorganic and organic contaminant photodegradation. In photocatalysis, a catalyst composed of semiconductor materials is first made sensitive by visible or ultraviolet light, which results in the creation of pairs of electrons and holes. Excited electrons in the valence band move towards the conduction band and form holes in valence band when the energy of the incident light reaches or exceeds the semiconductor's band gap energy. The hydroxyl radicals and hydrogen gas produced by these holes divide the water molecules and cause the oxidative destruction of organic contaminants. When dissolved oxygen reacts with the conductive band electrons, it produces superoxide anions leading to redox reaction. These holes and electrons eventually break down organic pollutants through a sequence of oxidation and reduction processes when exposed to light. These days, there is great interest in the biological synthesis of metal oxide NPs, like zinc, titanium, and silver, by means of photocatalysis for the elimination of environmental pollutants. The affordable, highly stable, and rapid oxidation process of photocatalytic degradation makes it superior [25].

Adsorption

The surface phenomenon known as adsorption is the binding of adsorbate to an adsorbent's surface. Surface and ionic interactions mediate this exothermic reaction. The term enthalpy of adsorption refers to the quantity of heat produced when one mole of adsorbate is adsorbed on an adsorbent. Because of the restriction of the adsorbate's freedom of movement during adsorption, entropy is decreased and the enthalpy (Δ H) of adsorption is always expressed as a negative number. Adsorption happens spontaneously at constant temperature and pressure, which lowers Gibb's free energy even more. Because they are environmentally acceptable and economically viable, biologically generated nanoscale adsorbents have become crucial for the exclusion of both inorganic and organic contaminants. Many biologically generated nanoadsorbents are currently used to eliminate pollutants from contaminated areas [25].

As illustrated in figure given, different adsorption mechanisms utilizing NMs can be explained. Adsorption takes place exclusively on the adsorbent's surface as a result of complex ion generation on its active sites. Adsorption takes place as a result of organic molecules joining together to form complexes. Adsorption is more likely when functional groups such as OH, COOH, NH_2, and SH are present in the adsorbent. Chemical interaction results in the transfer of cations and anions, the breaking down of adsorbent bonds, and the formation of new bonds with adsorbate. Π–Π stacking, hydrogen bonds, intermolecular interactions, etc. could be the cause. Adsorbent binding sites bind to metal ions by complexation and chelation (bonding with a metal ion that has a single centre atom). During the ion exchange process, NMs containing exchangeable ions bind to metal ions. Afterwards, the metal ion NMs are

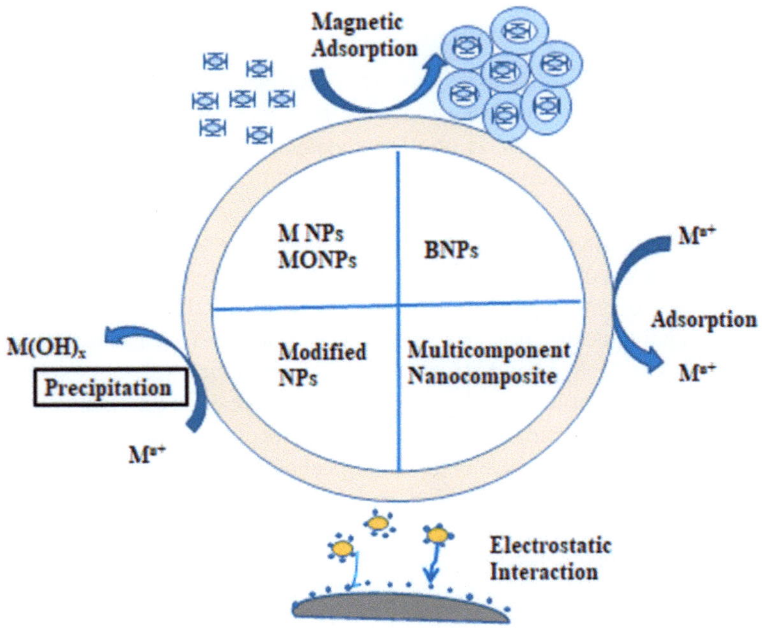

Fig. 4 Metal ion removal through adsorption mechanism

created with charge mobility and exchangeable ions separate out. In precipitation, metabolic products are formed [61] (Fig. 4).

3.1.2 NMs Used in Metal Ion Removal

Recently, a variety of NMs, including carbonaceous, ferrite, zeolite, polymer-based, chitosan, metal oxide, metallic, magnetic, and bimetallic, have been applied to eliminate heavy metals from water. The adsorption process on NMs' surfaces can be utilized to eliminate heavy metals from irrigation water. Multiple types of NMs are explored and analysed in detail, based on their primary function as absorbents and their ability to eliminate heavy metals [61].

Carbon-Based NMs

Functionalized carbon nanostructures are thought to be among the most capable adsorbents for potentially hazardous metals because of their special physical and chemical characteristics, which allow for large-scale application in water purification [61].

Graphene is a significant carbon-based nano medicine (NM) utilized in water treatment. One atom thick, planar sheet of sp2-bonded carbon atoms makes up this thinnest two-dimensional material. High strength, durability, and specific surface area are displayed by graphene. Similar in structure to graphene, graphene oxide (GO) contains more functional groups that contain oxygen [93]. Because graphene oxide (GO) surface contains oxygen functional groups, graphene has a high hydrophilic tendency and high dispersity in aqueous solution. Graphene oxide (GO) has enormous surface area and functional groups, which make it a material with great potential for water detoxification. Another oxidized form of graphene oxide (GO) is called exfoliated graphene oxide (EGO), and it has multiple surface functional groups, including carboxyl, epoxy, and hydroxyl, groups. Furthermore, reduced GO (rGO) exhibits enhanced interaction with metal pollutants and a high electron-transport capacity. By complexing the oxygen functional groups in graphene with the cationic metal, hazardous metals can be adsorbed from water using grapheme [134].

The dispersion of SiO_2 NPs on the GO surface prevents accumulation in addition to increasing the quantity of functional groups and removal efficiency. In addition, the silica/graphene oxide (SiO_2/GO) nanocomposite exhibits special qualities as wide surface area, functionality, resilience to acidity, and ease of separation. SiO_2/GO nanocomposites therefore present a viable option for removing heavy metal ions from water.

Using calcination and precipitation techniques, SiO_2 NPs were produced from rice husk ash (RHA). 3-Aminopropyltriethoxysilane (APS) was used as a coupling agent to create rice husk ash-silica/graphene oxide (RHA-SiO_2/GO) nanocomposites. The surface complex and electrostatic interactions were the focus of the RHA-SiO_2/GO adsorption mechanism proposal. It was suggested that Pb^{2+} ions and the functional groups of the nanocomposite interacted electrostatically. In addition, electron donor–acceptor complex was formed by linking of free electron pairs from the π system in the Gr carbon network with the electron-deficient Pb^{2+} cation atoms. Moreover, lone pairs of electrons on nitrogen and oxygen atoms (–OH, –C = O, –NH_2, –COOH) can interact with Pb element, which has an absolute electronegativity of 2.33, enhancing the adsorption capacity of RHA-SiO_2/GO [137].

$$\mathrm{GO-COOH+Pb^{2}} \rightleftharpoons \mathrm{GO-COOPb^{+}+H^{+}} \tag{1}$$

$$\mathrm{GO-OH+Pb^{2+}} \rightleftharpoons \mathrm{GO-OPb+H^{+}} \tag{2}$$

$$\mathrm{GO-COOH+Pb^{2+}+H_2O} \rightleftharpoons \mathrm{GO-COOPbOH+2H^{+}} \tag{3}$$

$$\mathrm{GO-OH+Pb^{2+}+H_2O} \rightleftharpoons \mathrm{GO-OPbOH+2H^{+}} \tag{4}$$

$$\mathrm{GO-OH+H_2O+Pb^{2+}} \rightleftharpoons \mathrm{(GO-O^{-})_2Pb^{2+}} \tag{5}$$

$$\mathrm{GO-COOH+H_2O+Pb^{2+}} \rightleftharpoons \mathrm{(GO-COO^{-})_2Pb^{2+}} \tag{6}$$

$$SiO_2 - NH_2 + Pb^{2+} \rightleftharpoons Si - NH - Pb^{2+} + H^+ \quad (7)$$

$$SiO_2 - NH_2 + H_2O + Pb^{2+} \rightleftharpoons (Si - NH_2)_2Pb_{2+} \quad (8)$$

Carbon Nanotubes (CNTs)

Another type of carbon material that environmental engineers find most appealing is carbon nanotubes (CNTs), due to their hollow cylindrical structure, electrical conductivity, small size, and enormous surface area. The two primary classifications of carbon nanotubes are single-walled and multi-walled carbon nanotubes [129]. The bulk adsorption of pollutants is caused by the primary surface functional groups of carbon nanotubes (CNTs), such as –COOH and –OH [134]. Carbon nanotubes (CNTs) exhibit strong interactions with other molecules or atoms through the hexagonal arrays of carbon atoms in their graphite sheets. By combining several processes, including sorption–precipitation, physical adsorption, chemical interaction, and electrostatic attraction, the CNTs may be able to extract heavy metal ions from the water. Functionalizing the surfaces of CNTs using a variety of techniques, including oxidation, nonmagnetic metal oxide coating, and magnetic iron oxide grafting, can improve their adsorption affinity [35],[130, 134].

Activated Carbon

One of the key components in the process of treating wastewater was activated carbon (AC). They are perfect for heavy metal cleanup because of their huge surface area, porous nature, affordability, and simplicity of preparation using a range of starting materials (Ethaib et al. 2020) [33]. The food processing industries generate a significant amount of bio waste, which is apple peels. Annual production of apple waste ranges from 17 to 21 million tons. Apple pomace, which primarily consists of apple peels, has high levels of polyphenols, lignin, pectin, and cellulose. These have functional groups like $-NH_2$, –OH, –CO, and –COO, which are mostly in charge of the metals' adsorption. In this work, chromium (VI) was extracted using activated carbon from an aqueous solution derived from apple peels (ACAP). Adsorbent dosage of 0.05 g/50 mL, pH of 2, contact time of 4 h, temperature of 28 °C, and Cr (VI) concentration of 50 mg/L were utilized to attain the maximum Cr (VI) adsorption of 36.01 mg/g. The Cr (VI) adsorption capacity provided by this ACAP was superior to specially designed products for the same purpose. The Cr (VI) adsorption capability of the prepared ACAP was superior to that of commercial products [33].

The amount of wastewater created and the quantity of sludge generated during the wastewater treatment process rise yearly in proportion to the country's growth and development. This sludge is primarily being disposed of in landfills. For the bio sorption or bioleaching of metal ions like Cu (II) from water, activated sludge is a useful biomaterial. Sludge-activated carbon (SAC) from water was used for the bio sorption of Cu (II) ions. The initial 100 mg/L metal concentration was kept constant.

The findings suggest that Cu (II) can be extracted from an aqueous solution using sludge-activated carbon as an adsorbent [117].

Zeolite NMs

Zeolites are crystalline alumino silicates consisting of AlO_4 and SiO_4 tetrahedra. These units provide a framework of pores and channels that can hold tiny molecules, cations, and/or water. Si and Al atoms are positioned in the centre, while O atoms are shared by SiO_4 and AlO_4 units at the corners, creating regular intra-crystalline cavities and channels. Alkaline earth and alkali metals can migrate inside the framework of silicon and aluminium ions to balance the negative charge between them [34]. Due to their larger surface area, high porosity, hydrophilic affinities, and ion exchange capabilities, zeolites can be used to eliminate harmful metals from water. Therefore, more accessible active site zeolite NMs are utilized for the effective adsorption of heavy metals because of their large and active surface areas [61].

Metal/Metal Oxide NMs (M/MONMs)

Metal/oxide NMs (M/MONMs) have been suggested to be a heavy metal adsorbent. Nonetheless, a number of recent investigations have also demonstrated the suitability of M/MONMs for the elimination of organic pollutants. Nowadays, magnetic NPs are one of the most important types of M/MONMs because they can easily separate magnetically after treating a lot of water [25]. A number of characteristics of nanometal oxides, including good selectivity and adsorption capability, make them potential heavy metal adsorbents. Metal-based NMs, like those based on iron, zinc, manganese, titanium, aluminium, and magnesium oxides, have a high BET surface area, little solubility, and solve the issue of secondary pollutants [98].

Silica-Based NMs

Due to their remarkable qualities, such as large surface area, distinct pore size, and controllable surface characteristics, silica-based NMs are now a valuable class of metal oxides for the heavy metal remediation in water. When these nanoadsorbents are treated with amino groups and thiol, they also exhibit enhanced selectivity and metal-adsorption capacity.

Their nontoxic nature and eco-friendliness have further expanded their applications in water remediation [134].

The most readily available and affordable method for removing toxic metals from a solution is iron. Iron oxide NPs were used for heavy metal adsorption because of their ease of usage, accessibility, and environmental friendliness. Moreover, it produces no secondary pollutants when applied. Many different types of iron nanoadsorbents were used extensively, such as amorphous hydrous iron oxides, magnetic magnetite (Fe_3O_4), magnetic maghemite (γ-Fe_2O_4), goethite (α-FeOOH), nonmagnetic hematite (α-Fe_2O_3), and iron/iron oxide (Fe/FexOy). The adsorption capacity can be increased by adding ligands, which make it easier for functional groups to attach. Polymers, on the other hand, improve dispersion, prevent aggregation, and act as a carrier and binder for metal ions. This prevents other metals from interfering with the process [98, 134].

Using *Syzygium cumini* leaf extract, TiO_2 NPs were created via a green method, and their morphological, structural, chemical, and physical properties were evaluated. Synthesized NPs were successfully used as a photocatalyst to remove 82.53% of the lead (Pb^{2+}) and 75.5% of the chemical oxygen demand (COD) from explosive industrial wastewater at a low cost. According to kinetic analysis, first-order kinetics was used for both the industrial wastewater's COD removal and the photocatalytic removal of lead. Photocatalytic activity is significantly enhanced by TiO_2 with a band gap energy of 3–3.5 eV. When TiO_2 is exposed to UV or visible light, the electron-transfer mechanism from valance band to conduction band causes a hole (h^+) to develop in the valance band. This hole then interacts with water to form an OH radical.

Through photo oxidation, this OH radical functions as a potent oxidizing agent, eliminating pesticides, heavy metals like lead (Pb), and organic pollutants from wastewater [122]:

$$TiO_2 + h\upsilon \rightarrow TiO_2 + h^+ + e^-$$
$$H_2O + 2h^+ \rightarrow OH^{\cdot} + H^+$$
$$TiOH + h^+ \rightarrow TiOH^{\cdot}$$

Mg $(NO_3)_2 \cdot 6H_2O$ was reduced by metabolites produced by *A. terreus* strain S1 to create MgONPs. The colour changing from colourless to turbid white and the maximum surface plasmon resonance being detected at 280 nm were the first indications that the MgONP synthesis was successful. The MgONPs demonstrated a significant level of chromium ion removal with a percentage of 97.5% [115].

Bimetallic NMs (BNMs)

BNMs are made from two metallic components that interact synergistically to produce features unique to each metal. In water, BNPs can remove a variety of pollutants [134]. One alternative, low-cost, and innovative green method to quicken the rate of decontamination of inorganic and organic pollutants in the environment is the use of biologically synthesized BNMs (including heavy metal pollutants). For instance, when compared to uncoated iron, the deposition of a discontinuous layer of noble metals has greatly enhanced the reactivity of nanoscale iron particles, increasing reaction rates [25].

Modified NMs

To decrease the tendency of nanoparticles to aggregate and to increase their stability in the system, their surfaces must be modified with either inorganic or organic materials, such as metals, biomolecules, polymers, and silica. More binding sites can be added to the surface of naturally occurring NMs and NPs using biodegradable polymers, enhancing their capacity to extract heavy metals and other impurities from water. Surface-modified NMs/NPs eliminate contaminants from water by surface-site binding, magnetic selective adsorption, electrostatic interaction, and modified ligand combination [135].

Multifunctional Nanocomposite

In order to build smart materials with two or more distinct actions, NMs are currently being investigated for their multifunctionality. Research has recently switched to concentrate on creating products that use nanotechnology to impart various functionalities, such as multitasking [25].

From the waste products of marine shrimp, chitosan was isolated. A multifunctional composite containing AgNPs, CNTs, and CNTs was made using equal weights of AgNPs, CuNPs, and chitosan at a total ratio of 5 weight percent of NMs with regard to chitosan. The removal of heavy metals from the multifunctional composite was then evaluated. Cd (II), Cu (II), and Pb (II) were selected as representative examples of heavy metal contamination in wastewater. The multifunctional composite shows a markedly improved heavy metal removal capability when compared to chitosan. The adsorption effectiveness of Cd (II), Cu (II), and Pb (II) on chitosan was 60% when 5 g L^{-1} was used; however, the multifunctional composite only required 1 g L^{-1} to achieve 100% adsorption efficiency. Almost all of the metal ions in the solution were eliminated after ten minutes of contact with the multifunctional composite. Regeneration of the multifunctional composite using EDTA solution was demonstrated to be possible in up to five effective cycles of adsorption/desorption processes. The multifunctional composite is recommended as a potential material for water purification in the submitted work [7].

Polymer-Based NMs

Polymer-based nanofibrous membranes packed with polymers (such as chitosan and cellulose) are gaining a lot of attention from researchers as an eco-friendly remediation technique. Their sorption efficacy is enhanced by their high gas penetrability, large specific surface area, and well-defined porous fibre skeleton. They make ideal adsorption alternatives due to their high skeletal strength, tuneable surface functional classes, and facile degradation behaviour. Because of their unique functional moieties—NH_2, COOH, and SO_3H—polymer-filled nano membranes have a high degree of selectivity and adsorption capacity [34].

Chitosan-Based NMs

Naturally occurring eco-friendly polymers with a lot of potential include chitin, and chitosan and its derivatives. The linear polysaccharide chitosan is made up of N-acetyl-D-glucosamine (acetylated unit) and randomly distributed (1 4)-linked D-glucosamine (de acetylated unit) [61]. Chitosan, a polysaccharide biopolymer, can be produced affordably from chitin and is also nontoxic, biocompatible, and has found extensive application in a variety of fields [56]. When its distinct qualities are altered chemically and mechanically, new properties are produced that make it easier to use in a variety of contexts. Because of its high $-NH_2$ and –OH group content and exceptional solubility in organic solvents and water, chitosan has drawn increased attention in the water filtration industry. This enables it to possibly chelate hazardous pollutants such as heavy metals and dyes [34]. Chitosan is well known for having good complexing ability due to the special interactions that its amino groups have with heavy metals. Because of its anticipated extraordinary features and uses, chitosan modification is of significant interest [137].

Cellulose-Based NMs

Cellulose has the ideal sorption properties of hydrophilicity, reactivity, and chirality because it is made up of repeating β-D-glucopyranose units that are covalently linked between the OH groups of the C4 and C1 carbon atoms through acetal functions. The world's most plentiful and renewable polymer resource is cellulose [56]. Because lignocellulosic materials have a higher surface area compared to nonporous materials, they offer excellent options for sorption materials, and nano cellulose fibres have a greater tendency to bind metal ions. Van der Waals forces and interactions between the surface of the cellulose fibre and the metals in the bath cause metal ions to be sorbed initially. The organic chemical moieties found in cellulose nanofibres are diverse, with significant amounts of OH– and COO– entities being among them. Cellulose's carboxyl groups have the ability to serve as sources of ion exchange capabilities [25]. Carboxylic acids can interact with cationic metal ions and are likely active sites for the sorption of metal ions due to the electrostatic attraction between their negatively charged groups and metallic cations.

Nano cellulose fibres (NCF) were prepared from seed pods of *Moringa oleifera* and applied for removal of toxic metals. XRD and FTIR measurements were used to characterize these cellulose fibres. The average grain size was found to be 10 nm using the Debye Scherer equation. According to nano remediation batch experiments for lead and cadmium metal ions, the manufactured NCF (0.500 gm) has removal efficiencies of 81.7% and 84.5% for Cd (II) and Pb (II) and from initial concentrations of 25 mg/l when compared to control native Moringa oleifera studies of biomass dose (2 gm) [56].

Magnetic NMs

One of the most sophisticated types of NMs is magnetic NM compounds, which efficiently eliminate dangerous metals through the use of nanotechnology and magnetic separation. Since the magnetic nanoparticles produced by this process have good recyclability, the magnetic separation approach is frequently applied to ecological issues and has several uses in environmental remediation. The characteristics of magnetic nanoparticles have altered significantly as their size has decreased. Nonmagnetic NMs have a smaller surface area and are less effective in treating water than magnetic NMs because of their hard separation from the aqueous phase. Several characteristics of the magnetic NMs include their larger surface area, biocompatibility, easy dispersion, lower toxicity, and inert chemical activity. The properties of magnetic nanoparticles allow for more dependable and affordable water treatment [34].

Bionanocomposite

Biopolymers and nanoparticles are combined to create hybrid materials called bionanocomposites, which have improved adsorption capabilities. These composites make use of the biocompatibility and regenerative nature of biopolymers, as well as the high surface area and functional characteristics of nanoparticles. In recent years, bionanocomposites as adsorbents have emerged as a potential strategy. Excellent mechanical strength and stiffness of the bionanocomposite point to adsorbents that may be used in high-pressure settings for applications involving the purification of water. Additionally, bionanocomposites typically exhibit excellent crystallinity, enabling the adsorbents to function as resistant to biological and chemical deterioration in water. However, there are a few things to take into account when using bionanocomposites in the water purification process, including their capacity for regeneration, agglomeration, long-term effectiveness, affordability, and immobilization. Bionanocomposites offer a novel, environmentally friendly solution to the issues brought on by the current heavy metal pollution because of their large specific area, which creates many active sites for metal ions to adhere to the surface of the biosorbent. The functional groups on the surface that act as biomass's metal-binding sites also facilitate the removal of heavy metal ions [67].

Regenerated micro fibrillated cellulose (R-MFC) was produced by dissolving native cellulose (nonwood cellulose), which was separated from jute fibres, using phosphoric acid. Then, dissolved cellulose chains were regenerated into micro fibrillated cellulose. This study revealed that the resulting R-MFC had strong adsorption properties, low crystallinity, and the cellulose II polymorph, which may effectively anchor the growth of ZnO nanocrystals. Arsenic ions (As (V)) were removed from water using a composite network made of R-MFC and ZnO nanocrystals (ZnO/R-MFC). The R-MFC scaffold firmly anchors ZnO nanoparticles, creating a sizable, positively charged ZnO surface that can interact electrostatically with the H2AsO4 ions (As (V) in pH 2–3). With a maximum adsorption capacity of 4,421 mg/g based on the Langmuir isotherm model, As (V) ion contaminants can be effectively removed from water with this ZnO/R-MFC composite adsorption system. As (V) contaminants can be safely removed from contaminated water using this bio nanomaterial, which also has an advantage over existing adsorbents in that stable nanocomposite networks lessen the issue of secondary contamination. This proved that the bio nanomaterial is inexpensive and environmentally friendly [125].

In a green synthesis procedure, eucalyptus extract was used as a reducing agent to produce metallic nanoparticles. Iron oxide nanoparticles were successfully synthesized with a well-formed structure. Under certain circumstances, produced iron oxide nanoparticles can be encapsulated in chitosan beads to form magnetic hybrid inorganic/organic materials. The sorbent qualities of the new hybrid materials were shown by a thorough analysis of arsenic removal (pH dependency, equilibrium, and kinetic studies), highlighting their positive traits and sorption capacity. Depending on the ratio of iron to extract, unique hybrid materials with distinct crystal phases were created when iron oxide nanoparticles were encapsulated in a chitosan matrix. The hybrid material utilized in arsenic adsorption trials was found to have the optimum magnetic and structural properties when it was manufactured as a 50:50 combination (INPB). It was found that INPB is an effective As (V) adsorbent due to its good adsorption capacity at natural pH and short equilibrium times. Column studies have validated the effective use of this new hybrid magnetic material for As (V) remediation under continuous conditions. Lastly, desorption investigations demonstrated that the hybrid material can be readily renewed with basic solutions at modest concentrations [87].

Bacteria in water pipelines produce iron oxide nanowires, which are being researched for potential use in removing arsenic from waterways. Research on the use of iron oxide material in batch adsorption to remove arsenic from water demonstrated that arsenic could be eliminated rapidly. This biofilm, which is produced as a waste product during pipeline cleaning and is freely available in huge amounts as a nanoscale material, is discharged into the environment, raising concerns about the ecosystem. This application demonstrates how to simultaneously treat two environmental issues: using a waste item to remove significant environmental issues like water contamination [8].

3.2 Dyes

Numerous contaminants have contaminated water due to the swift expansion of industry, population, intensive agriculture, and urbanization. Toxic textile dyes have received particular attention because they are frequently found in wastewater, contaminated irrigation water, industrial systems, and agricultural systems. Dyes are extremely toxic and have a strong propensity to cause eutrophication. Because toxic dyes and dyestuffs are broadly used in various industries, such as the paper, food processing, textile, cosmetics, and drug industries, water pollution has drawn a lot of attention internationally in the last few decades [105]. These dyes' great stability against heat, oxidizing agents, and light makes them resistant to degradation [48]. Specifically, azo dyes contain benzidine, the most common carcinogen, which needs to be properly treated before being released into the environment. They also produce highly carcinogenic and even explosive aromatic amines [118]. Nowadays, one of the most successful new technologies for removing pollutants from water is nano remediation. Biomass-based as well as chemically synthesized NMs have the ability to reduce and/or facilitate catalytic reactions that degrade, transform, and detoxify particular toxic dyes in wastewater. Table 2 shows the biomass-based NMs that are effective for the degradation of dyes. Applications of NMs supported on activated carbon, biochar, clay, and zeolites minerals have also enhanced the NMs' capacity to remove contaminants and their reactivity [86].

3.2.1 Mechanism Involved

NMs degrade the dye pollutant either through photocatalysis or through adsorption. Absorption, synthesis, and electron–hole pair transport are the three primary processes involved in photocatalysis. NMs have a powerful surface photoluminescence effect, which causes them to absorb photons from visible light and create an electron–hole pair. When oxygen and water are combined with the generated electrons (e^-) and holes (h^+), O^{2-} and OH radicals are created, which turn the dye molecules into colourless products. Dye degradation via adsorption involves the dye adsorption on NM surfaces and NMs acts as a relay centre for electron transfer from BH_4^- ions (e^- donor) to dye molecules (e^- acceptor).

The generated electrons (e^-) and holes ($h+$) combine with oxygen and water to form O^{2-} and OH radicals, which convert the dye molecules into colourless byproducts. In a process known as dye degradation by adsorption, dye is adsorbed, and NMs serve as a hub for the transfer of electrons from BH4-ions, or e-donor, to dye molecules, or e-acceptors (Fig. 5).

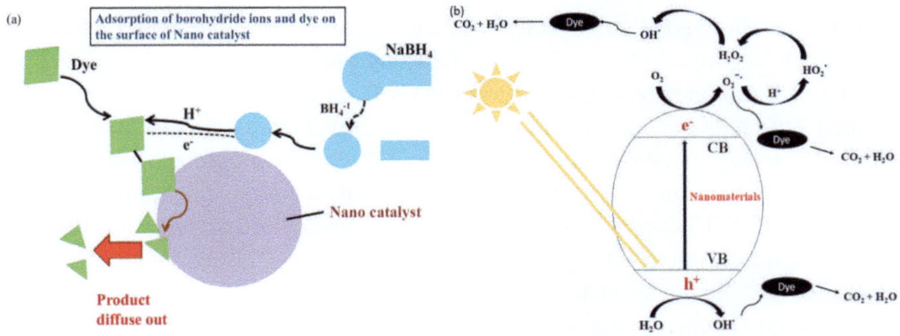

Fig. 5 Probable mechanism for dye degradation through **a** adsorption and **b** photocatalysis

3.2.2 NMs Used in the Removal of Dyes

Congo Red (CR)

Anionic dye CR is extensively used in the paper, leather, textile, plastic, and printing industries. Furthermore, it can be utilized to detect free hydrochloric acid in stomach contents to diagnose amyloidosis, as well as a histological stain and an amyloid pH indicator [126]. It contaminates the landmass and the water resources due to its non-biodegradable nature. AgNPs, chitosan nanoparticles, TiO_2 NPs, and undoped and Gd^{3+} doped cobalt ferrite nanoparticles show good catalytic efficiency for breaking down CR dye. *Amaranthus gangeticus* Linn (Chinese spinach)-mediated AgNPs demonstrated more than 50% catalytic efficiency within 15 min [68]. Chitosan nanoparticles (NPs) with a 98% removal efficiency were obtained by ionotropic gelation of chitosan and tripolyphosphates, which resulted in 5,107 mg of CR/g of adsorbent adsorbed in an aqueous solution. The removal of CR was influenced by tripolyphosphate concentration, encapsulation time, pH, and ionic strength. The mechanism involved a significant electrostatic interaction between anionic CR and protonated amino groups at low pH [4]. The ability of undoped cobalt ferrite nanoparticles to adsorb or remove Congo red was increased from 131 to 161 mg g^{-1} by Gd^{3+} doped cobalt ferrites ($CoFe_{2-x}Gd_xO_4$) [153].

Methyl Orange (MO)

Methyl orange is classified as an anionic synthetic dye and is linked to a number of health issues [20]. There are lots of publications available on the photocatalytic degradation of MO dye with nanomaterials. Various plant-based and chemically synthesized NMs including AgNPs, ZnO NPs, doped ZnO NPs, lead oxide NPs (PbO NPs), Iron NPs, TiO_2 NPs, Zinc aluminate, and $NiFe_2O_4$ NPs show good MO degradation efficiency. *Ulva lactuca* mediated AgNPs maximally degrade methyl orange in less than 12 h of incubation time (Kumar et al. 2013). Undoped ZnO nanoparticles show good photocatalytic efficiency as compared to doped ZnO NPs. While there was a significant drop in efficiency and dye degradation even after 150 min of radiation exposure for doped ZnO NPs, practically full deterioration for undoped ZnO NPs occurs after 80 min [60]. Al doped ZnO NPs (3% aluminium) achieve 95% degradation in 120 min [73]. When PbO NPs loaded with activated carbon are optimized to 22 min, pH2, and 0.02 g catalyst dosage, they can potentially degrade MO [39]. Undoped TiO_2 NPs effectively degrade 98% of the MO dye within 25 min when exposed to UV light [136], while doping of TiO_2 NPs with 5% molar solution of Cr^{3+} ion under visible light significantly increased the photodegradation of MO as compared to doping under UV-light (Hamadanian et al. 2014). Flower-like nanostructured rutile TiO_2 shows $<10\%$ degradation of MO, while the doping with H_2O_2 significantly accelerates the degradation rate of MO dye (Korosi et al. 2016). NPs like zinc aluminate and $NiFe_2O_4$ degrade dyes in aqueous solution. $NiFe_2O_4$ NPs, for instance, were able to degrade MO (at an original concentration of 10 mg/

L) by 77% in just 5 h when exposed to sunlight; in contrast, there was no discernible degradation in the absence of light [46].

Eosin Y

Eosin Y is a brominated anionic heterocyclic dye used for routine staining of histological tissue sections, including collagen, erythrocytes, and epithelial cells [37]. Several nanomaterials have been used in photodegradation studies to effectively remove eosin Y. The dye is totally degraded in 45 min by two-dimensional CuO nanoleaves and 160 min by silver-TiO_2 nanocomposites, respectively [6, 77]. Hesperdin-modified TiO_2 NPs exhibits a high level of catalytic efficiency (96%) when exposed for 180 min to an acidic pH of 4.0. The adsorption process is facilitated by an acidic pH, and dye is present in an aqueous solution in its anionic form [143]. Their results show that after three hours of exposure to visible light, the TiO_2/N-graphene nanocomposite efficiently degraded the dye due to the maximum surface contact between TiO_2 and the dye, yielding a degradation efficiency of 63.4% [77].

Rose Bengal

An aromatic group-containing xanthene nucleus acting as a chromophore distinguishes rose bengal as a noteworthy xanthene dye. It is extensively utilized in the printing, photochemical, dyeing, and textile industries. Xanthene dyes have been described as genotoxic, cytotoxic, mutagenic, and cytostatic [21, 113]. With a catalytic efficiency of 98.9%, hexagonal γ-Fe_2O_3 nanoparticles efficiently break down the dye under visible light in just three hours. Hexagonal nanoparticles effectively degrade the dye in visible light within 3 h with 98.9% catalytic efficiency [30]. ZnO NPs effectively degrade the dye with 98% degradation rate within 90 min [60]. Three-dimensional hierarchical SnO_2 nanoflowers degrade the dye with a degradation rate of 96% within 90 min [84].

Indigo Carmine

The highly poisonous indigoid dye known as indigo carmine is commonly utilized in biology as a microscopic stain, a redox indicator in analytical chemistry, and an additive in pharmaceuticals. The most popular textile dyeing agent harms farming operations and is a major contributor to water pollution [54].

The indigo carmine dye may be effectively removed by a variety of nanomaterials, including CdS nanofibres, grapheme sheet decorated platinum (Pt), and palladium (Pd) NPs [71] and palladium-coated zinc sulphide/reduced graphene oxide (Pd–ZnS/rGO) nanocomposites [3]. The dye is efficiently degraded by Sn/Al_2O_3 (60% ratio) and Pd-ZnS/rGO nanocomposites in 40 and 210 min, respectively [3, 22]. But within 5 min, Pt- and Pd-graphene nanocomposites roughly break down 75 and

70% of the indigo carmine dye, respectively [70]. CdS flowers and CdS nanofibres exhibit 10–19 times higher photoactivity for the reduction of Indigo carmine dye as compared to commercial CdS NPs which depends on their nanocrystalline nature and nanostructured shape [45].

Methylene Blue (MB)

The aquaculture industry uses methylene blue (thiazine dye) as a chemotherapeutic and anti-malarial [17]. Nanomaterials including iron and iron oxide NPs, ZnO NPs, graphene oxide-supported TiO_2/Co_3O_4 nanocomposite, TiO_2/BiO NPs, potassium metal hexacyanoferrate (KMHCF) NPs, ZVI-biochar-chitosan nanocomposite, and bismuth-doped zinc aluminate NPs were employed to degrade the MB dye from aqueous solutions. The degradation of MB dye was effectively achieved by the environment friendly synthesis of potassium metal hexacyanoferrate (KMHCF) nanoparticles using Cu, Ni, and Co metal ions, and Aegle marmelos as a natural surfactant. The maximum degradation efficiency (95%) is shown by KCuHCF nanoparticles, which are followed by KCoHCF (89.3%) and KNiHCF (91.4%) [49]. Through electrostatic attraction, *Punica granatum* mediated Cu-based NPs with an adsorption capacity of 166 mg/g MB efficiently remove the MB from an aqueous solution [141]. ZVI-biochar-chitosan nanocomposites also effectively degrade MB dye from aqueous sample and eliminate the initial concentration (20 mg/L) of dye up to 68% [154]. Similarly, MB is photodegraded by bismuth-doped zinc aluminate NPs by 98.28% in 150 min at a concentration of 10 mg/L. Bismuth doping significantly lowered the band gap energy of zinc aluminate, which in turn improved the photocatalytic degradation of MB [65].

Rhodamine B (RhB)

Rhodamine B, also called water tracer fluorescent, is a colorants that is frequently used in the food and textile industries and can contaminate water. Experiments have shown the detrimental effects, which include chronic toxicity, neurotoxicity, and carcinogenicity, to be detrimental to both humans and animals [63]. RhB is considerably degraded by Ag/ZnO/graphene oxide (Ag/ZnO/GO) and ZnO nanopowders under UV and solar light. Under UV irradiation, Ag/ZnO/GO degrades the dye with 100% catalytic efficiency in 60 min; however, degradation under solar light irradiation is superior, with over 95% degradation occurring in 8 min [96].

The dye's removal rate rose with the application of NiO NPs, reaching roughly 76% in 30 min [92]. The BiOBr/montmorillonite composite efficiently breaks down 98.96% of the dye in 120 min in an acidic environment [147].

Methyl Violet (MV)

Among the cationic dyes with the highest colour intensity and brilliance is methyl violet. Aquatic plants' ability to photosynthesize is inhibited by methyl violet [41]. Inhaling methyl violet or absorbing it through the skin can cause a variety of health problems, such as lightheadedness, headaches, diarrhoea, vomiting, pain, and respiratory tract injuries [100]. Since removing methyl violet from wastewater is an important task, many different types of nanoadsorbents have been employed. Due to its high negative charges, which promote the adsorbent's interaction with the cationic dye, nanographite/Fe_3O_4 composite is one such nanoadsorbent that functions well in basic conditions (pH > 10.0) [75]. At pH 11.0, Fe^{3+} doped ZnS quantum dots accelerate the degradation of UV radiation by over 95% [124]. ZnO NPs produced the best results, degrading all of the material in 60–80 min [51].

Crystal Violet (CV)

The triphenylmethane dye crystal violet serves as a coloured ligand, biological stain, and dermatological agent in the dextran sulphate sodium assay [42]. Several NMs, such as CdS, nickel dioxide, cobalt-doped titania, and zinc oxide nanorods, have been shown to effectively remove crystal violet dye [28, 44, 104]. One of the triphenylmethane dyes, crystal violet, is used in the dextran sulphate sodium assay as a coloured ligand, biological stain, and dermatological agent (Hao et al. 2009). Crystal violet dye has been effectively removed by a range of NMs, including CdS, cobalt-doped titania, nickel dioxide, and zinc oxide nanorods [132].

3.3 Pathogenic Microorganisms

In underdeveloped nations, untreated wastewater from home and municipal sources that contains high levels of pathogens is frequently utilized directly for irrigation due to its accessibility and nutritional content. When compared to other sources of irrigation water, untreated wastewater often has a higher pathogen burden. Although surface water sources are typically associated with pathogen contamination, groundwater can also be at risk if it receives recharging from wastewater sources. It has become difficult to monitor microbiological contamination in irrigation water due to its complexity [82]. Different pathogen species, including bacteria (Escherichia coli and Salmonella), protozoa (Giardia and Cryptosporidium), and viruses (like noroviruses), have been found in a variety of irrigation water sources [110], [79]. The largest risk to freshly eaten product is probably associated with surface water irrigation of food crops, which has received the most attention in recent years in terms of research and regulations. According to certain publications, it is possible to recover phyto pathogenic nematodes, and root and stem rot fungi including Pythium spp. and Phytophthora spp. [131] from different systems of irrigation. Reusing runoff or drainage water led to higher losses because of infections brought on by the contaminated organisms. To remove harmful bacteria from water, various treatment technologies are currently used. The most crucial ones are membrane separation, ultraviolet-based disinfection, ozonation, and chlorination [146, 145]. Nowadays, nanotechnology offers a new frontier in the domain of water disinfection due to the shortcomings of the previously described treatment methods. Researchers focusing on the synthesis, design, and structural manipulation of particles with sizes ranging from 1 to 100 nm for use in water purification, and treatment procedures have been drawn to this emerging field. Various biomass-based NMs (metallic and metal oxides NPs) chemically synthesized have been utilized to remediate the microbe-contaminated wastewater.

3.3.1 Mechanism Involved

Compared to their bulk counterparts, the positively charged NMs are more readily adsorbed on the surface and able to enter the cell. NMs work through a variety of mechanisms, including interaction, accumulation, membrane leakage, cell wall breakdown, inhibition of ATPase activity, inhibition of RNA and protein synthesis, generation of intracellular ROS, and increased SOD activity. The antiviral activity demonstrates the direct contact between capsid protein and NMs, leading to the destruction of virus particles [29] (Fig. 6).

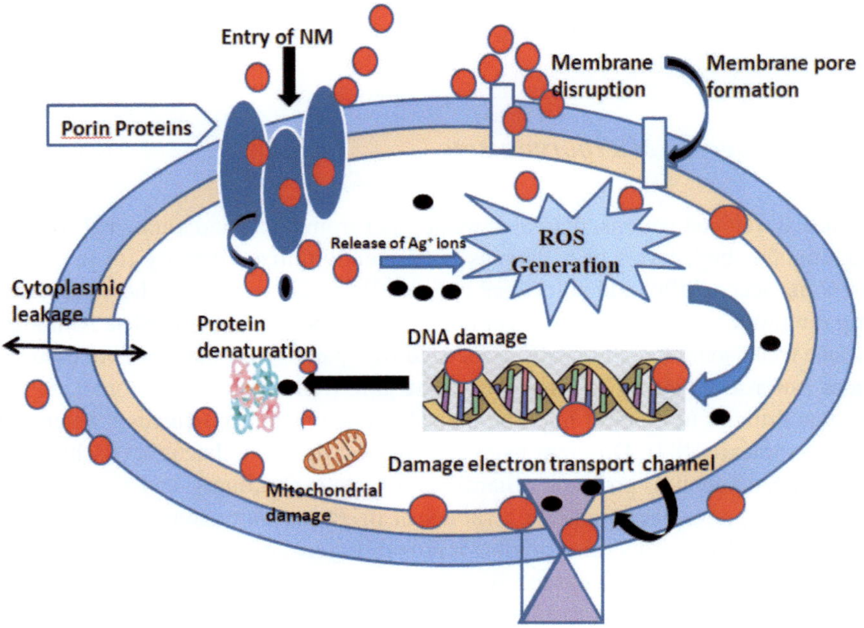

Fig. 6 Possible mechanism of action of NMs against pathogens

3.3.2 NMs Used for the Removal of Pathogenic Microorganisms

Table 1 shows the biomass-based NMs, effective against *E. coli*, Salmonella, Giardia spp. Ag-TiO_2 nanocomposite, Binary Mn/Co-doped TiO_2, Co–Mn, nanoporous cupper aluminosilicate, nanoparticulated film of TiO_2, and TiO_2/Ag, Fe_3O_4–SiO_2–NH_2 NPs are some of the chemically synthesized NMs that are effective against *E. coli* for wastewater remediation [66]. AgP-loaded clay with 90% antibacterial efficacy [59], QDs/g-C_3N_4, $BiVO_4$, and $AgVO_3$ QDs/g-C_3N_4 [146, 145] exhibit 90, 87.5, and 96.4% antibacterial efficacy respectively against Salmonella spp. When used in conjunction with proteinate-capped AgNPs, the excystation rate of Cryptosporidium parvum is reduced by 40%, and the arrangement of sporozoites within the oocyst is altered from its original banana-shaped form [2]. Silver nitrate and silver NPs at various concentrations effects norovirus surrogates (Murine norovirus, Feline calicivirus) with maximum $TCID_{50}$ (50% tissue culture infectious dose) at 2.1 mg/L for both surrogates [18]. In cool, moist conditions, the species of Pythium and Phytophthora found in irrigation water can cause damping off and root rot disease, which can affect 5–80% of the seedlings and result in large financial losses for the farmers [5]. One of the most destructive soil-borne diseases that affects tomatoes and potatoes globally is late blight, which is brought on by Phytophthora spp. MgONPs (500 μg/mL) and CuONPs (100 mg/L) are effective against Phytophthora nicotianae with 36.58 and 33.36% decline in tobacco black shank disease. Phytopthora infestans, causative agent for late blight of potato, gets affected by MgONPs due to the oxidative stress and cell membrane rupturing. Pythium spp. causes damping-off disease in cucumber seedlings, one of the foremost damaging disease to nd the S. fulvissimus Uts22 suspension to the sodium ALG solution [29]. From the discussion above, it is clear that the antimicrobial action of NPs against irrigation water pathogens, particularly fungal infections, didn't get much study (Tables 2 and 3).

3.4 Pesticides

Pesticides are extensively used worldwide to protect crops and food products as a result of modern agricultural practices and rapid industrialization. Their widespread presence, capacity to bioaccumulate, and long-term toxicity to living organisms have caused environmental alarms and prompted calls for their complete elimination, particularly from water. Owing to extensive agricultural practices in India, high concentrations of a variety of organochlorine and organophosphates have been found in groundwater, surpassing the country's drinking and irrigation water standards [144]. The accumulation of halogenated pesticides is being linked to an increase in cancer and neurological cases in the human population because they disrupt the nervous system and the cell cycle. However, pesticides, which are classified as organophosphates, effectively carbamylated acetyl cholinesterase at neuro-muscular junction. Exposition to organophosphate pesticides is frequently documented from developing and underdeveloped nations where agricultural products have higher than

Table 1 Major sources of toxic heavy metals and their permissible level

Heavy metal	Permissible level (mg/L)	Major source	References
Mercury	0.01	Batteries, mining, paint, and paper industries	[99, 121, 151]
Lead	0.1	Mining, burning of coal, electroplating, pigments, paint, manufacturing of batteries	
Cadmium	0.06	Refining, mining, plastic, pesticide, fertilizer, welding	
Nickel	0.2	Porcelain enamelling, electroplating, paint formulation, non-ferrous metal	
Zinc	15	Refineries, mining, plumping, brass manufacturing	
Copper	0.1	Plating, printing operations, paint, copper polishing	
Arsenic	0.02	Mining, smelting, pesticides, rock sedimentation	

allowed concentrations of these pesticides. In these nations, there is no industry oversight over pesticides, which allows very harmful pesticides to be used freely. Nanotechnology is a promising remedy for pesticide resistance and other problems. Nano formulations, mainly biopolymer-based ones, can resolve specific agricultural challenges due to their high surface-area-to-volume ratio, small size, and capacity to provide innovative crop protection approaches. NMs are effective adsorbents and photocatalysts for the elimination of pesticides because of their high surface area and semiconducting nature. Biopolymers, as nanoparticulate materials, have recently drawn interest because of their eco-friendly nature and capacity to provide a sustained release of the related active ingredients in pesticides [81].

3.4.1 Mechanism Involved

Adsorption

Using NMs for adsorption, pesticides can be successfully extracted from irrigation water. It is a surface phenomenon that depends on various types of interactions as well as the porosity, specific surface area, and number of available sites of the adsorbent. Its initial cost, adaptability, ease of operation, simplicity of design, and insensitivity to harmful pollutants are all positives. Adsorption doesn't lead to the creation of any hazardous materials [69].

Table 2 NMs based on biomass for the elimination of dyes

Sr. no.	Plant source	Type nanoparticles	Dye removed	References
1	Tea extract	Iron	MB, MO	[123]
2	*Viburnum opulus*	Silver	Tartrazine, carmoisine, and brilliant blue FCF dyes	[27]
3	*Artocarpus heterophyllus*	Iron	Fuchsin Basic dye	[47]
4	Amla extract	Iron oxide	MB	[11]
5	Lemon juice	ZnO	MO, MR, MB	[26]
6	Citrus Limetta	TiO_2	RhB	[95]
7	*Abelmoschus esculentus*	ZnO	Rhodamine B, MB	[102]
8	*Calotropis gigantea*	TiO_2	Congo red	[103]
9	*Artocarpus heterophyllus*	ZnO	Rose Bengal dye	[142]
10	Camellia japonica	AgNPs	Eosin Y	[58]
11	*Mangifera indica*	ZnO	MB, Eosin Y	[127]
12	*Typha latifolia* L	ZnO	Indigo carmine	[9]
13	Tragacanth gum	Titania nanomaterials	Crystal violet	[85]
14	*Shuteria involucrata*	Reduced graphene oxide-supported TiO_2/Co_3O_4 nanocomposite	MB, Crystal violet	[112]
15	*Sapindus mukorossi*	Titanium oxide coupled with bismuth oxide	MB	[111]
16	*Amaranthus gangeticus*	Silver Nanoparticles	Congo red	[68]
17	*Punica granatum*	Cu-based Nanoparticles	MB	[141]
18	*Moringa oleifera*	CeO_2	Crystal violet	[132]

Photocatalysis

An eco-friendly method for eliminating a variety of pesticides from irrigation water is photocatalytic oxidation. To improve the biodegradability of pesticides that are hazardous and not biodegradable, this is an appropriate pretreatment. In photocatalysis, photoexcited semiconductor solid surfaces are formed by exposure to solar or near-UV light. Consequently, positive surface charges and mobile electrons were produced. These electrons and excited sites accelerate the oxidation and reduction processes, which are crucial phases in the breakdown of pollutants. The reactivity and selectivity of semiconductor photocatalysts have changed as a result of advancements

Table 3 Plant-based NMs for the treatment of pathogenic microorganisms

Biomass used	Nanomaterials	Part of biomass	Targeted pathogen	Efficiency (MIC or ZOI)	References
Curcuma longa	AgNPs	Tuber	*E. coli*	50 mg/L (MIC)	[120]
Urtica dioica	AgNPs	Leaves	*E coli*	19 mm ZOI at 0.45 mg/100μL	[55]
Ocimum tenuiflorum	NiNPs	Leaves	*E. coli*	25 mm at 100 μg/mL	[50]
Phyllanthus reticulatus and *Conyza bonariensis*	CuONPs	Leaves	*E. coli*	16 ± 0.57 mm and 18 ± 0.20 mm ZOI respectively	[101]
Acasia nilotica	ZnO NPs	Pod	*E. coli*	17 ± 0.7 mm ZOI at 400 mg/mL	[36]
Portulaca oleracea	CuONPs	Leaves	*E. coli*	17.7 ± 1.2 mm ZOI	[31]
Capsicum annuum L	AuNPs	Leaves	*E. coli*	Maximum inhibition at 0.224 mg/mL	[13]
Zea mays and *Spomdias purpurea*	ZnO NPs	Seed	Salmonella typhimurium	>80%, >40% removal respectively	[23]
Pseudomonas aeruginosa and *E. coli*	AgNPs (both) ZnO (both)	–	Giardia trophozoites	91% (AgNPs) and 73% (ZnO) inhibition respectively	[12]

in nanotechnology. This idea has led to the application of photocatalytic degradation to the treatment of numerous pesticides [148].

Fenton Oxidation

Advanced oxidation techniques are frequently employed to degrade the inert or slowly biodegradable substances utilizing the oxidizing potential of oxidation radicals. Among advanced oxidation techniques, Fenton technology is an intriguing method of eliminating pesticides that involves adding hydrogen peroxide to a transition-metal catalyst, often ferrous iron. Hydroxyl radical generated in Fenton mechanism degrades the pesticides to harmless products:

$$H_2O_2 + Fe^{2+} \rightarrow HO^{\cdot} + HO^{\cdot} + Fe^{3+}$$
$$HO^{\cdot} + Fe^{2+} \rightarrow HO^{\cdot} + Fe^{3+}$$
$$HO^{\cdot} + H_2O_2 \rightarrow H_2O + HOO^{\cdot}$$
$$HOO^{\cdot} + H_2O_2 \rightarrow H_2O + HO^{\cdot} + O_2$$

3.4.2 NMs Used for the Removal of Pesticides

Organochlorines

Multi-walled carbon nanotubes (MWCN), nano alumina, and nano clay have the potential to effectively eliminate the organochlorine pesticide such as DDT. It reported that MWCN was a more effective adsorbent material than nano alumina and nano clay. 10% of MWCN, nano alumina, and nano clay adsorb 88.9, 68, and 67% of DDT respectively [133]. DDT is also effectively removed by magnetite NPs via an adsorption mechanism. DDT was degraded by the Ni–Fe bimetallic system of nanoparticles, and it was also demonstrated that the presence of ambient hydrogen ions, which are essential for DDT degradation, sped up the degradation process in an acidic solution [138]. Mg–Pd nanocomposite is able to decompose 100 μg/ml DDT in one hour [38]. Another significant organochloride that is frequently used in farming as a type of pesticide is lindane. Lindane is 100% degraded by Fe–S and Fe–Pd nanocomposites. These composites' high lindane degradation effectiveness is ascribed to their capacity to stimulate reduction processes [53]. Zerovalent Fe nanoparticles efficiently break down lindane in 24.0 h, producing smaller benzene and chloride ions as byproducts [32]. When airborne UV light exposes nano-TiO_2-coated sheets to organochlorine pesticides, photocatalytic destruction of the pesticides occurs rapidly, degrading all pesticides. Additionally, the use of a high-power UV lamp significantly accelerated the rate of degradation. Research has shown that using various nanotechnology techniques to remove pesticides has a large potential [149].

Organophosphates

Non-metallic carbon, or graphene, is used as an adsorbent to draw pesticides out of water. They are able to adsorb pollutants with aromatic rings through π–π interactions. When combined with other materials, it also improves the pesticide adsorption capacity. One very efficient sorbent that has been used to remove leftover organophosphorus pesticides from water is graphene-coated silica (GCS). It functions by means of the electron-donating qualities of nitrogen, phosphorus, and sulphur, atoms as well as the robust π-bonding network of benzene rings [78, 152].

Cadmium oxide nanoparticles (NPs) are categorized as n-type semiconductor materials among NMs due to their low electrical resistivity, ionic nature, and wide energy band gap [16]. These NPs are employed in the environmental field as a nanoadsorbent to extract pesticides from water. Hymexazol pesticide is removed from irrigation water using cadmium oxide nanoparticles (NPs) produced by the photoirradiation method [1]. Organic minerals like zeolites, when applied on the surface of ZnO-based metal oxide nanoparticles effectively remove monocrotophos. Additionally, it has been shown that a number of zeolite combinations containing metal oxide nanoparticles, such as FeO and TiO_2, help remove dichlorvos, monocrotophos, and methomyl [114].

4 Conclusion and Future Perspectives

Due to severe water constraint, a number of nations have started irrigating vegetable, fibre, and grain crops using wastewater. Although it is normal practice to use irrigation systems to irrigate crops without any prior or little treatment, it is unusual that both quantity and quality of wastewater have been improved from their raw state. Innovative and cutting-edge water technologies are desperately needed to improve industrial production processes by eliminating micropollutants and utilizing readily adjustable water treatment systems. New Mexico has the ability to develop water technologies that are easily customized to match the needs of individual clients. The ability of nanomaterials to combine different features to create multifunctional systems like nanocomposite membranes that allow for both particle retention and pollutant removal is one of their most significant advantages over conventional water technologies. Higher process efficiency is also made possible by nanomaterials because of their special qualities, such a fast reaction rate. This chapter has outlined the current status of research on NMs for water treatment. The key challenge is to effectively increase the reactive material's surface area while minimizing costs and without significantly compromising its mechanical qualities. To advance such materials, more investigation is needed into substrates, anchoring/impregnating methods, and post-synthesis treatments that can increase the nano-reactivity. Future research on the creation of materials needs to address a number of issues, including the development of hybrid compounds with enhanced stability and low toxicity, large-scale synthesis and reproducibility, identification of potential harmful effects on people and the environment, and long-term environmental effects.

References

1. Abdullah FM, Ismail AH, Rheima AM (2023) Effective removal of Hymexazol pesticide from polluted water using cadmium oxide nanoparticles prepared by photolysis methods: thermodynamic, kinetic, and isothermal studies. J Wildlife Biodivers 7:586–600
2. Abebe LS, Su YH, Guerrant RL, Swami NS, Smith JA (2015) Point-of-use removal of cryptosporidium parvum from water: independent effects of disinfection by silver nanoparticles

and silver ions and by physical filtration in ceramic porous media. Environ Sci Technol 49(21):12958–12967

3. Agorku ES, Mamo MA, Mamba BB, Pandey AC, Mishra AK (2015) Palladium-decorated zinc sulfide/reduced graphene oxidenanocomposites for enhanced visible light-driven photodegradation of indigo carmine. Mat Sci Semicon Proc 33:119–126
4. Alarcon-Payan DA, Koyani RD, Vazquez-Duhalt R (2017) Chitosan-based biocatalytic-nanoparticles for pollutant removal from wastewater. Enzyme Microbial Technol 100:71–78
5. Alcala AV, Paulitz TC, Schroeder KL, Porter LD, Derie ML, du Toit LJ (2016) Pythium species associated with damping-off of pea in certified organic fields in the Columbia Basin of central Washington. Plant Dis 100(5):916–925
6. Alfaro SO, Rodríguez-González V, Zaldívar-Cadena AA, Lee SW (2011) Sonochemical deposition of silver-TiO_2 nanocomposites onto foamed waste-glass: evaluation of Eosin Y decompositionunder sunlight irradiation. Catal Today 166:166–171
7. Alsabagh AM, Fathy M, Morsi RE (2015) Preparation and characterization of chitosan/silver nanoparticle/copper nanoparticle/carbon nanotube multifunctional nano-composite for water treatment: heavy metals removal; kinetics, isotherms and competitive studies. RSC Ad 5(69):55774–55783
8. Andjelkovic I, Azari S, Erkelens M, Forward P, Lambert MF, Losic D (2017) Bacterial iron-oxide nanowires from biofilm waste as a new adsorbent for the removal of arsenic from water. RSC Adv 7(7):3941–3948
9. Arthanareeswari M, Devikala S, Sridharan M (2019) Green synthesis of zinc oxide nanoparticles using *Typha latifolia.* L leaf extract for photocatalytic applications. Mater Today: Proc 14:332–337
10. Ashraf MA, Maah MJ, Yusoff I, Mehmood K (2010) Effects of polluted water irrigation on environment and health of people in Jamber, District Kasur, Pakistan. Int J Basic Appl Sci IJBAS-IJENS 10(03):31–48
11. Ashraf I, Singh NB, Agarwal A (2023) Green synthesis of iron oxide nanoparticles using Amla seed for methylene blue dye removal from water. Mater Today: Proc 72:311–316
12. Attiah BH, Obaid HM, AL-Salihi SS (2023) Assessment the Ag and ZnO biosynthesized nanoparticles effects on giardia lamblia trophozoites which grown in HSP-1 culture media. NTU J Pure Sci 2(1)
13. Baran MF, Acay H, Keskin C (2020) Determination of antimicrobial and toxic metal removal activities of plant-based synthesized (Capsicum annuum L. Leaves), ecofriendly, gold nanomaterials. Global Challenges 4(5):1900104
14. Bardos DC, Guillera-Arroita G, Wintle BA (2015) Valid auto-models for spatially autocorrelated occupancy and abundance data. Methods Ecol Evol 6(10):1137–49
15. Bardos P, Bone B, Daly P, Elliott D, Jones S, Lowry G, Merly C (2014) A risk/benefit appraisal for the application of nano-scale zero valent iron (nZVI) for the remediation of contaminated sites. In: WP9 NanoRem, pp 13355–13364
16. Bukhari A, Ijaz I, Zain H, Mehmood U, Iqbal MM, Gilani E, Nazir A (2023) Introduction of CdO nanoparticles into graphene and graphene oxide nanosheets for increasing adsorption capacity of Cr from wastewater collected from petroleum refinery. Arab J Chem 16(2):104445
17. Burhenne J, Riedl KD, Rengelshausen J, Meissner P, Muller O, Mikus G, Haefeli WE, Walter-Sack I (2008) Quantification of cationic anti-malaria agent methylene blue indifferent human biological matrices using cation exchange chromatography coupled to tandem mass spectrometry. J Chromatogr B 863:273–282
18. Castro-Mayorga JL, Randazzo W, Fabra MJ, Lagaron JM, Aznar R, Sánchez G (2017) Antiviral properties of silver nanoparticles against norovirus surrogates and their efficacy in coated polyhydroxyalkanoates systems. LWT-Food Sci Technol 79:503–510
19. Chauhan N, Jain U, Soni S (2019) Nanotools for irrigation water remediation. In: Nanoscience for sustainable agriculture, pp 233–263
20. Chen S, Zhang J, Zhang C, Yue Q, Li Y, Li E (2010) Equilibrium and kinetic studies of methyl orange and methyl violet adsorption on activated carbon derived from Phragmites australis. Desalination 252:149–156

21. Chequer FMD, de Vena^ncio PV, de Bianchi MLP, Antunes LMG (2012) Genotoxic and mutagenic effects of erythrosine B, axanthenes food dye, on HepG2 cells. Food Chem Toxicol 50:3447–3451
22. Coelho MG, de Lima GM, Augusti R, Maria DA, Ardisson JD (2010) New materials for photocatalytic degradation of indigo carminesynthesis, characterization and catalytic experiments of nanometric tin dioxide-based composites. Appl Catal B 96:67–71
23. Cruz GJ, Gómez MM, Solis JL, Rimaycuna J, Solis RL, Cruz JF, Rathnayake B, Keiski RL (2018) Composites of ZnO nanoparticles and biomass based activated carbon: adsorption, photocatalytic and antibacterial capacities. Water Sci Technol 2017(2):492–508
24. Dalal R, Singh S, Sangwan S, Moond M, Beniwal R (2022) Biochemical and molecular mechanism of plant-mediated synthesis of silver nanoparticles—a review. Mini-Rev Org Chem 19(8):939–954
25. Das S, Chakraborty J, Chatterjee S, Kumar H (2018) Prospects of biosynthesized nanomaterials for the remediation of organic and inorganic environmental contaminants. Environ Sci Nano 5(12):2784–2808
26. Davar F, Majedi A, Mirzaei A (2015) Green synthesis of ZnO nanoparticles and its application in the degradation of some dyes. J Am Ceram Soc 98(6):1739–1746
27. David L, Moldovan B (2020) Green synthesis of biogenic silver nanoparticles for efficient catalytic removal of harmful organic dyes. Nanomaterials 10(2):202
28. Dil EA, Ghaedi M, Ghaedi A, Asfaram A, Jamshidi M, Purkait MK (2016) Application of artificial neural network and responsesurface methodology for the removal of crystal violet by zinc oxide nanorods loaded on activate carbon: kinetics and equilibrium study. J Taiwan Inst Chem Eng 59:210–220
29. Dutta P, Kumari A, Mahanta M, Upamanya GK, Heisnam P, Borua S, Kaman PK, Mishra AK, Mallik M, Muthukrishnan G, Sabarinathan KG (2023) Nanotechnological approaches for management of soil-borne plant pathogens. Front Plant Sci 14:1136233
30. Dutta AK, Maji SK, Adhikary B (2014) C-Fe_2O_3 nanoparticles: an easily recoverable effective photo-catalyst for the degradation of rose Bengal and methylene blue dyes in the waste-water treatment plant. Mater Res Bull 49:28–34
31. Eid AM, Fouda A, Hassan SE, Hamza MF, Alharbi NK, Elkelish A, Alharthi A, Salem WM (2023) Plant-based copper oxide nanoparticles; biosynthesis, characterization, antibacterial activity, tanning wastewater treatment, and heavy metals sorption. Catalysts 13(2):348
32. Elliott DW, Lien HL, Zhang WX (2009) Degradation of lindane by zero-valent iron nano particles. J Environ Eng 135:317–324
33. Enniya I, Rghioui L, Jourani A (2018) Adsorption of hexavalent chromium in aqueous solution on activated carbon prepared from apple peels. Sustain Chem Pharm 7:9–16
34. Ethaib S, Al-Qutaifia S, Al-Ansari N, Zubaidi SL (2022) Function of nanomaterials in removing heavy metals for water and wastewater remediation: a review. Environments 9(10):123
35. Ethaib S, Omar R, Siti Mazlina MK, Dayang Radiah AB (2020) Evaluation of the interactive effect pretreatment parameters via three types of microwave-assisted pretreatment and enzymatic hydrolysis on sugar yield. Processes 8(7):787
36. Ezzat SM, Moustafa MT (2024) Efficient removal of E. coli from wastewater by novel phytofabricated nano-zinc using antibacterial potential, kinetic studies, and response surface methodology. Appl Water Sci 14(4):80
37. Fischer AH, Jacobson KA, Rose J, Zeller R (2008) Hematoxylin and eosin staining of tissue and cell sections. Cold spring harb protoc (5):4986
38. Gautam SK, Suresh S (2007) Studies on dechlorination of DDT (1,1,1-trichloro-2,2-bis(4 chlorophenyl)ethane) using magnesium/palladium bimetallic system. J Hazard Mater 139:146–153
39. Ghaedi M, Ghaedi AM, Mirtamizdoust B, Agarwal S, Gupta VK (2016) Simple and facile sonochemical synthesis of lead oxidenanoparticles loaded activated carbon and its application for methyl orange removal from aqueous phase. Mol Liq 213:48–57

40. Grey D, Garrick D, Blackmore D, Kelman J, Muller M, Sadoff C (2013) Water security in one blue planet: twenty-first century policy challenges for science. Phil Trans R Soc A 371(2002):20120406
41. Hameed BH (2008) Equilibrium and kinetic studies of methyl violet sorption by agricultural waste. J Hazard Mater 154:204–212
42. Hao TW, Xiang PY, Mackey HR, Chi K, Lu H, Chui HK, van Loosdrecht MC, Chen GH(2014) A review of biological sulfate conversions in wastewater treatment. Water res 65:1–21
43. Hassan NU, Mahmood Q, Waseem A, Irshad M, Pervez A (2013) Assessment of heavy metals in wheat plants irrigated with contaminated wastewater. Pol J Environ Stud 22(1)
44. He H, Yang S, Yu K, Ju Y, Sun C, Wang L (2010) Microwave induced catalytic degradation of crystal violet in nano-nickeldioxide suspensions. J Hazard Mater 173:393–400
45. Hernández-Gordillo A, Rodríguez-González V, Oros-Ruiz S, Gómez R (2015) Photodegradation of Indigo Carmine dye by CdS nanostructures under blue-light irradiation emitted by LEDs. Catal Today 266:27–35
46. Hirthna, Sendhilnathan S, Rajan PI, Adinaveen T(2018) Synthesis and characterization of NiFe 2 O 4 nanoparticles for the enhancement of direct sunlight photocatalytic degradation of methyl orange. J Supercond Novel Magn 31(10):3315–22
47. Jain R, Mendiratta S, Kumar L, Srivastava A (2021) Green synthesis of iron nanoparticles using Artocarpus heterophyllus peel extract and their application as a heterogeneous Fenton-like catalyst for the degradation of Fuchsin Basic dye. Curr Res Green Sustain Chem 4:100086
48. Jain R, Sikarwar S (2008) Removal of hazardous dye congored from waste material. J Hazard Mater 152(3):942–8
49. Jassal V, Shanker U, Kaith BS (2016) Aegle marmelos mediated green synthesis of different nanostructured metal hexacyanoferrates: activity against photodegradation of harmful organic dyes. Scientifica 2016:1–13
50. Jeyaraj Pandian C, Palanivel R, Dhanasekaran S (2016) Screening antimicrobial activity of nickel nanoparticles synthesized using Ocimum sanctum leaf extract. J Nanoparticles 2016:1–3
51. Jeyasubramanian K, Hikku GS, Sharma RK (2015) Photo-catalytic degradation of methyl violet dye using zinc oxide nano particles prepared by a novel precipitation method and its anti-bacterial activities. J Water Proc Eng 8:35–44
52. Jimoh OD, Ayodeji MA, Mohammed B (2003) Effects of agrochemicals on surface waters and groundwaters in the Tunga-Kawo (Nigeria) irrigation scheme. Hydrol Sci J 48(6):1013–1023
53. Joo SH, Zhao D (2008) Destruction of lindane and atrazine using stabilized iron nanoparticles under aerobic and anaerobic conditions: effects of catalyst and stabilizer. Chemosphere 70(3):418–25
54. Julkapli NM, Bagheri S, Hamid SBA (2014) Recent advances in heterogeneous photocatalytic decolorization of synthetic dyes. Sci World J 2014:1–25
55. Jyoti K, Baunthiyal M, Singh A (2016) Characterization of silver nanoparticles synthesized using Urtica dioica Linn. leaves and their synergistic effects with antibiotics. J Radiat Res Appl Sci 9(3):217–227
56. Kardam A, Raj KR, Srivastava S, Srivastava MM (2014) Nanocellulose fibers for biosorption of cadmium, nickel, and lead ions from aqueous solution. Clean Technol Environ Policy 16(2):385–93
57. Kardam A, Raj KR, Srivastava S (2018) Green nanotechnology for bioremediation of toxic metals from waste water. In: Chemistry of phytopotentials: health, energy and environmental perspectives, pp 373–377
58. Karthik R, Govindasamy M, Chen SM, Cheng YH, Muthukrishnan P, Padmavathy S, Elangovan A (2017) Biosynthesis of silver nanoparticles by using Camellia japonica leaf extract for the electrocatalytic reduction of nitrobenzene and photocatalytic degradation of Eosin-Y. J Photochem Photobiol, B 170:164–172
59. KassemA AyoubGM, Malaeb L (2019) Antibacterial activity of chitosan nano-composites andcarbon nanotubes: a review. Sci Total Environ 668:566–576

60. Kaur J, Singhal S (2014) Facile synthesis of ZnO and transition metal doped ZnO nanoparticles for the photocatalytic degradation of methyl orange. Ceramics Int 40:7417–7424
61. Kaur K, Jawa GK (2021) Use of nano-biomaterials for adsorption of heavy metals from wastewater–a review. In: AIP conference proceedings, vol 2352. AIP Publishing, p 2561
62. Khairy M, Zakaria W (2014) Effect of metal-doping of TiO_2 nanoparticles on their photocatalytic activities toward removal of organic dyes. Egypt J Pet 23(4):419–426
63. Khan TA, Sharma S, Ali I (2011) Adsorption of Rhodamine B dye from aqueous solution onto acid activated mango (Magniferaindica) leaf powder: equilibrium, kinetic and thermodynamic studies. J Toxicol Environ Health 3:286–297
64. Khandaker S, Hassan J, Saha GC (2013) Effect of textile dying wastewater on growth, yield and nutritional qualities of stem amaranth. Int J Agric Sci Res 3(3):171–178
65. Kirankumar VS, Sumathi S (2017) Structural, optical, magnetic and photocatalytic properties of bismuth doped copper aluminate nanoparticles. Mater Chem Phys 197:17–26
66. Kokkinos P, Mantzavinos D, Venieri D (2020) Current trends in the application of nanomaterials for the removal of emerging micropollutants and pathogens from water. Molecules 25(9):2016
67. Kolya H, Kang CW (2023) Next-generation water treatment: exploring the potential of biopolymer-based nanocomposites in adsorption and membrane filtration. Polymers 15(16):3421
68. Kolya H, Maiti P, Pandey A, Tripathy T (2015) Green synthesis of silver nanoparticles with antimicrobial and azo dye (Congo red) degradation properties using Amaranthusgangeticus Linn leaf extract. J Anal Sci Technol 6:33–39
69. Korotcenkov G (2023) II–VI wide-bandgap semiconductor device technology: stability and oxidation. In: Handbook of II–VI semiconductor-based sensors and radiation detectors: volume 1, materials and technology, pp 517–550
70. Kurt BZ, Durmus Z, Durmus A (2016) Preparation and characterization of platinum (Pt) and palladium (Pd) nanoparticle decorated graphene sheets and their utilization for the elimination of basic fuchsin and indigo carmine dyes. Solid State Sci 51:51–58
71. Kurt OK, Zhang J, Pinkerton KE (2016) Pulmonary health effects of air pollution. Current opinion in pulmonary medicine 22(2):138–43
72. Latheef S, Soundhirarajan K (2018) Heavy metal contamination in irrigation water and its effects on plants. Int Res J Eng Technol 5(5):3704–3710
73. Lee HJ, Kim JH, Park SS, Hong SS, Lee GD (2015) Degradation kinetics for photocatalytic reaction of methyl orange over Aldoped ZnO nanoparticles. Ind Eng Chem 25:199–206
74. Letshwenyo MW, Mokokwe G (2020) Accumulation of heavy metals and bacteriological indicators in spinach irrigated with further treated secondary wastewater. Heliyon 6(10)
75. Li C, Dong Y, Yang J, Li Y, Huang C (2014) Modified nano-graphite/ Fe_3O_4 composite as efficient adsorbent for the removal of methyl violet from aqueous solution. J Mol Liq 196:348–356
76. Li D, Zhang Y, Zhang Y, Zhou X, Guo S (2013) Fabrication of bidirectionally doped b-Bi_2O_3/TiO_2–NTs with enhanced photocatalysis under visible light irradiation. J Hazard Mater 258:42–49
77. Liu Y, Pei F, Lu R, Xu S, Cao S (2014) TiO_2/N-graphene nanocomposite via a facile in situ hydrothermal sol-gel strategyfor visible light photodegradation of eosin Y. Mater Res Bull 60:188–194
78. Liu X, Zhang H, Ma Y, Wu X, Meng L, Guo Y, Yu G, Liu Y (2013) Graphene-coated silica as a highly efficient sorbent for residual organophosphorus pesticides in water. J Mater Chem A 1(5):1875–1884
79. Lothrop N, Bright KR, Sexton J, Pearce-Walker J, Reynolds KA, Verhougstraete MP (2018) Optimal strategies for monitoring irrigation water quality. Agric Water Manag 199:86–92
80. Mahmood S (2006) Waste water irrigation: issues and constraints for sustainable irrigated agriculture. Ital J Agron 3:12–15
81. Mair L, Hill JK, Fox R, Botham M, Brereton T, Thomas CD (2014) Abundance changes and habitat availability drive species' responses to climate change. Nat Clim Change 4(2):127–31

82. Malakar A, Snow DD, Ray C (2019) Irrigation water quality—a contemporary perspective. Water 11(7):1482
83. Malik S, Kishore S, Shah MP, Kumar SA (2022) A comprehensive review on nanobiotechnology for bioremediation of heavy metals from wastewater. J Basic Microbiol 62(3–4):361–375
84. Malik R, Tomer VK, Rana PS, Nehra SP, Duhan S (2015) Surfactant assisted hydrothermal synthesis of porous 3-D hierarchical SnO_2 nanoflowers for photocatalytic degradation of rose Bengal. Mater Lett 154:124–127
85. Mallakpour S, Ramezanzade V (2021) Tragacanth gum mediated green fabrication of mesoporous titania nanomaterials: application in photocatalytic degradation of crystal violet. J Environ Manag 291:112680
86. Mandal S, Sarkar B, Mukhopadhyay R, Biswas JK, Manjaiah KM (2018) Microparticle-supported nanocomposites for safe environmental applications. In: Rai M, Biswas JK (eds) Nanomaterials: ecotoxicity, safety, and public perception, pp 305–317
87. Martínez-Cabanas M, López-García M, Barriada JL, Herrero R, de Vicente ME (2016) Green synthesis of iron oxide nanoparticles. Development of magnetic hybrid materials for efficient As (V) removal. Chem Eng J 301:83–91
88. Moond M, Singh S, Sangwan S, Devi R, Beniwal R (2022) Green synthesis and applications of silver nanoparticles: a systematic review. AATCC J Res 9(6):272–285
89. Moond M, Singh S, Sangwan S, Devi P, Beniwal A, Rani J, Kumari A, Rani S (2023) Biosynthesis of silver nanoparticles utilizing leaf extract of Trigonella foenum-graecum L. for catalytic dyes degradation and colorimetric sensing of Fe^{3+}/Hg^{2+}. Molecules 28(3):951
90. Moond M, Singh S, Sangwan S, Rani S, Beniwal A, Rani J, Kumari A, Rani I, Devi P (2023) Phytofabrication of silver nanoparticles using trigonella foenum-graceum L. Leaf and evaluation of its antimicrobial and antioxidant activities. Int J Mole Sci 24(4):3480
91. Moond M, Singh S, Sangwan S, Devi R, Rani J, Beniwal A, Beniwal R (2023) Synthesis, optimization and characterization of Silver nanoparticles using aqueous seed extract of Trigonella foenum-graceum L. for catalytic reduction of p-nitrophenol. Chem Select 8(29):e202301006
92. Motahari F, Mozdianfard MR, Salavati-Niasari M (2015) Synthesis and adsorption studies of NiO nanoparticles in the presence of H2acacen ligand, for removing Rhodamine B in wastewater treatment. Process Saf Environ Prot 93:282–92
93. Mukhopadhyay R, Sarkar B, Khan E, Alessi DS, Biswas JK, Manjaiah KM, Eguchi M, Wu KC, Yamauchi Y, Ok YS (2022) Nanomaterials for sustainable remediation of chemical contaminants in water and soil. Crit Rev Environ Sci Technol 52(15):2611–2660
94. Musico YLF, Santos CM, Dalida MLP, Rodrigues DF (2014) Surface modification of membrane filters using graphene and graphene oxide-based nanomaterials for bacterial inactivation and removal. ACS Sustain Chem Eng 2(7):1559–1565
95. Nabi G, Majid A, Riaz A, Alharbi T, Kamran MA, Al-Habardi M (2021) Green synthesis of spherical TiO_2 nanoparticles using Citrus Limetta extract: excellent photocatalytic water decontamination agent for RhB dye. Inorg Chem Commun 129:108618
96. Nagaraja R, Kottam N, Girija CR, Nagabhushana BM (2012) Photocatalytic degradation of Rhodamine B dye under UV/solarlight using ZnO nanopowder synthesized by solution combustion route. Powder Technol 215–216:91–97
97. Nair R, Varghese SH, Nair BG, Maekawa T, Yoshida Y, Kumar DS (2010) Nanoparticulate material delivery to plants. Plant Sci 179(3):154–163
98. Naseem T, Durrani T (2021) The role of some important metal oxide nanoparticles for wastewater and antibacterial applications: a review. Environ Chem Ecotoxicol 3:59–75
99. Nazir R, Khan M, Masab M, Rehman HU, Rauf NU, Shahab S, Ameer N, Sajed M, Ullah M, Rafeeq M, Shaheen Z (2015) Accumulation of heavy metals (Ni, Cu, Cd, Cr, Pb, Zn, Fe) in the soil, water and plants and analysis of physico-chemical parameters of soil and water collected from Tanda Dam Kohat. J Pharm Sci Res 7(3):89
100. Ofomaja AE (2008) Kinetic study and sorption mechanism of methylene blue and methyl violet onto mansonia (Mansoniaaltissima) wood sawdust. Chem Eng J 143:85–95

101. Potbhare AK, Chaudhary RG, Chouke PB, Yerpude S, Mondal A, Sonkusare VN, Rai AR, Juneja HD (2019) Phytosynthesis of nearly monodisperse CuO nanospheres using Phyllanthus reticulatus/Conyza bonariensis and its antioxidant/antibacterial assays. Mater Sci Eng, C 99:783–793
102. Prasad AR, Garvasis J, Oruvil SK, Joseph A (2019) Bio-inspired green synthesis of zinc oxide nanoparticles using Abelmoschus esculentus mucilage and selective degradation of cationic dye pollutants. J Phys Chem Solids 127:265–274
103. Prashanth V, Remya N (2021) Synthesis of TiO_2 using Calotropis gigantea for visible light excitation and degradation of congo red dye. J Hazard Toxic Radioactive Waste 25(4):021026
104. Preethi T, Abarna B, Vidhya KN, Rajarajeswari GR (2014) Sol–gel derived cobalt doped nanotitania photocatalytic system for solar light induced degradation of crystal violet. Ceram Int 40:13159–13167
105. Prevot AB, BaiocchiC Brussino MC, Pramauro E, Savarino P, Auguglliaro V, Marcı‘ G, Palmisano L (2001) Photocatalyticdegradation of acid blue 80 in aqueous solutions containing TiO_2 suspensions. Environ Sci Technol 35:971–976
106. Prica M, Dalmacija B, Dalmacija M, Agbaba J, Krčmar D, Tricković J, Karlovic E (2010) Changes in metal availability during sediment oxidation and the correlation with immobilization potential. Ecotoxicol Environ Saf 73(6):1370–1377
107. Qu X, Brame J, Li Q, Alvarez PJ (2012) Nanotechnology for a safe and sustainable water supply: enabling integrated water treatment and reuse. Acc Chem Res 46(3):834–843
108. Rad LR, Momeni A, Ghazani BF, Irani M, Mahmoudi M, Noghreh B (2014) Removal of Ni^{2+} and Cd^{2+} ions from aqueous solutions using electrospun PVA/zeolite nanofibrous adsorbent. Chem Eng J 256:119–127
109. Rai PK, Lee SS, Zhang M, Tsang YF, Kim KH (2019) Heavy metals in food crops: Health risks, fate, mechanisms, and management. Environ Int 125:365–385
110. Raj SR, Arnold AC, Barboi A, Claydon VE, Limberg JK, Lucci VE, Numan M, Peltier A, Snapper H, Vernino S (2021) Long-COVID postural tachycardia syndrome: an american autonomic society statement. Clin Auton Res 31(3):365–8
111. Rani M, Shanker U (2021) Efficient degradation of organic pollutants by novel titanium dioxide coupled bismuth oxide nanocomposite: green synthesis, kinetics and photoactivity. J Environ Manag 300:113777
112. Ranjith R, Renganathan V, Chen SM, Selvan NS, Rajam PS (2019) Green synthesis of reduced graphene oxide supported TiO_2/Co_3O_4 nanocomposite for photocatalytic degradation of methylene blue and crystal violet. Ceram Int 45(10):12926–12933
113. Ritchie EE, Princz JI, Robidoux PY, Scroggins RP (2013) Ecotoxicity of xanthenes dyes and a non-chlorinated bisphenol in soil. Chemosphere 90:2129–2135
114. Sahithya K, Das D, Das N (2015) Effective removal of dichlorvos from aqueous solution using biopolymer modified MMT–CuO composites: equilibrium, kinetic, and thermodynamic studies. J Mol Liq 211:821–830
115. Saied E, Eid AM, Hassan SE, Salem SS, Radwan AA, Halawa M, Saleh FM, Saad HA, Saied EM, Fouda A (2021) The catalytic activity of biosynthesized magnesium oxide nanoparticles (MgO–NPs) for inhibiting the growth of pathogenic microbes, tanning effluent treatment, and chromium ion removal. Catalysts 11(7):821
116. Saleh IA, Zouari N, Al-Ghouti MA (2020) Removal of pesticides from water and wastewater: chemical, physical and biological treatment approaches. Environ Technol Innov 19:101026
117. Salihi IU, Kutty SR, Ismail HH (2018) Copper metal removal using sludge activated carbon derived from wastewater treatment sludge. MATEC Web Conf EDP Sci 203:03009
118. San NO, Celebioglu A, Tumtas Y, Uyar T, Tekinay T (2014) Reusable bacteria immobilized electrospun nanofibrous webs fordecolorization of methylene blue dye in wastewater treatment. RSC Adv 4:32249–32255
119. Sandeep G, Vijayalatha KR, Anitha T (2019) Heavy metals and its impact in vegetable crops. Int J Chem Stud 7(1):1612–1621
120. Sathishkumar M, Sneha K, Yun YS (2010) Immobilization of silver nanoparticles synthesized using Curcuma longa tuber powder and extract on cotton cloth for bactericidal activity. Biores Technol 101(20):7958–7965

121. Sattar MU, Khan MA, Khalil AA, Amir RM (2013) Mitigation of heavy metals in vegetables through washing with house hold chemicals
122. Sethy NK, Arif Z, Mishra PK, Kumar P (2020) Green synthesis of TiO_2 nanoparticles from Syzygiumcumini extract for photo-catalytic removal of lead (Pb) in explosive industrial wastewater. Green Process Synth 9(1):171–181
123. Shahwan T, Sirriah SA, Nairat M, Boyacı E, Eroğlu AE, Scott TB, Hallam KR (2011) Green synthesis of iron nanoparticles and their application as a Fenton-like catalyst for the degradation of aqueous cationic and anionic dyes. Chem Eng J 172(1):258–266
124. Shamsipur M, Rajabi HR (2014) Study of photocatalytic activity of ZnS quantum dots as efficient nanoparticles for removal ofmethyl violet: effect of ferric ion doping. Spectrochim Acta Mol Biomol Spectrosc 122:260–267
125. Sharma SK, Sharma PR, Chen H, Johnson K, Zhan C, Wang R, Hsiao B (2020) Cellulose-supported nanosized zinc oxide: highly efficient bionanomaterial for removal of arsenic from water. In: Contaminants in our water: identification and remediation methods. American Chemical Society, pp 253–267
126. Shu J, Wang Z, Huang Y, Huang N, Ren C, Zhang W (2015) Adsorption removal of congo red from aqueous solution bypolyhedral Cu_2O nanoparticles: kinetics, isotherms, thermodynamics and mechanism analysis. J Alloys Compd 633:338–346
127. Silva MC, Castro-Lopes S, Jerônimo AG, Barbosa R, Lins A, Trigueiro P, Peña-Garcia RR (2024) Green Synthesis of Er-Doped ZnO Nanoparticles: an Investigation on the Methylene Blue, Eosin, and Ibuprofen removal by photodegradation. Molecules 29(2):391
128. Singh S, Sangwan S, Sharma P, Devi P, Moond M (2021) Nanotechnology for sustainable agriculture: an emerging perspective. J Nanosci Nanotechnol 21(6):3453–3465
129. Singh A, Sharma A, Verma RK, Chopade RL, Pandit PP, Nagar V, Aseri V, Choudhary SK, Awasthi G, Awasthi KK, Sankhla MS (2022) Heavy metal contamination of water and their toxic effect on living organisms. In: The toxicity of environmental pollutants. IntechOpen
130. Stafiej A, Pyrzynska K (2007) Adsorption of heavy metal ions with carbon nanotubes. Sep Purif Technol 58(1):49–52
131. Steadman JR, Maier CR, Schwartz HF, Kerr ED (1975) Pollution of surface irrigation waters by plant pathogenic organisms 1. JAWRA J Am Water Resour Assoc 11(4):796–804
132. Surendra TV, Roopan SM (2016) Photocatalytic and antibacterial properties of phytosynthesized CeO_2 NPs using Moringa oleifera peel extract. J Photochem Photobiol B: Biol 161:122–128
133. Taha MR, Mobasser S (2015) Adsorption of DDT and PCB by nanomaterials from residual soil. PLoS ONE 10(12):e0144071
134. Tahoon MA, Siddeeg SM, Salem Alsaiari N, Mnif W, Ben Rebah F (2020) Effective heavy metals removal from water using nanomaterials: a review. Processes 8(6):645
135. Tangahu BV, Sheikh Abdullah SR, Basri H, Idris M, Anuar N, Mukhlisin M (2011) A review on heavy metals (As, Pb, and Hg) uptake by plants through phytoremediation. Int J Chem Eng
136. Thapa R, Maiti S, Rana TH, Maiti UN, Chattopadhyay KK (2012) Anatase TiO_2 nanoparticles synthesis via simple hydrothermal route: degradation of orange II, methyl orange and Rhodamine B. J Mol Cat A: Chem 363:223–229
137. Thy LT, Hau DT, Lien LT, Hung ND, Tu TH, Nam HM, Hieu NH, ThanhPhong M (2021) Fabrication of RHA-silica/graphene oxide nanocomposites for removal of lead ions from water. Environ Technol 42(14):2202–2210
138. Tian T, Shu Y, Chen J, Hu Z, Xu L, Jin G, Liang J, Liu P, Zhou X, Miao R, Ma H (2009) A functional genetic variant in microRNA-196a2 is associated with increased susceptibility of lung cancer in Chinese. Cancer Epidemiol Biomarkers & Prev 18(4):1183–7
139. Tiwari A, Pandey K, Singh S, Chawla S (2023) Grass and their waste products for nanoparticles synthesis and applications. In: Husen A (eds) Nanomaterials from agricultural and horticultural products. Smart nanomaterials technology. Springer, Singapore
140. Ugya AY, Imam TS, Tahir SM (2015) The use of Pistiastratiotes to remove some heavy metals from Romi stream: a case study of Kaduna refinery and petrochemical company polluted stream. IOSR J Environ Sci Toxicol Food Technol 9:48–51

141. Vidovix TB, Quesada HB, Januario EFD, Bergamasco R, Vieira AMS (2019) Green synthesis of copper oxide nanoparticles using Punica granatum leaf extract applied to the removal of methylene blue. Mater Lett 257:126685
142. Vidya C, Prabha MC, Raj MA (2016) Green-mediated synthesis of zinc oxide nanoparticles for the photocatalytic degradation of Rose Bengal dye. Environ Nanotechnol Monit Manag 6:134–138
143. Vignesh K, Suganthi A, Rajarajan M, Sakthivadivel R (2012) Visiblelight assisted photodecolorization of eosin-Y in aqueous solution using hesperidin modified TiO_2 nanoparticles. Appl Surf Sci 258:4592–4600
144. Vinutha JS, Bhagat D, Bakthavatsalam N (2013) Nanotechnology in the management of polyphagous pest Helicoverpa armigera. J Acad Indus Res 1(10):606–608
145. Wang R, Kong X, Zhang W, Zhu W, Huang L, Wang J, Zhang X, Liu X, Hu N, Suo Y, Wang J (2018) Mechanism insight into rapid photocatalytic disinfection of Salmonella based on vanadate QDs-interspersed g-C_3N_4 heterostructures. Appl Catal B 225:228–237
146. Wang L, Wu W, Xie X, Chen H, Lin J, Dionysiou D (2018) Removing Escherichia coli from water using zinc oxide-coated zeolite. Water Res 141:145–151
147. Xu HJ, Wang XH, Li H, Yao HY, Su JQ, Zhu YG (2014) Biochar impacts soil microbial community composition and nitrogen cycling in an acidic soil planted with rape. Environ sci technol 48(16):9391–9
148. Yu B, Zeng J, Gong L, Zhang M, Zhang L, Chen X (2007) Investigation of the photocatalytic degradation of organochlorine pesticides on a nano-TiO_2 coated film. Talanta 72(5):1667–1674
149. Yu Q, Lian J, Siriponglert S, Li H, Chen YP, Pei SS (2008) Graphene segregated on Ni surfaces and transferred to insulators. Appl phys lett 93(11)
150. Zeid MA, Biswas AK (1990) Impacts of agriculture on water quality. Water Int 15:160–167
151. Zhang P, Yang M, Lan J, Huang Y, Zhang J, Huang S, Yang Y, Ru J (2023) Water quality degradation due to heavy metal contamination: health impacts and eco-friendly approaches for heavy metal remediation. Toxics 11(10):828
152. Zhang C, Zhang RZ, Ma YQ, Guan WB, Wu XL, Liu X, Li H, Du YL, Pan CP (2015) Preparation of cellulose/graphene composite and its applications for triazine pesticides adsorption from water. ACS Sustain Chem Eng 3(3):396–405
153. Zhao X, Wang W, Zhang Y, Wu S, Li F, Liu JP (2014) Synthesis and characterization of gadolinium doped cobalt ferrite nanoparticles with enhanced adsorption capability for congo red. Chem Eng J 250:164–174
154. Zhou Y, Gao B, Zimmerman AR, Chen H, Zhang M, Cao X (2014) Biochar-supported zerovalent iron for removal of various contaminants from aqueous solutions. Biores Technol 152:538–542

Hybrid Nanomaterials: Pioneering Sustainable Environmental Detection and Enhancing Agricultural Productivity

Faria Fatima and Manaal Zahera

Abstract The fields of materials science and nanotechnology have shown a great deal of potential in hybrid nanomaterials, a novel category originating from the combined action of organic and inorganic components at the nanoscale. This combination makes it easier to design buildings with enhanced functionality. Because of their cutting-edge and sustainable qualities, these materials play a crucial role in the field of sustainable agriculture and in environmental detection. This chapter explores their invaluable role in addressing agricultural and environmental issues through the utilization of nanomaterials' varied features. Due to their huge surface area, amazing sensitivity, and tuneable features, hybrid nanomaterials are promising options to act in the field of nano-agrochemicals. In the agricultural sector, hybrid nanomaterials are employed to increase crop yields, and towards pollutant detection, keeping pests away from plants and enhancing nutrient delivery. These chemicals offer effective, tailored treatments such as smart insecticides and slow-release fertilizers by combining organic and inorganic components. They promote sustainable agricultural practices and aid in lessening the negative environmental effects of conventional farming techniques.

Keywords Adsorption · Agriculture · Environment · Hybrid nanoparticles · Pollutant · Remediation

1 Introduction

It is more important than ever to address the most pressing environmental issues in our fast-paced world. Our world faces huge challenges that require creative answers, from pollution and climate change to depleting resources. Hazardous compounds

F. Fatima (✉)
Department of Agriculture, Integral Institute of Agricultural Science and Technology, Integral University, Lucknow, India
e-mail: fatimafaria45@gmail.com

M. Zahera
Department of Biotechnology, Era University, Lucknow, India

L. F. Fraceto et al. (eds.), *Emerging Nanotechnologies for Agroecosystem Management*, Smart Nanomaterials Technology, https://doi.org/10.1007/978-981-95-0187-8_10

are constantly being introduced into different environmental domains by industrial activity, which puts ecological balance and human health at risk [1]. Using nanotechnology—more specifically, hybrid nanomaterials—to solve these issues represents one of the most effective approaches. These tiny miracles of nature have become powerful instruments that have the potential to transform environmental management and monitoring [58]. By accurately modifying the size and shape of materials, devices and systems with unique features and functionalities can be created at the nanoscale (1–100 nm) scale thanks to nanotechnology [11].

Current issues include a limited range of measures, high operating costs, and the incapacity to identify new toxins quickly. Furthermore, traditional approaches often require large energy inputs and produce large amounts of trash. On the other hand, by providing accurate and efficient monitoring while reducing resource consumption and waste, hybrid nanomaterials provide a sustainable alternative. Nanotechnology harnesses the exceptional properties of materials—such as their high surface-to-volume ratio, enhanced reactivity, and superior efficiency—for a wide range of advanced applications [12, 25]. Unlike traditional techniques, nanomaterials can make use of unique surface chemistry. This makes it possible to alter them or mix them with functional groups in order to effectively remediate specific molecules of interest, like contaminants, with precision. After being used, the materials used in pollution repair cannot turn into new pollutants. Thus, it makes perfect sense to employ biodegradable material that provides a more secure substitute for cleaning up pollutants [42].

Nanotechnology has proven its adaptability in the environmental domain by helping to avoid and remediate contamination in addition to detecting and sensing pollutants. Hybrid nanomaterials are vital to protecting air, soil and water quality (Fig. 1).

2 Types of Hybrid Nanomaterials

A broad array of HNM are employed in environmental detection which include carbon-based nanomaterials, inorganic nanomaterials, and polymer-related nanomaterials (Fig. 2). These materials serve diverse functions such as membrane filtration, adsorption, photocatalysis, catalysis, and sensing, contributing significantly to the detection and remediation of environmental pollutants.

2.1 Carbon-Based Nanomaterials

Carbon nanotubes (CNTs) have extraordinary physiochemical characteristics, like high surface area and high temperature stability, by which they tend to attract a lot of interest in pollutants in remediation applications.

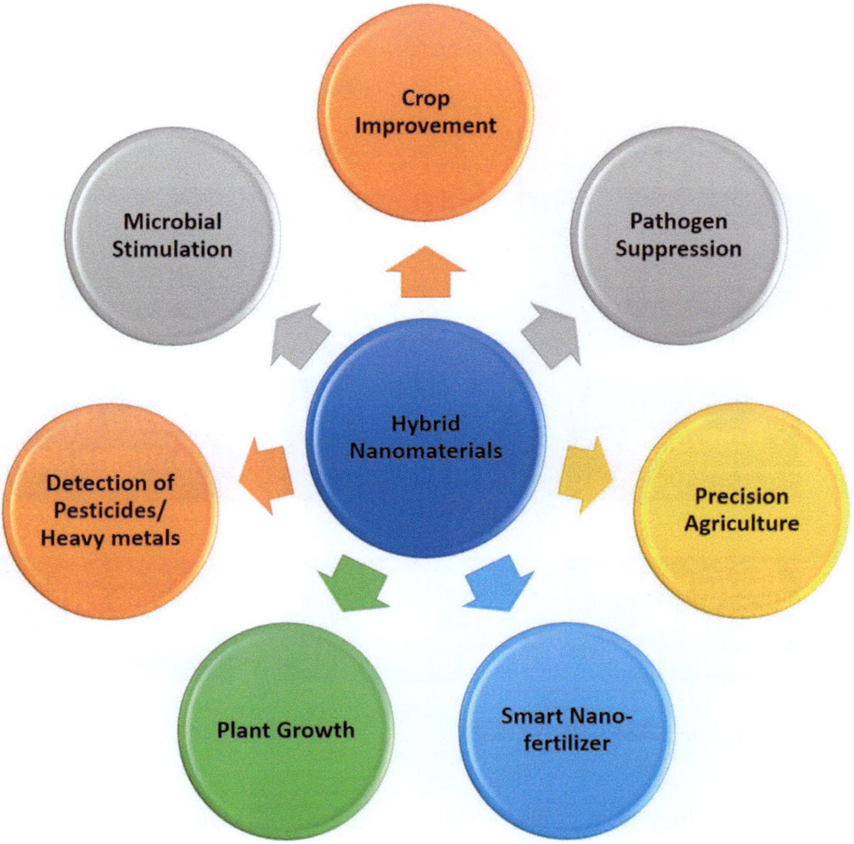

Fig. 1 Role of hybrid nanoparticles in agriculture

Furthermore, these nanoparticles are employed in photocatalytic methods for pollutant elimination. The size, structure, and properties of nanostructured carbon materials are used to classify them into different groups [31]. These categories comprise.

- **Zero-Dimensional Carbon**: At the nanoscale, these materials are essentially point formations with no discernible shape, e.g., nanodiamond (ND), carbon dots (CDs), fullerenes, and graphene quantum dots (GQDs).
- **One-Dimensional Carbon**: Nanotubes and nanofibers made of carbon are examples of one-dimensional materials made of carbon.
- **Two-Dimensional Carbon**: Graphene and its derivatives, graphitic carbon nitride (g-C_3N_4), and graphdiynes are examples of two-dimensional carbon compounds.

The distinctive optical, electrical, and structural characteristics of C-based nanomaterials of different sizes have drawn a lot of interest, as has their simplicity in functionalization. When various carbon-based matrices come into touch with

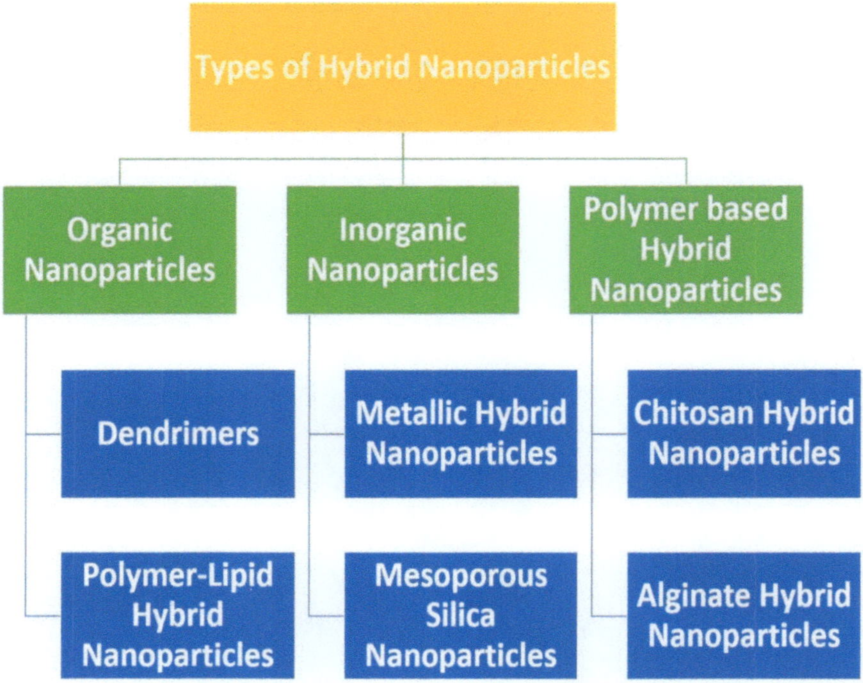

Fig. 2 Types of hybrid nanoparticles

these materials, they form effective heterojunctions that result in high interfaces. Heterostructures which have different morphologies are good support matrices. In catalytic applications, effective heterojunctions provide quick charge transfer, better catalytic dispersion, and increased catalytic activity [68]. For photocatalytic wastewater treatment, morphology-based heterojunctions are very beneficial. When coupled with their two-dimensional counterparts, zero-dimensional CNM's show remarkable potential because of their stable physicochemical features and good electrical conductivity, which facilitate effective charge-transfer kinetics. For wastewater treatment, for instance, composite materials show increased photoactivity. Due to their distinctive qualities, carbon nanotubes (CNTs) are useful additives in a variety of photocatalytic composites, particularly when paired with two-dimensional semiconductors like C_3N_4. The enhanced degradation of colourants is ascribed to the CNT/C_3N_4 hybrid catalyst's ability to absorb light and inhibit charge carrier exchange.

Growing interest has been shown in using carbon nanoparticles (CNPs) as a fertilizer synergist to improve crop growth. Researchers examined how CNPs applied at various rates affected the nutrient usage efficiency (NUE) of corn (*Zea mays* L.), plant development, and soil characteristics. The findings demonstrated that CNPs had a dose-dependent effect on maize growth, promoting and inhibiting it at low and high doses, respectively. Regardless of the type of soil, the application of CNPs significantly increased soil urease activity [55]. Similarly, despite harming the growth of seedlings, water-soluble carbon nanoparticles (CNPs) can greatly increase seed germination. CNP pretreatment greatly enhanced high temperature and seed germination. While CNP treatment encouraged the development of lateral roots and increased the amount of chlorophyll in seedlings growing under salt stress, it also marginally hindered the expansion of primary roots. The results of this study show that soluble nanoparticles can significantly improve lettuce seed germination under salinity stress, even though different varieties of lettuce respond differently to nanoparticle treatments. This provides substantial verification on the possible uses of nanoparticles in applications in agriculture to enhance quality and yield of crops under adverse circumstances [6].

2.2 *Inorganic Nanomaterials*

Heavy metal pollution is a persistent and pervasive environmental issue that poses serious threats to ecological health [3]. Due to their effectiveness in eliminating heavy metals from environmental matrices, such as soil and water, metals and metal oxide nanomaterials have become prominent subjects of study. These nanoparticles are interesting options for heavy metal cleanup because of their special qualities and surface chemistry [29]. Environmental remediation has shown a great deal of interest in nanoparticles made from metals as well as metal oxides. There increased attention has been attributed to their adaptability in wastewater purification, air quality enhancement, and soil pesticide identification [65]. These nanomaterials show a wide range of functions, including adhesion, antimicrobial properties, photocatalysts, dehalogenation, removal of sulphur, and catalytic reduction activities.

Metallic nanomaterials made from metal oxides are created via a different mode of synthesis including co-precipitation, hydrothermal production, and thermal breakdown. Because of their strong antibacterial capabilities against bacteria, fungus, and viruses, silver nanoparticles are widely used in water disinfection. Silver ions have photoactivity when exposed to UV radiation, which enhances the inactivation of viral and bacterial infections by UV light [30]. Silver nanoparticles preferentially identify Cd^{2+} ions in water samples and efficiently prevent bacterial biofilms. They are frequently used with metal oxides and polymers to increase their overall efficacy. Through chemical interactions, zero-valent iron nanoparticles (nZVI) effectively reduce and eliminate pollutants such as heavy metals, nitrate, and chlorinated organic substances in remediation processes. They usually possess core–shell morphology that is made up of mixed-valent oxides that include Fe (II) and Fe (III). Contaminants are easier to remove from the shell, especially heavy metals that have a larger standard reduction capability than the Fe^{2+}/Fe pair.

Natural zeolites and diatomaceous earth are examples of silica-based nanomaterials (SNNMs), which can adsorb polarized compounds or cations. This type of nanoporous characteristic controls the consequences of using SNNM as low-risk plant protectors and soil conditioners. These serve as transporters for foliar fertilizers that are sprayed on fruits like peach, apricot, and stone fruit plants. Their influence was shown on stone fruit trees' physiology, yield, and fruit quality. Implementation of SNNM produced effects that were particular to the development of particle films on leaves, such as decreased leaf temperature and increased water-use efficiency. These natural zeolites and diatomaceous earth encourage the build-up of polyphenols metabolites in the fruits [40].

However inorganic nanomaterials (INMs) have shown diverse and beneficial effects on various crops, significantly improving their growth, stress resilience, and yield. For instance, titanium dioxide nanoparticles (TiO_2 NPs) applied to wheat (*Triticum aestivum*) and spinach (*Spinacia oleracea*) enhance photosynthesis, resulting in increased biomass and larger leaf areas. Similarly, zinc oxide nanoparticles (ZnO NPs) in rice (*Oryza sativa*) improve root growth and nutrient uptake, especially zinc, leading to higher grain yields. Silver nanoparticles (Ag NPs) in tomato (*Solanum lycopersicum*) help boost resistance against bacterial pathogens, improving plant health and reducing disease incidence. In maize (*Zea mays*), silicon dioxide nanoparticles (SiO_2 NPs) increase drought and salinity tolerance, promoting growth in stressful conditions. Iron oxide nanoparticles (Fe_3O_4 NPs) in mustard (*Brassica juncea*) enhance iron uptake and photosynthetic efficiency, boosting seed yield. Similarly, manganese oxide nanoparticles (MnO_2 NPs) in soybean (*Glycine max*) improve nitrogen fixation and enzyme activity, leading to better crop performance. In barley (*Hordeum vulgare*), copper oxide nanoparticles (CuO NPs) increase plant height, root length, and chlorophyll content, improving overall growth. These examples demonstrate how INMs can significantly enhance agricultural productivity and sustainability across a variety of crops.

2.3 Polymer-Based Nanomaterial

Polymers are integrated as a fundamental ingredient at the nanoscale in polymer-based nanomaterials [51]. Nanoparticles are extracted from aqueous solutions, those made from chitosan, by solvent casting techniques. By directing particles through pores, these membranes promote chemical reactions that render contaminants immobile. Chitosan-mediated nanofiltration membranes are frequently used to cleanse textile wastewater because they reject negatively and electro neutrally charged dyes less frequently than positively charged dyes [39]. The size of the dyes also affects the effectiveness of nanofiltration.

Chitosan-based carbon nanofibers (CNFs), iron oxide nanoparticles, and polyvinyl alcohol can be combined to create hybrid films with exceptional metal absorption capacity that have been shown to be highly effective at eliminating Cr^{6+} from water sources. The system's potential for catalysis enhanced significantly when palladium NPs are supported by CNT's and chitosan, owing to the enhanced surface area, improved NPsdispersion and synergistic interaction between the support materials. [59]. The reduction of several nitroarenes, such as 4-nitrophenol, 2,4-dinitrophenol, and 2-nitroaniline, was effectively facilitated by this catalyst. In addition, this catalytic system demonstrated outstanding efficacy in the breakdown of many dyes, including methyl red, methyl orange, methylene blue, and Congo red.

Two key characteristics of polymeric nanocomposites—their intrinsic biocompatibility and biodegradability—are what propel their application: For example, a sustainable hybrid nanomaterial adsorbent that incorporates encased Fe_3SO_4 NPs with special property of the magnetic hydrogel. They can remove about 90% of the methylene blue dye. Similar to this, a nanogel made of sodium alginate loaded with Fe_3SO_4 and tetrasodium thiacalixarene tetra sulfonate showed a strong ability to adsorb a variety of ions from water-soluble solutions, including Ni^{2+}, Co^{2+}, Pb^{2+}, Cd^{2+}, Cu^{2+}, and Cr^{2+} [60].

Opportunities for nanocomposites in agriculture and the food sector can be developed using biopolymer-based nanocomposites that can serve as nanocarriers to transfer agrochemicals to crop plants in order to boost yields. Hydrogels or nanofilms based on biopolymers can also be applied directly to products to increase their shelf life, enhance seed germination, and provide defence against pests and diseases. The packaging of food has enormous promise for biopolymers. However, when paired with nanofillers like nanoparticles, their packing qualities—including their durability or gas, water, or microbial barriers—can be noticeably enhanced. This article provides an overview of the strategic applications of nanocarriers for active ingredients, polymer-based hydrogels, nanocoatings, and nanofilms in the agricultural and food sector [67].

2.4 *Bio-Nanocomposites*

Biopolymers are combined with nanoscale additional forces, usually in the form of nanoparticles, nanotubes, and nanofibers in the 1–100 nm range, to create bio-nanocomposites, which are novel materials. By combining the special qualities of the nanoscale additives with the biopolymer molecule matrix, these composites produce materials with enhanced thermal, mechanical, and barrier qualities. These materials, which are renowned for their hybrid nature, harmoniously blend biopolymers, nanoparticles, and other elements. Bio-nanocomposites differ from conventional nanocomposite and nanohybrid materials in a number of ways, including preparation methods, properties, functionalities, biocompatibility, biological degradation, and applicability. These materials have received significant interest as possible candidates for application in medical biomaterials due to their minimal toxicity, biocompatibility, intricately designed architectures, and biodegradability. These delve into a variety of biomedical applications, such as medication administration, wound dressings, tissue engineering, dental uses, anti-anaemia, antibacterial, and biosensors. This choice is motivated by these chemicals' well-established biocompatibility and intrinsic non-toxicity. For bio-nanocomposites to perform at their best, the filler and polymers need to get along well. Choosing manufacturing processes that are appropriate for the type of composite components is crucial to fostering such a positive relationship.

Biopolymeric nanocomposites have various significant benefits and roles in water remediation procedures. First off, biopolymeric nanocomposites are ecologically friendly since they are frequently made from natural sources. Their application in water remediation is in line with methods that are sustainable, supporting environmentally and socially responsible approaches to water treatment. Biopolymeric nanocomposites have the ability to efficiently eliminate contaminants from water due to their adsorption capability. Due to their large surface area and surface components containing functional groups, biopolymeric nanocomposites have the ability to either attract or adsorb impurities from water. To make it easier for contaminants to separate from the water, they either stick to the surface or interact with the structure of the nanocomposite. Chemicals such as heavy metals, chemical substances, pigments, antibiotics, and other contaminants fall under this category. Biopolymeric nanocomposites have the additional function of being filtering agents. They can be made with particular features to trap or filter out microbes, microplastics, particulates, and other unwanted elements from water, improving the quality of the water. Certain biopolymeric nanocomposites have the ability to exchange ions. This implies that they can efficiently lower the concentration of dangerous compounds in water by exchanging ions with impurities. Moreover, a noteworthy characteristic of several biopolymeric nanocomposites is their capacity for regeneration and reuse. These nanocomposites can go through a regeneration process after adsorbing pollutants, which enables them to keep their adsorption capability for several treatment cycles. It's interesting to note that certain biopolymeric nanocomposites can be engineered to bind selectively to particular pollutants, like antibiotics or other chemicals. This selectivity increases their effectiveness in focusing on specific pollutants without changing the water's overall composition.

Chitosan biopolymers are distinguished by their remarkable gelation characteristics, biodegradability, and noteworthy antibacterial capabilities. These characteristics set them out as a model material with a wide range of uses that go beyond wastewater treatment to a number of scientific and technological domains. Chitosan is a versatile biopolymeric nanocomposite contender due to its significant positive qualities, including non-toxicity, biodegradability, antibacterial, antioxidant, and biocompatibility. Chitosan interacts with negatively charged substances or ions, thanks to the positively charged amino groups on its surface. Because of this, chitosan is finding use in a variety of scientific and technological domains in addition to the environmental ones. With their unique ability to form gels, biopolymers like chitosan, especially when combined with nanoparticles like Ag ZnO, TiO_2, etc., to form bionanocomposites or nanocomposites, have a lot of potential applications in the packaging of food, biomedical, textiles, and environmental fields (Fig. 3). Chitosan has some disadvantages in addition to its intriguing properties, such as poor selectivity for specific pollutants.

Bio-nanocomposites have gained attention for their potential to enhance crop growth and sustainability. These materials offer benefits in agriculture

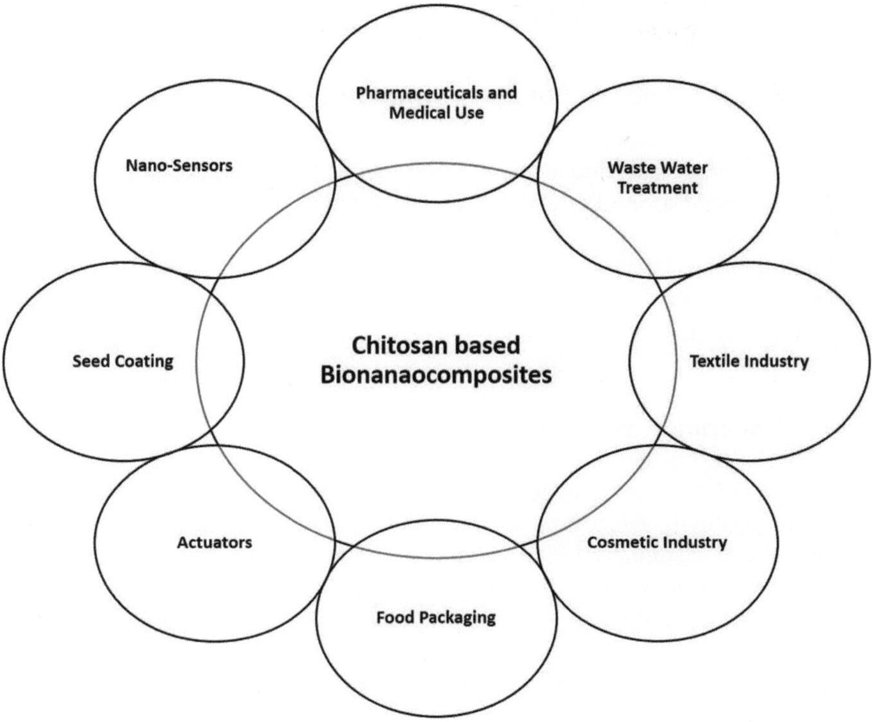

Fig. 3 Applications of role of bio-nanocomposites in different sectors

by improving nutrient delivery, protecting plants from pests and diseases, and enhancing plant resilience to environmental stresses. For instance, chitosan-based bio-nanocomposites combined with silver nanoparticles (Ag NPs) have been used in tomato (*Solanum lycopersicum*) to increase disease resistance [19], particularly against fungal infections, while also promoting better plant growth. Similarly, chitosan-silica bio-nanocomposites have shown promise in rice (*Oryza sativa),* improving plant growth, nutrient uptake, and stress tolerance under saline conditions [18]. Moreover, seaweed-derived alginate exhibits a strong affinity for metallic ions and organic contaminants due to its capacity to gel when exposed to divalent cations. It can be encapsulated and is biocompatible. Its performance may be impacted by its inability to sustain mechanical strength, stability, and the capacity to resist disintegration in harsh chemical conditions [37, 62]. Similar to this, there are advantages and disadvantages to using each type of biopolymer substrate in wastewater treatment.

3 Utility of Hybrid Nanomaterials

3.1 *Pollutant Detection in Environmental Components*

Pollution reduction and global sustainability are important concerns that need to be addressed right away. Pollutants are widely found in many different types of habitats, including energy fields, landfills, military stations, and industrial factories. This highlights the seriousness of the global situation. As a result, nanomaterials have a variety of uses, such as identifying air pollution, evaluating soil contamination, and monitoring water quality [69].

3.1.1 Monitoring of Water Quality

HNM powered by cutting-edge nanotechnological approach leads the way in improving sensitivity and accuracy in the monitoring of water quality. These hybrid materials create extremely effective and adaptable sensors by mixing different nanoparticles and nanoscale architectures [49]. Their wide range of capabilities make them possible to identify a variety of water pollutants, including organic chemicals, heavy metals, and microbes [70]. These sensors based on nanomaterials are essential for monitoring the environment, protecting aquatic ecosystems, and guaranteeing the safety of drinking water. An overview of all the metals and metal oxide nanoparticles that have been investigated by researchers for a range of environmental remediation uses is shown in Table 1.

Furthermore, the creation of nanoscale ceramic and polymer membranes is a key contribution of nanotechnology to the water treatment industry [5]. Their usage

Table 1 Applicability of hybrid nanomaterials in environmental remediation

S. no.	Nanomaterial	Application	References
1	Carbon nanotubes, carbon nanofibers	Electrode sensors for detecting heavy metal ions	[23])
2	Graphene quantum dot	Electrochemical biosensing	[9])
3	Quantum dot	Fluorescent-based sensing	[52])
4	Metal oxide NPs	VOCS	[57])
5	2D materials like MoS_2, WS2, black phosphorus	Gas sensing, chemicals, biomolecules	[14])
6	MOF, Nanocellulose-MOF	Sensor technologies	[4])
7	Carbon nanomaterials + MeNPs	PMS sensors	[17])
8	Carbon nanomaterials + Me oxides NPs + polymer	VOCS detection	[28])
9	QDs + organic dyes or plasmonic NPs	Optical sensors	[10])

during wastewater treatment is a noteworthy application in environmental detection. When compared to other approaches, the membranes employed in nanofiltration operations efficiently filter well or surface water while using less energy since they naturally resist big molecules. Many pollutants, such as bacteria, organic pesticides, and salts, can be removed from conventional water by using this procedure. Researchers have looked into the use of nanomaterials to remove oil, bacteria, and metals from contaminated water sources in an effort to address this problem. In particular, nanoparticles provide a great degree of adaptability in the remediation of pollutants. Applications for nanostructured particles are found in many environmental domains, such as liquid waste purification, water treatment, solid waste management, and soil and sediment remediation [47].

Iron-silver, iron-palladium, and zinc-palladium nanostructured bimetallic particulates have demonstrated remarkable efficacy in the treatment and purification of environmental contaminants, including organic pesticides that contain chlorination and halogenated organic solvents. Based on pore size, nano filters—which are divided into microfilters, ultrafilters, and nanofillers—are essential for energy conservation and environmental protection. Reverse osmosis operates at higher pressures than nanofiltration, which is more expensive but can remove bacteria and viruses from agricultural and human drinking water sources [56]. This protects the environment and encourages energy-saving water treatment methods. Although they offer certain improvements over conventional membranes, zeolite and catalytic membranes are not very useful for treating water. A study looked into treatment techniques based on nanotechnology, particularly membrane filtration and adsorption. Because it was easy to use and efficient at eliminating a wide range of pollutants, including poisons, bases, organic compounds, salts, and acids, adsorption was the method of choice. Adsorption efficiency is dependent on various parameters, including the pore structure of the adsorbent and how it interacts with pollutants. The investigation also looked into CNT, such as carbon nanotubes and graphene-based nanomaterials, which showed excellent removal capabilities for organic contaminants and heavy metals [54]. However, because of loose aggregation, CNT showed increased hydrophobicity and decreased surface area. These effects might be avoided by using metal oxide particles or functional groups. The study revealed that carbon nanoparticles might be repurposed without losing their adsorption ability by modifying the pH.

3.1.2 Air Pollution

Biodiversity and ecosystems are seriously threatened by air pollution, which is defined as particulates, biological components, or toxic compounds emitted within the atmosphere. It is classified as primary mode (directly released) and secondary mode (produced via chemical reactions). Activities like burning fossil fuels produce primary pollutants such CO, S, NO, and volatile organic compounds. A secondary pollutant *Peroxyacetyl nitrate* is created when main pollutants react chemically [27].

In order to address environmental issues, it is imperative to conduct ongoing monitoring of air quality. Nanotechnology presents creative answers in the form of novel materials with distinct qualities for long-term pollution control.

The increase in temperature and ecological climate change are largely caused by CO_2 emissions, and attempts are being made to create workable materials for this CO_2 capture. Carbonaceous adsorbents are seen to be promising for CO_2 capture due to their high thermal stability, hydrophobicity, adaptable pore shapes, and affordability. Toxic pollutants such as benzene, ethyl benzene, toluene, dioxin, and p-xylene present problems in industrial settings. The 2.9 nm-diameter porous wall of carbon nanotubes shows a high association with benzene dioxin rings, which increases the pore's capacity for adsorption [53]. They also exhibit strong resistance to oxidation, which is advantageous for regenerating adsorbents at high temperatures. Because of their distinct physicochemical properties, high specific surface areas, and increased reactivity, quantum dot nanomaterials show great promise as heavy metal detection tools. Robust detection systems are created by integrating them with different optical/chemical sensor converters that help to identify numerous metals.

Metal oxide nanoparticles of biological origins, such as silver, zinc, and titanium, exhibit photocatalytic activity to remove contaminants from the environment. For air purification, photocatalytic nanomaterials must be effectively immobilized [26]. This requires properties including mass transfer, stability in the presence of light, high surface area, efficient adsorption, and affordability. To summarize, the incorporation of nanotechnology in the detection of air pollution offers encouraging opportunities for effectively tackling environmental issues and advancing sustainable practises.

3.1.3 Soil Contamination Analysis

Excessive soil contamination is a result of human activity expanding due to urbanization. The investigation of nanotechnology as a vital remedy has been spurred by the widespread concern over soil remediation. One of the main causes of soil contamination is heavy metal accumulation, which occurs over time due to its resistance to biodegradation [34]. Common methods for remediating soil include oxidation, reduction processes, immobilization, adsorption, and creative combinations of bioremediation and nanotechnology. Persistent micro-pollutants can be more easily removed from soil thanks to the enhanced efficacy of adsorption and oxidation processes facilitated by nanoparticles, such as carbon nanomaterials, nanoscale zero-valent iron (nZVI), and different nanocomposites. Notably, nZVI is the method of choice for eliminating heavy metal contaminants and changing them into less dangerous forms.

3.1.4 Pesticide Detection

Colorimetric detectors may identify pesticides and bisphenol. A (phenolic substances) with relatively high analytical enactments due to biosynthesized metal-doped nanocomposites that comprise nanoparticles, nanotubes, and nanowires. They also possess applications in a variety of fields, such as optical and electrical nanodevices and sensors, as well as chemical and biological sensors. Many nanostructures have been investigated in order to learn more about their properties and their applications in biosensors. Some of the most promising near-term applications of nanotechnology can be identified in the interface between physical and biological phenomena. Therefore, a variety of synthesized metal infused nanocomposite-based colorimetric sensors were used to calorimetrically quantify chemicals and phenolic chemicals from various classes [8].

Metal-doped nanocomposites were used independently or in conjunction with additional modifiers, such as peptide nanotubes, electro spun carbon nanofibers (ECNF), covalently bonded organic framework, and AChE (acetylcholinesterase). Therefore, tests have been conducted at different metal-based nanocomposites to determine the pesticides and bisphenol. A detection in nM and pM amounts. In addition, a high degree of responsivity, measurement accuracy, and precision are seen in the recovered spiked samples. As a result, the benefits of using these nanocomposites to detect bisphenol A and pesticides are illustrated [45]. The well-established colorimetric sensing capabilities for pesticides and bisphenol. A (phenolic chemicals) will be used to examine and surpass a number of tasks in order to progress important research areas. The possibility of designing multiplexed arrays that utilize a micro-fluidic device that detected different analytes in different sensing points is a driving force for additional research. The challenge is associated with a relevant and contemporary trend in (electro) analytical chemistry, which is the lowering of excess production and chemical reagent use, the shrinking of analytical equipment, and the recommendation of portable apparatus for analysis within this field. Thus, the classes of biosynthesized metal-doped nanocomposites exhibit a high degree of promise for electro-analysis commitments, and further calorimetrical research on pesticide and bisphenol A (phenolic substances) sensing as well as biosensing may be permitted. Nonetheless, the extended storage durability of the sensors altered by metal-doped nanocomposite garners more interest from researchers, which may lead to further research in this connected topic in the future.

Similarly a sustainable preparation approach called microwave-assisted hydrothermal production was used to create CuS QD@ZnO hybrid nanocomposites, enhancing the ZnO NR photocatalyst's ability to degrade colours, medications, and pesticides—all of which are developing contaminants in water [41]. It is discovered that when the ratio of CuS is increased to 3%, the crystallization, absorption of light, and degradation efficiency all increase, and then they decrease with subsequent increases. The QDs@ZnO hybrid nanocomposite possesses the ability to reduce electron–hole recombination, thereby accelerating the degradation rate. For ZnO-NRs, the frequency gap drops from 3.1 to 2.90, improving the absorption of light and electron–hole pair formation. These HNC have shown effective photocatalytic capacity in the destruction of ciprofloxacin and ibuprofen and 2.4, 5 T pesticide. In a nutshell, CuS-QDs@ZnO nanocomposites proved to become promising and reasonably priced photocatalytic substances for decontaminating wastewater containing pesticides and medications when exposed to simulated sunlight (SL).

3.2 Development of Smart Fertilizers: Nanofertilizers

In comparison to conventional fertilizers, nanofertilizer provide a number of benefits due to their nano-sized particles. Their increased nutritional efficiency is a major advantage since the nanoscale facilitates higher plant nutrient absorption. In order to address these problems, nanofertilizers minimize nutrient losses into the environment, provide regulated nutrient release, and guarantee that plants receive the proper amount of nutrients at the right time. By improving nutrient use efficiency and lowering the amount of fertilizer needed, this tailored nutrient delivery helps to reduce pollution in the environment.

Unconventional materials can maximize agricultural input use and minimize environmental damage, providing sustainable solutions for the future [20]. Examples of these materials are nanoparticulate fertilizers and nanocomposites. By efficiently regulating nitrogen hydrolysis, lowering loss, and increasing utilization efficiency, a fertilizer made of nanocomposite materials can improve crop growth while supporting environmentally friendly and sustainable agriculture practices. Slow-release fertilizers made from NPK with improved efficiency in utilizing nutrients and ability to retain water can be created using cellulose nanocrystal-filled poly (vinyl alcohol) nanocomposites, which will help with growing crops sustainably.

For the delayed and persistent diffusion of fertilizers into soil and water, hybrid nanofertilizers (HNF) were employed. Urea-modified hydroxyapatite, a calcium, phosphate, and nitrogen-rich material, was created. To boost the effectiveness of the suggested fertilizer, urea-modified hydroxyapatite was mixed with nanomaterials of copper, iron, and zinc. Many techniques, such as Fourier-transform infrared spectroscopy (FTIR), powder X-ray powder diffraction, and scanning electron microscopy (SEM), were used to examine the size of the fertilizer produced and its morphology, The ladies' finger (*Abelmoschus esculentus*) plant was the subject of a field experiment using the created HNF [50]. Leaching investigations revealed the gradual release of HNF, which validated the presence of Ca^{2+}, Cu^{2+}, PO_4^{3-}, NO^{3-}, Zn^{2+}, and Fe^{2+}. Moreover, the inductively coupled plasma-optical emission spectrometry (ICP-OES) experiment verified the existence of Cu^{2+}, Fe^{2+}, and Zn^{2+} micronutrients in ladies' finger. The suggested fertilizer's physicochemical characteristics, such as its swelling ratio and water uptake and retention capacities, were found to have significantly increased. This makes the fertilizer more advantageous and appealing than commercial fertilizers. The suggested HNF's composition proved useful in terms of facilitating the steady and gradual delivery of nutrients to plants. *A. esculentus* received a weekly dosage of 50 mg of prepared HNF and a weekly dosage of 5 g of commercial fertilizer. The acquired data demonstrated that *A. esculentus* significantly increased its uptake of Cu^{2+}, Fe^{2+}, and Zn^{2+} nutrients as a result of gradual release from HNF.

3.3 Soil Health Monitoring

The dynamic balance of the soil's chemical, biological, and physical characteristics is known as soil health. Effective soil management necessitates a real-time assessment of the soil condition. Precision agriculture totally relies heavily on sensors for precise regulation of moisture and nutrients. Because of their higher specific surface and greater chemical reactivity, nanomaterials appear to be vital for biosensors creation [71]. Catalytic activity increases the use of nano biosensors because it produces results with high specificity, precision, stability, fast responsiveness, and unique physiochemical properties. Mercury elimination and detection are now possible with the introduction of multimodal nanosensors. A sensor device based on carbon nanodots and quantum dot particles has been created that can identify heavy metals. The use of nano biosensors to track soil enzymatic activity and urea detection was developed in a lab setting and implemented under carefully regulated conditions. Through efficient resource usage and eventually increased crop output, responsible and controlled application of nano biosensors can promote sustainable agriculture [7].

Plant growth-promoting rhizobacteria (PGPR) and nanomaterials are needed to enhance the growth of crops and soil health [63]. The impacts of two PGPR strains, nano zeolite and nano chitosan on fenugreek (*Trigonella foenum-graecum),* were studied where there was a discernible rise in morphological characteristics when fenugreek plants were treated with the nano-compounds and PGPR in comparison to the control. Moreover, it also showed highest concentrations of total chlorophyll, soluble leaf protein, sugar, and catalase. The high level of phenols, which are known to help plants cope with biotic and abiotic stressors, was discovered through GC–MS analysis of plant constituents. When nano zeolite and PS10 were coupled, the rate of activity of the enzyme fluorescein di-acetate hydrolase was 2.5 times more than in the control. In comparison to the control, there was an 11% rise in alkaline phosphatase activity in the same therapy [32]. The pot experiment results unmistakably show that PGPR and nano-compounds enhanced plant growth and soil health, suggesting the potential benefits of using these substances in agricultural practices to boost crop yields.

3.4 Role of Hybrid Nanocomposites in Enhancing Soil Structure

Enhancing the soil's composition is a major function of hybrid nanocomposites, which also improves soil health and agricultural output. These cutting-edge materials have special qualities that favourably affect the physical properties of soil because they blend organic and inorganic components at the nanoscale. Improving soil aggregation is one of the main ways hybrid nanocomposites improve soil structure. By serving as binding agents, they encourage the development of stable soil aggregates that are resistant to compaction and erosion [72]. Improved root penetration and water absorption are made possible by the increased aeration and porosity of the soil caused by this improved aggregation. Increased availability to nutrients and water benefits crops result in stronger development and higher yields. Additionally, hybrid nanocomposites can aid in absorbing and releasing of moisture from the soil. Because of their large surface area and hydrophilic qualities, which allow them to hold onto water and release it gradually, they can provide plants with a steady supply of moisture even in dry spells. This feature is especially helpful in dry and semi-arid areas where a main problem is a lack of water. Such nanocomposites can also improve the soil's cation exchange capacity (CEC), which increases the soil's capability to hold onto vital nutrients and increases their availability to plants [44]. Because of this nutrient retention, less frequent fertilization is required, encouraging more environmentally friendly farming methods.

3.5 Pathogen Suppression

Hybrid nanoparticles are demonstrating themselves to be an innovative instrument in combating against soil-borne pathogens, providing novel approaches to the control of agricultural illnesses and improvement of crop health [48]. These nanoparticles have special qualities that allow for targeted and effective disease control because they blend organic and inorganic elements at the nanoscale. The antibacterial property of hybrid nanoparticles is one of the main ways in which they inhibit infections. A lot of hybrid nanoparticles are made to release metal ions, such as copper or silver, which are known to have antibacterial qualities. These ions cause microbial cellular functions to malfunction, which renders them inactive or kills them. This focused intervention lowers the prevalence of infections brought on by fungus, bacteria, and viruses that live in the soil. Additionally, antimicrobial drugs can be delivered to infection locations directly by means of hybrid nanoparticles, thereby reducing the negative effects on beneficial soil microbes and preserving ecological equilibrium [24]. The modification techniques enable nanoparticles to detect and bind to certain diseases, enabling selective targeting. After adhering, the nanoparticles can gradually release their antibacterial payload, maintaining long-term infection suppression. Furthermore, because of their unique chemical compositions, certain hybrid nanoparticles have built-in antifungal and antibacterial qualities, which increases their efficacy in pathogen management.

The capacity of hybrid nanoparticles to pierce biofilms—protective coatings created by microbial communities—is another important benefit [16, 46]. Pathogens are difficult to eliminate because biofilms frequently protect them from standard therapies. By breaking up these biofilms and exposing the bacteria to antimicrobial chemicals, hybrid nanoparticles can increase the treatment's overall effectiveness [35]. Hybrid nanoparticles can also strengthen plants' natural defence against diseases by causing systemic resistance in plants through soil application. This generated resistance lessens the overall effect of illnesses on crop output by enabling plants to endure infections and recover more quickly.

Hybrid nanoparticles encourage ecologically friendly farming because they reduce the need for synthetic pesticides, which have detrimental effects on both the ecosystem and human health. They also promote crop safety and environmental sustainability in integrated pest control techniques by providing a more focused and environmentally friendly method of disease suppression. The complete potential in pathogen suppression is still being investigated through ongoing research and field testing, with the goal of perfecting their formulations and application techniques for optimal effectiveness and least possible negative environmental impact.

3.6 Microbial Stimulation

In agricultural soils, hybrid nanoparticles are becoming more and more potent stimulants of microbiological activity, providing potentially beneficial ways to improve crop productivity and soil fertility. These nanoparticles have special qualities that can favourably affect microbial activity and community dynamics because they combine organic and inorganic components at the nanoscale. Essential nutrients are provided by hybrid nanoparticles, which is one of the main ways they promote microbial activity [66]. Hybrid nanoparticles can function as slow-release fertilizers by integrating nutrients like K, P, and N into their structure. This allows the nutrients to be released over time and provide a consistent supply of nutrients that support bacteria development and growth. The growth of advantageous soil microorganisms, such as nitrogen-fixing bacteria and mycorrhizal fungi, which are crucial to the cycling of nutrients and the health of the soil, is aided by this continuous release of nutrients. Hybrid nanoparticles can also increase the nutrients' absorption in the soil, increasing their accessibility to microbial populations [66]. These nanoparticles' large surface area and sites of reaction aid in absorbing and subsequent release of nutrients, supplying soil microbes with the materials they require for growth and metabolic processes. The increased microbial biomass and diversity that can result from this improved nutrient availability are important markers of productive and healthy soils. The capacity of hybrid nanoparticles to alter the physicochemical characteristics of the soil is a noteworthy additional method by which they promote microbial activity. These nanoparticles enhance the soil's structure and porosity, which makes the environment more conducive to microbial colonization and activity. Hybrid nanoparticles' improved soil permeability and water retention create ideal conditions for microbial development and function, which raises soil fertility overall.

Additionally, hybrid nanoparticles have the ability to come into contact with microbial cells to improve their metabolic processes. Certain nanoparticles can enhance microbial respiration and energy production by serving as electron donors or acceptors. This interaction stimulates the growth and activity of specific microbial populations, leading to increased rates of organic matter decomposition, nutrient mineralization, and overall soil fertility [15]. Hybrid nanoparticles can also trigger advantageous microbial reactions, such as the synthesis of hormones that stimulate plant growth and enzymes that improve plant uptake of nutrients and stress tolerance. Although hybrid nanoparticles reduce the need for chemical pesticides and fertilizers, which can damage beneficial microbial ecosystems, they promote sustainable agriculture. Hybrid nanoparticles encourage the development and activity of advantageous soil microbes, which strengthens and balances the soil ecology. Hybrid nanoparticles have an inclusive range of uses in microbial stimulation, and research is still being done to better understand these uses and develop formulations and application techniques that will maximize agricultural advantages [21]. This novel method has enormous potential for boosting crop output, maintaining agricultural systems' long-term sustainability, and increasing soil health.

4 Hybrid Nanomaterials' Sustainable Advantages Compared to Conventional Detection Techniques

Using hybrid nanomaterials to address a range of resource efficiency and environmental issues is a major step towards achieving sustainability goals. These nanoparticles offer clear benefits over conventional detection techniques, resulting in lower resource usage, increased energy economy, and decreased waste production.

4.1 Decreased Use of Resources

4.1.1 Material Efficiency

By precisely controlling composition and characteristics at the nanoscale, hybrid nanomaterials minimize the demand for bulk materials [36]. Comparing smaller amounts of nanoparticles to greater amounts of traditional materials can yield outcomes that are either equal to or better than the former.

4.1.2 Utilization of Raw Material

Hybrid nanomaterial synthesis typically uses readily available, low-impact components, as opposed to the resource-intensive procedures needed to extract and synthesize many conventional elements.

4.2 Efficiency in Energy Use

4.2.1 Low Energy Synthesis

In order to fabricate hybrid nanomaterials, energy-efficient processes like chemical vapour deposition or sol–gel synthesis are usually used, which use less energy than conventional production procedures.

4.2.2 Enhanced Properties

High surface area and improved conductivity are two examples of hybrid nanomaterials' superior qualities, which can result in more energy-efficient procedures and equipment like catalysts and sensors.

4.3 Minimal Production of Waste

4.3.1 Accurate Manufacturing

By decreasing flaws or unwanted byproducts, the accurate synthesis of nanomaterials reduces waste and improves the usage of raw materials.

4.3.2 Recyclability

By making a lot of hybrid nanomaterials recyclable, they help to further reduce waste output and advance a circular economy.

4.4 *Better Selectivity and Sensitivity*

4.4.1 Decreased Sample Size

Smaller amounts of sample are often required for hybrid nanomaterial-based detectors, which reduces the quantity of material or substance needed. This is especially helpful in situations where sample availability is limited, such as monitoring the environment applications.

4.4.2 Enhanced and Early Detection

High surface-to-volume ratio and customized surface chemistry are two distinctive features of hybrid nanomaterials that improve sensitivity and selectivity and reduce false positives and negatives for more precise readings.

4.5 *Durability and Extended Life*

4.5.1 Improved Life Expectancy

Eco-friendly devices utilizing hybrid nanomaterials, like catalysts and sensors, usually have a greater lifespan and involve less repair and replacement. By increasing source efficiency, preserving energy, and improving environmental routine in a variety of applications, hybrid nanomaterials help to achieve sustainability goals [38]. They are essential to creating a more sustainable future because they offer a viable path forward for technological advancement while reducing the negative effects of scientific and industrial operations on the environment [61]. The development, application, and possible effects on several facets of society and the environment are all factors to be taken into account while discussing the ethics of hybrid nanoparticles. Important ethical factors to think about include the following:

Safety

It is vital to studying the safety of hybrid nanoparticles, particularly in relation to possible harm to people and ecosystems. Thorough testing and risk evaluations are part of ethical practices in order to reduce potential harm.

Environmental Impact

Due to their persistence in ecosystems, capacity for bioaccumulation, and possible influence on biodiversity, hybrid nanoparticles may have a detrimental effect on the environment. Ethical concerns entail evaluating and reducing these effects in order to reduce environmental harm.

Regulation

To control the creation, manufacturing, and application of hybrid nanoparticles, ethical standards and laws are required. Throughout their existence, regulations should guarantee openness, responsibility, and adherence to safety requirements.

Equity and Access

Ensuring equitable access to hybrid nanoparticles and their advantages for various people and geographical areas is a key component of equity considerations. Taking care of potential gaps in access to technology and its advantages is a component of ethical practices.

Privacy and Data Security

Safeguarding individuals' rights to privacy and guaranteeing the secure processing of data gathered from individuals or communities are ethical considerations in applications employing hybrid nanoparticles, such as sensors and monitoring systems.

Social Impacts

The social effects of hybrid nanoparticles, such as their capacity to worsen already-existing inequities or generate new ones, are likewise covered by ethical issues. In order to further justice and fairness, ethical practices entail evaluating and addressing these social ramifications.

Responsible Innovation

In the creation and application of hybrid nanoparticles, ethical principles encompass responsible innovation. This entails taking into account the wider societal ramifications of their use and including stakeholders in the decision-making process.

Thus, in order to ensure the ethical creation and use of hybrid nanoparticles in a variety of applications, it is necessary to address the ethics of these particles in a comprehensive manner that takes into account their safety, environmental impact, regulation, equity, privacy, and social impacts.

5 Challenges and Considerations

Hybrid nanomaterials are revolutionizing a number of industries, such as environmental monitoring, the agricultural sector, and medicine. To optimize advantages and minimize dangers, a number of issues and concerns related to their deployment must be taken into account. Knowing the possible effects, these materials have an adverse effect on environment and human health (Table 2). Hybrid nanomaterials might interact with biological processes in unexpected ways because of their unique features and nanoscale size, which raises questions about their toxicity and potential long-term impacts on the environment and human health [43]. According to studies, certain nanoparticles can build up in the environment and within living things, which may have negative consequences like inflammatory responses, cellular damage, and disturbance of natural ecosystems. Thus, thorough toxicological evaluations and ecological impact studies are crucial for their safe use [33].

Another major obstacle is the expense and scalability of using hybrid nanomaterials. Despite the well-established advantages of these materials, their widespread acceptance may be hampered by their high production costs and the difficulty of scaling up manufacturing methods. The development of economical synthesis techniques and enhancement of production efficiency are crucial measures in enabling the use of hybrid nanomaterials in large-scale applications [2]. Furthermore, for these materials to be successfully integrated into a variety of industries, it is essential that they can be produced reliably and in large enough quantities to meet demand. The creation and application of hybrid nanomaterials are significantly influenced by ethical and regulatory factors as well. Regulations must change as nanotechnology develops to take into account the special qualities and hazards of nanoparticles.

This entails defining standards for their safe manufacture, handling, and disposal in addition to imposing upper bounds on the amount of exposure that the general

Table 2 Challenges associated with hybrid nanoparticles

Aspect	Current state	Future directions
Scalability	Limited large-scale production	Scalable manufacturing processes and cost-effectiveness
Environmental impact	Potential toxicity and environmental risks	Eco-friendly materials and sustainable production practices
Regulatory and safety	Emerging regulations and safety assessments	Comprehensive regulations and standardized safety protocols

public and employees are permitted to experience. Ensuring that the benefits of nanotechnology are distributed equitably—without unfairly favouring certain groups or disadvantaging others—is an important ethical concern. This calls for a well-rounded strategy that encourages responsible innovation while accounting for the possible societal effects of these materials.

Hybrid nanomaterials' acceptance and perception by the general population are also essential to their effective use. Active involvement with stakeholders and open communication about the advantages and hazards of nanotechnology can help allay fears and foster confidence. Notifying the public and policymakers on the possibilities of hybrid nanomaterials and the significance of reliable and ethical methods in their development can also be accomplished through outreach efforts.

6 Future Prospects and Innovations

With their integration with precision agriculture, use in innovative technologies, and benefits to long-term sustainability, hybrid nanomaterials are at the cutting edge of technological innovation and present exciting opportunities for the near future. These materials are anticipated to propel major breakthroughs in a variety of fields as research continues, transforming sectors and solving global issues.

One important area where hybrid nanomaterials are expected to grow is in emerging technologies. In order to produce materials with improved qualities and capabilities, scientists are investigating cutting-edge synthesis methods and functionalization tactics. The creation of multifunctional hybrid nanomaterials, which combine many characteristics like targeted medication delivery, environmental sensing, and pollution removal, is one exciting avenue. More sophisticated methods of nanofabrication, such as template-assisted and self-assembly, are making it possible to precisely create complex nanostructures with customized features [13]. Moreover, nanomaterials that mimic natural processes are being developed for novel uses through biomimetic design, which draws inspiration from natural systems. In this area, artificial intelligence and machine learning are also revolutionizing the field by analysing large datasets to anticipate material behaviours, expedite the design process, and find new uses for hybrid nanomaterials. In line with the increasing focus on environmentally conscious innovation, green chemistry techniques are being used to create more environmentally friendly synthesis processes that lessen their influence on the environment and dependency on dangerous materials.

The use of hybrid nanomaterials into precise agriculture has encouraging prospects for contemporary farming methodologies. Hybrid nanomaterials are at the forefront of precision agriculture, a sector that optimizes crop yield and resource utilization using data-driven methodologies. Hybrid nanomaterial-based smart sensors are able to measure crop health, monitor soil conditions, and identify environmental changes in real time. With the precise data these sensors provide on variables like soil moisture, nutrient levels, and pest infestations, farmers are better equipped to make decisions and implement solutions. Furthermore, a notable breakthrough in controlled release fertilizers is the creation of nano-enabled fertilizers, which permit the release of nutrients gradually in response to environmental cues such soil moisture and pH [22]. Crop yields are increased and nutrient waste is decreased with this focused approach. In order to minimize the influence on beneficial soil bacteria and non-target creatures while delivering pesticides selectively to pest populations, hybrid nanomaterials are also being investigated for use in targeted pest management. Hybrid nanomaterials are also essential for soil remediation, where they aid in the detoxification of contaminated soils by decomposing organic contaminants and adhering heavy metals, thereby improving the fertility of the soil and health.

In the long run, hybrid nanomaterials will play a significant role in sustainability and have a significant worldwide influence. These resources provide cutting-edge answers to some of the most important social and environmental problems. Sustainable development objectives are supported by hybrid nanomaterials because they enhance resource efficiency in industry, agriculture, and environmental management. They are additionally utilized in climate change mitigation technologies, such as carbon capture and storage systems, which aid in the fight against global warming by lowering greenhouse gas emissions. Furthermore, hybrid nanomaterials hold great promise for improving food security and global health. Applications include water purification, illness detection, and raising agricultural yield using cutting-edge insect and fertilizer techniques. These developments contribute to the global health goals of ensuring access to clean water and boosting food security. Another major area of concentration is sustainable development, with research going towards creating environmentally friendly nanomaterials that are biodegradable and have negative consequences on the environment. Future success of hybrid nanomaterials depends on ensuring that they support sustainability without endangering ecosystems or human health [64]. This entails creating materials with the minimal negative impact on the environment, and encouraging moral behaviour in the process of creation and use. Thus a combination of opportunity and innovation will define the future of hybrid nanomaterials. The capabilities of these materials are being expanded by research and emerging technologies, and their use into precision agriculture marks a significant development in sustainable agricultural methods. Hybrid nanomaterials are essential to long-term sustainability initiatives because of their potential to address global issues like resource efficiency, food security, and climate change. Hybrid nanomaterials are expected to be crucial in forming a more robust and sustainable future for society and industry as the field develops. Continuous research, innovative problem-solving, and a dedication to moral behaviour will be necessary for their development and implementation in order to guarantee that their advantages are achieved in a way that promotes environmental wellness as well as global well-being.

7 Conclusion

Use of hybrid nanomaterials is becoming a promising approach in enhancing crop yield, monitoring pollutants, toxins, and pathogens by improving sensitivity, selectivity, and efficiency. In agriculture, hybrid nanomaterials—which include organic and inorganic components—are being utilized more and more to boost crop development, keep pests away from plants, and enhance soil health. By acting as slow-release fertilizers, reducing the necessity for chemical pesticides, and improving nutrient delivery efficiency, these materials can support sustainable farming techniques. They can also be designed to recognize and react to developments in the environment, which will maximize agricultural output. Due to their adaptability to a variety of detection methods, such as sensors and biosensors, they can be used in a variety of environmental contexts and provide real-time data for early warning and monitoring systems. Furthermore, these nanomaterials are made to be affordable, non-toxic, and environmentally benign. With the ability to create materials with many functions for all-encompassing monitoring, hybrid nanomaterials have great potential for agricultural and environmental identification in future generations. Sensing performance can be further improved by investigating novel combinations, such as combining 2D materials like graphene and carbon nanotubes. Environmental effect concerns are addressed by research on biodegradable nanomaterials; hybrid nanomaterials have the ability to both enhancing the plant growth and detecting and remediate environmental problems. As the sector develops, it will be crucial to addressing ethical and regulatory issues, such as data security and privacy.

References

1. Ahmad F, Saeed Q, Shah SMU, Gondal MA, Mumtaz S (2022) Environmental sustainability: challenges and approaches. In: Natural resources conservation and advances for sustainability, pp 243–270
2. Ahmed SF, Mofijur M, Rafa N (2022) Green approaches in synthesising nanomaterials for environmental nanobioremediation: technological advancements, applications, benefits and challenges. Environ Res 204:111967
3. Ali H, Khan E, Ilahi I (2019) Environmental chemistry and ecotoxicology of hazardous heavy metals: environmental persistence, toxicity, and bioaccumulation. J Chem
4. Ashour RM, Abdel-Magied AF, Wu Q, Olsson RT, Forsberg K (2020) Green synthesis of metal-organic framework bacterial cellulose nanocomposites for separation applications. Polymers 12(5):1104
5. Asif MB, Zhang Z (2021) Ceramic membrane technology for water and wastewater treatment: a critical review of performance, full-scale applications, membrane fouling and prospects. Chem Eng J 418:129481
6. Baz H, Creech M, Chen J, Gong H, Bradford K, Huo H (2020) Water-soluble carbon nanoparticles improve seed germination and post-germination growth of lettuce under salinity stress. Agronomy 10(8):1192
7. Bharti A, Jain U, Chauhan N (2024) From lab to field: nano-biosensors for real-time plant nutrient tracking. Plant Nano Biol 100079

8. Boher S, Ullah R, Tuzen M, Saleh TA (2023) Metal doped nanocomposites for detection of pesticides and phenolic compounds by colorimetry: trends and challenges. OpenNano 13:100168
9. Campuzano S, Yáñez-Sedeño P, Pingarrón JM (2019) Carbon dots and graphene quantum dots in electrochemical biosensing. Nanomaterials 9(4):634
10. Caroleo F, Magna G, Naitana ML (2022) Advances in optical sensors for persistent organic pollutant environmental monitoring. Sensors 22(7):2649
11. Chaudhary P, Fatima F, Kumar A (2020) Relevance of nanomaterials in food packaging and its advanced future prospects. J Inorg Organomet Polym Mater 30(12):5180–5192
12. Cooley K, Lynn V (2018) Nanotechnology: Principles and practices. scientific e-resources
13. Das PP, Chaudhary V, Singh RK, Singh D, Bachchan AA (2021) Advancement in hybrid materials, its applications and future challenges: a review. Mater Today: Proc 47:3794–3801
14. Donarelli M, Ottaviano L (2018) 2D materials for gas sensing applications: a review on graphene oxide, MoS_2, WS_2 and phosphorene. Sensors 18(11):3638
15. Enerijiofi K, Musa S, Igiebor F (2024) Nanotechnological approaches for enhanced microbial activities, sustainable agriculture, and the environment. Sustain Agric Nanotechnol Biotechnol Crop Prod Protect 153
16. Fatima F, Siddiqui S, Khan WA (2021) Nanoparticles as novel emerging therapeutic antibacterial agents in the antibiotics resistant era. Biol Trace Elem Res 199(7):2552–2564
17. García-Aguilar J, Miguel-García I, Berenguer-Murcia Á, Cazorla-Amorós D (2014) Single wall carbon nanotubes loaded with Pd and NiPd nanoparticles for H_2 sensing at room temperature. Carbon 66:599–611
18. Gatahi D (2017) Use of bio-nanocomposites in enhancing bacterial wilt plant resistance, and water conservation in greenhouse farming tomato production
19. Giri VP, Pandey S, Kumari M (2022) Hybridization of chitosan and biosynthesized silver nanoparticles to enhance antimicrobial activity against phytopathogens in tomato (Solanum lycopersicum). ACS Agric Sci Technol 2(4):719–733
20. Gomiero T, Pimentel D, Paoletti MG (2011) Environmental impact of different agricultural management practices: conventional versus. organic agriculture. Crit Rev Plant Sci 30(1–2):95–124
21. Goud BR, Raghavendra M, Prasad PS (2022) Sustainable management and restoration of the fertility of damaged soils. Agric Issues Policies 113
22. Goyal V, Rani D, Ritika MS, Deng C, Wang Y (2023) Unlocking the potential of nano-enabled precision agriculture for efficient and sustainable farming. Plants 12(21):3744
23. Han X, Meng Z, Zhang H, Zheng J (2018) Fullerene-based anodic stripping voltammetry for simultaneous determination of Hg (II), Cu (II), Pb (II) and Cd (II) in foodstuff. Microchim Acta 185:1–9
24. Harish V, Tewari D, Gaur M (2022) Review on nanoparticles and nanostructured materials: bioimaging, biosensing, drug delivery, tissue engineering, antimicrobial, and agro-food applications. Nanomaterials 12(3):457
25. Hashim A, Fatima F (2024) Exploring the transformative potential of hybrid nanoparticles in biomedical applications: relevance of hybrid nanoparticles technological applications of nano-hybrid composites. IGI Global, pp 226–246
26. Jaison A, Mohan A, Lee Y-C (2023) Recent developments in photocatalytic nanotechnology for purifying air polluted with volatile organic compounds: effect of operating parameters and catalyst deactivation. Catalysts 13(2):407
27. Kang Y, Wang Y, Cheng M (2024) Peroxyacetyl nitrate can be used as a comprehensive indicator of air pollution complex. Environ Pollut 123905
28. Kaushik A, Kumar R, Arya SK, Nair M, Malhotra B, Bhansali S (2015) Organic–inorganic hybrid nanocomposite-based gas sensors for environmental monitoring. Chem Rev 115(11):4571–4606
29. Khan I, Saeed K, Khan I (2019) Nanoparticles: properties, applications and toxicities. Arab J Chem 12(7):908–931

30. Kumar A, Nath K, Parekh Y, Enayathullah MG, Bokara KK, Sinhamahapatra A (2021) Antimicrobial silver nanoparticle-photodeposited fabrics for SARS-CoV-2 destruction. Colloid Interface Sci Commun 45:100542
31. Kumar R, Joanni E, Sahoo S (2022) An overview of recent progress in nanostructured carbon-based supercapacitor electrodes: from zero to bi-dimensional materials. Carbon 193:298–338
32. Kumari S, Sharma A, Chaudhary P, Khati P (2020) Management of plant vigor and soil health using two agriusable nanocompounds and plant growth promotory rhizobacteria in Fenugreek. 3 Biotech 10(11):461
33. Lebre F, Chatterjee N, Costa S (2022) Nanosafety: an evolving concept to bring the safest possible nanomaterials to society and environment. Nanomaterials 12(11):1810
34. Li C, Zhou K, Qin W (2019) A review on heavy metals contamination in soil: effects, sources, and remediation techniques. Soil Sediment Contam Int J 28(4):380–394
35. Li X, Chen D, Xie S (2021) Current progress and prospects of organic nanoparticles against bacterial biofilm. Adv Coll Interface Sci 294:102475
36. Liu M, Zhao Z, Duan X, Huang Y (2019) Nanoscale structure design for high-performance Pt-based ORR catalysts. Adv Mater 31(6):1802234
37. Matos GS, Pereira SG, Genisheva ZA, Gomes AM, Teixeira JA, Rocha CM (2021) Advances in extraction methods to recover added-value compounds from seaweeds: sustainability and functionality. Foods 10(3):516
38. Mazari SA, Ali E, Abro R (2021) Nanomaterials: applications, waste-handling, environmental toxicities, and future challenges–a review. J Environ Chem Eng 9(2):105028
39. Mijinyawa AH, Durga G, Mishra A (2022) Preparation and application of chitosan-based membrane: focusing on dye removal. Membrane based methods for dye containing wastewater: recent advances, pp 121–179
40. Moale C, Ghiurea M, Sîrbu CE (2021) Effects of siliceous natural nanomaterials applied in combination with foliar fertilizers on physiology, yield and fruit quality of the apricot and peach trees. Plants 10(11):2395
41. Mohammed R, Ali MEM, Gomaa E, Mohsen M (2022) Copper sulfide and zinc oxide hybrid nanocomposite for wastewater decontamination of pharmaceuticals and pesticides. Sci Rep 12(1):18153
42. Moshood TD, Nawanir G, Mahmud F, Mohamad F, Ahmad MH, Abdulghani A (2022) Biodegradable plastic applications towards sustainability: a recent innovations in the green product. Clean Eng Technol 6:100404
43. Pérez-Hernández H, Pérez-Moreno A, Sarabia-Castillo C et al (2021) Ecological drawbacks of nanomaterials produced on an industrial scale: collateral effect on human and environmental health. Water Air Soil Pollut 232:1–33
44. Rahimzadeh S, Ghassemi-Golezani K (2022) Biochar-based nutritional nanocomposites altered nutrient uptake and vacuolar H+-pump activities of dill under salinity. J Soil Sci Plant Nutr 22(3):3568–3581
45. Rajabnejad S-H, Badibostan H, Verdian A, Karimi GR, Fooladi E, Feizy J (2020) Aptasensors as promising new tools in bisphenol a detection-an invisible pollution in food and environment. Microchem J 155:104722
46. Rameez M, Khan N, Ahmad S, Ahmad MM (2024) Bionanocomposites: a new approach for fungal disease management. Biocatalysis Agric Biotechnol 103115
47. Rani M, Yadav J, Chaudhary S, Shanker U (2021) An updated review on synthetic approaches of green nanomaterials and their application for removal of water pollutants: current challenges, assessment and future perspectives. J Environ Chem Eng 9(6):106763
48. Ray MK, Mishra AK, Mohanta YK (2023) Nanotechnology as a promising tool against phytopathogens: a futuristic approach to agriculture. Agriculture 13(9):1856
49. Rebber M, Willa C, Koziej D (2020) Organic–inorganic hybrids for CO_2 sensing, separation and conversion. Nanoscale Horizons 5(3):431–453

50. Reshma Anjum M, Maheswari J, Anusha K, Sravya B, Narasimha G, Abd-Elsalam KA (2023) maximizing crop yield with macro and micro nano enhanced fertilizers nanofertilizers for sustainable agroecosystems: recent advances and future trends. Springer, pp 1–33
51. Rofeal M, Abdelmalek F, Steinbüchel A (2022) Naturally-sourced antibacterial polymeric nanomaterials with special reference to modified polymer variants. Int J Mol Sci 23(8):4101
52. Sargazi S, Fatima I, Kiani MH (2022) Fluorescent-based nanosensors for selective detection of a wide range of biological macromolecules: a comprehensive review. Int J Biol Macromol 206:115–147
53. Sebai W, Ahmad S, Brun N (2023) Treatment of wastewater containing pharmaceutical micropollutants by adsorption under flow in highly porous carbon monoliths. Chem Mater 35(20):8464–8482
54. Sheoran K, Kaur H, Siwal SS, Saini AK, Vo D-VN, Thakur VK (2022) Recent advances of carbon-based nanomaterials (CBNMs) for wastewater treatment: synthesis and application. Chemosphere 299:134364
55. Shojaei TR, Salleh MAM, Tabatabaei M (2019) Applications of nanotechnology and carbon nanoparticles in agriculture Synthesis, technology and applications of carbon nanomaterials. Elsevier, pp 247–277
56. Singh R, Bhadouria R, Singh P, Kumar A, Pandey S, Singh VK (2020) Nanofiltration technology for removal of pathogens present in drinking water Waterborne pathogens. Elsevier, pp 463–489
57. Stucchi M, Bianchi CL, Pirola C (2016) Copper NPs decorated titania: a novel synthesis by high energy US with a study of the photocatalytic activity under visible light. Ultrason Sonochem 31:295–301
58. Sudha PN, Sangeetha K, Vijayalakshmi K, Barhoum A (2018) Nanomaterials history, classification, unique properties, production and market Emerging applications of nanoparticles and architecture nanostructures. Elsevier, pp 341–384
59. Sun W, Wu H, Xu Z, Li C, Qian X, Chen L (2018) Adsorption of heavy metal ions by carbon-nanofibers-blended carbon nanotubes. ChemistrySelect 3(44):12410–12414
60. Talwar A, Anand S, Nayyar A, Fatima F, Zahera M (2024a) Hybrid Nanomaterials: a sustainable tool to detect environmental problems technological applications of nano-hybrid composites. IGI Global, pp 63–96
61. Talwar A, Nayyar A, Anand S, Zahera M, Fatima F (2024b) A futuristic approach on the multifunctionality of nanomaterials: relevance of nanoparticles technological applications of nano-hybrid composites. IGI Global, pp 1–36
62. Tiwari A, Pandey K, Singh S, Chawla S (2023) Grass and their waste products for nanoparticles synthesis and applications. In: Husen A (eds) Nanomaterials from agricultural and horticultural products. Smart nanomaterials technology. Springer, Singapore
63. Upadhyay SK, Rajput VD, Kumari A (2023) Plant growth-promoting rhizobacteria: a potential bio-asset for restoration of degraded soil and crop productivity with sustainable emerging techniques. Environ Geochem Health 45(12):9321–9344
64. Ur Rahim H, Qaswar M, Uddin M, Giannini C, Herrera ML, Rea G (2021) Nano-enable materials promoting sustainability and resilience in modern agriculture. Nanomaterials 11(8):2068
65. Vidu R, Matei E, Predescu AM (2020) Removal of heavy metals from wastewaters: a challenge from current treatment methods to nanotechnology applications. Toxics 8(4):101
66. Wang Z, Yue L, Dhankher OP, Xing B (2020) Nano-enabled improvements of growth and nutritional quality in food plants driven by rhizosphere processes. Environ Int 142:105831
67. Wypij M, Trzcińska-Wencel J, Golińska P, Avila-Quezada GD, Ingle AP, Rai M (2023) The strategic applications of natural polymer nanocomposites in food packaging and agriculture: chances, challenges, and consumers' perception. Front Chem 10:1106230
68. Xiao J, Han J, Zhang C, Ling G, Kang F, Yang QH (2022) Dimensionality, function and performance of carbon materials in energy storage devices. Adv Energy Mater 12(4):2100775
69. Yadav N, Garg VK, Chhillar AK, Rana JS (2021) Detection and remediation of pollutants to maintain ecosustainability employing nanotechnology: a review. Chemosphere 280:130792

70. Zahera M, Mishra K, Fatima F (2024) Exploiting the bioprospecting potential of nanoparticles with micro-organisms the nanotechnology driven agriculture. CRC Press, pp 245–263
71. Zain M, Ma H, Chaudhary S (2023) Nanotechnology in precision agriculture: advancing towards sustainable crop production. Plant Physiol Biochem 108244
72. Zhou L, Monreal CM, Xu S (2019) Effect of bentonite-humic acid application on the improvement of soil structure and maize yield in a sandy soil of a semi-arid region. Geoderma 338:269–280

Carbon-Based Nanocomposites for Environmental Cleaning

Djadjiti Namla, Mohammad Oves, Mohammad Omaish Ansari, Majed Al-Shaeri, and George Mangse

Abstract Human activities continuously introduce pollutants into the environment, posing risks to human health, as well as that of microorganisms and animals. These pollutants, pervasive in various environments, stem from industries like pharmaceuticals, textiles, and thermal power generation, yielding hazardous chemicals that persist and accumulate. Consequently, there arises a pressing need for a reliable, cost-effective, and environmentally sustainable solution to mitigate these harmful substances from air, soil, and water, ensuring human health. One promising strategy for environmental remediation involves employing carbon-based nanocomposites (CNCs). These materials have gained attention due to their exceptional efficiency in pollutant removal, owing to their unique properties such as high surface area, adjustable porosity, and remarkable adsorption capacity. Moreover, their versatility and adaptability make them suitable for diverse environmental conditions and various pollutants. Notably, recent advancements in synthesis processes have made these nanocomposites more cost-effective and scalable for production. This chapter delves into the multifaceted applications of CNCs in environmental remediation. It discusses various synthesis approaches, emphasizing their efficacy in soil, water, and air pollutant remediation. Furthermore, it explains the photocatalytic degradation mechanism and explores the applications of CNCs in oil spill cleanup, wastewater treatment, and air filtration, addressing both organic and inorganic contaminant removal. Finally, the chapter provides an overview of the opportunities, challenges, and future outlooks for using carbon-based nanocomposites in environmental cleanup efforts.

D. Namla · M. Al-Shaeri
Department of Biological Sciences, Faculty of Sciences, King Abdulaziz University, Jeddah, Saudi Arabia

M. Oves (✉)
Center of Excellence in Environmental Studies, King Abdulaziz University, Jeddah, Saudi Arabia
e-mail: mmateenuddin@kau.edu.sa; owais01m@gmail.com

M. O. Ansari
Center of Nanotechnology, King Abdulaziz University, Jeddah, Saudi Arabia

D. Namla · G. Mangse
Department of Biotechnology, Faculty of Sciences, Nile University of Nigeria, Abuja, Nigeria

L. F. Fraceto et al. (eds.), *Emerging Nanotechnologies for Agroecosystem Management*, Smart Nanomaterials Technology, https://doi.org/10.1007/978-981-95-0187-8_11

Keywords Pollution · Nanocomposites · Carbon · Nanotechnology · Environmental remediation

1 Introduction

Human activities across various industries introduce several pollutants into the environment [70], posing significant risks to human health and ecosystem stability [102]. These pollutants' persistent and accumulative nature exacerbate environmental degradation, necessitating urgent remediation measures [83]. In response to this pressing need, many techniques have emerged as promising approaches for efficient and sustainable environmental cleanup including carbon-based nanocomposites (CNCs). The inherent properties of CNCs, such as their tunable surface chemistry and nanoscale morphology, enable tailored approaches for pollutant capture and degradation [64]. They possess unique properties, including high surface area, adjustable porosity [27, 49, 93, 103], and remarkable adsorption capacity, making them highly effective in pollutant removal from air, soil, and water [10, 14, 81]. Moreover, recent advancements in synthesis techniques have enhanced the scalability and cost-effectiveness of CNC production [34], further bolstering their applicability in environmental remediation efforts [13].

The versatility of CNCs extends beyond pollutant removal to encompass applications in diverse environmental matrices and conditions [32]. These materials are used in environmental surveillance, healthcare, forensic science, and the identification of minute antibiotics and diverse environmental pollutants [87]. Also, they are used in cancer identification, glucose detection, and the creation of optical biosensors for forensic investigations [107]. Whether deployed in contaminated soils, polluted water bodies, or airborne particulate matter [32], CNCs demonstrate versatility and effectiveness in addressing various pollution scenarios [69]. From adsorbing organic contaminants to catalyzing the degradation of hazardous chemicals, CNCs exhibit unparalleled potential in mitigating environmental pollution across multiple domains [51].

This chapter explores the multifaceted applications of CNCs in addressing environmental pollution, delving into synthesis approaches, pollutant remediation mechanisms, and various applications in soil, water, and air cleanup.

2 Sources and Types of Pollution

The process of globalization has amplified industrial activities and supply chain complexities, leading to increased pollution across various environmental domains [134]. Consequently, this phenomenon has heightened pollution levels, exacerbated climate change, and increased the adverse impacts on human health associated with these industries [15]. Pollutants can be categorized into several types based on their

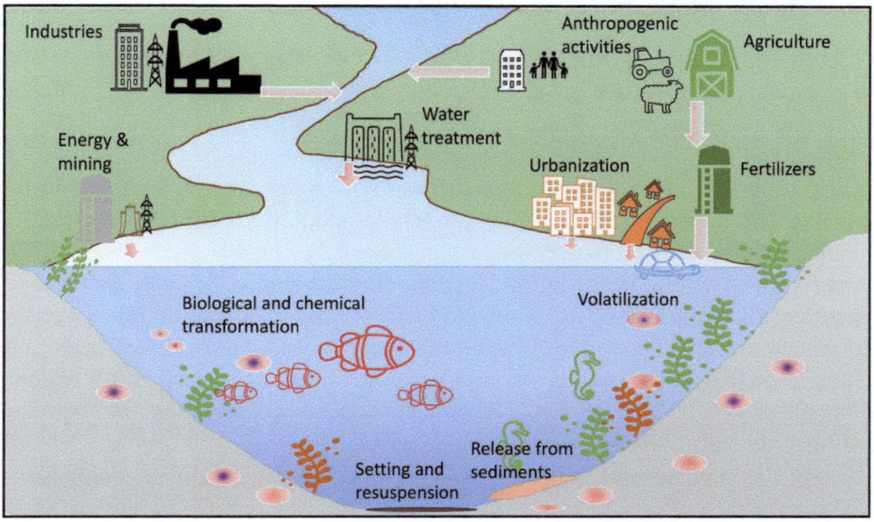

Fig. 1 Sources and sinks of pollutants in the environment

sources, characteristics, and environmental and human health impacts. Based on the environmental effects of pollutants the present chapter discusses the environmental risks associated with air, water, and soil pollutants. These contaminants significantly cause damage to ecosystems differently [56]. Nevertheless, air and water pollution remain constant threats in developing countries with fragile medical infrastructures. Previous studies have linked the proportion of the disease burden attributable to environmental pollution to 23–30% [53, 89, 133]. This includes waterborne infections, sanitation, and food hygiene diseases, respiratory diseases stemming from severe indoor air pollution due to biomass combustion; and protozoal diseases with considerable dimensions in environmental ecology such as malaria [53]. In their 2021 report, the Global Burden Disease (GBD) stated that the cohort life expectancy for both sexes is about 56 years, and each of these diseases' accounts for about 6% of the total disabled-adjusted life years.

WHO states outdoor air pollution causes 0.6 to 1.4% of non-communicable diseases in developing regions and at the same time, contaminated water, air, and soil that contain lead may contribute 0.9 percent [131]. These numbers might appear quite low, but the majority of the risks that are not listed in the "top 10" contribute between 0.5 and 1% [131]. As shown in Fig. 1, environmental pollutants have diverse sources including agricultural, mining, industrial, and urban.

2.1 Air Pollutants

Air pollutants represent the gravest form due to their significant effects on human health and global warming. They include toxicants such as carbon dioxide (CO_2), methane, and particulate matter. According to a recent report by the World Health Organization [131], air pollutants are associated with approximately 7 million deaths annually. The impact of air pollutants extends to infant mortality rates, respiratory ailments, mental health disorders, and numerous other health concerns [132]. At concentrations as low as 30 μg/m^3, epidemiological studies indicate that $PM_{2.5}$, containing significant amounts of soluble toxic metals, could exhibit redox cycling and induce oxidative stress [130]. Upon inhalation of particulate matter, lung cells release reactive oxygen species, which subsequently attack and oxidize other cellular components within the lungs [5]. This process triggers an inflammatory response, leading to damage to tissues, proteins, and DNA [21].

Using a comprehensive analysis involving atmospheric dynamics, combustion byproducts, and aerosol characteristics, Sun et al. [108] analyzed the emission behaviors of carbon and aerosol co-emissions around power plants using satellite-based atmospheric observations, to understand the interplay between combustion processes, aerosol characteristics, and coal composition for better air quality management. They observed strong correlations and synchronous peaks in Aerosol Optical Depth (AOD) and column-averaged dry-air mole fraction of carbon dioxide (XCO_2) emission patterns. They also revealed that variations in CO/XCO_2 and black carbon extinction and total aerosol extinction (BCEXT/TOTEXT) ratios indicate differences in combustion efficiency and aerosol characteristics. Different pollutants behave and interact based on the combustion process and the type of coal used—and it has been proven that NO_2 distributions consistently mirror XCO_2, while SO_2 distributions vary due to differences in coal sulfide content [108].

Particulate matters (PM) are also important air pollutants. These contaminants are a combination of a large number of small particles and liquid droplets that contain acids, organic chemicals, metals dust or other materials. PMs occur naturally—and often as a result of human activities. Mechanical and Industrial combustion, vehicle emission, and tobacco smoking are all anthropogenic sources of PM emissions; while naturally occurring, PMs originate from phenomena such as volcanic activity, fire, dust storms, and aerosolized sea salt. Generally, PMs are described based on aerodynamic equivalent diameter (AED) with particles of the same AED having similar settling velocity patterns. They are grouped as AED fractions based on their generation and deposition in human airways—with varying particle sizes as <10, <2.5, and <0.1 μm (PM_{10}, $PM_{2.5}$, and $PM_{0.1}$), respectively. Particles larger than 10 μm in size mainly have short suspension half-life and the major part of them are removed by the nose and upper respiratory tract. One category is the "fine" particles which are defined as greater than 2.5 and 10 μm ($PM_{2.5\text{–}10}$), termed coarse of about 5 μm as "fine," and any particle size greater than 20 μm as "coarse." However, anything smaller particle is not to be referred to as "ultrafine." A noteworthy point is that PM_{10} encompasses ultra-fine or $PM_{0.1}$; fine or $PM_{2.5\text{–}10}$. Another important key point is that

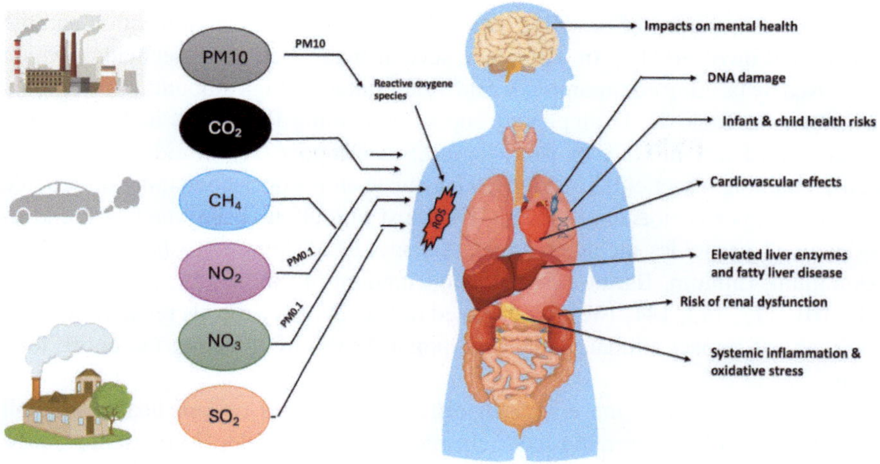

Fig. 2 Airborne pollutants

the total number of particles and the surface area in the mixed environmental sample increases with the reduction of the particle diameter, though the total particulate mass is normally reduced exponentially with the smaller particle diameters. Industries typically involving combustion processes, chemical reactions, or physical disturbances that release both coarse and fine particles into the atmosphere contribute the most to particulate matter pollution. However, governments and industries are increasingly adopting technologies such as electrostatic precipitators, fabric filters, and scrubbers to reduce particulate emissions. As shown in Fig. 2, the type of particulate matter originated from industries.

2.2 *Soil Pollutants*

Soil contaminants originate from various sources, including agricultural runoff (fertilizers and pesticides), industrial discharge (heavy metals), and urban waste [6, 99]. These pollutants result in detrimental effects such as biodiversity decline, water reservoir contamination, and nutrient exhaustion [2]. The application of pesticides can be detrimental to the soil, and the numerous living organisms including both the microscopic and macroscopic of the soil. Pesticides are insecticides, fungicides, rodenticides, nematicides, and herbicides containing organic or inorganic toxic compounds to the specific target. However, some of these such as the organophosphate pesticides have been identified to be more harmful to soil organisms than others. Also, the main chemicals contained in some pesticides degrade faster in the soils than others. Organophosphates and other types of pesticides might be important in agriculture but it has been found that these chemicals have significant effects on the soil profile [111].

The concentrations of pesticide residuals vary by geographical location and the type of soil involved [17]. In past years, several fungicides and insecticides, which were used by farming communities worldwide, contained high amounts of metalloids (as shown in Table 1). Examples of copper-containing fungicides include copper sulfate (Bordeaux mixture or Bordo® mix) and copper oxychloride. In contrast, lead-containing insecticides include lead arsenate and copper-containing insecticide includes copper acetoarsenite [123]. The most ubiquitous heavy metals found as residuals in pesticides include Copper, Arsenic, Lead, Mercury, Chromium, Zinc, Aluminum, Lithium, Barium, Boron, and Titanium [4, 36, 50, 52, 65, 76, 79, 92, 100, 101, 112, 113, 144, 147]. As depicted in Fig. 3, heavy metals released into the environment impact animal, plant, and human health by affecting the entire food chain.

Table 1 summarizes the effects of different pesticides on soil health and soil organisms. It includes the main active chemical of each pesticide, its pesticide family, the specific impact on soil, and references for further reading.

Heavy metal concentrates are known to have negative inputs to the quality of the soil. The World Health Organization (WHO) has set generally tolerable concentrations of these metals in soil expressed in $mg{\cdot}Kg^{-1}$ as Maximum Residual Limits (MRLs). For two metals, the metal with a higher permissible MRL is considered safer. Concerning the permissible limit found in soils, lead (85 $mg{\cdot}Kg^{-1}$) is the highest, while zinc (50 $mg{\cdot}Kg^{-1}$) and copper (36 $mg{\cdot}Kg^{-1}$) are in the middle cadre and cadmium (0.8 $mg{\cdot}Kg^{-1}$) is the lowest [6, 129]. This goes a long way in explaining why cadmium accumulation is even more dangerous than copper, zinc, and lead—albeit in lesser concentration. Such substances affect plants and alter the quality of the soils thus leading to poor crop production.

Again, most metals get into the ground and end up in edible parts of most crops. They often evaporate into the air or remain sorbed in the soil because they can form volatile residuals through natural biogenic processes. These heavy metals significantly affect many soil microorganisms including the total heterotrophic bacteria (THB) and rhizobia [76]. The level of heavy metal toxicity in crops depends on certain factors such as the type and variety of crops, environmental conditions, susceptibility of the elements, crop developmental stage, characteristics of the soil, physical and chemical properties and concentrations of accessibility of these metalloids in the soil solution [9]. In soil microorganisms, heavy metals have the potential to impact a range of metabolic and developmental procedures [37], as displayed in Tables 1 and 2.

2.3 *Water Pollutants*

Water pollutants encompass various substances that contaminate water bodies, including heavy metals, pesticides, fertilizers, industrial chemicals, pharmaceuticals, and microbial pathogens [412]. Common sources of water pollutants include agricultural runoff carrying nutrients and chemicals, leachates from landfills, discharge

Table 1 Impact of various pesticides on soil and soil organisms

Pesticide name	Main active chemical	Pesticide family	Effecton soil	References
Chlorpyrifos	O, O-Diethyl O-(3,5,6-trichloro-2-pyridyl) phosphorothioate	Organophosphate	Microbial communities, nutrient cycling	[48]
Atrazine	6-chloro-N-ethyl-N'-(1-methylethyl)-1,3,5-triazine-2,4-diamine	Triazine	SOM	[20]
Glyphosate	N-(phosphonomethyl)glycine	Glyphosate	Inhibits specific enzymes in plants, and plant residue composition	[84]
2,4-Dichlorophenoxyacetic Acid (2,4-D)	2,4-Dichlorophenoxyacetic acid	Phenoxycarboxylic acid	Plant hormone balance	[120]
Imidacloprid	1-{(6-chloro-3-pyridinyl)methyl}-7-nitroimidazolo[4,5-d] pyridine	Neonicotinoid	Soil organisms like earthworms	[43]
Metalaxyl	(2-methoxy-N-phenylacetamide)-6-methylpyrimidin-4-ol	Acylaminoacid	Soil fungi involved in nutrient cycling	[43]
Fipronil	1-(2,6-difluorophenyl)-5-trifluoromethylpyridin-4-ylaminosulfonyl cyanide	Phenylpyrazole	Soil biota	[120]
Dicamba	3,6-dichloro-2-methoxybenzoic acid	Chlorophenoxy	Non-target vegetation in soil	[120]

(continued)

Table 1 (continued)

Pesticide name	Main active chemical	Pesticide family	Effecton soil	References
Propazine	6-chloro-N-isopropyl-N’-propyl-1,3,5-triazine-2,4-diamine	Triazine	Leaching and impact on soil fertility	[20]
Tebuthiuron	N-[1,1-dimethyl-3-(trifluoromethyl)phenyl]ethylsulfonyl]-1-methyl-3-methoxyurea	Substituted Urea	Impact soil health	[120]

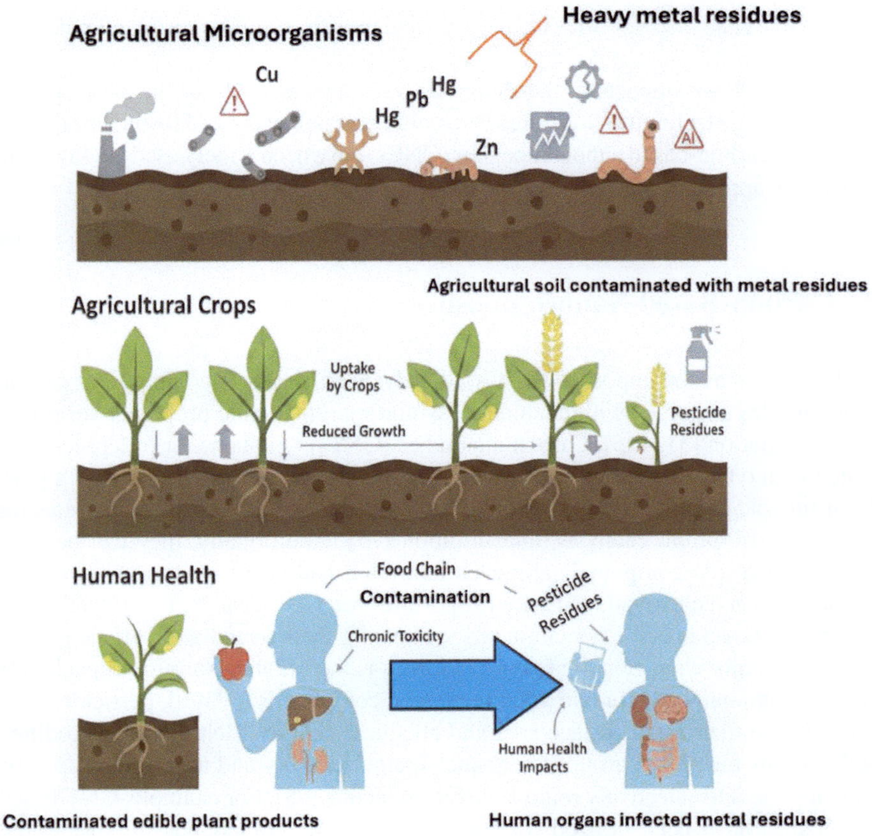

Fig. 3 Sources and impact of heavy metal on the food chain

Table 2 Impact of certain heavy metals on soil and soil organisms

S/N	Metal	Impact on soil health	References
1	Cadmium (Cd)	• Disrupts nutrient uptake • Reduces soil fertility and plant growth	[76, 79, 113]
2	Lead (Pb)	• Inhibits microbial activity • Reduces decomposition and nutrient cycling	[11, 28, 123, 147]
3	Arsenic (As)	• Disrupts soil microbial communities	[36, 65, 71, 100]
4	Mercury (Hg)	• Disrupts soil biota and nutrient cycling	[4, 52]
5	Copper (Cu)	• Disrupts soil microbial balance	[60, 113, 144]
6	Zinc (Zn)	• Disrupts microbial activity and nutrient cycling	[23, 92]
7	Chromium (Cr)	• Disrupts soil microbial communities	[2, 50, 101]

of toxic dyes and microplastics from waste, and industrial spills causing oil pollution [33, 40, 47].

The significant impacts of these various pollutants have led to the development of cutting-edge techniques to address the pollution challenges and foster sustainable practices. Carbon-based nanocomposites (CNCs) have been widely used among these innovative approaches.

3 Carbon-Based Nanocomposites

Carbon-based nanocomposites (CNCs) are advanced materials composed of carbon nanoparticles dispersed within a matrix, exhibiting remarkable properties for pollutant remediation [31, 115]. Carbon nanotubes (CNTs), graphene, or fullerenes integrated with other substances like polymers or metals are all classified as CNCs [54]. Their tunable porosity and exceptional mechanical strength are ideal candidates for pollutant adsorption, catalysis, and filtration [75]. Additionally, they can be used to adsorb organic compounds and even microbial pathogens, owing to their large surface area and chemical reactivity [98]. Moreover, CNCs can be functionalized to selectively target specific contaminants, enhancing their remediation efficacy [142, 144]. The major characteristic feature of CNCs is their high adsorption capacity for organic compounds, including volatile organic compounds (VOCs), pesticides, and dyes, due to their large surface area and chemical activity. CNCs can be modified with various functional groups to enhance their selectivity and adsorption capacity, allowing them to selectively remove target pollutants [98]. For example, CNT-based composites have been utilized for capturing phenolic compounds in wastewater, while graphene-based CNCs can capture heavy metals like lead and cadmium [75]. This versatility of design and operation underscores the suitability of CNCs to be optimized for solving various environmental issues and serving as a sustainable option in eradicating various pollutants within ecosystems.

Generally, CNCs are composed of carbon nanoscale materials with high surface-to-volume ratios, which greatly enhance their interaction with external species. The nanoscale dimension also leads to quantum effects, contributing to their remarkable electronic, optical, and mechanical properties while graphene is a honeycomb lattice in two dimensions which have the highest standards in thermal conductivity and mechanical properties [55]. CNTs, in particular, exhibit exceptional electrical conductivity due to their one-dimensional structure, allowing electrons to travel with minimal resistance. These characteristics are central to the enhanced performance of CNCs in various fields of material science and engineering [54]. In addition, the hybrid nature of CNCs allows for the integration of different types of nanomaterials, resulting in a synergistic combination of their properties. For instance, the incorporation of metal nanoparticles into CNCs can introduce new functionalities including magnetic and catalytic properties, while preserving the inherent advantages of the carbon nanomaterial. This hybridization offers the opportunity to design multifunctional composites that can perform multiple tasks simultaneously, such as pollutant detection and degradation. The ability to integrate multiple materials into a single nanocomposite structure is a major advantage that sets CNCs apart from other conventional materials [75].

Additionally, the structural versatility of CNCs offers a wide range of possibilities for customization at the molecular level. Functionalization can be achieved by attaching various chemical groups or molecules to the surface of the carbon nanoparticles, significantly expanding the scope of CNC applications. This approach allows researchers to fine-tune the surface properties of CNCs, such as their hydrophobicity or hydrophilicity, to better match specific environmental conditions [106]. The ability to control the surface chemistry of CNCs makes them highly adaptable to complex and dynamic systems, further enhancing their relevance in addressing environmental challenges [98]. From a theoretical standpoint, the electronic structure of CNCs has been a subject of intense study, particularly regarding their interaction with other materials. Density functional theory (DFT) has been employed to model the electronic behavior of carbon nanomaterials and predict their responses to external stimuli. By simulating the electronic density distribution and the energy band structure of CNCs, researchers have been able to design more efficient nanocomposites with tailored properties [114]. Theoretical models have provided insights into the charge transfer mechanisms between carbon nanoparticles and other materials, which are crucial for optimizing the performance of CNCs in various applications, such as pollutant adsorption, catalysis, and energy storage [142, 144]. Figure 4 represents the main groups of carbon-based nanocomposites.

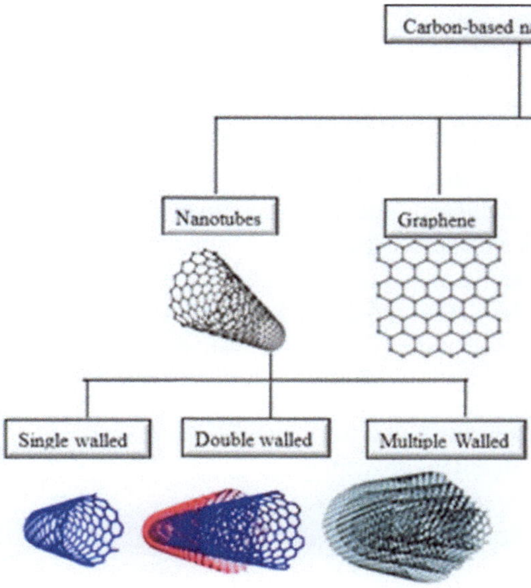

Fig. 4 Classification of carbon-based nanocomposites

4 Synthesis Approaches of CNCs

The fabrication of CNCs involves various synthesis methods including chemical vapor deposition (CVD), solution-based methods like hydrothermal synthesis, and mechanical exfoliation techniques such as liquid-phase exfoliation of graphene [1]. Each of these methods offers distinct advantages and challenges. Recent advancements have focused on enhancing the scalability, cost-effectiveness, and precision of CNC synthesis. Innovations like microwave-assisted synthesis, plasma-enhanced techniques, and template-directed growth have emerged to streamline production processes and improve material properties [86]. These methods enable precise control over CNC morphology, composition, and functionality, facilitating tailored design for specific applications [57]. Additionally, efforts to enhance the cost-effectiveness and scalability of CNC production have led to the development of continuous synthesis techniques and the utilization of renewable carbon sources. These novel improvements aim to reduce production costs and environmental impact while increasing the commercial viability of CNCs for various industrial applications [35, 68, 104]. CNC synthesis is followed by a growth mechanism, mainly vapor–liquid-solid (VLS) and vapor–solid-solid (VSS). These mechanisms depend on the metal catalyst condition and carbon dispersion at high temperatures. VLS involves catalyst-mediated decomposition of a carbon precursor, leading to CNT formation at the catalyst-substrate interface, while VSS comprises three stages involving carbon particle deposition and crystallization [126]. Thus, as shown in Table 3, the growth mechanism can directly impact factors like the electrical conductivity, tensile strength, and reactivity of carbon nanocomposites.

Following the growth of CNCs through VLS or VSS mechanisms, further processing is often required to enhance their properties and prepare them for specific applications [88]. This can involve techniques such as purification, functionalization, and assembly. Purification steps remove impurities and defects from the CNCs, ensuring high-quality materials. Functionalization involves modifying the surface of the CNCs with functional groups, which can improve their compatibility with other materials or impart specific properties like electrical conductivity, biocompatibility, or catalytic activity [35]. Assembly techniques, such as filtration, centrifugation, or self-assembly, are used to organize the CNCs into desired structures or composites.

Once processed, CNCs can be incorporated into a wide range of applications, including:

- Energy storage: Batteries, supercapacitors, and fuel cells
- Electronics: Transistors, sensors, and flexible displays
- Materials: Composites, coatings, and membranes
- Biomedical: Drug delivery, tissue engineering, and diagnostics.

Table 3 Properties of the different types of carbon-based nanocomposites

Carbon nanomaterial	Dimensions	Diameter (nm)	Experimental specific surface area ($m^2\ g^{-1}$)	Thermal conductivity ($W\ m^{-1}\ K^{-1}$)	Electrical conductivity ($S\ cm^{-1}$)	Tenacity	Hardness	References
Graphene	2	0.1–0.3	1500–2630	4840–5300	~2000	Flexible, elastic	Uppermost (for single layer)	[16]
Carbon nanotube	1	0.5–2.0	~1300	3500	Structure-dependent	Flexible, elastic	High	[74]
Fullerene	NA	0.7	80–90	0.4	10^{-10}	Elastic	High	[78]
Activated charcoal	NA	<0.4–50	500–1500	0.04–0.08	10^{-8}–10^{-6}	NA	High	

5 Characterization of CNCs

Efforts to characterize and understand CNCs involve various methods. These analytical techniques confirmed successful surface modification, with FTIR, XRD, TEM, SEM, and EDX indicating functionalization on the surface [44]. Transmission electron microscopy (TEM) and scanning electron microscopy (SEM) are key for morphology assessment. Raman spectroscopy (RS) evaluates purity and quality, while techniques like thermogravimetric analysis (TG), X-ray photoelectron spectroscopy (XPS), and infrared spectroscopy confirm CNT functionalization [25]. Optimization of CVD parameters enhances CNT properties for diverse applications [1, 44]. The major characterization techniques used to detect carbon nanocomposites are summarized in Fig. 5.

5.1 *Fourier Transform Infrared Spectroscopy (FTIR)*

FTIR is the most preferred method to study CNCs and especially the identification of functional groups on the surface of carbon nanomaterials. FTIR technique concerns the vibrational changes taking place in molecules and can determine the presence of functional groups, such as hydroxyl (–OH), carboxyl (–COOH) or amine ($-NH_2$), which are usually formed during the surface modification processes. In the case of

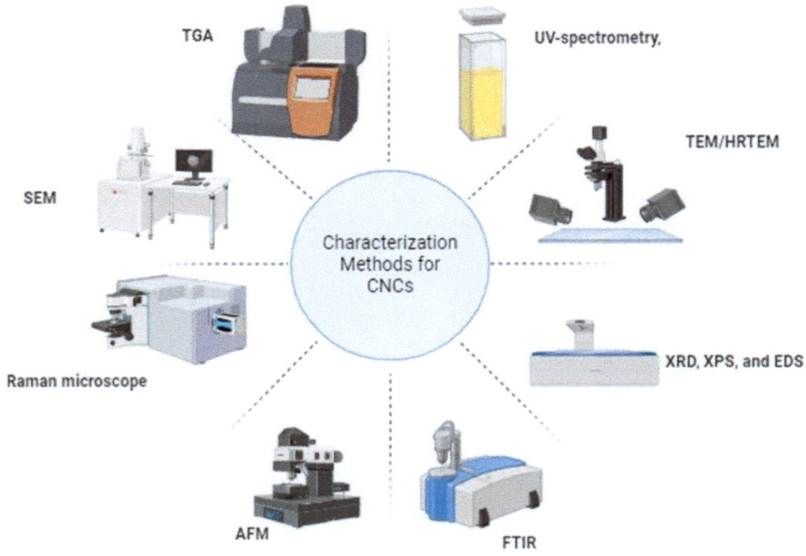

Fig. 5 Characterization techniques for carbon-based nanocomposites

FTIR spectroscopy, the presence or the absence of several peaks in the spectrum justifies whether the corresponding functionalization was achieved—which is important in cases where specific interaction with pollutants is desirable [25].

For identification of carbon nanotubes (CNTs) changes in the molecular structure are identified using the FTIR. In pristine CNTs, peaks at 1580–1530 cm^{-1} correspond to sp^2 hybridized carbon and 1200–1100 cm^{-1} resulting from disordered structures [85]. Thus, oxygenated functional groups, particularly carboxyl groups (COOH) in functionalized CNTs, induce new peaks such as those corresponding to carbonyl (C = O) stretching frequency ranging between 1800 and 1700 cm^{-1} upon oxidation. Other functional groups that are observed by FTIR are functional groups that are incorporated as a result of functionalization such as ether (C–O–C), hydroxyl (–OH), and C = O functional groups. However, for nitrogen-doped CNTs, new peaks at 1375 cm^{-1} and 1250 cm^{-1} are linked to the N–N bonds, confirming doping.

Characterization of some functional groups in graphene oxide (GO) samples is made using FTIR spectroscopy. For pure graphene, there are no major characteristic FTIR peaks that correspond to the presence of functional groups while in the case of GO, there are several characteristic peaks that appear, such as stretching peaks at 1740–1720 cm^{-1} suggesting the presence of C = O groups, and peaks ranging in between 1230 and 1040 cm^{-1} due to epoxy, alkoxy, and carbonyl. The stretching vibration of hydroxyl groups appears between 4000 and 3000 cm^{-1} as a result of GO being hydrophilic [8]. When reducing to pristine graphene or RGO structure, many of the oxygen functional groups' characteristic peaks decrease or vanish and new characteristic peaks also appear, specifically around 1585–1565 cm^{-1}. The appearance of these new peaks is assigned to the C = C bonds which may indicate the restoration of the graphene structure after the reduction.

5.2 *Electron Microscopy*

Transmission electron microscopy (TEM) and scanning electron microscopy (SEM) are essential tools for assessing the morphology and size distribution of CNCs. TEM, which uses electron beams to create high-resolution images, provides detailed insights into the internal structure of the CNCs at the nanoscale. This allows researchers to examine the alignment and dispersion of carbon nanotubes (CNTs) or graphene layers within the composite matrix [90]. In addition to distinguishing various morphologies of nanocarbon, TEM plays a significant role in designing the electronic, chemical, mechanical, and magnetic properties at the nano or atomic level by allowing defect control at such levels. These defects include point defects of the nanocomposites, especially graphene sheet and cluster defects, edge, bond rotations, grain boundaries, doping-induced defects, non-sp^2 carbon defects and high-strain folding of graphene sheets. These defects are associated with natural ones, growth-induced defects and the ones intentionally produced, and enhances the knowledge of carbon-based phenomena [72]. High-resolution TEM (HRTEM) gives important information on the pure samples' atomic arrangement and the identities and positions of defects in the samples defects in nanocarbon. Since defect studies in nanocarbon are an area of particular interest, one major outcome in the past year which ought to be mentioned is the utilization of HRTEM by [125] to study the motion of dislocation pairs in graphene, and the use of HRTEM to monitor the actual growth process of single-wall carbon nanotubes (SWNTs) and to determine their structural properties under controlled-environmental conditions by [110].

SEM, on the other hand, offers surface morphology images that reveal the overall shape and texture of the CNCs. These microscopy techniques are often complemented by energy-dispersive X-ray spectroscopy (EDX), which provides elemental analysis, confirming the composition and detecting any impurities present on the CNC surface [44]. SEM provides detailed morphological information, allowing researchers to observe the surface structure of CNCs at high magnification. It confirmed the homogeneity and dispersion of nanoparticles within the CNCs, ensuring proper integration of components for hybridization. The use of SEM in the visualization of the carbon nanocomposite morphology is also effective due to its high lateral resolution, large depth of focus, and ability to perform X-ray microanalysis to establish the fact that the structure of the carbon nanocomposite appeared to be nonuniform with unequal thickness and density [12]. SEM examination requires samples to be coated with a thin layer of gold or platinum, for nonconductive nanomaterial. At times, there is a necessity to apply any surface treatment processes such as ion sputtering or chemical etching to improve structural features. SEM images have revealed that carbon nanocomposites possess various roughness of surfaces and inhomogeneous density of chemical components at the nanoscale [97]. These images often display surface imperfections such as cracks, etching residues, differential swelling, depressions, and perforations.

5.3 Raman Spectroscopy (RS)

This is another technique commonly employed to evaluate the purity and quality of carbon-based nanomaterials within CNCs. Raman spectroscopy is particularly sensitive to the carbon–carbon bond vibrations in materials like carbon nanotubes and graphene, making it ideal for assessing defects, crystallinity, and the overall structural integrity of CNCs. Analyzing the D-band (disorder) and G-band (graphitic) ratios in the Raman spectrum helps determine the level of functionalization or defects introduced during the synthesis process. A high G-band intensity relative to the D-band suggests higher purity and fewer defects, while an increase in the D-band signals indicates more significant functionalization [1].

5.4 Thermogravimetric Analysis (TGA)

It is employed to assess the thermal stability of CNCs and to quantify the functional groups attached to their surface. TGA involves heating a sample in a controlled environment while monitoring the mass loss as a function of temperature. This technique provides valuable data on the degradation temperature of the nanocomposite and helps identify the amount of organic or polymeric material present [82]. In CNCs, a sharp weight loss at lower temperatures can indicate the presence of volatile organic functional groups, while stable weight retention at higher temperatures confirms the robustness of the carbon framework [25]. This thermal analysis is particularly important when designing CNCs for high-temperature applications or environments.

5.5 X-Ray Diffraction (XRD)

X-ray diffraction (XRD) is also an effective method used to analyze the crystalline structure of carbon nanocomposites. It provides information on the arrangement of atoms within the material and measures the degree of crystallinity of the carbon nanomaterials, such as graphene or carbon nanotubes, incorporated into the composite [18]. The XRD technique makes it possible to analyze the orientation and dispersion of carbon nanomaterials within the matrix of CNCs. This information can help determine how it influences such properties as mechanical strength and chemical activity. The height and the position of the peaks within the XRD pattern give information about the crystallinity and phase distribution thus enabling researchers to control the synthesis of CNC for a specific application [44].

5.6 X-Ray Photoelectron Spectroscopy (XPS)

X-ray photoelectron spectroscopy (XPS) is an effective instrument for surface analysis of materials for relative concentration of elements up to 10 nm depth. It has been extensively used to analyze surface chemistry and follow the time-dependent changes in the nanostructure of biochar [105]. For instance, the oxidation of C groups and formation of carboxylic C in black carbon particles incubate at high temperatures (70 °C). It gives a precise report about the binding energies of surface atoms which gives an understanding of the interaction between carbon nanomaterials and the functional groups attached to them. This technique is most valuable in ascertaining the presence of heteroatoms, for example, oxygen, nitrogen or sulfur added during surface modification. Changes in the binding energy peaks in the XPS spectrum can also indicate the formation of successful chemical bonding of the pollutants or catalytic materials [146]. Hence, XPS plays a crucial role in the enhancement of pollutant adsorption CNCs or improved catalytic performance [1]. Furthermore, the process of XPS peak calibration makes it possible to estimate the surface concentration of functional groups, which is the determining factor in assessing the applicability of CNC's surface chemistry [30].

5.7 Other Characterization Techniques

In addition to these analytical techniques, researchers have been focusing on optimizing the chemical vapor deposition (CVD) parameters to enhance the properties of carbon nanotubes (CNTs) for use in CNCs. CVD is a popular method for synthesizing high-quality CNTs by depositing carbon atoms onto a substrate in the form of a thin film. The growth conditions, such as temperature, pressure, and precursor concentration, play a crucial role in determining the size, shape, and quality of the CNTs produced. By fine-tuning these parameters, researchers can improve the electrical conductivity, mechanical strength, and surface area of the CNTs, making them more effective when integrated into CNCs for specific environmental applications [44]. While the techniques mentioned above focus on analyzing CNC structure and functionalization, other advanced methods are being developed to enhance their understanding further. For example, atomic force microscopy (AFM) has been employed to measure the surface roughness and mechanical properties of CNCs at the nanoscale. This technique allows for the detailed mapping of surface topography and provides insights into the interaction forces between CNCs and pollutants. Furthermore, AFM can measure the mechanical properties, such as stiffness and elasticity, of individual carbon nanomaterials within the composite, providing a more comprehensive understanding of how these properties impact the performance of CNCs in real-world applications [91].

Ultraviolet–visible (UV–Vis) spectroscopy is another technique applied in CNC characterization, particularly for studying their optical properties. UV–Vis spectroscopy helps in analyzing the electronic transitions within the CNCs and provides information about the absorption of light by the material [116]. This technique is particularly useful in studying photocatalytic CNCs, where light absorption plays a critical role in pollutant degradation. The absorption spectra obtained from UV–Vis spectroscopy can be correlated with the efficiency of CNCs in catalytic and sensing applications, providing a better understanding of their interaction with light and pollutants [1].

6 Applications of CNCs in Environmental Remediation

Compared to traditional remediation methods such as chemical precipitation or activated carbon filtration, CNCs offer several advantages that allow for tailored design and application in diverse environmental settings, promising innovative solutions to pollution challenges.

6.1 *Air Purification*

Air pollutant remediation stands as a pressing environmental concern, with CNCs offering innovative solutions for air purification. The porous structure and high surface area of CNCs allow them to adsorb a wide range of airborne contaminants, including nitrogen oxides (NOx), sulfur dioxide (SO_2), and carbon dioxide (CO_2). CNCs can be modified to enhance their affinity for specific gases, further improving their performance in air filtration systems [19, 75]. [26] synthesized S–BiOBr nano-flowers via isomorphous replacement combined self-assembly methods and deposited Bi2Sn2O7 in-situ to construct a 3D/0D interfacial Z-scheme S–BiOBr/Bi2Sn2O7 heterojunction. The photocatalyst exhibited high activity in degrading refractory pollutants including rhodamine B, and tetracycline, while activating Spirulina. Fluorescence spectroscopy tracked the dynamic degradation process—and revealed the involvement of h+ and radical dotO_2 – radicals, with radical dotOH playing a supporting role. CNCs have proven to be useful in the photocatalytic degradation of gaseous pollutants. [66] synthesized $Bi_2Sn_2O_7$ nanocrystalline material hydrothermally using stannic chloride pentahydrate and bismuth citrate as starting materials for the photocatalytic elimination of nitrogen oxide (NO). The efficiency was assessed under simulated solar light irradiation, with the as-prepared $Bi_2Sn_2O_7$ samples tested in a continuous reactor for the photocatalytic degradation of NO. The $Bi_2Sn_2O_7$ sample synthesized for BSO-12 (12 h) exhibited a 37% removal rate for NO, surpassing the removal rates achieved by $Bi_2Sn_2O_7$ samples synthesized for BSO-24 (24 h) and BSO-36 (36 h). Particulate matter (PM) and volatile organic compounds (VOCs) have been efficiently removed using a polyester-based air filter by coating it with Sericin [117]. Results showed that the Sericin-coated filter effectively reduced PM2.5 and PM10 levels from 1000 to 5 $\mu g/m^3$ within 27 and 23 min of operation, respectively [117]. Table 4 summarizes nanomaterials used for gaseous pollutant photodegradation from the past few years.

Furthermore, these nanomaterials can be incorporated into filter media to enhance their capabilities—and remove higher pollutants including fine particulates at the same time. Studies have shown that these filters have longer useful life and higher efficiency when compared to conventional filters [118]. These composites can be combined with appropriate photocatalysts for use in splitting solar energy to break down hazardous substances into relatively non-toxic forms, as noted by Wang et al. [123]. It is an innovative and eco-friendly way of carrying out the process of air

Table 4 Summary of typical nanomaterials used for gaseous pollutant photodegradation

Photocatalyst	Synthesis	Reaction conditions			Performances	References
		Target pollutant	Initial concentration	Light source		
B-Doped g-C_3N_4	Hydrothermal reaction and calcination in N_2	NO	400 ppb (continuous flow mode)	300 W Xenon lamp ($\lambda >$ 420 nm)	NO removal efficiency (30.4%)	[124]
Ag-AgBr-TiO_2	Stirring under room temperature	Benzene/acetone	250 ppm (fixed-bed mode)	Visible light ($\lambda >$ 400 nm)	Removal efficiency: benzene (47.2%) acetone (70.4%)	[141]
Au-TiO_2	Calcination in air	HCHO	50 ± 2 ppmv (continuous flow mode)	Visible light	Conversion rate: 83.3% (RH = 44.0%)	[145]
F-TiO_2/Pt	Photodeposition for Pt and NaF soaking (pH = 3.5) for surface fluorination	Toluene	50 ppmv (closed reactor)	370 nm UV light	Removal efficiency: 53.0% (after the fifth cycle)	[128]
Carbon nanodots/$ZnFe_2O_4$	Hydrothermal reaction	NO	400 ppb (continuous flow mode)	Visible light ($\lambda >$ 420 nm)	Removal efficiency: 38.0%	[38]
Bi/$ZnWO_4$	Hydrothermal reaction	NO	400 ppb (continuous flow mode)	Visible light ($\lambda >$ 420 nm)	Removal efficiency: 63.0%	[29]
$CuInS_2$/Bi_2WO_6	Solvothermal and hydrothermal synthesis	toluene	1 μL (fixed-bed mode)	Visible light ($\lambda >$ 420 nm)	Removal efficiency: 63.0%	[67]
$Bi_2O_2CO_3$/$ZnFe_2O_4$	Hydrothermal reaction	NO	400 ppb (continuous flow mode)	Visible light ($\lambda >$ 420 nm)	Removal efficiency: 32.0%	[39]
Ag-decorated WO_3/Bi_2WO_6	Hydrothermal reaction	Chlorobenzene	2 μL in a closed 120 mL reactor	Xe lamp	Removal efficiency: 80.0%	[143]

(continued)

Table 4 (continued)

Photocatalyst	Synthesis	Reaction conditions			Performances	References
		Target pollutant	Initial concentration	Light source		
BiOCl (001) and (010)	Hydrothermal reaction	NO	600 ppb (continuous flow mode)	Visible light ($\lambda >$ 420 nm)	Removal efficiency: BiOCl (001): 50.1% BiOCl (010): 60.4%	[59]
(001) TiO_2 nanotubes	Electrochemical anodization and NaF treatment	Toluene; HCHO; Aldehyde	10 ppmv (closed reactor)	370 nm UV light	Removal efficiency: toluene (32.7%) HCHO (97.2%) aldehyde (58.7%)	[128]
Sr-doped defective TiO_2	Hydrothermal reaction	HCHO	1 ppm (continuous flow mode)	UV light ($\lambda =$ 365 nm, 90 mW cm^{-1})	Removal efficiency: 50.85%, QE: 5.53%	[62]
Er^{3+}-Doped TiO_2	Sol–gel method	Acetaldehyde; o-xylene; ethylene	o-xylene and ethylene (25 ppm flow mode); acetaldehyde (25 and 500 ppm flow mode)	Visible light	Removal efficiency: Acetaldehyde (99.2%) o-xylene (84.6%) ethylene (22.4%)	[94]
Graphene-TiO_2 composite mats	Hydrothermal reaction and electrospinning method	Gas-phase methanol	4000 ± 200 ppm (static mode)	UV light	Removal efficiency: 100%	[96]

(continued)

Table 4 (continued)

Photocatalyst	Synthesis	Reaction conditions			Performances	References
		Target pollutant	Initial concentration	Light source		
Fe-based metal–organic-frameworks	Solvothermal method	Toluene	460 ppm (closed-circulation reactor)	Simulated sunlight (100 mW cm^{-2})	100% adsorption and photodegradation removal efficiency	[61]
Core–shell CeO_2@LDHs	Reverse precipitation and hydrothermal method	Formaldehyde	8.0, 16.0, 24.0, 32.0, and 40.0 mg m^{-3} (continuous flow mode)	Visible light	Removal efficiency: 86.9%	[135]

purification. Also, these materials have been studied in an attempt to identify the applicability of advanced oxidation processes (AOPs). Therefore, CNCs can facilitate the generation of reactive species when used as catalyst support, inducing the degradation of many pollutants [139]. As concerns over air pollution and climate change continue to grow, CNCs offer a promising solution for reducing harmful emissions and mitigating their environmental impact.

6.2 Soil Pollutant Removal

CNCs have exceptional adsorption capabilities in soil pollutant remediation due to their large surface area and tunable properties, making them promising candidates for the removal of heavy metals, organic contaminants, and other pollutants from soil environments. Yang et al. [137] investigated the capability of multi-walled carbon nanotubes (MWCNTs) to alter the extractability of antimony (Sb) and cadmium (Cd) in contaminated soil. They discovered that MWCNTs exhibited selective impacts on the extractability of soil heavy metals, particularly Sb and Cd. At low Sb content levels, the addition of MWCNTs reduced Sb bioavailability, attributed to increased adsorption sites. However, excess Sb(OH)6– was not efficiently adsorbed. Additionally, the mild acid-soluble fraction increased maximally only at high content levels and 0.1% MWCNTs addition, indicating the influence of effective specific surface area differences. In an experiment to remove Pb^{+2} and Cd^{+2} ions [7] synthesized eco-friendly mesoporous magnetic activated carbon-based nanosorbents with incorporated antimicrobial AgNPs (Mag@AC1–Ag and Mag@AC1–Ag). The prepared Mag@AC1–Ag and AgNPs (Mag@AC1–Ag) exhibited rapid and excellent uptake of Pb^{+2} and Cd^{+2}. They stressed that pseudo-second-order kinetics and the Langmuir isotherm are more suitable for the explanation of their findings. Moreover, the high reactivity of CNCs facilitates in-situ chemical degradation of pollutants, enhancing soil quality and fostering sustainable agriculture practices.

In addition to their use in adsorption, CNCs have been employed in catalysis to enhance pollutant degradation. Carbon nanoparticles, such as graphene oxide, can be integrated with catalytic metals like titanium dioxide or silver to create photocatalytic nanocomposites. These composites can break down harmful pollutants under light irradiation, further improving the efficiency of the remediation process [140]. The catalytic properties of CNCs are particularly advantageous for the degradation of persistent organic pollutants (POPs), which are resistant to conventional treatment methods. CNCs can also be utilized in advanced oxidation processes, where they facilitate the generation of reactive oxygen species that degrade pollutants more effectively than traditional catalysts [31].

Carbon-based nanocomposites also play a significant role in microbial remediation, where they can adsorb and neutralize microbial pathogens in contaminated water sources. This capability is especially relevant in addressing waterborne diseases in regions with inadequate water treatment infrastructure. Studies have demonstrated that CNCs functionalized with antimicrobial agents, such as silver nanoparticles, can effectively eliminate bacterial pathogens like *Escherichia coli* and *Staphylococcus aureus* from water samples [115]. Combining antimicrobial properties with pollutant adsorption makes CNCs a powerful tool in water purification and public health protection.

6.3 Water Treatment

Carbon nanomaterials, such as nanoporous activated carbon (NAC), fullerenes, graphene-based CNCs, and carbon nanotubes (CNTs), are widely found incorporated in wastewater treatment methods such as membrane processes, photocatalysis, disinfection, and adsorption. Ahmad et al. [2] used four distinct chemical moieties (amidoxime, hydrazone, ethylenediamine (EDA), and diethylenetriamine (DETA)) as CNCs to enhance the adsorption and photocatalytic performance in a water treatment system. They discovered that the surface-functionalized halloysite nanotubes (HNTs) effectively adsorbed Cr(VI) ions from water, with adsorption rates ranging from 58 to 94% under optimized conditions. Even materials made with low concentrations of nanocomposites can efficiently remove pollutants from the environment [41]. For instance, it was demonstrated that small amounts of multi-walled carbon nanotubes (MWCNTs), as low as 0.7 g, for a short stirring time of 5 min, resulted in the highest percentage of removal of unleaded gasoline from water [63]. Functionalized graphene oxide membranes exhibit selective permeability, enabling the efficient removal of heavy metals, pathogens, and emerging contaminants from wastewater streams [148]. Furthermore, CNCs-based electrochemical sensors facilitate real-time monitoring of water quality parameters, enabling proactive intervention and resource management in water treatment facilities [138]. The different environmental applications of carbon nanocomposites are summarized in Fig. 6.

7 Challenges and Opportunities

However, despite their promising attributes, the utilization of CNCs for environmental remediation is not without challenges. One major limitation is the potential for unintended consequences such as the release of nanomaterials into the environment during their production, application, or end-of-life disposal, raising concerns about their long-term environmental impacts. Additionally, the scalability of CNC synthesis processes and their cost-effectiveness at large scales remain areas of concern. Furthermore, the stability and reusability of CNCs in real-world environmental conditions

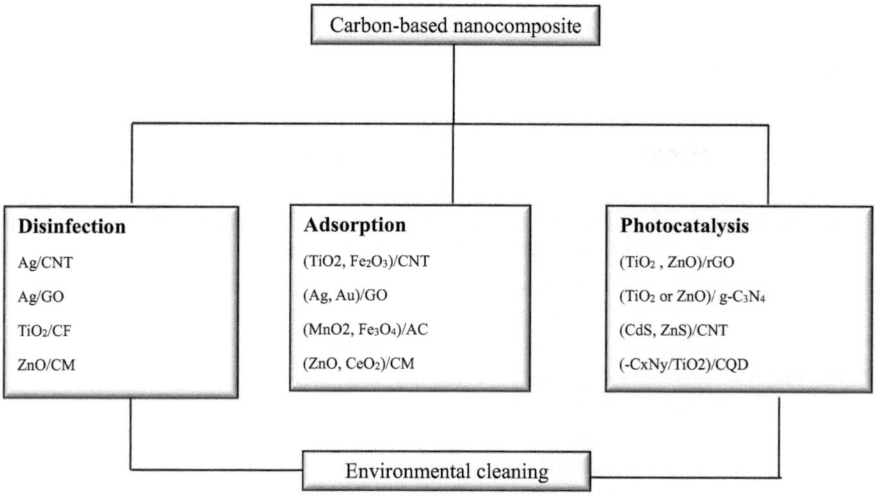

Fig. 6 Applications of carbon-based nanocomposites for environmental cleaning

need further investigation to ensure their long-term effectiveness. Addressing these limitations will be crucial for realizing the full potential of CNCs in sustainable environmental remediation efforts. As shown in Table 5, carbon nanocomposites have diverse environmental impacts such as toxicity to aquatic and terrestrial organisms, potential for bioaccumulation, and environmental release during production and disposal. In addition to environmental concerns, the safety of carbon-based nanocomposites (CNCs) in human health applications must be carefully considered. While CNCs hold promise for various biomedical applications, including drug delivery and medical imaging, their potential cytotoxicity and long-term effects on human health require thorough evaluation (Chen et al. 2020). Understanding the interactions between CNCs and biological systems is essential to mitigate any adverse effects and ensure their safe use in medical and healthcare settings. Addressing these safety concerns through comprehensive toxicity studies and regulatory frameworks will be essential to harness the full potential of CNCs in both environmental remediation and biomedical applications while safeguarding human health and the environment.

Table 5 Summary of the environmental impact of typical carbon-based nanocomposites

Type of carbon-based nanocomposite	• Impact on environment	References
Graphene Oxide	• Potential toxicity to aquatic organisms with environmental persistence • Difficulty in separation and recovery	[136]
Carbon nanotubes	• Potential toxicity to human health and environment • Low solubility, difficulty in dispersion	[42]
Activated carbon	• Leaching of heavy metals leading to secondary pollution • Low selectivity for some pollutants	[73, 119]
Carbon aerogels	• High cost of production • Low mechanical strength that could lead to secondary pollution	[95]
Carbon dots	• Limited understanding of toxicity that could lead to accumulation • Low specific surface area compared to other carbon materials	[58]
Graphene oxide-metal oxide composites	• Potential toxicity of metal oxide components • Environmental persistence • Difficulty in separation and recovery	[45]
Carbon nanotubes-metal oxide composites	• Potential toxicity of metal oxide components and environmental persistence • Difficulty in separation and recovery	[46]
Activated carbon-polymer composites	• Leaching of polymer components that leads to microplastic pollution • Reduced adsorption capacity compared to pure activated carbon	[121]
Carbon aerogel-polymer composites	• High cost of production • Low mechanical strength • Potential for microplastic pollution	[80]
Carbon dots-polymer composites	• Limited understanding of toxicity • Potential for microplastic pollution • Reduced photocatalytic activity compared to pure carbon dots	[109]

8 Conclusion

Carbon-based nanocomposites (CNCs) emerge as versatile solutions in addressing environmental pollution, offering tailored approaches for pollutant capture and degradation across air, soil, and water matrices. Their tunable properties, high surface area, and scalability enhance their effectiveness in diverse pollution scenarios. Despite their promise, challenges like environmental release during production and scalability issues persist. Moreover, ensuring the safety of CNCs in both environmental and human health applications remains paramount. However, advancements in synthesis techniques and regulatory frameworks present opportunities to overcome these challenges. With further research and development, CNCs hold immense potential for sustainable environmental remediation, offering innovative solutions to mitigate pollution and safeguard human health and ecosystems in the future.

References

1. Abbas Q, Shinde PA, Abdelkareem MA, Alami AH, Mirzaeian M, Yadav A, Olabi AG (2022) Graphene synthesis techniques and environmental applications. Materials (Basel, Switzerland) 15(21)
2. Ahmad MF, Ahmad FA, Alsayegh AA, Zeyaullah M, AlShahrani AM, Muzammil K et al (2024) Pesticides impacts on human health and the environment with their mechanisms of action and possible countermeasures. Heliyon 10(7)
3. Ahmad SS, Zulfiqar S, Ruipérez F, Rafique M, Iqbal M, Forrester MJ et al (2024) An integrated experimental and theoretical approach to probe Cr(vi) uptake using decorated halloysite nanotubes for efficient water treatment. RSC Adv 14(5):2947–2960
4. Ahmad W, Alharthy RD, Zubair M, Ahmed M, Hameed A, Rafique S (2021) Toxic and heavy metals contamination assessment in soil and water to evaluate human health risk. Sci Rep 11(1)
5. Ahn Y, Yim YH, Yoo HM (2024) Particulate matter induces oxidative stress and ferroptosis in human lung epithelial cells. Toxics 12(2)
6. Alengebawy A, Abdelkhalek ST, Qureshi SR, Wang MQ (2021) Heavy metals and pesticides toxicity in agricultural soil and plants: ecological risks and human health implications. Toxics 9(3)
7. Ali OI, Zaki ER, Abdalla MS, Ahmed SM (2023) Mesoporous Ag-functionalized magnetic activated carbon-based agro-waste for efficient removal of Pb(II), Cd(II), and microorganisms from wastewater. Environ Sci Pollut Res Int 30(8):53548–53565
8. Aliyev E, Filiz V, Khan MM, Lee YJ, Abetz C, Abetz V (2019) Structural characterization of graphene oxide: surface functional groups and fractionated oxidative debris. Nanomaterials (Basel, Switzerland) 9(8)
9. Angon PB, Islam MS, Kc S, Das A, Anjum N, Poudel A, Suchi SA (2024) Sources, effects and present perspectives of heavy metals contamination: soil, plants and human food chain. Heliyon 10(7)
10. Anjaneyulu B, Rana R, Versha AM, Carabineiro SA (2024) The Use of magnetic porous carbon nanocomposites for the elimination of organic pollutants from wastewater. Surfaces 7:120–142
11. Arao T, Ishikawa S, Murakami M, Abe K, Maejima Y, Makino T (2010) Heavy metal contamination of agricultural soil and countermeasures in Japan. Paddy Water Environ 11(9)
12. Asano N, Lu J, Asahina S, Takami S (2021) Direct observation techniques using scanning electron microscope for hydrothermally synthesized nanocrystals and nanoclusters. Nanomaterials (Basel, Switzerland) 11(4)
13. Asghar N, Hussain A, Nguyen DA, Ali S, Hussain I, Junejo A, Ali A (2024) Advancement in nanomaterials for environmental pollutants remediation: a systematic review on bibliometrics analysis, material types, synthesis pathways, and related mechanisms. J Nanobiotechnol 22(1)
14. Attia Y, Ezet AE, Saeed S, Galmed AH (2024) Nano carbon-modified air purification filters for removal and detection of particulate matters from ambient air. Sci Rep 14
15. Bolan S, Padhye LP, Jasemizad T, Govarthanan M, Karmegam N, Wijesekara H et al (2024) Impacts of climate change on the fate of contaminants through extreme weather events. Sci Total Environ 909

16. Bonaccorso F, Colombo L, Yu G, Stoller M, Tozzini V, Ferrari AC et al (2015) Graphene, related two-dimensional crystals, and hybrid systems for energy conversion and storage. Science 347(6217)
17. Bondareva L, Fedorova N (2021) Pesticides: behavior in agricultural soil and plants. Molecules (Basel, Switzerland) 26(17)
18. Cao A, Xu C, Liang J, Wu D, Wei B (2001) X-ray diffraction characterization on the alignment degree of carbon nanotubes. Chem Phys Lett 344(1–2):13–17
19. Cao J-J, Huang Y, Zhang Q (2021) Ambient air purification by nanotechnologies: from theory to application. Catalysts 11(11)
20. Chaplain V, Mamy L, Vieuble-Gonod L, Mougin C, Benoit P, Barriuso E, Nelieu S (2011) Fate of pesticides in soils: toward an integrated approach of influential factors. In: Stoytcheva M (ed) Pesticides in the modern world—risks and benefits. InTech
21. Chen L, Deng H, Cui H, Fang J, Zuo Z, Deng J et al (2017) Inflammatory responses and inflammation-associated diseases in organs. Oncotarget 9(6):7204–7218
22. Chen L, Moon J-H, Ma X, Zhang L, Chen Q, Chen L et al (2018) High performance graphene oxide nanofiltration membrane prepared by electrospraying for wastewater purification. Carbon 130:487–494
23. Cheng SC (2003) Heavy metals in plants and crops. Encycl Soils Environ 1:1–10
24. Collaborators G (2024) Global incidence, prevalence, years lived with disability (YLDs), disability-adjusted life-years (DALYs), and healthy life expectancy (HALE) for 371 diseases and injuries in 204 countries and territorie and 811 subnational locations, 1990–2021: a systemat. Lancet (London, England) 403(10440):2133–2161
25. Dasi G, Lavanya T, Suneetha S, Vijayakumar S, Shim J-J, Thangaraju K (2022) Raman and X-ray photoelectron spectroscopic investigation of solution processed Alq_3/ZnO hybrid thin films. Spectrochimica Acta Part A: Mole Biomole Spectro 265
26. Fan G, Zhou J, Ruan F, Li Y, Tian H, Fan D et al (2023) The Z-scheme photocatalyst S-BiOBr/Bi_2Sn2O_7 with 3D/0D interfacial structure for the efficient degradation of organic pollutants. Sep Purif Technol 309
27. Feng X, Han G, Cai J, Wang X (2022) Au@Carbon quantum Dots-MXene nanocomposite as an electrochemical sensor for sensitive detection of nitrite. J Colloid Interface Sci 607(Pt 2):1313–1322
28. Gadd GM (2010) Metals, minerals and microbes: geomicrobiology and bioremediation. Microbiology (Reading, England) 156(pt 3):609–643
29. Gao Y, Huang Y, Li Y, Zhang Q, Cao J, Ho W, Lee S (2016) Plasmonic Bi/$ZnWO_4$ microspheres with improved photocata-lytic activity on NO removal under visible light. ACS Sustain Chem Eng 4:6912–6920
30. Geißler D, Nirmalananthan-Budau N, Scholtz L, Tavernaro I, Resch-Genger U (2021) Analyzing the surface of functional nanomaterials-how to quantify the total and derivatizable number of functional groups and ligands. Mikrochimica Acta 188(10)
31. Giannakoudakis DA, Zormpa FF, Margellou AG, Qayyum A, Colmenares-Quintero RF, Len C et al (2022) Carbon-based nanocatalysts (CnCs) for biomass valorization and hazardous organics remediation. Nanomaterials (Basel, Switzerland) 12(10)
32. Godja NC, Munteanu FD (2024) Hybrid nanomaterials: a brief overview of versatile solutions for sensor technology in healthcare and environmental applications. Biosensors 14(2)
33. Guabloche A, Alvariño L, Acioly TM, Viana DC, Iannacone J (2024) Assessment of essential and potentially toxic elements in water and sediment and the tissues of sciaena deliciosa (Tschudi, 1846) from the Coast of Callao Bay, Peru. Toxics 12(1)
34. Guerra FD, Attia MF, Whitehead DC, Alexis F (2018) Nanotechnology for environmental remediation: materials and applications. Molecules (Basel, Switzerland) 23(7)
35. Hashemi M, Khodaei MM, Teymouri M, Rashidi A, Mohammadi H (2016) Preparation of NiO nanocatalyst supported on MWCNTs and its application in reduction of nitrobenzene to aniline in liquid phase. Synth React Inorg Met-Org Nano-Met Chem 46(7)
36. Hoque MM, Islam A, Islam AR, Pal SC, Mahammad S, Alam E (2023) Assessment of soil heavy metal pollution and associated ecological risk of agriculture dominated mid-channel bars in a subtropical river basin. Sci Rep 13(1)

37. Hu X, Wang J, Lv Y, Liu X, Zhong J, Cui X et al (2021) Effects of heavy metals/metalloids and soil properties on microbial communities in farmland in the vicinity of a metals smelter. Front Microbiol 12
38. Huang Y, Liang Y, Rao Y, Zhu D, Cao J-J, Shen Z et al (2017) Environment-friendly carbon quantum dots/znfe2o4 photocatalysts: characterization, biocompatibility, and mechanisms for NO removal. Environ Sci Technol 51:2924–2933
39. Huang Y, Zhu D, Zhang Q, Zhang Y, Cao J-J, Shen Z et al (2018) Synthesis of a $Bi_2O_2CO_3$/$ZnFe_2O_4$ heterojunction with enhanced photocatalytic activity for visible light irradiation-induced NO removal. Appl Catal B 234:70–78
40. Hunt CF, Voulvoulis N (2020) Chemical pollution of the aquatic environment and health. In: Harrison RM (ed) Environmental pollutant exposures and public health, pp 39–69
41. Hussain E, Ahtesham A, Shahadat M, Ibrahim MN, Ismail S (2024) Recent advances of Clay/polymer-based nanomaterials for the treatment of environmental contaminants in wastewater: a review. J Environ Chem Eng 12(3)
42. Jackson P, Jacobsen NR, Baun A, Birkedal R, Kühnel D, Jensen KA et al (2013) Bioaccumulation and ecotoxicity of carbon nanotubes. Chem Central J 7(1)
43. Jat M, Dohling PN, Ahuja A, Singh J (2022) Effect of pesticides on soil ecosystem services and processes. Indian J Entomol 84(4):981–990
44. Jayawardena HS, Liyanage SH, Rathnayake K, Patel U, Yan M (2021) Analytical methods for characterization of nanomaterial surfaces. Anal Chem 93(4):1889–1911
45. Joel EF, Lujanienė G (2022) Progress in graphene oxide hybrids for environmental applications. Environments 9(12)
46. John JP, Mary-Nancy TE, Bindu-Sharmila TK (2021) A comprehensive review on the environmental applications of graphene–carbon nanotube hybrids: recent progress, challenges and prospects. Mater Adv 2(21):6816–6838
47. John J, Nandhini AR, Chellam PV, Sillanpää M (2022) Microplastics in mangroves and coral reef ecosystems: a review. Environ Chem Lett 20:397–416
48. Kalia A, Gosal SK (2011) Effect of pesticide application on soil microorganisms. Arch Agron Soil Sci 57(6):569–596
49. Kannan P, Maduraiveeran G (2023) Carbon nanocomposites-based electrochemical sensors and biosensors for biomedical diagnostics. Curr Med Chem
50. Kanwal H, Raza SH, Ali S, Iqbal M, Shad MI (2024) Effect of riboflavin on redox balance, osmolyte accumulation, methylglyoxal generation and nutrient acquisition in indian squash (Praecitrullus fistulosus L.) under chromium toxicity. Environ Sci Pollut Res Int 31(14):20881–20897
51. Kaur H, Devi N, Siwal SS, Alsanie WF, Thakur MK, Thakur VK (2023) Metal-organic framework-based materials for wastewater treatment: superior adsorbent materials for the removal of hazardous pollutants. ACS Omega 8(10):9004–9030
52. Kinuthia GK, Ngure V, Beti D, Lugalia R, Wangila A, Kamau L (2020) Levels of heavy metals in wastewater and soil samples from open drainage channels in Nairobi, Kenya: community health implication. Sci Rep 10(1)
53. Kjellstrom T, Lodh M, McMichael T, Ranmuthugala G, Shrestha R, Kingsland S (2006) Air and water pollution: burden and strategies for control. In: Jamison DT, Breman JG, Measham AR, Alleyne G, Claeson M, Evans DB et al (eds) Disease control priorities in developing countries, 2nd edn. Washington (DC)
54. Kolahdouz M, Xu B, Nasiri AF, Fathollahzadeh M, Manian M, Aghababa H et al (2022) Carbon-related materials: graphene and carbon nanotubes in semiconductor applications and design. Micromachines 13(8)
55. Krasley AT, Li E, Galeana JM, Bulumulla C, Beyene AG, Demirer GS (2024) Carbon nanomaterial fluorescent probes and their biological applications. Chem Rev 124(6):3085–3185
56. Lasek-Bal A, Rybicki W, Student S, Puz P, Krzan A, Derra A (2024) Direct exposure to outdoor air pollution worsens the functional status of stroke patients treated with mechanical thrombectomy. J Clin Med 13(3)

57. Leong SL, Tiong SI, Siva SP, Ahamed F, Chan C-H, Lee CL et al (2022) Morphological control of cellulose nanocrystals via sulfuric acid hydrolysis based on sustainability considerations: an overview of the governing factors and potential challenges. J Environ Chem Eng 10(4)
58. Li H, Kang Z, Liu Y, Lee S-T (2012) Carbon nanodots: synthesis, properties, and applications. J Mater Chem 22(46):24230–24253
59. Li J, Dong X, Sun Y, Cen W, Dong F (2018) Facet-dependent interfacial charge separation and transfer in plasmonic photo-catalysts. Appl Catal B 226:269–277
60. Li J, Wang JT, Hu HW, Ma YB, Zhang LM, He JZ (2016) Copper pollution decreases the resistance of soil microbial community to subsequent dry-rewetting disturbance. J Environ Sci (China) 39:155–164
61. Li P, Kim S, Jin J, Chun DH, Park JH (2020) Efficient photodegradation of volatile organic compounds by iron-based met-al-organic frameworks with high adsorption capacity. Appl Catal B 263:118284–118293
62. Li X, Li H, Huang Y, Cao J, Huang T, Li R et al (2022) Exploring the photocatalytic conversion mechanism of gaseous formaldehyde degradation on TiO_2-x-OV surface. J Hazard Mater 424:127217–127226
63. Lico D, Vuono D, Siciliano C, Nagy JB, De-Luca P (2019) Removal of unleaded gasoline from water by multi-walled carbon nanotubes. J Environ Manag 237:636–643
64. Liu J, Li R, Yang B (2020) Carbon dots: a new type of carbon-based nanomaterial with wide applications. ACS Cent Sci 6(12):2179–2195
65. Liu Z, Li J, Luo YF, Christie P, Guo ZW (2013) Enhanced phytoextraction of soil arsenic by Miscanthus sacchariflorus. Environ Sci Technol 47(23):13745–13753
66. Lu Y, Huang Y, Cao J-J, Ho W, Zhang Q, Zhu D, Lee SC (2016) Insight into the photocatalytic removal of NO in air over nanocrystalline $Bi_2Sn_2O_7$ under simulated solar light. Ind Eng Chem Res 55(40):10609–11061
67. Luo S, Ke J, Yuan M, Zhang Q, Xie P, Deng L, Wang S (2018) $CuInS_2$ quantum dots embedded in Bi_2WO_6 nanoflowers for enhanced visible light photocatalytic removal of contaminants. Appl Catal B 221:215–222
68. Mahajan H, Mohanan KU, Cho S (2022) Facile synthesis of biocarbon-based MoS_2 composite for high-performance supercapacitor application. Nano Lett 22(20):8161–8167
69. Mallakpour S, Azadi E, Hussain CM (2023) Chapter 19—Innovative progress in functionalized carbon nanomaterials, their hybrids, and nanocomposites: fabrication, antibacterial, biomedical, bioactivity, and biosensor applications. In: Mallakpour S, Hussain CM (eds) Functionalized carbon nanomaterials for theranostic applications. Elsevier, Amsterdam, pp 439–479
70. Manisalidis I, Stavropoulou E, Stavropoulos A, Bezirtzoglou E (2020) Environmental and health impacts of air pollution: a review. Front Public Health 8(14)
71. Meharg AA, Rahman MM (2003) Arsenic contamination of Bangladesh paddy field soils: implications for rice contribution to arsenic consumption. Environ Sci Technol 37(2):229–234
72. Menaa F, Fatemeh Y, Vashist SK, Iqbal H, Sharts ON, Menaa B (2021) Graphene, an interesting nanocarbon allotrope for biosensing applications: advances, insights, and prospects. Biomed Eng Comput Biol 12
73. Mishra RK, Singh B, Acharya B (2024) A comprehensive review on activated carbon from pyrolysis of lignocellulosic biomass: an application for energy and the environment. Carbon Resour Convers 7(4)
74. Murjani BO, Kadu PS, Bansod M, Vaidya SS, Yadav MD (2022) Carbon nanotubes in biomedical applications: current status, promises, and challenges. Carbon Lett 32(5):1207–1226
75. Muttil N, Jagadeesan S, Chanda A, Duke M, Singh SK (2023) Production, types, and applications of activated carbon derived from waste tyres: an overview. Appl Sci 3(1)
76. Namla D, Mangse G, Koleoso PO, Ogbaga CC, Nwagbara OF (2022) Assessment of heavy metal concentrations of municipal open-air dumpsite: a case study of gosa dumpsite, Abuja. In: Mambo A, Gueye A, Bassioni G (eds) Innovations and interdisciplinary solutions for underserved areas, vol 449. Springer, Cham, pp 165–174

77. National Center for Biotechnology Information (2024) PubChem Substance Record for SID 347827832, Activated charcoal, DrugBank. Accessed from www.pubchem.ncbi.nlm.nih.gov/substance/347827832
78. Navaneethan RD, Thangaiyan S, Sounderrajan V, Rathish RJ, Nguyen TA, Rajendra S (2022) Animal health monitoring using nanosensor networks. In: Denizli A, Nguyen TA, Rajendran S, Yasin G, Nadda AK (eds) Nanosensors for smart agriculture. Elsevier, pp 573–608
79. Navarro-Torre S, Garcia-Caparrós P, Nogales A, Abreu MM, Santos E, Cortinhas AL, Caperta AD (2023) Sustainable agricultural management of saline soils in arid and semi-arid Mediterranean regions through halophytes, microbial and soil-based technologies. Environ Exp Botany 212
80. Naveen P, Jayakumar P, Haritha S, Pavithra M, Palanisamy K, Rathinavel T (2024) Carbon-polymer composites for environmental applications. Phys Sci Rev
81. Noamani SN, Sadrzadeh M (2019) Carbon-based polymer nanocomposite membranes for oily wastewater treatment. NPJ Clean Water 2(20)
82. Nurazzi NM, Asyraf MR, Rayung M, Norrrahim MN, Shazleen SS, Rani MS et al (2021) Thermogravimetric analysis properties of cellulosic natural fiber polymer composites: a review on influence of chemical treatments. Polymers 13(16)
83. Padhye LP, Srivastava P, Jasemizad T, Bolan S, Hou D, Shaheen SM et al (2023) Contaminant containment for sustainable remediation of persistent contaminants in soil and groundwater. J Hazard Mater 455
84. Padilla JT, Selim HM (2020) Environmental behavior of glyphosate in soils. In: Sparks DL (ed) Advances in agronomy, vol 159. Academic, pp 1–34
85. Park SK, Mahmood Q, Park HS (2013) Surface functional groups of carbon nanotubes to manipulate capacitive behaviors. Nanoscale 5(24):12304–12309
86. Pawelski D, Plonska-Brzezinska ME (2023) Microwave-assisted synthesis as a promising tool for the preparation of materials containing defective carbon nanostructures: implications on properties and applications. Materials 16(19)
87. Potes-Lesoinne HA, Ramirez-Alvarez F, Perez-Gonzalez VH, Martinez-Chapa S, Gallo-Villanueva RC (2022) Nanomaterials for electrochemical detection of pollutants in water: a review. Electrophoresis 43:249–262
88. Prekob Á, Szegedi MP, Muránszky G, Kristály F, Nagy M, Halasi G et al (2023) Development of magnetizable, nickel-ferrite-decorated carbon nanocomposites as hydrogenation catalyst for aniline synthesis. Int J Mole Sci 24(24)
89. Prüss-Ustün A, Wolf J, Corvalán C, Bos R, Neira M (2016) Preventing disease through healthy environments: a global assessment of the burden of disease from environmental risks. Accessed from WHO: https://iris.who.int/bitstream/handle/10665/204585/9789241565196_eng.pdf
90. Qian H (2021) Major factors influencing the size distribution analysis of cellulose nanocrystals imaged in transmission electron microscopy. Polymers 13(19)
91. Raimondo M, Naddeo C, Vertuccio L, Lafdi K, Sorrentino A, Guadagno L (2019) Carbon-based aeronautical epoxy nanocomposites: effectiveness of atomic force microscopy (AFM) in investigating the dispersion of different carbonaceous nanoparticle. Polymers 11(5)
92. Rajput P, Singh A, Agrawal S, Ghazaryan K, Rajput VD, Movsesyan H et al (2024) Effects of environmental metal and metalloid pollutants on plants and human health: exploring nano-remediation approach. Stress Biol 4(1)
93. Ranjan P, Yadav S, Sadique MA, Khan R, Chaurasia JP, Srivastava AK (2021) Functional ionic liquids decorated carbon hybrid nanomaterials for the electrochemical biosensors. Biosensors 11(11)
94. Rao Z, Xie X, Wang X, Mahmood A, Tong S, Ge M, Sun J (2019) Defect chemistry of Er3+-doped TiO_2 and its photocata-lytic activity for the degradation of flowing gas-phase VOCs. J Phys Chem C 123:12321–12334
95. Rathi K, Kim D (2023) Super-compressible and mechanically stable reduced graphene oxide aerogel for wearable functional devices. Sci Technol Adv Mater 24(1)

96. Roso M, Boaretti C, Pelizzo MG, Lauria A, Modesti M, Lorenzetti A (2017) Nanostructured photocatalysts based on different oxidized graphenes for VOCs removal. Ind Eng Chem Res 56:9980–9992
97. Rydz J, Šišková A, Andicsová EA (2019) Scanning electron microscopy and atomic force microscopy: topographic and dynamical surface studies of blends, composites, and hybrid functional materials for sustainable future. Mater Sci Eng (1)
98. Salama E, Mohamed S, Samy M, Mensah K, Ossman M, Elkady MF, Hassan SH (2024) Catalytic fabrication of graphene, carbon spheres, and carbon nanotubes from plastic waste. RSC Adv 14(3):1977–1983
99. Sarkar D, Shikha Rakesh S, Ganguly S, Rakshit A (2017) Management of increasing soil pollution in the ecosystem. Adv Res 12(2):1–9
100. Shahzad A, Aslam U, Ferdous S, Qin M, Siddique A, Billah M et al (2024) Combined effect of endophytic Bacillus mycoides and rock phosphate on the amelioration of heavy metal stress in wheat plants. BMC Plant Biol 24(1)
101. Sharma A, Kapoor D, Wang J, Shahzad B, Kumar V, Bali AS et al (2020) Chromium bioaccumulation and its impacts on plants: an overview. Plants (Basel, Switzerland) 9(1)
102. Shetty SS, Deepthi D, Harshitha S, Sonkusare S, Naik PB, Kumari NS, Madhyastha H (2023) Environmental pollutants and their effects on human health. Heliyon 9(9)
103. Shin M, Lim J, Park Y, Lee JY, Yoon J, Choi JW (2024) Carbon-based nanocomposites for biomedical applications. RSC Adv 14(10):7142–7156
104. Shrestha S, Sapkota KP, Lee I, Islam MA, Pandey A, Gyawali N, et al (2022) Carbon-based ternary nanocomposite: bullet type ZnO-SWCNT-CuO for substantial solar-driven photocatalytic decomposition of aqueous organic contaminants. Molecules (Basel, Switzerland) 27(24)
105. Singh B, Fang Y, Cowie BC, Thomsen L (2014) NEXAFS and XPS characterisation of carbon functional groups of fresh and aged biochars. Org Geochem 77:1–10
106. Singh B, Lohan S, Sandhu PS, Jain A, Mehta SK (2016) Functionalized carbon nanotubes and their promising applications in therapeutics and diagnostics. Nanobiomater Med Imaging 455–478
107. Speranza G (2021) Carbon nanomaterials: synthesis, functionalization and sensing applications. Nanomaterials 11(967)
108. Sun L, Yu S, Liu D (2024) Characterization of aerosol and CO_2 co-emissions around power plants through satellite-based synergistic observations. Remote Sens 16(9)
109. Szczepankowska J, Khachatryan G, Khachatryan K, Krystyjan M (2023) Carbon dots-types, obtaining and application in biotechnology and food technology. Int J Mole Sci 24(19)
110. Takeda S, Yoshida H (2004) HRTEM image simulation of carbon nanotubes under actual growth environment. Microsc Microanal 10(S03):18–19
111. Tudi M, Ruan DH, Wang L, Lyu J, Sadler R, Connell D et al (2021) Agriculture development, pesticide application and its impact on the environment. Int J Environ Res Public Health 18(3)
112. Ullah A, Lin YJ, Zhang H, Yu XZ (2023) Identification of the key genes involved in proline-mediated modification of cell wall components in rice seedlings under trivalent chromium exposure. Toxics 12(1)
113. Vácha R (2021) Heavy metal pollution and its effects on agriculture. Agronomy 11(9)
114. van-Mourik T, Bühl M, Gaigeot MP (2014) Density functional theory across chemistry, physics and biology. Philos Trans Ser A Math Phys Eng Sci 372(2011)
115. Veeman D, Shree MV, Sureshkumar P, Jagadeesha T, Natrayan L, Ravichandran M, Paramasivam P (2021) Sustainable development of carbon nanocomposites: synthesis and classification for environmental remediation. J Nanomater
116. Venkatachalam S (2016) Ultraviolet and visible spectroscopy studies of nanofillers and their polymer nanocomposites. In: Thomas S, Rouxel D, Ponnamma D (eds) Spectroscopy of polymer nanocomposites. William Andrew, pp 130–157
117. Verma VK, Subbiah S, Kota SH (2019) Sericin-coated polyester based air-filter for removal of particulate matter and volatile organic compounds (BTEX) from indoor air. Chemosphere 237

118. Viswanathan G, Kane DB, Lipowicz PJ (2004) High efficiency fine particulate filtration using carbon nanotube coatings. Adv Mater 16(22):2045–2049
119. Wang B, Lan J, Bo C, Gong B, Ou J (2023) Adsorption of heavy metal onto biomass-derived activated carbon: review. RSC Adv 13(7):4275–4302
120. Wang CZ (2008) Detecting invasive sericea lespedeza (Lespedeza cuneata) in Mid-Missouri pastureland using hyperspectral imagery. Environ Manag 41(6):853–862
121. Wang J, Wang R, Ma J, Sun Y (2022) Study on the application of shell-activated carbon for the adsorption of dyes and antibiotics. Water 14(22)
122. Wang X, Chen Y, Li Z (2021) Photocatalytic degradation of air pollutants using carbon-based nanocomposites. Nano Today 35
123. Wang Y, Tan R, Zhou L, Lian J, Wu X, He R et al (2021) Heavy metal fixation of lead-contaminated soil using Morchella mycelium. Environ Pollut (Barking, Essex: 1987) 289
124. Wang Z, Chen M, Huang Y, Shi X, Zhang Y, Huang T et al (2018) Self-assembly synthesis of boron-doped graphitic carbon nitride hollow tubes for enhanced photocatalytic NOx removal under visible light. Appl Catal B 239:352–361
125. Warner JH, Margine ER, Mukai M, Robertson AW, Giustino F, Kirkland AI (2012) Dislocation-driven deformations in graphene. Science 337(6091):209–212
126. Weissker U, Hampel S, Leonhardt A, Büchner B (2010) Carbon nanotubes filled with ferromagnetic materials. Materials 3(8):4387–4427
127. Weon S, Choi E, Kim H, Kim J, Park H-J, Kim S-M et al (2018) Active 001 facet exposed TiO_2 nanotubes pho-tocatalyst filter for volatile organic compounds removal: from material development to commercial indoor air cleaner application. Environ Sci Technol 52:9330–9340
128. Weon S, Kim J, Choi W (2018) Dual-components modified TiO_2 with Pt and fluoride as deactivation-resistant photocatalyst for the degradation of volatile organic compound. Appl Catal B 220:1–8
129. WHO (1996) Permissible limits of heavy metals in soil and plants. World Health Organization, Geneva, Switzerland
130. WHO (2020) Mortality and burden of disease from ambient air pollution. Accessed 2024, from World Health Organisation: https://www.who.int/data/gho/data/indicators/indicator-details/GHO/ambient-and-household-air-pollution-attributable-death-rate-(per-100-000-population)
131. WHO (2022) Compendium of WHO and other UN guidance on health and environment. World Health Organization, Geneva
132. WHO (2024).Air pollution. Accessed from World Health Organization: https://www.who.int/health-topics/air-pollution#tab=tab_1
133. Wolf J, Johnston RB, Ambelu A, Arnold BF, Bain R, Brauer M et al (2023) Burden of disease attributable to unsafe drinking water, sanitation, and hygiene in domestic settings: a global analysis for selected adverse health outcomes. Lancet (London, England) 401(10393):2060–2071
134. Wu CF, Chang T, Wu TP, Leng KJ, Lin MC, Huang SC (2022) Impact of globalization on the environment in major CO_2-emitting countries: evidence using bootstrap ARDL with a Fourier function. Front Public Health 10
135. Xia S, Zhang G, Meng Y, Yang C, Ni Z, Hu J (2020) Kinetic and mechanistic analysis for the photodegradation of gaseous formaldehyde by core-shell CeO_2@LDHs. Appl Catal B 278:119266–119279
136. Yadav S, Raman SA, Meena H, Goswami AG, Kumar BV, Jain P et al (2022) An update on graphene oxide: applications and toxicity. ACS Omega 7(40):35387–35445
137. Yang Z, Xu Z, Geng L, Shu W, Zhu T (2020) Effect of multi-walled carbon nanotubes on extractability of Sb and Cd in contaminated soil. Ecotoxicol Environ Saf 205
138. Yaroshenko I, Kirsanov D, Marjanovic M, Lieberzeit PA, Korostynska O, Mason A et al (2020) Real-time water quality monitoring with chemical sensors. Sensors (Basel, Switzerland) 20(12)
139. Zhang BT, Yan Z, Liu Y, Chen Z, Zhang Y, Fan M (2022) Nanoconfinement in advanced oxidation processes. Crit Rev Environ Sci Technol 53(12):1197–1228

140. Zhang P, Yang M, Lan J, Huang Y, Zhang J, Huang S et al (2023) Water quality degradation due to heavy metal contamination: health impacts and eco-friendly approaches for heavy metal remediation. Toxics 11(10)
141. Zhang Y, Tang Z-R, Fu X, Xu Y-J (2011) Nanocomposite of Ag-AgBr–TiO_2 as a photoactive and durable catalyst for degradation of volatile organic compounds in the gas phase. Appl Catal B 106:445–452
142. Zhang Y, Zhang Y, Xu W, Wu H, Shao Y, Han X et al (2023) Preparation methods of cellulose nanocrystal and its application in treatment of environmental pollution: a mini-review. Colloid Interface Science Commun 53
143. Zhou H, Wen Z, Liu J, Ke J, Duan X, Wang S (2019) Z-scheme plasmonic Ag decorated WO_3/Bi_2WO_6 hybrids for enhanced photocatalytic abatement of chlorinated-VOCs under solar light irradiation. Appl Catal B 242:76–84
144. Zhu D, Wang Z, Zhang Z (2023) Effects of heavy metal pollution and soil physicochemical properties on the Sphagnum farmland soil microbial community structure in Southern Guizhou China. Int J Phytorem 25(13):1762–2177
145. Zhu X, Jin C, Li X-S, Liu J-L, Sun Z, Shi C et al (2017) Photocatalytic formaldehyde oxidation over plasmonic Au/TiO_2 under visible light: moisture indispensability and light enhancement. ACS Catal 7:6514–6524
146. Zorn G, Liu LH, Arnadóttir L, Wang H, Gamble LJ, Castner DG, Yan M (2014) X-ray photoelectron spectroscopy investigation of the nitrogen species in photoactive perfluorophenylazide-modified surfaces. J Phys Chem 118(1):376–383
147. Zunaidi AA, Lim LH, Metali F (2024) Heavy metal tolerance and accumulation in the Brassica species (Brassica chinensis var. parachinensis and Brassica rapa L.): a pot experiment. Heliyon 10(8)
148. Zunita M (2021) Graphene oxide-based nanofiltration for Hg removal from wastewater: a mini review. Membranes 11(4)

Risk Assessment and Regulation

Potential Effects of Nanomaterials on Honey Bees: Impacts on Survival, Physiology, and Behaviour

R. Karthik, R. Loganathan, Sabuj Ganguly, Rahul Datta, and Mukul Kumar Deka

Abstract Honey bees play a crucial role in ecosystems, offering essential ecosystem services, such as provisioning, regulating, and cultural services. The application of nanotechnology is expanding in various fields, including agriculture and apiculture. While the use of nanomaterials in these sectors holds promise, it can have negative consequences for honey bees, potentially affecting their pollination efficiency. These bees are exposed to nanomaterials through various routes, viz., application to crops, beekeeping activities, and environmental deposition. Such exposure can affect their biology, physiology, survival, and behaviour. This chapter explores the potential effects of nanomaterials on honey bees, drawing on relevant studies.

Keywords Nanoparticle exposure · Honey bees · Digestive system · Detoxification system · Behavioural aspects

1 Introduction

Among insects, honey bees are the most fascinating species, providing vital ecosystem services such as provisioning (honey, pollen, wax, propolis, royal jelly, and bee venom), regulating (crop pollination), and cultural services (apicultural tourism) [6]. Additionally, honey bees act as bioindicators for environmental monitoring due to their sensitivity to changes in their surroundings [7]. While more than 20,000

R. Karthik (✉) · M. K. Deka
Department of Entomology, Assam Agricultural University, Jorhat, Assam, India
e-mail: karthikagr1996@gmail.com

R. Loganathan
Division of Entomology, ICAR-Indian Agricultural Research Institute, New Delhi, India

S. Ganguly
BSM&TC (BTSSO), Central Silk Board, Baripada, Odisha, India

R. Datta
Department of Geology and Pedology, Faculty of Forestry and Wood Technology, Mendel University in Brno, Brno, Czech Republic

L. F. Fraceto et al. (eds.), *Emerging Nanotechnologies for Agroecosystem Management*, Smart Nanomaterials Technology, https://doi.org/10.1007/978-981-95-0187-8_12

bee species have been reported worldwide, only eleven species of honey bees are predominantly recognized, with the Western honey bee (*Apis mellifera*) and Indian honey bee (*Apis cerana*) being the most significant [37, 47]. Beekeeping with these two species offers livelihood opportunities for many people around the globe. The other major wild bee species include *Apis dorsata, Apis laboriosa, Apis florea, Apis andreniformis, Apis koschevnikovi,* and *Apis nuluensis* [12].

Recently, the potential applications of nanotechnology in agriculture have gained significant attention. The utilization of nano fertilizers in agricultural crops has been shown to enhance yield, while nanoparticles and nanoparticle-based pesticides effectively manage biotic stresses such as insect pests, plant diseases, and weeds [29, 38, 54]. These nanomaterials offer several advantages, including reduced application doses, improved solubility due to higher surface area-to-volume ratio, stable and controlled release, and increased deposition on target surfaces [39].

The application of nanotechnology extends to the beekeeping sector as well. For example, providing artificial feed during the dearth periods is critical for effective colony management, as inappropriate feeding can increase susceptibility of bees to diseases and mortality. Recent studies have found that feeding honey bee colonies with chitosan nanoparticles can enhance colony performance, including increased bee populations, enhanced hygienic behaviour, and comb-building abilities [1]. Similarly, incorporating zinc oxide nanoparticles into honey syrup has been shown to improve the antioxidant status and body weight of newly emerged bees, boosting the performance of both queens and workers and enhancing their heat resistance [4]. Moreover, nanotechnology offers promising solutions for managing pests and diseases affecting honey bees. For instance, glycerol monolaurate nano capsules were found to be effective against American foulbrood disease caused by the bacteria *Paenibacillus larvae* [41]. Silver nanoparticles derived from camphor extract and tea tree oil have demonstrated effectiveness against bacterial infections [44], while zinc oxide (ZnO) and silver (Ag) nanoparticles synthesized from *Withania coagulans* fruit extract significantly inhibit chalkbrood disease (*Ascosphaera apis*) [28]. The combination of bee venom and silver nanoparticles enhances antibacterial efficacy and extends the lifespan of honey bees [18]. Furthermore, thymol nanoemulsions effectively reduce *Varroa* mite infestation [20], and nanoformulations of essential oils like citronella and mustard exhibit significant effect on the greater wax moth (*Galleria mellonella*) [34].

Despite the numerous advantages of applying nanomaterials, there are concerns regarding their impacts on non-target organisms, including honey bees [31]. Therefore, this chapter aims to provide comprehensive information regarding the potential effects of nanomaterials on the various physiological, behavioural, and survival aspects of honey bees.

2 Nanomaterials

Nanomaterials are natural or manufactured materials with extremely small-sized particles (ranges from 1 to 100 nm), with different dimensions of nanoscales from zero to three dimensions, and available as unbound or aggregated with different shapes [55]. The nanomaterials can be solid, colloid, aerogels, or ferrofluids. Solid nanomaterials are either in single phase (crystalline and amorphous) or multiphase (coated particles and matrix composites) while colloid, aerogels, and ferrofluids nanomaterials are under multiphase system and these nanomaterials can be synthesized by mechanical methods or gas or liquid phase reactions [19]. Based on construction, nanomaterials are classified as carbon-based, metal-based, metalloid-based, composites, and dendrimers [52]. Carbon-based nanomaterials are hollow tubes or sphere-shaped while metal-based nanomaterials can be quantum dots, nano silvers, and oxides with metal bases [10]. Silicon nanoparticles are metalloid-based nanomaterials and in composites, a combination of nanoparticles with other materials such as nano-sized clays. Some of the most commonly used materials are zinc oxide, titanium dioxide, copper oxide, silicon dioxide, silver, and fullerene for the synthesis of nanomaterials.

3 Exposure Route of Nanomaterials to Honey Bees

In agricultural ecosystems, honey bees primarily facilitate the pollination of cross-pollinated crops [33]. Honey bee pollination involves both active and passive collection of pollen, known as melittophily, and these pollens are the predominant protein source of bees [11]. During the active pollination process, bees intentionally collect and gather pollen using their corbicula or pollen basket [45]. In contrast, passive pollination occurs through electrostatic forces, whereby the positive charge is generated by the wings of foraging bees attracts the pollen from the flowers having a negative charge [22]. Apart from pollen, bees also collect nectar, the primary carbohydrate source that is essential for the growth and development of their colonies [51].

It is important to note that while foraging for pollen, honey bees can unintentionally collect non-pollen particles from plants [13]. Nanoparticles or nanoparticle-based pesticides may settle on plant and flower surfaces during foliar applications, leading to their accumulation in pollen and on flowers. Because these particles are much smaller than pollen grains, bees may inadvertently gather them while pollinating [26]. Furthermore, bees can be exposed to nanomaterials through beekeeping practices, as recent studies indicate a potential usage of nanomaterials in apiculture (Fig. 1).

Fig. 1 Exposure routes of nanomaterials to honey bees. Illustration created with BioRender.com (accessed on 14 October 2024)

4 Potential Effects of Nanomaterials on Honey Bees

4.1 Effect of Nanomaterials on Survival of Honey Bees

A honey bee colony consists of three different castes, namely queen, drone, and worker. Usually, a colony contains a queen bee, a few drone bees, and many worker bees. The total population per colony varies from 20,000 to 40,000 bees. Honey bees are eusocial organisms and they have the characteristics such as overlapping generations, cooperative brood care, and division of labour. The honey bee is a holometabolous insect, its life cycle consists of egg, larva, pupa, and adult. The duration of each life stage varies with the honey bee castes. The developmental period of queen, worker, and drone is 16, 21, and 24 days, respectively. Three days after mating and fertilization, the queen bee will start laying eggs. The egg period is three days for all the castes. After the egg hatches, the larva develops by obtaining the major food sources such as bee bread (honey and pollen) and royal jelly. The larvae are determined to become queen bee receives royal jelly throughout their larval duration while the worker and drone bees receive only for the first three days. The feeding process to larva is done by nursing bees through mouth-to-mouth trophallaxis. After the larval period, the cells are sealed after 8, 9, and 10 days for queen, worker, and drone, respectively [58]. The adult queen bee lives up to 2–4 years based on environmental conditions. In contrast, the worker bees live from 6 weeks to 3 months, while the drone bees live up to 6 months.

Several experiments have been conducted to study the effect of various nanoparticles and nanoparticle-based pesticides on honey bee survival. The survival of honey bees varies depending on the type and formulation of nanomaterials used (Table 1).

Table 1 Potential effects of nanomaterials on survival of honey bees

Sl. No	Experiment	Effect on bees	References
1	Evaluation of zinc oxide (ZnO) nanoparticles synthesized from *Petroselinium crispum* on *A. mellifera*	A significant reduction in the survival of worker bees has been observed in nanoparticle treatments compared to control. The highest toxicity was recorded in the treatment with ZnO nanoparticles at a concentration of 500 mg/100 mL	[32]
2	Effect of co-exposure to titanium dioxide (TiO_2) nanoparticles and deltamethrin on the toxicity of *A. mellifera*	The LD_{50} values for the combined exposure of TiO_2 nanoparticles and deltamethrin at 24, 48, and 72 h (h) were 0.101, 0.09, and 0.02 μg/bee, respectively	[53]
3	Influence of dinotefuran using nanomaterial based on 1,3,5-triformylphloroglucinol and benzidine or melamine monomers covalent organic framework carrier on honey bees	Reduced acute toxicity on honey bees was noticed in the neonicotinoid insecticide dinotefuran loaded with the nanomaterial, i.e. 1,3,5-triformylphloroglucinol and benzidine or melamine monomers covalent organic framework	[42]
4	Study on the effect of Chlorfenapyr nanopesticide based on O-carboxymethyl chitosan on honey bees	The LC_{50} values of chlorfenapyr suspension (CF SC) and chlorfenapyr encapsulated with O-carboxymethyl chitosan nanoparticles (CF@O-CMCS) were calculated. At 96 h, the LC_{50} value for CF@O-CMCS was 11.41 mg/L, compared to 2.71 mg/L for CF SC, indicating that CF@O-CMCS has four times lower toxicity than CF SC. This study establishes the relative safety of using chlorfenapyr nanopesticide on honey bees	[27]
5	Evaluation of nano emulsion of peppermint oil on the toxicity of *A. mellifera*	This study evaluates the toxicity of peppermint oil in its nano emulsion form compared to its crude form. The nano emulsion form exhibited greater toxicity than the crude form. The LC_{50} for nano emulsion form was 2629.85 ppm/mL for oral exposure and 4246.84 ppm/mL for contact exposure. In contrast, the crude oil recorded oral and contact LC_{50} values of 5471.13 and 11,895.65 ppm/mL, respectively	[59]
6	Effect of sublethal concentrations (20% LC_{50}) of cadmium oxide (CdO) and lead oxide (PbO) nanoparticles on *A. mellifera*	The lowest survival was recorded in the treatment of CdO nanoparticles, followed by the PbO nanoparticles and control	[3]

(continued)

Table 1 (continued)

Sl. No	Experiment	Effect on bees	References
7	Study on the toxicity of fullerenol nanoparticles on *A. mellifera carnica*	The maximum tested dose of fullerenol nanoparticles is 68.8 μg/bee. However, after 48 h of feeding, none of the tested dose of fullerenol nanoparticles showed toxic effects on worker bees	[35]
8	Effect of silica nano particles and gallic acid nanohybrids using silica nanoparticles on the survival of *A. mellifera*	The nano hybrids of gallic acid show lower toxicity to bees compared to gallic acid alone. The LD_{50} values are as follows: 9700 ppm for gallic acid, >18,000 ppm for silica nanoparticles, 30,400 ppm for hybrid of silica nanoparticles and gallic acid via alkoxide bonding and 91,600 ppm for hybrid of silica nanoparticles and gallic acid via amide bonding. This study concludes that using silica nanoparticles and their hybrids with gallic acid is advantageous for bee survival	[17]
9	Effect of chronic exposure to cerium (IV) oxide (CeO_2) nanoparticles on *A. mellifera carnica*	The survival of the bees was not significantly impacted by the chronic exposure to CeO_2 nanoparticles	[36]
10	Impact of zinc oxide (ZnO) nanomaterials on *A. mellifera carnica*	Two days after chronic oral exposure to ZnO nanoparticles, no significant effects on bee mortality were observed	[21]
11	Evaluation of acute toxicity of poly (vinylferrocenium) with platinum nanoparticles on *A. mellifera*	The LC_{50} values for 48 and 96 h after exposure to poly (vinylferrocenium) and poly (vinylferrocenium) supported with platinum nanoparticles were determined. For 48 and 96 h after exposure, the LC_{50} values of poly (vinylferrocenium) were 148.15 and 0.344, respectively. In contrast, for 48 and 96 h after exposure, the LC_{50} values of the poly (vinylferrocenium) supported with platinum nanoparticles were 12,458,374 and 178.262, which reveals the lower toxicity of poly (vinylferrocenium) supported with platinum nanoparticles compared to poly (vinylferrocenium) alone	[16]
12	Effect of carbon black nanomaterial and titanium dioxide nanoparticles on survival of *A. mellifera carnica*	The tested dose of 1 mg/L of nanosized carbon black and titanium dioxide nanoparticles did not affect the survival of bees	[30]

(continued)

Table 1 (continued)

Sl. No	Experiment	Effect on bees	References
13	Comparative toxicity study of nano boron and non-nano boron particles on *A. mellifera*	Four different concentrations such as 0.001, 0.01, 0.1, and 1 mg L of nano boron and non-nano boron particles were investigated. After 48 and 96 h of exposure, the LC_{50} values of non-nano boron particles were found to be 62.330 and 4.694, respectively, while it was 229.09 and 0.339 for 48 and 96 h after exposure, respectively for nano boron particles. According to these findings, for 48 h after exposure, non-nano boron particles have higher toxicity compared to nano boron particles, while for 96 h after exposure, nano boron particles have higher toxicity compared to non-nano boron particles	[15]
14	Acute toxicity studies of TiO_2, composite nanoparticles viz. TiO_2 with silver nanoparticles (Ag–TiO_2) and TiO_2 with zinc nanoparticles (Zn–TiO_2) on *A. mellifera*	The LC_{50} values for a 96-h exposure was reported as follows: 5.865 mg/L for TiO_2, 312.845 mg/L for Ag–TiO_2, and 6.315 mg/L for Zn–TiO_2 nanoparticles, respectively	[48]

4.2 *Effect of Nanomaterials on Physiology of Honey Bees*

The physiological functions of honey bees are very essential for their growth, development, and survival. Digestive system, excretory system, respiratory system, circulatory system, reproductive system, and endocrine system are the crucial physiological systems in honey bees.

4.2.1 Effect of Nanomaterials on Digestive System of Honey Bees

The digestive system of honey bees plays a critical role and the midgut epithelial cells are responsible for digestion and absorption of food (bee bread) as they secrete various digestive enzymes [43]. Nanomaterials can impact the digestive processes of honey bees by affecting the midgut epithelial cells (Fig. 2). For instance, a nine-day exposure to sublethal concentrations (20% LC_{50}) of CdO and PbO nanoparticles (0.01 mg/L for CdO and 0.65 mg/L PbO nanoparticles) resulted irregular distribution of epithelial cells, chromatin condensation in the nucleus, mitochondrial swelling and dilation of rough endoplasmic reticulum in the midgut epithelial cells of *A. mellifera*, demonstrating the adverse effects of these nanoparticles on midgut [14]. Similarly, O-carboxymethyl chitosan nanoparticles, which act as a carrier for nano pesticides, also affect bee digestion [40].

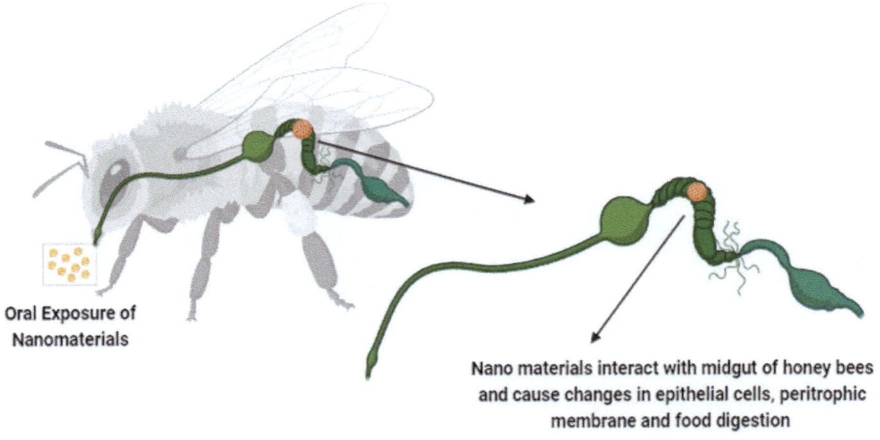

Fig. 2 Oral exposure of nanomaterials and their interaction with the gut of honey bees. *Illustration created with BioRender.com* (accessed on 16 October 2024)

Nanoparticles interact directly with the midgut epithelial cells and cause cytotoxicity. However, when the nanoparticles are loaded with botanicals or insecticide, their toxicity to midgut cells is reduced, as the chance of direct interaction of nanoparticles with cells is minimized. For example, a study found that the cytotoxicity of epithelial cells increased with oral exposure to empty solid lipid nanoparticles, leading to greater cell elimination, whereas solid lipid nanoparticles loaded with pyrethrum, exhibited lower cytotoxicity [46]. Similarly, abamectin nano capsules did not cause apparent damage on midgut tissue but resulted in a significant reduction in digestive enzymes activity [24].

The midgut epithelial cells lack cuticular intima. Instead of cuticular intima, midgut of honey bee contains peritrophic membrane. The peritrophic membrane surrounds the food bolus and acts as a barrier between lumen and epithelial cells thereby protecting the midgut epithelial cells from mechanical damage [25]. The application of nanomaterials affects peritrophic membrane, which can be supported by the study of [14]. The results revealed that the worker bees of *A. mellifera* had normal peritrophic membrane when they were fed with sugar syrup alone whereas the sugar syrup consisted of CdO nanoparticles causing the complete rupture of peritrophic membrane.

Gut microbiota plays a vital role in honey bees. The microorganisms present in the gut of honey bees aid in food digestion, growth, and development [9]. Additionally, gut microbiota enhances innate immunity and provides resistance against pathogens [50]. It was found that lambda-cyhalothrin nano capsules increased the diversity of microbial communities in the gut of *A. mellifera,* although they also reduced the network complexity of gut microbes [24]. Nano plastics also affect honey bee gut microbiota. For instance, oral exposure to 100 nm polystyrene nano plastics significantly reduced the abundance of beneficial gut microbiota, such as Lactobacillus and Bifidobacterium. Moreover, bees exposed to 100 nm polystyrene nano plastics became more susceptible to the pathogen *Hafnia alvei* [56].

4.2.2 Effect of Nanomaterials on Nervous, Reproductive, and Circulatory Systems of Honey Bees

The nervous system of bees plays a crucial role in coordinating their activities. In honey bees, the brain is a primary component of the central nervous system. A recent study has shown that worker bees of *A. mellifera* exposed to TiO_2 nanoparticles and deltamethrin exhibit alterations in brain physiology [53]. The impact of nanomaterials on nerve transmission of honey bees can be explained by the cholinergic nerve transmission, which is primarily mediated by the enzyme acetylcholinesterase (AchE). Dynamic responses in AchE activity were noted following chronic exposure to CeO_2 nanoparticles for nine days [36]. However, [21] found no significant changes in AchE in *A. mellifera carnica* exposed to ZnO nanoparticles. [30] also found that the tested dose of carbon black nanoparticles and TiO_2 nanoparticles did not affect enzymes such as AchE, catalase, and GST in the brain of honey bees *A. mellifera carnica.*

The reproductive ability of honey bees is also influenced by nanomaterials. Research indicated that artificial feeding with zinc nanoparticles (nZn) and selenium nanoparticles (nSe) mixed with sugar syrup enhances the egg-laying area of queen bees compared to control [4]. The observed egg laying areas for the control, nZn at 100 μg/L, nZn at 200 μg/L, nSe at 50 μg/L, and nSe at 100 μg/L were 2389, 2575, 2606, 2421, and 2439 cm^2, respectively. This study also highlights the beneficial effects of Zn on immune integrity in bees, as Zn deficiency can lead to pycnosis and loss of hemocytes, ultimately resulting in increased mortality. However, limited research exists on the effect of nanomaterials on the reproductive system and cellular immune responses of honey bees. Likewise, the studies on the effect of nanomaterials on respiratory tract and functions of honey bees are also remains unexplored.

4.2.3 Effect of Nanomaterials on Antioxidant and Detoxification System of Honey Bees

The detoxification system of honey bees is crucial for their survival, as detoxification enzymes play a major role in detoxifying xenobiotic substances, including nanomaterials. The detoxification process involves converting lipophilic substances into hydrophilic, excretable forms through three phases. In Phase I, functionalization of xenobiotic substances is mediated by enzymes such as carboxylesterases (CCE) and cytochrome P450 monooxygenases (P450), altering the structure of toxins, so they cannot interact with lipophilic target sites. Phase II entails conjugation reactions facilitated by glutathione *S*-transferase (GST) for solubilization, while Phase III manages the transport of conjugated products for excretion [8]. The detoxification of nanoparticles can be studied with the activity of CCE, P450, and GST. For example, exposure to ZnO nanoparticles for two days did not alter the activity of GST in *A. mellifera carnica* [21]. Conversely, exposure to CeO_2 nanoparticles elevated GST activity in the head and thorax of honey bees, indicating GST's role in mitigating oxidative damage caused by these nanoparticles [36]. Research regarding the effects of nanomaterials on the activity of P450 and CCE in honey bees is limited, highlighting a need for further studies in this area.

Honey bees are aerobic organisms. They produce reactive oxygen species (ROS) through the process of oxidative metabolism. These ROS involve in oxidation of DNA, RNA and proteins, subsequently leads to cell death [49]. To protect from these adverse effects, antioxidant system plays a major role in honey bees. These include the activity of antioxidant enzymes such as superoxide peroxidase (SOD), catalase (CAT), peroxidase (POD), and GST [57]. A few studies have been carried out to study the influence of nanomaterials on antioxidant system of honey bees. For instance, heat stress poses a significant threat to honey bees, making antioxidant status crucial during such conditions. Zinc (Zn) helps maintain the stability of antioxidant enzymes, while selenium (Se) is a well-known antioxidant. Recent findings suggest that artificial feeding with sugar and water syrup (1:1) containing 200 μg/L of Zn nanoparticles or 50 μg/L of Se nanoparticles enhances the antioxidant status of honey bees during heat stress, demonstrating the beneficial effects of utilizing nanoparticles under these conditions [5]. Similarly, the antioxidant capacity of honey bees was significantly enhanced with the treatment of fulleronol nanoparticles + paraquat when compared to sucrose + paraquat treatment [35]. Additionally, this study has also reported that the fulleronol nanoparticles + paraquat significantly enhanced the expression of antioxidant enzyme, manganese SOD.

4.3 Effect of Nanomaterials on Behaviour of Honey Bees

Honey bees are highly evolved insects and exhibits many distinct behaviours such as mating, festooning, thermoregulation, hygiene, foraging, and communication behaviour. The mating behaviour involves queen and drone bees in a typical nuptial flight. Five days after the new queen emerges from the pupa, it will make a nuptial flight and mate with drones. Queen bee maintains the cohesion of a colony through queen substance secreted by its mandibular glands. Festooning is another crucial for comb construction, while thermoregulation behaviour is important for temperature maintenance within the colony. Bees perform clustering to accumulate heat during winter and fanning to reduce excess heat during summer. The communication behaviour is distinct among bees and is mediated by both dancing behaviour and through pheromones. Foraging nectar and pollen is the major behaviour that bees perform outside the colony.

A very few studies have been examined the impact of nanoparticles on the behavioural aspects of honey bees. For example, the nanoparticle application affects feeding behaviour, as shown by previous studies such as increased concentrations of ZnO nanomaterials have been exhibited to reduce honey bee feeding [21], and

CdO nanoparticles have exhibited malaise effects [3]. Conversely, chronic exposure of 1 mg/L nanosized carbon black and titanium dioxide nanoparticles did not affect the feeding of *A. mellifera carnica* [30]. However, further research is needed to gain a comprehensive understanding of how nanomaterials affect other behavioural aspects of honey bees. For instance, application of external agrochemicals affects the foraging behaviour of bees. Although a previous study shows that the application of Proteck Calbor nanoparticles of 2 mL/L and the inclusion of pollinators enhanced the yield of sunflower [2], more studies can be conducted in future to find out the impact of nanoparticles on foraging behaviour of honey bees which provides clear idea for better understanding the pollination ecology and crop production.

5 Conclusion and Future Prospects

The majority of investigations referenced here primarily focused on the effect of nanomaterials on the survival, physiology, pests, and diseases affecting honey bees. To our knowledge, the influence of nanomaterials on many behavioural aspects and products of honey bees remains largely unexplored. Therefore, future studies should investigate the effects of nanomaterials on honey bees emphasizing various behaviours, pollination efficiency in different crops, and quality of bee products. Moreover, long-term molecular-level studies are necessary to understand the effects of nanomaterials on various physiological and biochemical mechanisms in honey bees.

The application of nanotechnology holds considerable promise in agriculture and scientific beekeeping. Honey bees are vital for ensuring food security, and any decline in their populations can adversely impact food production. Ensuring the safety of nanomaterials for honey bees through future studies will pave the way for their sustainable use in both agricultural and apicultural ecosystems. In this context, the green synthesis of nanoparticles emerges as a particularly valuable approach, offering environmentally friendly and less toxic alternatives to conventional methods.

References

1. Abou-Shaara HF (2022) Testing the ability of chitosan nanoparticles to improve development and performance of honey bee colonies during early spring. Entomol Res 52(4):214–222
2. Al-Husani AH, Alkinani LQ, Al-Sultany AK, Almosawy AN, Al-Amery AA, Kadhim ZK, Abbas ZF, Razak A, Kthem ARA (2021) Role of honey bee visits (*Apis mellifera*) and the treatment with nanoparticale (Proteck Calbor) on some growth characteristics for six varieties of sunflower (*Helianthus annuus* L.). Int J Agric Stat Sci 17:1807–1813
3. Al Naggar Y, Dabour K, Masry S, Sadek A, Naiem E, Giesy JP (2020) Sublethal effects of chronic exposure to CdO or PbO nanoparticles or their binary mixture on the honey bee (*Apis millefera* L.). Environ Sci Pollut Res 27:19004–19015

4. Amini MB, Sadeghi AA, Shawrang P, Chamani M, Aminafshar M (2020) Nano-selenium and nano-zinc oxide supplementation in syrup on laying area, population size and hsp gene expression of honey bees in hot climate. Acta Sci Anim Sci 43:e48574
5. Amini-Esfidvajani MB, Sadeghi AA, Shawrang P, Chamani M Aminafshar M (2022) Effect of nano-particles of zinc oxide and selenium on antioxidant status, aminotransferase enzymes activities and genes expression of *sod-1* and *vg* in honey bee during the hot season. J Trace Elem Min 2:100034
6. Aryal S, Ghosh S, Jung C (2020) Ecosystem services of honey bees; regulating, provisioning and cultural functions. J Apic 35(2):119–128
7. Barganska Z, Slebioda M, Namiesnik J (2016) Honey bees and their products: Bioindicators of environmental contamination. Crit Rev Env Sci Tech 46(3):235–248
8. Berenbaum MR, Johnson RM (2015) Xenobiotic detoxification pathways in honey bees. Curr Opin Insect Sci 10:51–58
9. Bonilla-Rosso G, Engel P (2018) Functional roles and metabolic niches in the honey bee gut microbiota. Curr Opin Microbiol 43:69–76
10. Bratovcic A (2019) Different applications of nanomaterials and their impact on the environment. Int J Mater Sci Eng 5(1):1–7
11. Clarke D, Morley E, Robert D (2017) The bee, the flower, and the electric field: electric ecology and aerial electroreception. J Comp Physiol A 203:737–748
12. Crane E (2009) *Apis* species:(Honey Bees). In: Encyclopedia of insects. Academic, pp 31–32
13. Connor LJ, Muir RG (2013) Pollen and bee bread. Am Bee J 153(7):727–729
14. Dabour K, Al Naggar Y, Masry S, Naiem E, Giesy JP (2019) Cellular alterations in midgut cells of honey bee workers (*Apis millefera* L.) exposed to sublethal concentrations of CdO or PbO nanoparticles or their binary mixture. Sci Total Environ 651:1356–1367
15. Dağlioğlu Y, Kabakçi D, Akdeniz G (2015) Toxicity of nano and non-nano boron particles on *Apis mellifera* (honey bee). Res J Chem Environ Sci 3(3):06–13
16. Dağlioğlu Y, Kabakcı D, Akdeniz G, Çelebi MS (2016) Determining the acute toxic effects of poly (vinylferrocenium) supported platinum nanoparticle (Pt/PVF+ NPS) on *Apis mellifera.* Mugla J Sci Tech 2(2):1–8
17. Domínguez E, Moliné MP, Churio MS, Arce VB, Mártire DO, Mendiara SN, Alvarez B, Gende LB, Damiani N (2019) Bioactivity of gallic acid–conjugated silica nanoparticles against *Paenibacillus larvae* and their host, *Apis mellifera* honeybee. Apidologie 50:616–631
18. El-Sayied Ali S, El-Ghannam G, Hashish MES, Elsayed H, Ali AK, Marzouk WM Attia Y (2024) Exploring bee venom and silver nanoparticles for controlling foulbrood pathogen and enhancing lifespan of honeybees. Sci Rep 14(1):19013
19. Findik F (2021) Nanomaterials and their applications. Period Eng Nat Sci 9(3):62–75
20. Gamal Eldin NK, Ebeid AA, Sallam AERM, Basuny NK Elaidy WK (2024) Efficacy of thymol nanoemulsion against *Varroa destructor* mites infesting *Apis mellifera* colonies under the stress of abiotic factors in Egypt. Egyptian J Basic Appl Sci 11(1):626–643
21. Glavan G, Milivojevic T, Bozic J, Sepcic K, Drobne D (2017) Feeding preference and sub-chronic effects of ZnO nanomaterials in honey bees (*Apis mellifera carnica*). Arch Environ Contam Toxicol 72:471–480
22. Greggers U, Koch G, Schmidt V, Durr A, Floriou-Servou A, Piepenbrock D, Gopfert MC, Menzel R (2013) Reception and learning of electric fields in bees. Proc R Soc B: Biol Sci 280(1759):20130528
23. Guo D, Wang Y, Li Z, Zhang DX, Wang C, Wang H, Liu Z, Liu F, Guo X, Wang N, Xu B, Gao Z (2024) Effects of abamectin nanocapsules on bees through host physiology, immune function, and gut microbiome. Sci Total Environ 930:172738
24. Guo D, Li Z, Zhang Y, Zhang W, Wang C, Zhang DX, Liu F, Gao Z, Xu B, Wang N (2024) The effect of lambda-cyhalothrin nanocapsules on the gut microbial communities and immune response of the bee elucidates the potential environmental impact of emerging nanopesticides. J Hazard Mater 479:135650
25. Hegedus DD, Toprak U, Erlandson M (2019) Peritrophic matrix formation. J Insect Physiol 117:103898

26. Hooven LA, Chakrabarti P, Harper BJ, Sagili RR, Harper SL (2019) Potential risk to pollinators from nanotechnology-based pesticides. Mol 24(24):4458
27. Hou R, Li C, Tan Y, Wang Y, Huang S, Zhao C, Zhang Z (2023) Eco-friendly O-carboxymethyl chitosan base chlorfenapyr nanopesticide for effective pest control and reduced toxicity to honey bees. Int J Biol Macromol 224:972–983
28. Hussain R, Hasan M, Iqbal KJ, Zafar A, Tariq T, Saif MS, Anjum SI (2023) Nano-managing silver and zinc as bio-conservational approach against pathogens of the honey bee. J Biotech 365:1–10
29. Jalil SU, Ansari MI (2020) Role of nanomaterials in weed control and plant diseases management. In: Husen A, Jawaid M (eds) Micro and nano technologies. Elsevier, Nanomaterials for agriculture and forestry applications, pp 421–434
30. Jemec A, Milivojevic T, Drobne D, Sepcic K, Glavan G (2016) No chronic effects on biochemical biomarkers, feeding and survival of carnolian honeybees (*Apis mellifera carnica*) after exposure to nanosized carbon black and titanium dioxide. Acta Biologica Slovenica 59(1):45–55
31. Kannan M, Elango K, Tamilnayagan T, Preetha S, Kasivelu G (2020) Impact of nanomaterials on beneficial insects in agricultural ecosystems. In: Thangadurai D, Sangeetha J, Prasad R (eds) Nanotechnology for food, agriculture, and environment. Nanotechnology in the life sciences. Springer, Cham, pp 379–393
32. Karim AS, Abdullah FO, Mohammad NZM (2024) Green synthesis and evaluation of ZnO NPs and study the effect of their toxic on honey bee (*Apis mellifera*). Baghdad Sci J 21(6):2124–2124
33. Khalifa SAM, Elshafiey EH, Shetaia AA, El-Wahed AAA, Algethami AF, Musharraf SG, AlAjmi MF, Zhao C, Masry SHD, Abdel-Daim MM, Halabi MF, Kai G, Naggar YA, Bishr M, Diab MAM, El- Seedi HR (2021) Overview of bee pollination and its economic value for crop production. Insects 12(8):688
34. Khalil MS (2023) Larvicidal activity of certain natural essential oils and their nanoemulsions against *Galleria mellonella* L. (Lepidoptera: Pyralidae). Ann Agric Sci Moshtohor 61(1):237–250
35. Kojic D, Purac J, Celic TV, Jovic D, Vukasinovic EL, Pihler I, Borisev I, Djordjevic A (2020) Effect of fullerenol nanoparticles on oxidative stress induced by paraquat in honey bees. Environ Sci Pollut Res 27:6603–6612
36. Kos M, Kokalj AJ, Glavan G, Marolt G, Zidar P, Bozic J, Novac S, Drobne D (2017) Cerium (IV) oxide nanoparticles induce sublethal changes in honeybees after chronic exposure. Environ Sci Nano 4(12):2297–2310
37. Kremen C (2018) The value of pollinator species diversity. Science 359(6377):741–742
38. Kumar M, Shamsi TN, Parveen R, Fatima S (2017) Application of nanotechnology in enhancement of crop productivity and integrated pest management. In: Prasad R, Kumar M, Kumar V (eds) Nanotechnology. Springer, Singapore, pp 361–371
39. Kumar S, Nehra M, Dilbaghi N, Marrazza G, Hassan AA, Kim KH (2019) Nano-based smart pesticide formulations: emerging opportunities for agriculture. J Control Release 294:131–153
40. Li SN, Wu H, Hou RQ, Zhou JT, Lin SK, Zhao KY, Wang R, Huang S, Zhang ZX (2023) Nano-pesticide carrier O-Carboxymethyl chitosan is indigestible in *Apis cerana cerana* and affects intestinal flora. Sci Total Environ 885:163769
41. Lopes LQ, Santos CG, de Almeida VR, Gende L, Raffin RP, Santos RC (2016) Evaluation of antimicrobial activity of glycerol monolaurate nanocapsules against American foulbrood disease agent and toxicity on bees. Microb Pathog 97:183–188
42. Ma X, Liu M, Chen Y, Huang Q, Cao C (2024) Adsorption of dinotefuran using TpBDMe COF nanoparticles for soil leaching reduction and bee safety. Chem Eng J 497:155082
43. Martins GF, Neves CA, Campos LAO, Serrao JE (2006) The regenerative cells during the metamorphosis in the midgut of bees. Micron 37(2):161–168
44. Masry SHD, Taha TH, Botros WA, Mahfouz H, Al-Kahtani SN, Ansari MJ, Hafez EE (2021) Antimicrobial activity of camphor tree silver nano-particles against foulbrood diseases and finding out new strain of *Serratia marcescens* via DGGE-PCR, as a secondary infection on honeybee larvae. Saudi J Biol Sci 28(4):2067–2075

45. Milne CP Jr, Pries KJ (1984) Honeybee corbicular size and honey production. J Apic Res 23(1):11–14
46. Oliveira CR, Caio EC, Domingues NFS, de Melo RTC, Malaspina O, Monica JC, Elaine CM, Zacarin S, Fraceto LF (2019) Nanopesticide based on botanical insecticide pyrethrum and its potential effects on honeybees. Chemosphere 236:124282
47. Osterman J, Aizen MA, Biesmeijer JC, Bosch J, Howlett BG, Inouye DW, Jung C, Martins DJ, Medel R, Pauw A, Seymour CL, Paxton RJ (2021) Global trends in the number and diversity of managed pollinator species. Agriculture Ecosyst Environ 322:107653
48. Özkan Y, Irende İ, Akdeniz G, Kabakçi D, Sökmen M (2014) Evaluation of the comparative acute toxic effects of TiO_2, Ag–TiO_2 and ZnO–TiO_2 composite nanoparticles on honey bee (*Apis mellifera*). J Int Environ Appl Sci 10(1):26–36
49. Pardini RS (1995) Toxicity of oxygen from naturally occurring redox-active pro-oxidants. Arch Insect Biochem Physiol 29(2):101–118
50. Raymann K, Moran NA (2018) The role of the gut microbiome in health and disease of adult honey bee workers. Curr Opin Insect Sci 26:97–104
51. Schmid-Hempel P (1987) Efficient nectar-collecting by honeybees I. Economic models. J Anim Ecol 209–218.
52. Saleh TA (2016) Nanomaterials for pharmaceuticals determination. Bioenergetics 5(1):1000226
53. Shahzad K, Manzoor F (2024) Evaluation of Apis mellifera brain morphophysiology in response to titanium dioxide nanoparticles and deltamethrin co-exposure. Physiol Entomol 49(3):189–201
54. Singh MD (2017) Nano-fertilizers is a new way to increase nutrients use efficiency in crop production. Int J Agric Sci 9(7):0975–3710
55. Trotta F, Mele A (2019) Nanomaterials: classification and properties. Nanosponges: synthesis and applications. Wiley-VCH Verlag GmbH & Co. KGaA, Weinheim, pp 1–26
56. Wang K, Zhu L, Rao L, Zhao L, Wang Y, Wu X, Zheng H, Liao X (2022) Nano-and micro-polystyrene plastics disturb gut microbiota and intestinal immune system in honeybee. Sci Total Environ 842:156819
57. Weirich GF, Collins AM, Williams VP (2002) Antioxidant enzymes in the honey bee, *Apis mellifera.* Apidologie 33(1):3–14
58. Yadav S, Kumar Y, Jat BL (2017) Honeybee: diversity, castes and life cycle. In: Entomology I (ed) Omkar. Springer, Singapore, pp 5–34
59. Youssef DAA, Abdelmegeed SM (2021) Polymer-based encapsulation of peppermint oil (*Mentha piperita*) nanoemulsion and its effects on life and some physiological activities of honeybees *Apis mellifera* (Hymenoptera: Apidae). Egyptian Pharm J 20(4):313–322

Interaction, Toxicity, and Fate of Nanoparticles in Agroecosystems

Pragati Khare, Noopur Khare, Raviraj Barot, Nitish R. Dave, Rahul Datta, Nisha Singh, and Sachidanand Singh

Abstract Nanotechnology has emerged as a promising field with significant applications in agriculture, offering potential benefits such as enhanced crop protection, improved nutrient delivery, and soil remediation. Due to their unique physicochemical properties, nanoparticles (NPs) have the potential to improve agricultural productivity and sustainability. However, their introduction into agroecosystems raises concerns regarding their interactions with plants, soil, microbial populations, and environmental sustainability. Understanding these interactions is crucial for evaluating both the advantages and risks associated with the use of NPs in agriculture. Nanoparticles influence plant physiology through root uptake and foliar absorption, impacting germination, growth, photosynthesis, and nutrient assimilation. While some NPs, such as silver (Ag) and titanium dioxide (TiO_2), may enhance plant growth at lower concentrations, excessive exposure can induce phytotoxicity, oxidative stress, and membrane damage. Additionally, the potential bioaccumulation of NPs in edible plant tissues raises food safety concerns. In soil ecosystems, NPs interact with soil particles and microbial communities, affecting soil fertility, water

P. Khare
Department of Pharmacy, BPS Educational Institution of Pharmacy, Etah, Uttar Pradesh, India

N. Khare
Bhai Gurdas Institute of Engineering and Technology, Sangrur, Punjab, India

Shri Ramswaroop Memorial University, Barabanki, Uttar Pradesh, India

R. Barot · N. Singh · S. Singh (✉)
Department of Biotechnology, School of Energy and Technology, Pandit Deendayal Energy University, Gandhinagar, Gujarat, India
e-mail: drsachinbioinfo@gmail.com

N. R. Dave
Department of Science and Technology (GoG), Gujarat State Biotechnology Mission, Gandhinagar, Gujarat, India

R. Datta
Department of Geology and Pedology, Faculty of Forestry and Wood Technology, Mendel University in Brno, Brno, Czech Republic

L. F. Fraceto et al. (eds.), *Emerging Nanotechnologies for Agroecosystem Management*, Smart Nanomaterials Technology, https://doi.org/10.1007/978-981-95-0187-8_13

retention, and nutrient cycling. While some NPs enhance soil properties and plant–microbe interactions, others, such as copper (Cu) and AgNPs, exhibit antimicrobial effects that may disrupt beneficial microbial populations. Furthermore, NPs' mobility and persistence in soil raise concerns regarding groundwater contamination and bioaccumulation in higher trophic levels. Addressing these challenges requires multidisciplinary research to assess the long-term ecological risks and develop regulatory frameworks for the safe use of nanotechnology in agriculture. A balanced approach that maximises the benefits of NPs while mitigating potential environmental and health risks is essential for ensuring sustainable agricultural practices.

Keywords Nanoparticles · Interactions · Toxicity · Transport · Risk assessment

1 Introduction

The field of nanotechnology has shown great promise and has a wide range of possible uses, especially in agriculture. Because of their special physicochemical characteristics, nanoparticles (NPs) have attracted a lot of interest because of their potential to improve agricultural sustainability and production [1]. Concerns about NPs' effects on plants, soil, and the environment are brought up by their introduction into agroecosystems. Evaluating the potential risks and benefits of nanoparticle applications in agriculture necessitates a clear understanding of their interactions with soil, plants, and the environment [2]. The interactions between nanoparticles in agroecosystems are examined in this essay, along with their effects on plant development, soil characteristics, microbial populations, and environmental sustainability.

2 Interactions with Plants

When applied to the soil or foliage of plants, nanoparticles can directly interact with the plants and affect a variety of physiological processes. It has been demonstrated that engineered nanoparticles, such as titanium dioxide nanoparticles (TiO_2NPs) and silver nanoparticles (AgNPs), influence root elongation, seed germination, and total plant growth. These effects are frequently dose-dependent, with greater doses causing phytotoxicity and lower concentrations perhaps encouraging plant growth. According to research, AgNP exposure, for example, can promote root elongation and seed germination in certain plant species but, because of oxidative stress and membrane damage, can hinder growth in other plant species [3].

Moreover, plants' absorption and transport of nanoparticles are important interactions that affect their physiological reactions. Plant tissues can be penetrated by nanoparticles via a variety of routes, such as foliar absorption and root uptake. After being ingested, nanoparticles have the potential to gather in many plant parts, including the roots, stems, leaves, and even fruits. Plant metabolism, photosynthesis,

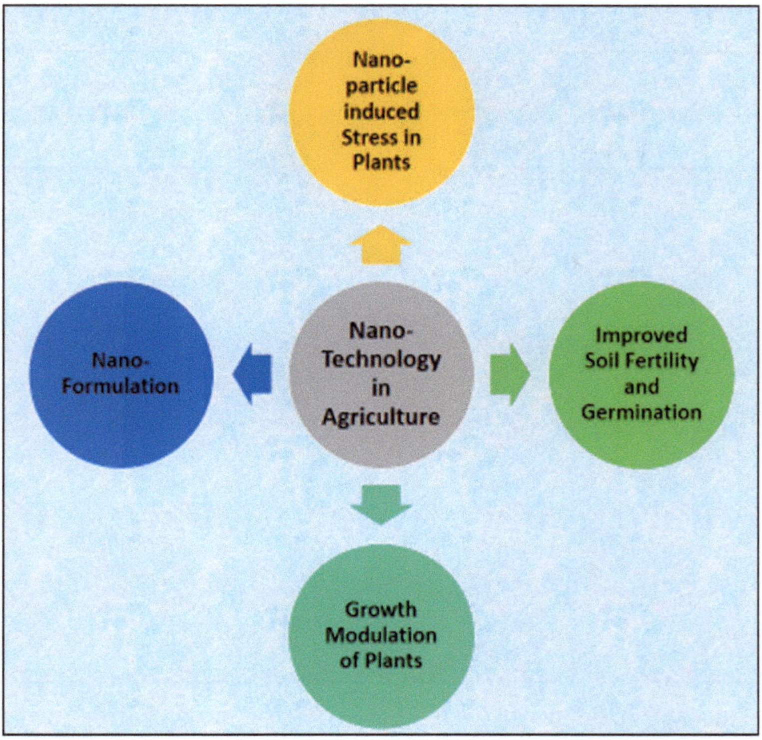

Fig. 1 Nanotechnology in agriculture

and nutrient uptake can all be impacted by the translocation of nanoparticles, which in turn affects crop yield and quality. On the other hand, issues with food safety and human health are raised by the possible bioaccumulation of nanoparticles in edible plant components [4] (Fig. 1).

3 Interactions with Soil

In addition to directly affecting plants, nanoparticles can interact with other elements of the soil to change its characteristics and the microbial communities that live there. The structure, porosity, and ability of the soil to retain water can all be impacted by nanoparticles that adsorb onto soil particles. For instance, it has been suggested that applying nanoscale additions, such as nano-silica or nano-clay, will increase soil fertility and improve nutrient retention, both of which will encourage plant development. Further research is necessary to determine the long-term effects of nanoparticle additions on soil health and microbial diversity [5].

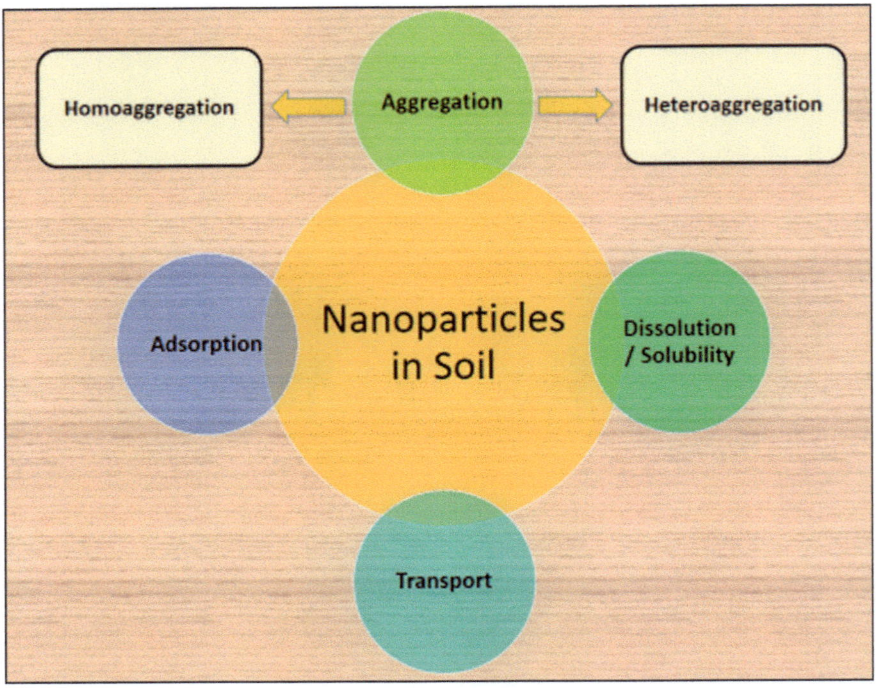

Fig. 2 Nanoparticles in soil

Additionally, soil microbial communities—which are vital to the decomposition of organic matter, the cycling of nutrients, and the interactions between plants and microbes—can be impacted by nanoparticles. According to studies, some nanoparticles, like AgNPs and CuNPs, have antibacterial qualities that can stop soil microorganisms from growing and functioning. This could reduce the impact of some soil-borne diseases, but it also has the potential to disturb beneficial microbial communities that are vital to the health of the soil ecosystem. Furthermore, accidental outcomes from the introduction of nanoparticles into the soil environment could include the emergence of microbial strains resistant to the particles or the encouragement of horizontal gene transfer among soil bacteria [6] (Fig. 2).

4 Environmental Implications

Significant environmental consequences are raised by the interactions of nanoparticles in agroecosystems, including worries about the persistence, mobility, and ecological risk of these particles. When applied to agricultural soils, nanoparticles may stay in the environment for lengthy stretches of time, which could cause them to bioaccumulate in soil organisms and then move up the food chain. Furthermore, there is a risk of groundwater pollution due to the mobility of nanoparticles within soil profiles and their tendency to leach, particularly in areas where nanoparticle use is intensive [7].

Comprehensive risk evaluations are also required due to the ecotoxicological impacts of nanoparticles on non-target animals, such as beneficial insects, aquatic organisms, and soil fauna. Agrochemicals based on nanoparticles may provide focused nutrient delivery or pest control, but their unintentional effects on non-target species may upset ecosystem dynamics and biodiversity. Furthermore, the sustainability of nanoparticle-enabled agriculture as a whole is called into question by the possibility of phytotoxicity or genotoxicity in plants [8].

Comprehensive risk evaluations are also required due to the ecotoxicological impacts of nanoparticles on non-target animals, such as beneficial insects, aquatic organisms, and soil fauna. Agrochemicals based on nanoparticles may provide focused nutrient delivery or pest control, but their unintentional effects on non-target species may upset ecosystem dynamics and biodiversity. Furthermore, the sustainability of nanoparticle-enabled agriculture as a whole is called into question by the possibility of phytotoxicity or genotoxicity in plants [18].

5 Toxicity of Nanoparticles in Agroecosystems

The field of nanotechnology has shown great promise and has the potential to be applied in many different fields, including agriculture. Because of their small size, high surface area-to-volume ratio, and quantum effects, NPs have special physicochemical features that make them appealing for use in fertilisers, agrochemicals, and plant growth regulators [10]. Nonetheless, questions have been raised about the toxicity that nanoparticles in agroecosystems may pose. In order to evaluate nanoparticles' environmental impact and guarantee sustainable agricultural practices, it is essential to comprehend how they interact with agricultural ecosystems [11].

5.1 Nanoparticles in Agroecosystems

In order to improve their efficacy and target delivery, agrochemicals like fertilisers, herbicides, and pesticides are increasingly include nanoparticles. Because of their antibacterial qualities and capacity to manage diseases and pests, metal-based nanoparticles, such as those based on silver, copper, zinc oxide, and titanium dioxide, are among the most often utilised in agricultural applications. Furthermore, the potential of carbon-based nanoparticles, such graphene and carbon nanotubes, to enhance soil fertility and plant growth is being investigated [12].

5.2 *Toxicity Mechanisms*

Because nanoparticles can interact with both the environment and living things, they may be hazardous in agroecosystems. Plants can absorb nanoparticles through a variety of pathways, such as soil uptake, foliar uptake, and root uptake. Once inside the plant, nanoparticles may disrupt physiological functions including water transport, nutrient uptake, and photosynthesis, which could have a negative impact on the growth and development of the plant. Additionally, through the food chain, nanoparticles can endanger human health and the environment by accumulating in various plant tissues [13].

5.3 *Impact on Soil Microorganisms*

Plant health, soil fertility, and nutrient cycling are all greatly influenced by soil microbes. However, microbial communities may be upset and their functions may be impacted by the introduction of nanoparticles into the soil. According to studies, some nanoparticles—like those of silver and copper—have antibacterial qualities that can prevent the growth of helpful soil microbes while encouraging the spread of harmful ones. The health of the soil and agricultural productivity may suffer as a result of this imbalance in the microbial communities of the soil [14].

5.4 *Ecological Implications*

Nanoparticles have the ability to be hazardous not just to individual species but also to entire agroecosystems. Long-term environmental persistence of nanoparticles can result in bioaccumulation and biomagnification throughout the food chain. Therefore, by consuming tainted food and water, creatures at higher trophic levels—including humans—may be exposed to elevated amounts of nanoparticles. Moreover, pollinators, aquatic life, and soil fauna are examples of non-target creatures that may be impacted by the release of nanoparticles into the environment during agricultural activities. This can upset ecosystem dynamics and biodiversity [15].

6 Risk Assessment and Regulation

The creation of regulatory frameworks for nanotechnology in agriculture is still in its early phases, and there are currently few standards for the safe application and disposal of agrochemicals based on nanoparticles [16]. Comprehensive risk evaluations that take into account the possible ecological effects of nanoparticles at various geographical and temporal scales are therefore desperately needed. Regulatory bodies also need to work with scientists, business leaders, and legislators to develop policies and guidelines for the sustainable application of nanotechnology in agriculture [17].

There are chances for innovation and advancement in agricultural methods due to the increasing application of nanoparticles in agroecosystems. However, the sustainability of the environment and human health are seriously threatened by the potential toxicity of nanoparticles [18]. To tackle these obstacles, multidisciplinary research endeavours are necessary to get a deeper comprehension of the interplay between nanoparticles and agroecosystems, establish dependable risk assessment techniques, and execute efficacious regulatory strategies. We can maximise the advantages of nanoparticles in agriculture while reducing their negative impacts on the environment and biodiversity by supporting responsible nanotechnology practices and taking a cautious approach [19].

7 The Fate of Nanoparticles in Agroecosystems: Understanding Implications for Sustainable Agriculture

Applications for nanotechnology can be found in many industries, including agriculture. It has become a very promising field. Because of their special physicochemical characteristics, NPs present fresh possibilities for raising agricultural output, such as better fertiliser delivery, pest control, and soil remediation. On the other hand, questions about NPs' behaviour, fate, and possible ecological effects are raised by their introduction into agroecosystems. In order to evaluate the advantages and disadvantages of NPs and maintain sustainable agricultural practices, it is essential to comprehend their fate in agroecosystems [20].

Following application, NPs may interact with organic matter, minerals, and microbes in the soil, affecting their movement and behaviour. In soil environments, the destiny of nanoparticles is determined by factors like coating, surface charge, and size [21]. Research has demonstrated that NPs can change through aggregation, dissolution, and surface alteration, which impacts their environmental persistence and bioavailability. Furthermore, NP–soil interactions and the ensuing transport mechanisms are significantly influenced by soil characteristics including pH, texture, and moisture content [22].

7.1 Transport and Mobility in Soil

In soil, NPs can display a variety of complex transport behaviours, including as retention, diffusion, and advection. Numerous variables, including soil porosity, hydraulic conductivity, and interactions with soil surfaces, affect their movement

[23]. Certain NPs have the ability to move through soil pores and enter nearby ecosystems or groundwater, whereas other NPs may agglomerate or adhere to soil particles, rendering them immobile. Predicting the possible leaching or accumulation of NPs in the environment requires an understanding of the mechanisms driving their mobility [24].

7.2 Uptake and Translocation by Plants

Nanoparticles have the ability to enter plants through their roots and go to aboveground tissues, where they may have an effect on the physiology, growth, and uptake of nutrients by the plants. A number of variables, including NP size, surface chemistry, and plant species, affect how well plants absorb NPs. Once internalised, NPs may have an impact on plant health and biochemical processes by interacting with cellular components. Moreover, NPs have the ability to move via vascular tissues within plants, possibly reaching edible sections and joining the food chain [25].

7.3 Impact on Soil Microorganisms

The breakdown of organic matter, the cycling of nutrients, and the general fertility of the soil are all greatly aided by soil microbes. Changes in microbial populations and functions brought about by NP introduction into soil ecosystems may have an impact on ecosystem services and soil health [26]. Certain NPs have the potential to be antimicrobial, but others can have an impact on gene expression, enzyme activity, and microbial metabolism. Determining the long-term consequences of NPs for the sustainability of agroecosystems requires an understanding of their impacts on soil microbiota [27] (Fig. 3).

8 Ecological Risks and Mitigation Strategies

Concerns concerning the possible ecological problems posed by NPs, such as soil contamination, biodiversity loss, and negative impacts on non-target organisms, are raised by the extensive use of NPs in agriculture [28]. To guide regulatory choices and risk management tactics, it is essential to examine the environmental destiny and toxicity of NPs using sound risk assessment frameworks. Environmentally friendly formulations, tailored distribution methods, and NP encapsulation are examples of mitigation strategies that might reduce possible dangers and maximise the advantages of NP-enabled agricultural technologies [29].

Agro ecosystems fate for nanoparticles is determined by intricate interactions between biological activities, soil features, and NP properties. In order to ensure

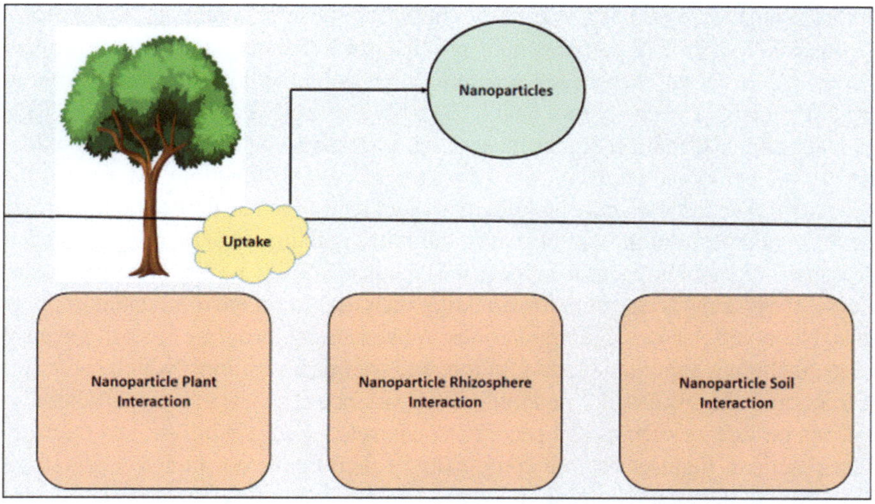

Fig. 3 Impact on soil microorganisms

sustainable agricultural operations and reduce potential hazards to soil health, plant production, and ecosystem integrity, it is imperative to comprehend the environmental behaviour of NPs. It will be easier to make wise decisions and create secure and practical agricultural nanotechnology solutions if research on NP destiny, transport, and ecological effects is conducted indefinitely [30].

9 Conclusion

Interactions, toxicity, and fate of nanoparticles in agroecosystems have become topics of significant concern and study in recent years due to the increasing utilisation of nanotechnology in agriculture. Nanoparticles, typically ranging from 1 to 100 nm in size, exhibit unique physicochemical properties that make them attractive for various agricultural applications, such as crop protection, nutrient delivery, and soil remediation. However, their introduction into agro ecosystems raises questions about their potential impacts on the environment, organisms, and ecosystem processes [31]. One of the primary focuses of research has been on understanding the interactions between nanoparticles and components of agro ecosystems, including soil, plants, microbes, and other organisms. Nanoparticles can interact with soil constituents, altering soil structure, nutrient availability, and microbial activity. These interactions may influence plant uptake of nanoparticles, potentially affecting plant growth, development, and overall health. Furthermore, nanoparticles may interact with soil-dwelling organisms, such as earthworms and beneficial microbes, with potential consequences for soil fertility and ecosystem functioning [32].

Assessing the toxicity of nanoparticles in agroecosystems is another crucial aspect of research. While nanoparticles hold promise for enhancing agricultural productivity, concerns have been raised regarding their potential adverse effects on non-target organisms and ecosystem health [33]. Nanoparticles can induce toxic effects in plants, animals, and soil organisms through various mechanisms, including oxidative stress, membrane damage, and disruption of cellular functions. Moreover, the toxicity of nanoparticles may be influenced by factors such as particle size, surface chemistry, concentration, and exposure duration, making it challenging to predict their environmental impacts accurately [34]. Understanding the fate of nanoparticles in agro ecosystems is essential for assessing their long-term environmental implications. Nanoparticles can undergo various transformation processes in soil environments, including aggregation, dissolution, and chemical reactions, which can affect their mobility, bioavailability, and ultimate persistence. The fate of nanoparticles also depends on factors such as soil properties, environmental conditions, and management practices, highlighting the complexity of their behaviour in agro ecosystems [35]. Overall, the interactions, toxicity, and fate of nanoparticles in agro ecosystems represent a multifaceted research frontier with significant implications for sustainable agriculture and environmental stewardship. While nanoparticles offer potential benefits for improving agricultural practices, their deployment must be approached cautiously to minimise potential risks to ecosystems and human health. Future research efforts should focus on elucidating the mechanisms underlying nanoparticle interactions, toxicity, and fate in agro ecosystems, as well as developing effective risk assessment strategies and mitigation measures to ensure the safe and sustainable deployment of nanotechnology in agriculture [36]. By integrating interdisciplinary approaches and engaging stakeholders, we can advance our understanding of nanoparticle-agro ecosystem interactions and foster the responsible development and deployment of nanotechnology for agricultural innovation [37].

References

1. Shang Y, Hasan MK, Ahammed GJ, Li M, Yin H, Zhou J (2019) Applications of nanotechnology in plant growth and crop protection: a review. Molecules 24(14):2558
2. Remédios C, Rosário F, Bastos V (2012) Environmental nanoparticles interactions with plants: morphological, physiological, and genotoxic aspects. J Botany
3. Junedi MA, Mukhopadhyay R, Manjari KS (2023) Alleviating salinity stress in crop plants using new engineered nanoparticles (ENPs). Plant Stress 29:100184
4. Wang X, Xie H, Wang P, Yin H (2023) Nanoparticles in plants: Uptake, transport and physiological activity in leaf and root. Materials 16(8):3097
5. Khan ST, Adil SF, Shaik MR, Alkhathlan HZ, Khan M, Khan M (2021) Engineered nanomaterials in soil: their impact on soil microbiome and plant health. Plants 11(1):109
6. Khanna K, Kohli SK, Handa N, Kaur H, Ohri P, Bhardwaj R, Yousaf B, Rinklebe J, Ahmad P (2021) Enthralling the impact of engineered nanoparticles on soil microbiome: a concentric approach towards environmental risks and cogitation. Ecotoxicol Environ Saf 1(222):112459
7. Lavicoli I, Leso V, Beezhold DH, Shvedova AA (2017) Nanotechnology in agriculture: opportunities, toxicological implications, and occupational risks. Toxicol Appl Pharmacol 15(329):96–111

8. Ali S, Ahmad N, Dar MA, Manan S, Rani A, Alghanem SM, Khan KA, Sethupathy S, Elboughdiri N, Mostafa YS, Alamri SA (2023) Nano-Agrochemicals as substitutes for pesticides: prospects and risks. Plants 13(1):109
9. Chaud M, Souto EB, Zielinska A, Severino P, Batain F, Oliveira-Junior J, Alves T (2021) Nanopesticides in agriculture: Benefits and challenge in agricultural productivity, toxicological risks to human health and environment. Toxics 9(6):131
10. Saritha GN, Anju T, Kumar A (2022) Nanotechnology-Big impact: how nanotechnology is changing the future of agriculture? J Agric Food Res 1(10):100457
11. Paramo LA, Feregrino-Pérez AA, Guevara R, Mendoza S, Esquivel K (2020) Nanoparticles in agroindustry: applications, toxicity, challenges, and trends. Nanomaterials 10(9):1654
12. Hazarika A, Yadav M, Yadav DK, Yadav HS (2022) An overview of the role of nanoparticles in sustainable agriculture. Biocatal Agric Biotechnol 1(43):102399
13. Takeshita V, Campos EV, Rodrigues JS, Fraceto LF (2023) Opinion: hybrid nanoparticle systems–two-way delivery approach for agriculture. Plant Nano Biology 1(6):100053
14. Rajput VD, Kumari A, Upadhyay SK, Minkina T, Mandzhieva S, Ranjan A, Sushkova S, Burachevskaya M, Rajput P, Konstantinova E, Singh J (2023) Can nanomaterials improve the soil microbiome and crop productivity? Agriculture 13(2):231
15. Ray PC, Yu H, Fu PP (2009) Toxicity and environmental risks of nanomaterials: challenges and future needs. J Environ Sci Health Part C 27(1):1–35
16. Yousef HA, Fahmy HM, Arafa FN, Abd Allah MY, Tawfik YM, El Halwany KK, El-Ashmanty BA, Al-Anany FS, Mohamed MA, Bassily ME (2023) Nanotechnology in pest management: advantages, applications, and challenges. Int J Trop Insect Sci 43(5):1387–1399
17. Allan J, Belz S, Hoeveler A, Hugas M, Okuda H, Patri A, Rauscher H, Silva P, Slikker W, Sokull-Kluettgen B, Tong W (2021) Regulatory landscape of nanotechnology and nanoplastics from a global perspective. Regul Toxicol Pharmacol 1(122):104885
18. Zain M, Ma H, Chaudhary S, Nuruzaman M, Azeem I, Mehmood F, Rahman SU, Aiwang D, Sun C (2023) Nanotechnology in precision agriculture: advancing towards sustainable crop production. Plant Physiol Biochem 30:108244
19. Mittal D, Kaur G, Singh P, Yadav K, Ali SA (2020) Nanoparticle-based sustainable agriculture and food science: recent advances and future outlook. Front Nanotechnol 4(2):579954
20. Jiang M, Song Y, Kanwar MK, Ahammed GJ, Shao S, Zhou J (2021) Phytonanotechnology applications in modern agriculture. J Nanobiotechnol 19:1–20
21. Chen H (2018) Metal based nanoparticles in agricultural system: behavior, transport, and interaction with plants. Chem Speciat Bioavailab 30(1):123–134
22. Wu J, Zhai Y, Liu G, Bosker T, Vijver MG, Peijnenburg WJ (2021) Dissolution dynamics and accumulation of Ag nanoparticles in a microcosm consisting of a soil–lettuce–rhizosphere bacterial community. ACS Sustain Chem Eng 9(48):16172–16181
23. Doro KO, Stoikopoulos NP, Bank CG, Ferris FG (2022) Self-potential time series reveal emergent behavior in soil organic matter dynamics. Sci Rep 12(1):13531
24. Sun C, Hu K, Mu D, Wang Z, Yu X (2022) The widespread use of nanomaterials: the effects on the function and diversity of environmental microbial communities. Microorganisms 10(10):2080
25. Rizwan M, Ali S, ur Rehman MZ, Riaz M, Adrees M, Hussain A, Zahir ZA, Rinklebe J (2021) Effects of nanoparticles on trace element uptake and toxicity in plants: a review. Ecotoxicol Environ Saf 221:112437
26. Daunoras J, Kačergius A, Gudiukaitė R (2024) Role of soil microbiota enzymes in soil health and activity changes depending on climate change and the type of soil ecosystem. Biology 13(2):85
27. Ameen F, Alsamhary K, Alabdullatif JA, ALNadhari S (2021) A review on metal-based nanoparticles and their toxicity to beneficial soil bacteria and fungi. Ecotoxicol Environ Saf 213:112027
28. Singh D, Gurjar BR (2022) Nanotechnology for agricultural applications: facts, issues, knowledge gaps, and challenges in environmental risk assessment. J Environ Manag 15(322):116033

29. Schwirn K, Voelker D, Galert W, Quik J, Tietjen L (2020) Environmental risk assessment of nanomaterials in the light of new obligations under the REACH regulation: which challenges remain and how to approach them? Integr Environ Assess Manag 16(5):706–717
30. Dilnawaz F, Misra AN, Apostolova E (2023) Involvement of nanoparticles in mitigating plant's abiotic stress. Plant Stress 3:100280
31. Neme K, Nafady A, Uddin S, Tola YB (2021) Application of nanotechnology in agriculture, postharvest loss reduction and food processing: food security implication and challenges. Heliyon 7(12)
32. Mgadi K, Ndaba B, Roopnarain A, Rama H, Adeleke R (2024) Nanoparticle applications in agriculture: overview and response of plant-associated microorganisms. Front Microbiol 6(15):1354440
33. Mwaanga P (2018) Risks, uncertainties, and ethics of nanotechnology in agriculture. New Vis Plant Sci 22(3):143
34. Patani A, Patel A, Prajapati D, Khare N, Singh S (2023) Various agriculture crop plant-based bioactive compounds and their use in nanomaterial synthesis and applications. In: Nanomaterials from agricultural and horticultural products. Springer Nature Singapore, pp 223–241
35. Martínez G, Merinero M, Pérez-Aranda M, Pérez-Soriano EM, Ortiz T, Villamor E, Begines B, Alcudia A (2020) Environmental impact of nanoparticles' application as an emerging technology: a review. Materials 14(1):166
36. Rai PK, Kumar V, Lee S, Raza N, Kim KH, Ok YS, Tsang DC (2018) Nanoparticle-plant interaction: implications in energy, environment, and agriculture. Environ Int 1(119):1–9
37. Cummings CL, Kuzma J, Kokotovich A, Glas D, Grieger K (2021) Barriers to responsible innovation of nanotechnology applications in food and agriculture: a study of US experts and developers. NanoImpact 1(23):100326

Regulatory Aspects and Risk Assessment of Nanoplatforms for Agricultural Uses

Vanessa Takeshita, Jéssica de Souza Rodrigues, Estefânia Vangelie Ramos Campos, Gustavo Vínicios Munhoz-Garcia, Jhones Luiz de Oliveira, Felipe Franco de Oliveira, Giovanna Moura Silva, and Leonardo Fernandes Fraceto

Abstract The present chapter discusses the development and innovation of nanopesticides aimed at agriculture, emphasizing sustainable platforms that seek to reduce environmental impacts and promote more efficient and ecological farming. It addresses the environmental and health risks of using these nanopesticides, considering their ecotoxicity and potential effects on biodiversity and human health. Additionally, the chapter analyzes the complex regulatory challenges specific to these nanoplatforms, highlighting the urgent need for clear guidelines and new evaluation methods to handle the novelty and complexity of nanopesticides. In the context of risk assessment, it emphasizes the importance of rigorous scientific reports and reliable data. Finally, it discusses the prioritization and reuse of existing data as crucial

V. Takeshita · J. de Souza Rodrigues · E. Vangelie Ramos Campos · J. Luiz de Oliveira · F. Franco de Oliveira · G. Moura Silva · L. Fernandes Fraceto (✉)
Institute of Science and Technology, São Paulo State University (UNESP), Sorocaba, SP, Brazil
e-mail: leonardo.fraceto@unesp.br

V. Takeshita
e-mail: vanessatakeshita@usp.br

J. de Souza Rodrigues
e-mail: rodrigues.jessicad@gmail.com

E. Vangelie Ramos Campos
e-mail: estefaniavan@gmail.com

J. Luiz de Oliveira
e-mail: jhones.luiz@unesp.br

F. Franco de Oliveira
e-mail: felipe_oliveira@usp.br

G. Moura Silva
e-mail: silvam.giovanna29@gmail.com

E. Vangelie Ramos Campos · J. Luiz de Oliveira
B. Nano Soluções Tecnológicas Ltda, São Miguel Arcanjo, São Paulo, Brazil

G. Vínicios Munhoz-Garcia
Center of Nuclear Energy in Agriculture, University of São Paulo, Piracicaba, SP, Brazil
e-mail: gvmgarcia@usp.br

L. F. Fraceto et al. (eds.), *Emerging Nanotechnologies for Agroecosystem Management*, Smart Nanomaterials Technology, https://doi.org/10.1007/978-981-95-0187-8_14

strategies to accelerate the regulatory process and reduce costs, while ensuring the safety and efficacy of the nanoplatforms used in agriculture.

Keywords Nanopesticides · Sustainable agriculture · Environmental risks · Regulatory challenges · Risk assessment

1 Introduction

Conventional pesticides and agrochemicals have long played a pivotal role in enhancing food production, ensuring safety, and maximizing crop yields by effectively managing pests [69]. According to the Food and Agriculture Organization of the United Nations (FAO), the judicious use of agrochemicals has restored approximately 30% of agricultural output globally [72]. However, in the twenty-first century, the utilization of pesticides has encountered significant challenges, including issues related to the low solubility of the active ingredient (a.i.), non-selective behavior, and uncontrolled environmental release [18, 58].

The agricultural sector finds itself at a critical juncture in response to the staggering inefficiencies and environmental repercussions plaguing conventional pesticide use. Nanotechnology emerges as a transformative force poised to revolutionize agricultural practices. By harnessing the unique properties of nanoscale materials, nanotechnology offers a promising avenue to rectify conventional pesticides' shortcomings while mitigating their adverse environmental impacts [61]. Indeed, the potential of nanotechnological innovations spans a wide spectrum of agricultural applications, from the precise detection of toxic agrochemicals to the development of advanced seed treatments, efficient fertilizer delivery systems, and sophisticated pesticide formulations [35, 78, 80]. Nanoformulations of pesticides exemplify nanotechnology's potential, offering advantages such as heightened effectiveness, enhanced durability, biodegradability, and reduced toxicity [103]. These formulations can be target-specific, minimizing collateral damage and environmental contamination. Nanotechnology revolutionizes agriculture, providing sustainable pest management solutions [40, 44, 78].

A diversity of nanoplatforms has been meticulously studied and developed, encompassing nanoemulsions, nanocapsules, and nanosuspensions [32, 33, 45, 47, 95, 107, 119]. Each formulation variant offers distinct mechanisms and benefits, such as sustained release, improved stability, and enhanced solubility of the a.i. in water. Nanocapsules, for instance, feature a unique shell or membrane structure that encloses a.i., providing protection against early degradation and improving formulation stability. This encapsulation also facilitates an efficient delivery system with controlled release of the a.i., thereby extending the longevity of the pesticide's effectiveness. In contrast, nanoemulsions present a different structural arrangement, consisting of two immiscible phases—organic or oil and water—separated by surfactants. These emulsions can take on various forms, including oil in water (O/W), water in oil (W/O), and bi-continuous, each offering specific advantages in pesticide delivery and solubility enhancement [8]. Similarly, nanosuspensions represent another innovative formulation approach characterized by nanocolloidal dispersions of the a.i. particles stabilized with surfactants. These suspensions enhance the solubility of insoluble a.i. in water, thereby improving formulation stability and efficacy.

Nanostructured platforms can be developed (i) from natural polymers including cellulose, chitosan, pectin, and zein [6, 9, 48, 62, 84], (ii) from inorganic structures such as synthetic silica, titania, alumina, copper, and silver [7], and (iii) from synthetic polymers such as polylactic acid (PLA), polylactic-co-glycolic acid (PGA), polyethylene glycol (PEG), polyvinyl alcohol (PVA), and poly-ε-caprolactone (PCL) [42]. These materials, renowned for their environmental compatibility, biodegradability, and cost-effectiveness, are considered ideal carriers for encapsulating a.i. Noteworthy examples include zein-based nanoparticles (NP) loaded with botanical pesticides, which have demonstrated prolonged efficacy and minimal phytotoxicity in crop applications [77].

While exploring the exciting possibilities of nanotechnology in agriculture, it is crucial to acknowledge and tackle the specific regulations-related challenges. Nanotechnology offers great potential to revolutionize farming practices but raises concerns about safety, environmental impact, and how it fits within existing regulations. Many nanoplatforms are developed for plant protection; however, comparative criteria with conventional a.i. or robust efficacy and risk assessment must be reported more cautiously. As we progress, it is important to understand and work within the rules and regulations designed to keep people safe and protect the environment. This means carefully assessing potential risks and ensuring that any new nanotechnology applications in agriculture are safe and sustainable. By shedding light on these regulatory challenges and conducting thorough risk assessments, this chapter aims to help stakeholders navigate the complexities and promote the responsible use of nanotechnology in agriculture.

2 Nanopesticides Developed for Agricultural Applications

According to Wu et al. [110], 314 studies have been conducted to evaluate the efficacy of nanopesticides, including metal-based and polymeric NP, as carriers for fungicides, insecticides, and herbicides. The state of knowledge in nanopesticide development and examples of nanoplatforms developed for crop protection are presented in Table 1. In the table, the type of material, a.i. loaded, nanoparticle properties, target species, application mode, and agronomic efficacy are the main factors described for these nanoplatforms, separated by fungicides, insecticides, and herbicides.

Some interesting cases of fungicide nanoplatforms (present in Table 1) deserve to be highlighted. For example, using zeolitic imidazolate to deliver prochloraz or benzoylated lignin sulfonates to deliver difenoconazole resulted in a reduction in a.i. degradation under light conditions [64, 65]. This approach increased fungicide residual effect, which proved to be more effective than the commercial product.

Table 1 Nanoplatforms developed as carriers for pesticide agricultural uses reported on literature

Nanoparticle		Active ingredient	Application mode	Efficacy	Nanoparticle effects on the active ingredient	Potential practical application	References
Material (type)	Characteristics						
Fungicides							
Mesoporous selenium (Metal)	150 nm, 18 mV, 23.5% of loading, pH-responsive particle	Carbendazin	Foliar spray	Reduced damage severity of *Sclerotinia sclerotiorum*	Not evaluated	More studies are necessary to prove agronomic applications	[50]
Benzoylated lignin sulfonates (Modified biopolymer)	92 nm, –45 mV	Difenoconazole	Foliar spray	Increased residual period of fungicide in plant leaves	Reduced by 6.6-times the degradation time in UV light	Control of disease in plants exposed to *Colletotrichum gloeosporioides*	[65]
Hydrotalcite (Mineral)	314 nm, 31 mV, 25% of loading	Fenhexamid + salicylic acid	Foliar spray	Increased in 65% of control	Increased in 3.9-times the half-life time in photodegradation	1. Increase the active stability during the day, reducing losses 2. Application in crop areas to control *S. sclerotiorum*	[37]

(continued)

Table 1 (continued)

Nanoparticle		Active ingredient	Application mode	Efficacy	Nanoparticle effects on the active ingredient	Potential practical application	References
Material (type)	Characteristics						
Glycine methyl ester (Aminoacid) + polysuccinimide (Biopolymer)	615 nm, –23 mV, 28% of loading, pH responsive	Fludioxonil	Foliar spray	50% more effective against *Fusarium oxysporum* f. sp. *cubense*	Increased in 2-times the half-life time in photodegradation	1. Reduces degradation by light effects 2. Promotes control for a root-soil disease, by foliar application due higher transport to plant stem and roots	[109]
Zeolitic imidazolate (Metal–organic frameworks)	130 nm, 30 mV, light and pH responsive	Prochloraz	Foliar spray	Increased residual affect protecting plant leaves	Reduced in 40% the degradation time under light conditions	1. Reduces fungicide degradations and increase retention in leaves 2. Promotes a better control of *Sclerotinia* sp. after 14 days after sprayed	[64]
Star polycations (Synthetic polymer)	86 nm, positive charge, 15% of loading	Thiophanate-methyl	Seed treatment	30% more efficient than conventional fungicide	Not evaluated	Enable *Verticillium dahlia* control in cotton plants	[100]

(continued)

Table 1 (continued)

Nanoparticle		Active ingredient	Application mode	Efficacy	Nanoparticle effects on the active ingredient	Potential practical application	References
Material (type)	Characteristics						
β-cyclodextrin (Biopolymer) + Synthetic ester*	190 nm, –27 mV, loading 22% and encapsulation efficiency 69%, responsive to ROS accumulation	Pyraclostrobin	Localized application on stem	No differences within commercial pyroclostrobin	Not evaluated	More studies are needed to prove agronomic efficacy	[113]
Insecticides							
Mesoporous silica (Mineral)	380 nm, 80% of insecticide retention	Matrine	Localized application to stem	30% more toxic to *Monochamus alternatus*	Not evaluated	Promotes more distribution inside plant vessels, however, more studies need to be done do prove the agronomic efficacy	[31]
	140–300, 33% of loading	Rotenone	Foliar spray	20% more control of larvae of *Pieris rapae*	Increased in 73% half-life time in photodegradation in tomato leaves	Possibility of using nanocarriers for bioinsecticides delivery in field conditions	[111]

(continued)

Table 1 (continued)

Nanoparticle		Active ingredient	Application mode	Efficacy	Nanoparticle effects on the active ingredient	Potential practical application	References
Material (type)	Characteristics						
	180 nm, –33 mV, 29% of encapsulation, redo responsive	Acetamiprid	Foliar spray	No differences found	Not specified	Increases in 17% the adhesion in plant leaves, however, no data supports better insect control than commercial acetamiprid	[30]
Mesoporous silica (Mineral) coated with synthetic polymers**	220–340 nm, 27 mV, 12% of loading	Imidacloprid	Foliar spray	LC_{50} was similar to conventional imidacloprid, no improvements were found	Reduced in 90% the degradation of imidacloprid under UV light	More studies are needed to confirm the agronomic efficacy	[67]
Lipidic Nanoemulsions (Organic materials)	43–58 nm, –6 to –4 mV,	Eucalyptus and nutmeg essential oils	Not specified	Increase in 12–35-times LC_{90} of oils against *Odontotermes formosanus*	Not evaluated	Can contribute to pest management but need to confirm the agronomic efficacy	[79]
Chitin (Biopolymer)	37–140 nm, 27 mV,	Not used	Foliar spray	In mixture with conventional insecticides enable the dose reduction and a rapid aphid control in 24 h	Not evaluated	Can be used in *Rhopalosiphum padi* control, but it needs field studies to confirm agronomic efficacy	[63]

(continued)

Table 1 (continued)

Nanoparticle		Active ingredient	Application mode	Efficacy	Nanoparticle effects on the active ingredient	Potential practical application	References
Material (type)	Characteristics						
Poly(ethylene glycol) (PEG) + poly(D, L-lactide) (PDLLA) (Synthetic polymers)	120 nm, 99% of encapsulation and 49% of loading	Lambda-cyhalothrin	Foliar spray	No differences with the commercial insecticide	Not evaluated	Promotes *Aphis craccivora* control, at the same intensity of commercial active	[21]
Chitosan (Biopolymer)	100 nm	Chlorfenapyr	Foliar spray and root application	Control 7–17-times higher than commercial through root	Reduced degradation in leaves	Efficient against *Spodoptera frugiperda*, being tested at field conditions	[118]
Herbicides							
Poly-Ɛ-caprolactone (Synthetic polymer)	260 nm, –30 mV, 87% of encapsulation	Atrazine	Foliar spray	Better control than conventional atrazine at the same dose (2 kg a.i. ha^{-1}) and similar control when the dose of nanoatrazine is 1/10 of the recommended dose	Not evaluated	Effective against *Amaranthus viridis* and *Bidens pilosa* in post-emergente	[97]

(continued)

Table 1 (continued)

Nanoparticle		Active ingredient	Application mode	Efficacy	Nanoparticle effects on the active ingredient	Potential practical application	References
Material (type)	Characteristics						
	Not specified	Atrazine	Soil application	Increase in twofold the mortality of *Bidens pilosa* at 1/10 of the recommended dose and no phytotoxicity was observed to soybean plants	Not evaluated	Control of *B. pilosa* in pre-emergence of soybean	[89]
	200–289 nm, –32 mV, 82% of encapsulation	Metribuzin	Foliar and soil application	Reduced twofold more the fresh weight of *Ipomoea grandifolia* when diluted tenfold, than commercial herbicide	Not evaluated	No changes in a.i. environmental behavior, more effective in reduced doses, can be used to weed management	[101]
Copper-benzene-1,4-dicarboxylate (Metal – organic framework) + mesoporous silica (Mineral)	140 nm, –7 mV, pH responsive	Quizalofop-p-ethyl	Foliar spray	11% more control than commercial product (both at 1/8 of the recommended dose)	Not evaluated	Strong evidence are needed	[28]
Poly-lactic-co-glycolic Acid (Synthetic polymer)	110 nm, 50% of encapsulation	Atrazine	Not specified	No improvement	Not evaluated	Need tests with target weed species	[93]

(continued)

Table 1 (continued)

Nanoparticle		Active ingredient	Application mode	Efficacy	Nanoparticle effects on the active ingredient	Potential practical application	References
Material (type)	Characteristics						
Covalent Organic Framework (Porous and synthetic polymers) ****	150–200 nm, 57% of loading	Cyalofop-buthyl	Foliar spray	3–6 times more control in 1/4 of the recommended dose	Increase in 25% the a.i. half-life time under UV light	Can be used to control grasses in rice fields, with good effect at reduced doses	[29]
Synthetic polymer	200 nm, 16% of loading, light responsive	2,4-D	Foliar spray	LC_{50} of 2,4-D in nanoformulation increased 10-folds	Not evaluated	No evidence of improvement	[94]
Attapulgite (Mineral) + Polyvinyl alcohol (Synthetic polymer)	Not specified, temperature responsive	Glyphosate	Foliar spray	No improvements at normal conditions. with rainfall after application, increased sevenfold the weed control	Not evaluated	Several studies need to be conducted to prove the field applicability	[22]
Zein (Biopolymer)	130–170 nm, 12 mV, 90% of encapsulation	Atrazine	Soil application	Reduced doses at 1/80 presented the same efficacy than commercial atrazine at full dose against *Brassica juncea*	Not evaluated	Efficient weed control without compromising maize, but field studies are needed	[17]

(continued)

Table 1 (continued)

Nanoparticle		Active ingredient	Application mode	Efficacy	Nanoparticle effects on the active ingredient	Potential practical application	References
Material (type)	Characteristics						
Polydopamine (Synthetic polymer) + Metal – organic framework*	782 nm, –2.6 mV, 14% of loading	Thiachloprid + 2,4-D	Foliar spray	Not evaluated for 2,4-D and no improvement for Thiachloprid	Reduced both a.i. degradation under UV light	Not demonstrated	[94]
Polyvinyl alcohol (Synthetic polymer)	15 nm, 21% of encapsulation	2,20-disulfanediyldianiline	Foliar and soil application	Increase in tenfold the IC_{50} in coleoptile bioassays and reduced 13-fold more the germination, compared to the non-formulated compound	Not evaluated	Need further studies to demonstrate the efficacy against weeds at least at greenhouse conditions	[106]

*4-(hydroxymethyl) phenylboronic acid pinacol ester (PBAP). **Polyethylene glycol diacrylate (PEGDA) and sulfhydryl-modified polyethylenimide (PEI-SH). ***Fe (III) coordinated with carboxylate salts. ****1,3,5-tris(4-aminophenyl)-benzene and -divinylterephthalaldehyde)

Another advantage of nanoplatform uses is to deliver two or more compounds. For example, using hydrotalcite in NP design to deliver fenhexamid associated with salicylic acid (an immune inducer) in plant leaves promoted 65% more effective control than commercial fungicides [36]. The nanoplatforms can also be used for fungicide transport in long distances for the control of root diseases. The glycine methyl ester (an amino acid) + polysuccinimide (a biopolymer) was used to deliver fludioxonil from leaves to roots [109]. This approach resulted in 50% more control of *Fusarium oxysporum* f. sp. *cubense* infecting plant roots. For seed treatment, synthetic polymers (star polycations) combined with thiophanate-methyl is more effective than the commercial formulation to *Verticillium dahlia* control (30% higher) [100]. However, not all nanoplatforms are effective for the increase efficacy of the a.i.. For example, the use of mesoporous selenium acting as carbendazim carrier in plant leaves or β-cyclodextrin (biopolymer) + synthetic ester for delivery of pyraclostrobin in plant stems [50, 113].

The scenario for insecticides is a bit more complex. Several studies have developed nanoplatforms, but in some cases, the evidence for their efficacy is inconclusive. For example, for mesoporous silica systems, the toxicity of the a.i. was improved for matrine but not for acetamiprid [31]. In a more complex approach using mesoporous silica coated with polymer, imidacloprid toxicity was not improved and no evidence of agronomic efficiency was presented [67]. Synthetic polymers-loaded lambda-cyhalothrin also demonstrated this phenomenon [21]. In contrast, the efficacy of chlorfenapyr against *Spodoptera frugiperda* was increased by up to 17-fold when biopolymers like chitosan were used, both in foliar and root applications [118]. Additionally, approaches using natural compounds with insecticides were tested for delivery by nanoplatforms. Using mesoporous silica-loaded rotenone (a bioinsecticide) resulted in a 20% increase in the control of *Pieris rapae* in tomato plants under field conditions [111]. In contrast, lipidic nanoemulsions as carriers for essential oils (eucalyptus and nutmeg) demonstrated toxicity to *Odontotermes formosanus* ranging from 12 to 35 times that of the oils alone [79]. A different approach was employed by [63], in which chitin was combined with conventional insecticides, resulting in an effective control of *Rhopalosiphum padi* in reduced doses. However, further efficacy studies are necessary to confirm the true potential of these nanoplatforms for insecticides.

The development of nanoplatforms for herbicide delivery it is also a challenge. There are platforms from a simple encapsulation strategy, such as poly-Ɛ-caprolactone for the delivery of atrazine, to more complex formulations, such as polydopamine and metal–organic frameworks to encapsulate 2,4-D and insecticide thiacloprid [43, 94]. Furthermore, nanoherbicides demonstrate considerable potential for use in weed control, contingent on the nanoplatform employed. For instance, encapsulating atrazine and metribuzin with poly-Ɛ-caprolactone enhances the a.i. toxicity to target weed species, enabling a dose reduction (up to 80-fold) without compromising efficacy [17, 101]. A comparable outcome was observed with zein as an atrazine carrier. These results were observed in both foliar and soil applications, indicating the versatility of nanoplatforms. Factors such as light, temperature, and pH are explored in the design of responsive nanoplatforms, yet there is a lack of evidence regarding their field applicability. For example, the herbicide quizalofop-p-ethyl was encapsulated in a copper-benzene-1,4-dicarboxylate + mesoporous silica formulation and demonstrated only a 11% increase in control efficacy compared to the conventional formulation in in vitro studies [28]. Similarly, the herbicide 2,4-D was encapsulated in a synthesized polymer and exhibited reduced toxicity to plants [94]. A tenfold increase in efficacy was observed for 2,4-D when it was combined with thiacloprid and encapsulated in polydopamine and a metal–organic framework. However, the efficacy of this formulation against weeds was not evaluated. Glyphosate was entrapped into structures of attapulgite coated with polyvinyl alcohol [22]. However, under normal environmental conditions, no improvement was observed.

Besides classical nanomaterials (synthetic polymers and metal NP) for pesticides, an innovative and sustainable approach in developing nanopesticides is the adoption of natural molecules and biodegradable nanoplatforms [60]. This strategy aims to improve pest control efficacy and mitigate the environmental impacts associated with synthetic formulation ingredients. A crucial aspect of this advancement is using natural platforms such as chitosan, zein, lignin, and others, which offer additional advantages due to their biodegradability and low toxicity.

The chitosan matrix encapsulates pesticides due to its polymeric structure, forming a three-dimensional network around the active compounds. This network retains the pesticides and gradually releases them in response to environmental factors such as pH, temperature, and humidity. [60] developed alginate-chitosan nanocapsules using ionic pre-gelation and polyelectrolyte complexation, resulting in pH-dependent release for acetamiprid with reduced side effects. Campos et al. [15] developed chitosan nanoparticle carriers to transport carvacrol and linalool. These compounds, which have low aqueous solubility, were encapsulated in chitosan NP functionalized with β-cyclodextrin. The results revealed that these complexes exhibited favorable physicochemical properties, with high encapsulation efficiency for both compounds. Biological assays demonstrated that NP containing carvacrol and linalool were significantly more effective in acaricidal activity and oviposition, while non-encapsulated compounds exhibited better repellent activity. In another study, Campos et al. [16] observed a reduction in compound toxicity when nanoencapsulated, and the NP demonstrated insecticidal efficacy against *Helicoverpa armigera* and *Tetranychus urticae*, as well as repellent and oviposition reduction activities in mites. These findings suggest that the nanoformulation of these botanical compounds may be a promising strategy for sustainable pest control in agriculture, reducing environmental contamination and improving pest control efficacy.

In the same way, several studies highlight the innovation in the nanoencapsulation of botanical compounds by zein aiming to enhance efficacy and sustainability in agricultural pest control. The study by [77] describes the preparation of zein NP containing limonene and carvacrol, with a responsive approach to proteolytic enzymes present in the intestines of insects. These NP demonstrated stability, efficacy against *Spodoptera frugiperda* larvae, and controlled degradation under biological conditions. Meanwhile, the work by [86] focused on the use of zein NP to encapsulate the main constituents of citronella essential oil (geraniol and R-citronellal), aiming for protection against premature degradation, reduction of toxicity, and efficacy as repellents against the *Tetranychus urticae* mite. These NP exhibited high encapsulation efficiency, physicochemical stability, and reduced toxicity, emerging as an effective option for using botanical repellents in agriculture. Another study by de Oliveira et al. [85] investigated the encapsulation of a mixture of botanical compounds (geraniol, eugenol, and cinnamaldehyde) in zein NP. These formulations demonstrated stability, protection against degradation, and reduced toxicity, as well as superior efficacy in repellence and insecticidal activity compared to emulsified compounds.

In addition, lignin NP has been increasingly recognized at the nanoscale due to their enhanced properties, such as antibacterial and antioxidant activities. Recent studies have demonstrated the potential of lignin NP in encapsulating different types of pesticides. [68] developed chemically cross-linked lignin nanocarriers to encapsulate fungicides, offering a promising plant drug delivery platform. These authors prepared the NC through lignin cross-linking via aza-Michael addition in miniemulsion, demonstrating high encapsulation efficiency (70–99%) for various fungicides. The NC successfully inhibited the growth of lignase-producing fungi associated with the vine trunk disease Esca in planta studies, highlighting their effectiveness as biodegradable delivery vehicles. [38] proposed an innovative approach for lignin utilization by combining it with brochantite to create organic–inorganic hybrid materials (lignin@Cu) as an alternative to copper-based pesticides. They developed a greener synthesis method using mechanochemistry and demonstrated the efficacy of lignin@Cu materials against pathogens in greenhouse-grown strawberry and tomato plants. These materials exhibited synergistic activity between lignin and copper ions, enabling efficient pathogen control with significantly lower copper content than commercial copper-based pesticides.

These natural platforms represent a significant advancement in creating ecofriendlier and more efficient nanopesticides. These innovations promote sustainable agriculture by leveraging renewable resources and advanced nanotechnology techniques, ensuring crop protection and environmental preservation. It is worth mentioning that these materials can be used combining with other nanomaterials in hybrid NP to enhance the delivery of a.i. [103], turning the nanoformulations more complex and specific for agricultural uses.

The interaction between NP, pesticides, and plants is a complex process involving many factors, including nanomaterial, NP characteristics, plant physiology, pesticide mode of action, target species, and environmental conditions. Nanoplatforms can enhance the efficacy of pesticides by increasing absorption rates and routes [12, 102], facilitating long-distance translocation [110], and allowing for substance-triggered accumulation [113]. Besides, can presents the ability to protect the a.i. against photodegradation is frequently investigated, particularly in the case of fungicides and insecticides (Table 1). In addition, natural nanoplatforms can be adopted to reduce environmental risk as responsive and protector materials to deliver conventional and botanical compounds. However, NP can contribute to the increased toxicity of the a.i. to the target and the non-target effects must be evaluated. From an agronomic perspective, the advantages described in this section result in a more efficient use of pesticides, reducing environmental losses and promoting a more sustainable form of agriculture, contingent on the materials used in nanoplatforms synthesis. From an environmental risk perspective, nanoplatforms have been investigated worldwide. However, it has been challenging for new products to reach the global market.

3 Environmental and Health Risks

Nanoplatforms aim to increase solubility, provide slow or targeted release, and protect a.i. from degradation. These properties can reduce application rates and potentially lower environmental impact [112]. However, they also pose new contamination risks due to enhanced transportation, persistence, and toxicity.

[56] studied the environmental fate of three nanoformulations of the insecticide clothianidin, finding that these nanoformulations increased photodegradation half-life in water by up to 21% and soil sorption by up to 51% compared to commercial formulations. This suggests that nanoformulations significantly affect the fate of pesticides a.i., underscoring the need to understand their durability in the environment to properly evaluate nano-based agrochemicals.

Interactions between nanopesticides and microorganisms are critical, as microbes play key roles in organic matter decomposition, nutrient recycling, and growth enhancement [13, 91, 112]. The solubility of pesticides compounds affects their transportation and biodegradation, emphasizing the need to study nanoformulations designed to increase solubility.

The interactions between NP and environmental matrices are important to determine the nanoformulation dynamics. NP in nano-based formulations may undergo aggregation, affecting their environmental fate and toxicity. This includes homoaggregation (between NP) and heteroaggregation (NP with natural colloids), especially in aquatic environments where they accumulate in sediments [24, 71]. Natural NP in these environments, such as colloidal clays and organic matter, contribute to this aggregation process.

Several studies suggest that nano-based pesticides may be environmentally safer due to their lower application rates and sustained release of a.i. [4, 14, 70]. These properties are believed to result in fewer environmental impacts than conventional formulations [66, 116]. However, the limited knowledge about the environmental fate and effects of nano-based pesticides complicates accurate risk and benefit assessment [87]. Some studies indicate that the unique properties of nanomaterials, such as particle size and crystalline structure, influence their toxicity to non-target species [25, 83, 117].

[34] examined four model nanopesticides, including tebuconazole and chlorpyrifos loaded on polymeric and lipid NP, focusing on their fate and bioaccumulation in soil microcosms with earthworms (*Eisenia fetida*) and lettuce (*Lactuca sativa*). The study found that the soil half-lives of the a.i. loaded in NPs were significantly longer

than conventional formulations due to slow release from NC. Additionally, nano-based formulations showed enhanced bioavailability and greater bioaccumulation on earthworms compared to non-encapsulated compounds. Notably, nanoencapsulated tebuconazole exhibited significant translocation from the roots to the edible parts of lettuce, posing a potential threat to human health through the food chain.

These findings indicate that the fate and bioavailability of a.i. in nanopesticides differ from those in conventional pesticides. However, results may vary depending on environmental factors such as soil type and nanopesticide composition. Another study by [57] demonstrated that copper oxide nanopesticides applied to edible plants like collard greens, kale and lettuce could be absorbed by the leaves, leading to the bioaccumulation of copper oxide NP and posing a potential risk to consumers [55].

Another important aspect is those related to soil microorganisms. Those agents are crucial for plant growth as they help transform compounds into usable forms. Nanopesticides entering the soil may have unintended effects on these microorganisms, such as mycorrhizal communities and nitrogen-fixing bacteria [51, 73, 96]. For example, $Cu(OH)_2$ NPs were shown to inhibit urease activity and alter microbial community structure, differing in impact from $CuSO_4$ [114]. The response of soil enzyme activity and bacterial community to $Cu(OH)_2$ NPs also varied depending on application modes, unlike $CuSO_4$, which had consistent effects regardless of application frequency [104]. Additionally, nanopesticides doses influence soil microorganisms, with low concentrations inhibiting and high concentrations stimulating soil microbial enzyme activity [96].

It is known that soil microorganisms play a vital role in nutrient cycles and mitigating organic pollutants [108]. $Cu(OH)_2$ NPs altered soil NO3– –N and NH_4^+–N concentrations compared to ionic copper salts, affecting nitrogen pools [104]. Another study showed that applying $Cu(OH)_2$ NPs changed the microbial communities and reduced the nitrile hydratase activity, which resulted in a decreased biodegradation of thiacloprid in soil [115]. [81] evaluated the toxicity of copper nanoformulation to soil invertebrates using *Folsomia candida* as a model. The results showed that the nanoformulations were more toxic to the organism when compared to the macro counterparts, with EC_{50} of 156 mg Cu kg^{-1} and 3030 mg Cu kg^{-1}, respectively. This toxicity was more pronounced in soils with low pH and organic content, enhancing copper solubility and bioavailability.

These studies highlight that the environmental risks of nano-based formulations can differ significantly from conventional pesticides. Furthermore, some nano-based formulations might have lesser impacts on non-target organisms than conventional pesticides [76, 85], indicating that it is not always possible to generalize the effects of nanopesticides on non-target organisms.

Apart from soil microorganisms, the increased development and application of nano-based formulations raise the risk of exposure to non-target aquatic organisms. Permethrin nanoemulsion's ecotoxicity was tested on zebrafish species, revealing abnormal behaviors such as unstable swimming and loss of reflexes at concentrations above 6.4 mg L^{-1} [76]. However, at concentrations below 6.4 mg L^{-1}, it was non-toxic to freshwater algae and zebrafish, highlighting the importance of dosage in environmental risks, like other toxicants.

[96] studied the effects of a commercial $Cu(OH)_2$ nanopesticide on aquatic primary producers in wetland mesocosms. Chronic exposure at low doses led to algal blooms and prolonged eutrophication, lasting over 50 days longer than nutrient-only treatments. The same nanopesticide caused no developmental toxicity at 1 mg L^{-1} for zebrafish embryos and larvae, but higher doses (2 to 8 mg L^{-1}) resulted in morphological abnormalities. Increased transcription levels of immune system-related genes were also observed [2]. In *Daphnia magna*, $Cu(OH)_2$ nanopesticide changed gene expression related to detoxification and the reproductive system after 24–48 h of exposure, indicating genotoxic responses [1]. However, it is worth mentioning that in both studies did not compare impacts with macro counterparts, potentially biasing interpretations. Bullfrog tadpoles (*Lithobates catesbeianus)* exposed to solid lipid NP with pyrethrum extract for 48 h showed increased leukocytes and erythrocytes, higher DNA damage frequency, abnormal erythrocyte nuclei, and cell death [85]. Conversely, nanoencapsulated bifenthrin was less toxic to rainbow trout than conventional bifenthrin, with a 96 h LC_{50} values of 11.2 μg L^{-1} and 6.2 μg L^{-1}, respectively [11].

In summary, the studies have shown that the environmental risks of nano-based formulations for agriculture applications depend heavily on the organism tested, dosage and environmental conditions. A comprehensive understanding requires further research, collaboration between developers and environmental scientists to balance the risks and benefits and promoting sustainable application of nanotechnology in the agriculture sector.

Only few studies make the comparison in the impact of nano-based pesticides and conventional ones on the same species at environmentally relevant concentration, leaving significant gaps in understanding their safety. Nanopesticides' higher persistence and lower mobility in soil and water allow them to move through protozoans and non-target organisms, potentially reaching humans [41]. This raises concerns about their safety, especially since nanomaterials can enter organisms through inhalation, ingestion, and absorption [3]. Therefore, further studies are needed to evaluate the impact of nanopesticides under realistic environmental conditions and exposure scenarios.

Understanding the behavior and fate of nanopesticides in the environment is crucial. The lack of robust studies, combined with the complex properties of nanopesticides and varying environmental conditions, complicates this issue [59]. Developing reliable methods to measure the environmental impact of nanopesticides is essential for establishing regulations to minimize risks to natural water, soil, and biota.

4 Regulatory Aspects for Pesticides and Agrochemicals Nanoplatforms

Size classification is the first step in regulatory decisions about nanomaterial risk assessment, environmental fate, and food security. Over the last 15 years, researchers and regulatory organizations have focused on developing a framework that regulates nanomaterials, mainly for food and cosmetic applications. The reports are often present size definitions for nanomaterials from 1 to 100 nm, determined by ISO/TS 27,687, "Nanotechnologies—Terminology and definitions for nano-objects—Nanoparticle, nanofibre and nanoplate", in 2008 (2008). The size definition depends on the technology sectors, usually indicated as a natural or manufactured product in these dimensions [74]. However, nanomaterials are not necessarily defined by size but by properties and functionalities that differ from the origin materials on the macrometric scales [39].

It is worth mentioning that [74] carefully discuss the legal and practical challenges in the regulatory definitions, highlighting that in the Novel Food Regulation defined engineering nanomaterials may have a size above the order of 100 nm, if they have properties that are characteristic of the nanoscale. These characteristics must be related to the large specific surface area of the materials considered nanomaterials and specific physicochemical properties that differ from non-nanomaterials. The same authors presented a double challenge in resolving the definitions of "nano", such that lawmakers must clearly define the term. At the same time, appropriate techniques are needed to identify nanomaterials that are based on previously determined size definitions for the identification of nanomaterials in products.

Currently, nanotechnology has evolved into complex platforms with higher dimensions than those defined by regulatory agencies, adding a certain difficulty in distinguishing between what is or is not a nanomaterial/nanoplatform [54]. Then, a complete understanding of the current state of nanoplatforms may be affected by disagreement in size definitions [44]. Importantly, some nanoplatforms with sophisticated functionalities and structures, such as hybrid nanomaterial combinations [67, 92], can make classifications regarding size and composition even more difficult.

After all, what is important to consider when defining the term "nano" to regulate these materials adequately? [74] suggested that size and structure, size distribution, origin, persistence, and characteristic properties can compose the information necessary to define a nanomaterial. This distribution range was unclear for food regulation Regulation (EU) 2015/2283 but contained an implicit requirement, instead, a single particle could define a material. Additionally, according to [74], for food, 50% as a limit of particles in the nanoscale range was proposed by EFSA in 2012, such as in the Biocide's Regulation Regulation (EU) 528/2012. For nanomaterials, it is important to provide origin information by manufacturing to address the associated risk of these materials. [74] suggested caution on including some terms in regulation definitions, such as "intentionally produced materials" proposed by novel food regulation to not exclude from natural origins materials. The term "manufactured" can be seen in other regulatory documents and sectors, addressing intentional and not intentional purposes.

In 2022, the European Commission reported the last nanomaterial definitions for legislative and policy purposes. The recommendations reinforce that 1–100 nm size, the 50% particle limitation, other terminology definitions such as "agglomerates" and "aggregates", and specific surface area (< 6 m^2/cm^3 shall not be considered as nanomaterial, equivalent to <1000 nm). With the revised guidelines, it is possible that nanoplatforms for agricultural use will be properly classified and can be evaluated for registration after 20 years of evolution of nanomaterials guidelines. However, many products are available on the market (see the 4.1 section below), and the classification and risk assessment of these products must be discussed with attention by countries' authorities. An effort to follow the registration process framework, following the guidelines and recommendations available, is strongly necessary, with local policies and restrictions.

According to [82] which compiled the European guidelines reported over the last 20 years, discussions advanced in the cosmetics (2009), food (2015), and biocides (2012) sectors, with the establishment of rules for labels (2009) and chemical analysis (2018). The authors pointed out that progress is still needed: (i) in environmental risk assessment and analysis instruments consistent with the guidelines,(ii) when implementing the guidelines, an effort is required to apply them; (iii) and that updating regulations that do not include nanomaterials in their scope need to be reevaluated. Based on these advances, the nanoplatforms, as nanomaterials in formulations for agricultural uses, will be allowed to follow the registration process in different countries to reach the market as soon as possible.

4.1 Challenges in the Regulatory Process

4.1.1 Risk Assessment—Regulatory Aspects and Scientific Reports for Nanopesticides

The theme of nanotechnology holds vast potential, as described in Sect. 2, encompassing a wide range of technologies to modernize the agricultural industry. However, despite all these advances, understanding these materials' mechanisms and toxic properties is essential [5]. In this context, risk assessment and regulatory aspects of nanomaterials, including those for agricultural applications, are being addressed worldwide, as presented in Sect. 3. Many countries are concerned with evaluating and adjusting their governmental plans to handle nanotechnology. This also includes regulatory aspects, including environmental issues, health and safety risks, socio-ethical issues, consumer acceptance and market demands [19].

We can highlight that the assessment of risks associated with the impact of agricultural chemicals on human health is complex due to the vast diversity of substances used, the different mixtures applied in the field, the variations in exposure dose, and the geographical and meteorological characteristics of the agricultural areas where these agrochemicals are employed [52]. Considering all these points for conventional formulations, it is even more challenging to deal with the specific risk assessment of nanomaterials. This is because often, the assumptions, test methodologies, and modeling paradigms for environmental behavior and potential human absorption may not be adequate for products containing nanomaterials [54]. For example, when considering the fate and behavior of molecules in the environment, if a nanoplatform only protects that active ingredient from degradation, we can consider that these characteristics will not be altered. However, other molecules, often inorganic, where there is a significant size reduction, these characteristics can be drastically altered, requiring a comprehensive life cycle analysis from the insertion of the nanoplatform in the field to the disposal of residues [10].

Another important point to be explored, which has significantly contributed to regulatory uncertainties regarding nanoplatforms for agricultural applications, is the variations in definitions and limits of concern established by different regulatory bodies [105], as described previously. For instance, even when presenting certain alignment points in the definitions, regulatory bodies such as the European Commission, Environmental Protection Agency, and Food and Drug Administration still have distinct points regarding size variations and other physicochemical parameters. Such facts are crucial for the initial decision process regarding the classification of a nanoplatform and its approval process [20]. A striking example of this is the so-called nanopesticides. These formulations represent an emerging technological development offering various benefits, such as greater efficacy, durability, and reduction in the amounts of a.i. compared to conventional formulations [46].

In their chapter, [27] address issues on assessing these materials' environmental risk for regulatory purposes. The authors also discuss the problem of definitions (size and physicochemical properties). Moreover, the authors discuss potential adaptations to existing environmental risk assessment tests and procedures for use with nanopesticides, indicating that field urges for researchers, funding and organized international collaborative initiatives to thrive. Despite the significant challenges within the regulatory theme for nanoplatforms for agricultural applications, regulatory agencies and other public entities worldwide are concerned with updating testing guidelines and orientations [52]. In this scenario, we can highlight the role of the Organization for Economic Co-operation and Development (OECD), which has updated to address relevant regulatory parameters. This includes, for example, guidelines on sample preparation and dosimetry, estimation of worker exposure, and guidelines for inhalation toxicity, agglomeration and dissolution, aquatic toxicity, bioaccumulation, and retention in activated sludge [90].

For nanopesticides or agrochemical nanocarriers, [53] discussed an operational approach for the regulatory framework. They suggested that the nanoplatforms as active ingredient carriers or as directly a.i. also need to be considered to classify the workflow and regulation processes. The same authors suggest that this simple classification can guide decisions about the requirement for specific risk assessment tests in each case. In some cases, the adverse effects, degradability, and environmental interactions of nanostructured formulations and components need to be investigated, in addition to the active ingredient information. In summary, the nanomaterial, the active ingredient, and the combination of both materials are essential for reports for physical–chemical characterization, efficacy, and environmental fate investigations. It is important to highlight that comparing nanoformulations with conventional formulations can improve more details about these new agricultural products' efficacy and environmental risk.

Considering all these points, we present in Table 2 the main regulatory endpoints for nanoplatforms aimed at agricultural applications. As mentioned earlier, regulatory processes differ between countries; however, we focused the searches on relevant publications and reports from regulatory entities in the EU, the USA, and the OECD.

4.1.2 Prioritization and Data Reuse in the Registration Process of Nanoformulations

Besides all the benefits and products exhaustively studied and described in the literature, one question remains: why are not many commercial nanoproducts available to growers? The survey of the number of patents (Table 3) shows a significant number of products focused on agriculture. However, those patents still do not reflect availability in the market (1180 patents *versus* 135 available products). This absence of available products seems to be related to regulation products.

Due to their distinctive characteristics, regulatory agencies still lack a clear pathway for nanoformulations. Despite some recent advances, mainly in Europe, the discussion about specific rules is ongoing. Due to their specificities, nanoformulations must meet criteria that consider not only their source material's physicochemical characteristics when on the macroscale but also when submitted to the nanoscale.

From a safety point of view, the regulation must be rigorous. However, the challenge is to analyze the important aspects of the nanoformulations without adding impractical time to all registration processes, mainly in cases where no other solutions are available for the growers. Using the product can be a game changer for that community. Some problems related to pests, mainly to monocultures and commodities such as soybean, corn, cotton, etc., have many efficient technologies that only need to be refined using nanoplatforms, which means no urgency on new products. However, several small farmers, especially from developing countries, have been suffering from problems with no solutions available that could be minimized by new technologies, and extensive time product evaluation can heavily affect those communities, being determinant for their subsistence and even food security, which leads to the necessity of a creation of a priority list based on a well-delimited target/problem of the new product.

It is known that when required, the regulation agencies can analyze and release pesticides in urgency regimes. In 2012/2013, a *Helicoverpa armigera* outbreak, a polyphagous bollworm, was reported in Brazil, causing severe damage to soybean, corn, and cotton [23, 98]. The outbreak and the consequently economic loss led to the farmers' massive use of insecticides [99]. As a response to growers, in the same year, the Brazilian Ministry of Agriculture released in an emergency regime the use

Table 2 Regulatory endpoints for agricultural nanoplatforms

Regulatory endpoint	Tests and required information	Commonly used techniques	Importance of information in the regulatory process
Physical–chemical properties and material characterization	Particle size distribution	Dynamic Light Scattering (DLS), Transmission Electron Microscopy (TEM), Scanning Electron Microscopy (SEM), Nanoparticle Tracking analysis (NTA)	Size affects how NP interact with biological systems and their stability, solubility, and dispersibility
	Surface charge (zeta potential)	Electrophoretic Light Scattering (ELS)	Indicates stability in suspension; influences aggregation behavior and interaction with cell membranes
	Shape and morphology	TEM, SEM, Atomic Force Microscopy (AFM)	Shape can impact the biological activity and toxicity of NP
	Chemical composition	Energy Dispersive X-ray Spectroscopy (EDS), Fourier Transform Infrared Spectroscopy (FTIR), X-ray Diffraction (XRD)	Essential to understand the composition for predicting behavior and interaction with the environment and organisms
	Stability studies	Accelerated Stability Testing, Thermogravimetric Analysis (TGA), Differential Scanning Calorimetry (DSC)	Under varied storage circumstances, the nanopesticides will continue to be safe and effective
Toxicological studies	Dermal exposure/ toxicity	ATA for skin corrosion and irritation, Skin absorption (In vitro method), In vitro skin corrosion (Reconstructed human epidermis (RHE) test method)	NP can pass through the skin more readily, increasing the risk of local or systemic exposure. Testing makes sure that these special qualities don't result in toxicity
	Eye irritation	Bovine corneal opacity and permeability test method	NP can cause different profiles of irritation in the eye. Testing can help to clarify any elevated danger of irritability or harm

(continued)

Table 2 (continued)

Regulatory endpoint	Tests and required information	Commonly used techniques	Importance of information in the regulatory process
	Genotoxicity	In vitro mammalian chromosomal aberration test, In vitro mammalian cell gene mutation tests, In vitro mammalian cell micronucleus test	Determining the possibility for genotoxicity is essential for evaluating long-term health hazards, such as cancer
	Cytotoxicity	TP Cell Titer-Glo, neutral red uptake, LDH release, MTT, XTT, cell impedance, trypan blue, BrdU, Alamar Blue, WST-1, live/dead cell counting, colony-forming efficiency	Cytotoxicity assays help to understand the mode of action and potential dangers of the nanoformulation, providing information about how it causes damage or death to cells
	Inhalation exposure	Acute inhalation toxicity (short-term exposure); Acute inhalation toxicity: acute toxic class method	NP can be inhaled into the respiratory system and get into the deeper tissues of the respiratory system and thus may have a more profound impact compared to larger particles. Testing also helps to identify and control these risks. Inhalation is one of the main pathways of the workers. It assists in determining standard exposure levels, and precautionary measures

(continued)

Table 2 (continued)

Regulatory endpoint	Tests and required information	Commonly used techniques	Importance of information in the regulatory process
Ecotoxicological studies	Aquatic toxicity	Aquatic test (fresh water algae, cyanobacteria and growth inhibition test), *Daphnia sp.* acute immobilization test, Fish acute toxicity test, Daphnia magna reproduction test, Fish embryo acute toxicity (FET) test on embryonic stages of fish (*Daniorerio*)	Nanoformulation can enter the pulmonary system profoundly, NP may have more profound impacts than bigger particles. Testing guarantees that these hazards are recognized and controlled. One common way that workers handling nanoformulations are exposed is through inhalation. Testing aids in the establishment of protective measures and safe exposure limits
	Terrestrial toxicity	Terrestrial plant test (seedling emergence and seedling growth test), *Allium* and *Vicia root* tip chromosome breaks, *Tradescantia* chromosome break, *Tradescantia* micronucleus, *Tradescantia* stamen-hair mutation, and *Arabidopsis* mutation bioassays	For the purpose of soil aeration, nutrient cycling, and general soil fertility, earthworms and other soil organisms are essential. It is essential to ensure that nanoformulations do not damage these organisms in order to preserve soil health. ensures that beneficial insects' ecological services and non-target plants are safeguarded, preventing crop health from being adversely affected by nanoformulations

(continued)

Table 2 (continued)

Regulatory endpoint	Tests and required information	Commonly used techniques	Importance of information in the regulatory process
	Effects on soil microbiota and macrobiota	Earthworm reproduction, Long-term effects after a single exposure of chemicals, on nitrogen transformation activity of soil microbiota and carbon transformation activity of soil microorganisms, Bacterial Reverse Mutation Test, Alteration of biotic and abiotic soil properties	The breakdown of organic matter, soil fertility, and nutrient cycling are all significantly influenced by the soil bacteria. Sustainable agriculture requires careful consideration to make sure nanoformulations don't interfere with these procedures. Plant development and agricultural productivity are impacted by the macrobiota and microbiota of the soil, which are markers of the general health of the soil. Regulatory evaluations guarantee that these species are not adversely affected by nanoformulations
Environmental fate	Persistence in soil and water	Degradation studies, half-life (DT50), mineralization	Understanding persistence helps predict the duration of environmental exposure and potential long-term effects. Determines the potential for accumulation in the environment, which can impact non-target organisms and ecosystems

(continued)

Table 2 (continued)

Regulatory endpoint	Tests and required information	Commonly used techniques	Importance of information in the regulatory process
	Mobility in soil	Adsorption/desorption studies, column leaching studies, partition coefficient (Kd/Koc)	Assist in determining how likely groundwater is to get polluted or contaminated thus posing risks to sources of drinking water. Mobility determines the extent to which the nanoformulation is dispersed over different parts of the environment and condensed in specific areas
	Bioavailability and bioaccumulation	Uptake studies, bioaccumulation factor (BAF), biomagnification	Key to understanding potential negative impacts on other species and ecosystems. It ensures that living organisms accumulate certain substances deciding the degree of harm that is likely to affect the human population through contaminated food chains and even water
	Transformation and metabolites	High-performance liquid chromatography (HPLC), gas chromatography (GC), and mass spectrometry (MS), NMR spectroscopy	This makes sure that all breakdown products in the safety evaluation have been taken into account. Substances can have metabolites that are more persistent or toxic than the parent compound, thus influencing the entire environmental risk

(continued)

Table 2 (continued)

Regulatory endpoint	Tests and required information	Commonly used techniques	Importance of information in the regulatory process
	Sorption and desorption	Sorption coefficients (Kd/Koc), desorption rates	Affects the probability of the nanoformulation dissolving into water or phytotoxicity if ingested by plants. High sorption can limit transport but at the same time can result in long-term soil reservoirs
	Photodegradation	Rate of photodegradation	Can be used to forecast the stability of nanoformulations on other surfaces which are exposed to the sun light like the leaves and soil surfaces. Makes sure that the photodegradation products are recognized and the toxicity of each product is assessed

Table 3 Number of patents with agricultural use potential and products available worldwide depending on the searched term

Searched term	N° of patents*	N° of products available**
"nano" + "insecticide"	258	3
"nano" + "fungicide"	320	7
"nano" + "herbicide"	200	5
"nano" + "nematicide"	56	–
"nano" + "acaricide"	61***	–
"nano" + "fertilizer"	285	120
Total	1180	135

*Number of patents available on "patents.google.com" containing the combination of terms searched
**Number of products available for the market according to "statnano.com"
***Despite targeting mites, some of the products can be focus on cosmetics

of five a.i. that are not allowed so far (MAPA 2013). The urgency of the response can be attributed to the high economic damage caused by the insect. However, there are not a few cases in which small farmers are also affected by pests, and the response is slow or absent due to the low economic representation. We can assume that the same energy spent on the celerity on the release process of some products when economically important could be used to create preference analysis categories when the new product aims to solve small farmers problems, especially when no alternative is available.

The registration of new pesticides in Brazil is a shared duty. The bureaucratic pathway must follow strict guides on the constitution through analyzing three independent agencies, each with its responsibilities. The Brazilian Health Regulatory Agency (Anvisa) demands data regarding toxicological effects. Regarding environmental effects, the data analysis was conducted by the Brazilian Institute of Environment and Renewable Natural Resources (IBAMA). The Brazilian Ministry of Agriculture is responsible for the agronomic data containing the efficacy data of the product. As already described, efforts on prioritizing products that focus on solving or at least mitigating emergency problems are possible, which can only be made by the agronomic analysis, considering its ability to judge the priority based on agronomic issues. Considering that the nanoplatforms can act as an improver of agricultural product effects, these characteristics should be considered when the guidelines of a priority list are set up.

One celerity strategy that is possible is based on the use of shared characteristics of the materials. Studies focus on seeing the behavior differences of matrices when macro and nano should be considered to gain celerity in the evaluation process. Recently, [88] reported that some soil microbiome responses to Cu-based pesticides are the same when treated with bulk or nanoforms of Cu. However, functional and compositional responses on those non-target organisms are formulation dependent. Those types of studies can act as guides to skip some steps, which can be time and resource-consuming, and at the same time, can show where the evaluations of a new nanoformulation can focus on.

5 Conclusion and Perspective

[49], based on data available on the maturity of the technology, demonstrated the evolution of nanoplatforms for agricultural uses from proof of concept to validation and commercialization. Most of these "nano" solutions are between greenhouse trials and field tests (seed coating, NC for gene editing, growth stimulators, macro and micronutrients, sensors, RNAi, a.i., microbiome manipulation, and soil conditioning). The next milestone in agricultural nanotechnology will be validating and commercializing these products, in the initial stages mainly for nanopesticides. Therefore, it is still necessary for discussions in terms of nanomaterials and nanoplatforms classification, based on nanotechnology science evolution, to clarify what aspects of materials can determine the term "nano", considering that it will guide the associated risks to be assumed for each material, required tests, and phases in product registration. In Fig. 1, we presented the steps for the regulation of pesticide nanoplatforms, from design to availability in the market.

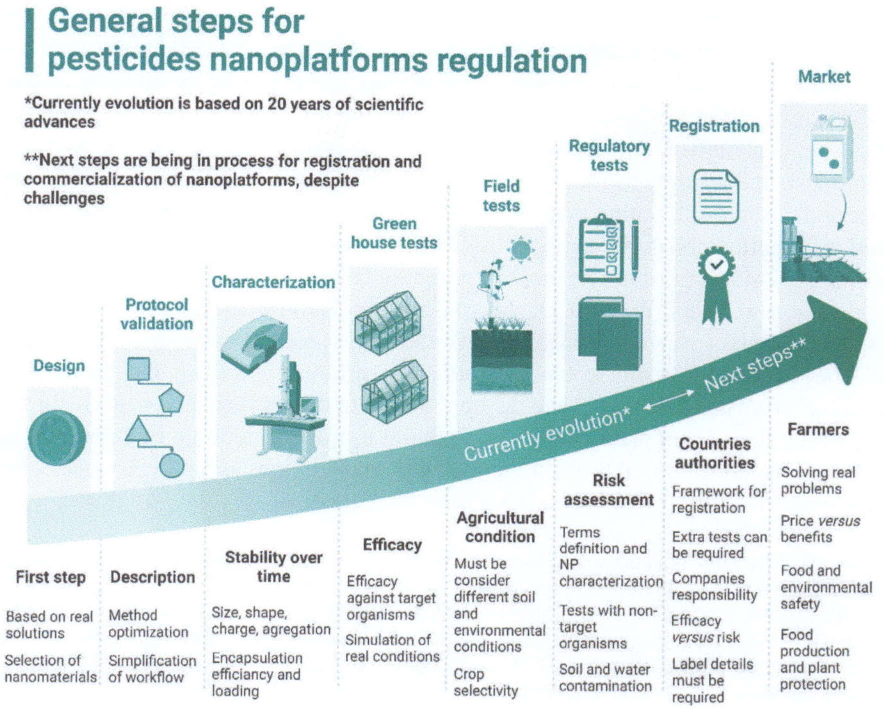

Fig. 1 Steps for pesticide nanoplatforms regulation from design to availability in the market. Created with BioRender.com

As presented in this chapter, the scientific advances in the last 20 years present many potential nanoplatforms for pesticides to delivery a.i., protect them and reduce the environmental risk associated with these compounds. The nanoplatforms can be constituted by simple or combined nanomaterials, presenting complex systems and interactions with target and non-target organisms. Some reports present some gaps that must be considered in the design and experimental process, such as the simplification of methods, simulations of real agricultural conditions, comparison with conventional products to emphasize the robustness of efficacy data, and, importantly, always consider the potential to increase in damage for non-target organisms. In general, the great potential of pesticides delivered by nanoplatforms for solving real agricultural problems in plant protection was reported, and the next steps for nanopesticides to reach the market are regulation, registration, and commercialization.

The regulatory aspects are still a challenge for commercializing nanoplatforms for agricultural uses. To meet society's urgent demands for safer, more modern and effective products, the scientific community and regulatory bodies have made efforts to place nanomaterials on the market. Some guidelines and definitions were addressed for products related to food production, but for environmental risks of chemicals, such as pesticides, these regulatory frameworks are still in discussion and elaboration. For a clear registration process of safer products compared to the conventional tools, the regulatory process must be considered a robust framework encompassing efficacy and risk tests, requiring additional information (such as nanomaterials characteristics and interactions with environment and organisms), and requiring the addition of this information in the leaflets of new products. Despite the extensive time required for agrochemical registration, accelerating this process is important to offer modern and more sustainable products for farmers and attend to current social and environmental demands. It is worth noting that this must be done considering the associated risks, and information about the composition, structure, and ingredients contained in the nanoformulations must not be omitted or ignored as has been observed.

In the next years, the OECD will finish the guidelines for nanomaterials, and complete characterization, toxicology, ecotoxicology, and environmental risk assays will be required for pesticide nanoplatforms registration. Until there, the responsible agencies from countries will have the opportunity to discuss legislation allowing for registering these nanoplatforms for agricultural use. It is important to mention that these efforts are essential to meet the global demand of agricultural producers who need safer and more effective tools for crop protection.

Acknowledgements Authors would like to thanks São Paulo State Research Support Foundation (FAPESP, grants, #2024/00869-1, #2022/00718-8, #2023/00335-4, #2023/16519-7 #2022/00509-0, #2022/03399-0, CBioClima #2021/10639-5, #2023/07905-0, and #2017/21004-5) and Brazilian National Council for Scientific and Technological Development (CNPq-MCTI-INCT NanoAgro #405924/2022-4, #308439/2021-0 L.F.F), and Coordenação de Aperfeiçoamento de Pessoal de Nível Superior—Brasil (CAPES-MEC INCTNanoAgro #88887.953443/2024-00).

References

1. Aksakal FI, Arslan H (2020) Detoxification and reproductive system-related gene expression following exposure to $Cu(OH)_2$ nanopesticide in water flea (Daphnia magna Straus 1820). Environ Sci Pollut Res 27:6103–6111. https://doi.org/10.1007/S11356-019-07414-X/METRICS
2. Aksakal FI, Sisman T (2020) Developmental toxicity induced by $Cu(OH)_2$ nanopesticide in zebrafish embryos. Environ Toxicol 35:1289–1298. https://doi.org/10.1002/TOX.22993
3. Ale A (2023) Nanomaterials and safety concerns to end users. Mater Res Found 135:341–365. https://doi.org/10.21741/9781644902172-14
4. Ali S, Ahmad N, Dar MA et al (2023) Nano-agrochemicals as substitutes for pesticides: prospects and risks. Plants 13:109. https://doi.org/10.3390/PLANTS13010109
5. Altabbaa S, Mann NA, Chauhan N et al (2023) Era connecting nanotechnology with agricultural sustainability: issues and challenges. Nanotechnol Environ Eng 8:481–498. https://doi.org/10.1007/S41204-022-00289-3/METRICS
6. Anderson TJ, Lamsa BP (2011) REVIEW: Zein extraction from corn, corn products, and coproducts and modifications for various applications: a review. Cereal Chem 88:159–173. https://doi.org/10.1094/CCHEM-06-10-0091
7. Athanassiou CG, Kavallieratos NG, Benelli G et al (2017) Nanoparticles for pest control: current status and future perspectives. J Pest Sci 91:1, 91:1–15. https://doi.org/10.1007/S10340-017-0898-0
8. Barradas TN, de Holanda e Silva KG (2020) Nanoemulsions as optimized vehicles for essential oils, pp 115–167. https://doi.org/10.1007/978-3-030-41842-7_4

9. Berardi A, Bisharat L, AlKhatib HS, Cespi M (2018) Zein as a pharmaceutical excipient in oral solid dosage forms: state of the art and future perspectives. AAPS PharmSciTech 19:2009–2022. https://doi.org/10.1208/S12249-018-1035-Y/METRICS
10. Bleeker EAJ, Swart E, Braakhuis H et al (2023) Towards harmonisation of testing of nanomaterials for EU regulatory requirements on chemical safety—a proposal for further actions. Regul Toxicol Pharmacol 139. https://doi.org/10.1016/J.YRTPH.2023.105360
11. Blewett TA, Qi AA, Zhang Y et al (2019) Toxicity of nanoencapsulated bifenthrin to rainbow trout (Oncorhynchus mykiss). Environ Sci Nano 6:2777–2785. https://doi.org/10.1039/C9EN00598F
12. Bombo AB, Pereira AES, Lusa MG et al (2019) A mechanistic view of interactions of a nanoherbicide with target organism. J Agric Food Chem 67:4453–4462. https://doi.org/10.1021/ACS.JAFC.9B00806/ASSET/IMAGES/LARGE/JF-2019-008062_0007.JPEG
13. Calder AJ, Dimkpa CO, McLean JE et al (2012) Soil components mitigate the antimicrobial effects of silver nanoparticles towards a beneficial soil bacterium, Pseudomonas chlororaphis O6. Sci Total Environ 429:215–222. https://doi.org/10.1016/J.SCITOTENV.2012.04.049
14. Camara MC, Campos EVR, Monteiro RA et al (2019) Development of stimuli-responsive nano-based pesticides: emerging opportunities for agriculture. J Nanobiotechnol 17:1–19. https://doi.org/10.1186/S12951-019-0533-8/FIGURES/4
15. Campos EVR, Proença PLF, Oliveira JL et al (2018a) Chitosan nanoparticles functionalized with β-cyclodextrin: a promising carrier for botanical pesticides. Sci Rep 8:1, 8:1–15. https://doi.org/10.1038/s41598-018-20602-y
16. Campos EVR, Proença PLF, Oliveira JL et al (2018b) Carvacrol and linalool co-loaded in β-cyclodextrin-grafted chitosan nanoparticles as sustainable biopesticide aiming pest control. Sci Rep 8:1, 8:1–14. https://doi.org/10.1038/s41598-018-26043-x
17. Carvalho LB, Godoy IS, Preisler AC et al (2023) Pre-emergence herbicidal efficiency and uptake of atrazine-loaded zein nanoparticles: a sustainable alternative to weed control. Environ Sci Nano 10:1629–1643. https://doi.org/10.1039/D2EN01064J
18. Chatterjee S, Bag S, Biswal D et al (2023) Neem-based products as potential eco-friendly mosquito control agents over conventional eco-toxic chemical pesticides-a review. Acta Trop 240:106858. https://doi.org/10.1016/J.ACTATROPICA.2023.106858
19. Chávez-Hernández JA, Velarde-Salcedo AJ, Navarro-Tovar G, Gonzalez C (2024) Safe nanomaterials: from their use, application, and disposal to regulations. Nanoscale Adv 6:1583–1610. https://doi.org/10.1039/D3NA01097J
20. Chelliah R, Madar IH, Sultan G et al (2023) Risk assessment and regulatory decision-making for nanomaterial use in agriculture. In: Engineered nanomaterials for sustainable agricultural production, soil improvement and stress management, pp 413–430. https://doi.org/10.1016/B978-0-323-91933-3.00009-X
21. Chen K, Fu Z, Wang M et al (2018) Preparation and characterization of size-controlled nanoparticles for high-loading λ-Cyhalothrin delivery through flash nanoprecipitation. J Agric Food Chem 66:8246–8252. https://doi.org/10.1021/ACS.JAFC.8B02851/SUPPL_FILE/JF8B02851_SI_001.PDF
22. Chi Y, Zhang G, Xiang Y et al (2017) Fabrication of a temperature-controlled-release herbicide using a nanocomposite. ACS Sustain Chem Eng 5:4969–4975. https://doi.org/10.1021/ACSSUSCHEMENG.7B00348/ASSET/IMAGES/MEDIUM/SC-2017-00348X_0010.GIF
23. Czepak C, Albernaz KC, Vivan LM, et al (2013) Primeiro registro de ocorrência de Helicoverpa armigera (Hübner) (Lepidoptera: Noctuidae) no Brasil. Pesqui Agropecu Trop 110–113
24. Dai H, Han T, Cui J et al (2022) Stability, aggregation, and sedimentation behaviors of typical nano metal oxide particles in aqueous environment. J Environ Manag 316:115217. https://doi.org/10.1016/J.JENVMAN.2022.115217
25. de Albuquerque FP, de Oliveira JL, dos Santos Machado L, et al (2021) Use of nontarget organism Chironomus sancticaroli to study the toxic effects of nanoatrazine. Ecotoxicology 30:4, 30:733–750. https://doi.org/10.1007/S10646-021-02400-X

26. de Oliveira JL, Campos EVR, Germano-Costa T et al (2019) Association of zein nanoparticles with botanical compounds for effective pest control systems. Pest Manag Sci 75:1855–1865. https://doi.org/10.1002/PS.5338
27. de Oliveira Pereira FS, Agarrayua DA, Quines CB, Ávila D (2020) Risk assessment of nanofertilizers and nanopesticides. nanopesticides: from research and development to mechanisms of action and sustainable use in agriculture, pp 299–316. https://doi.org/10.1007/978-3-030-44873-8_10
28. Le DX, Qing LJ, Ming YJ et al (2023a) A pH-responsive MOF-functionalized hollow mesoporous silica controlled herbicide delivery system exhibits enhanced activity against ACCase-herbicide-resistant weeds. Pest Manag Sci 79:5237–5249. https://doi.org/10.1002/PS.7729
29. Deng X, Zhao P, Xie Y, Bai L (2023) Self-assembled sphere covalent organic framework with enhanced herbicidal activity by loading Cyhalofop-butyl. J Agric Food Chem 71:1417–1425. https://doi.org/10.1021/ACS.JAFC.2C07616/SUPPL_FILE/JF2C07616_SI_001.PDF
30. Ding Y, Xiao Z, Chen F et al (2023) A mesoporous silica nanocarrier pesticide delivery system for loading acetamiprid: effectively manage aphids and reduce plant pesticide residue. Sci Total Environ 863:160900. https://doi.org/10.1016/J.SCITOTENV.2022.160900
31. Ding Z, Wei K, Zhang Y et al (2024) "One-Pot" method preparation of dendritic mesoporous silica-loaded matrine nanopesticide for noninvasive administration control of monochamus alternatus: the vector insect of Bursapherenchus xylophophilus. ACS Biomater Sci Eng 10:1507–1516. https://doi.org/10.1021/ACSBIOMATERIALS.3C01270/ASSET/IMAGES/MEDIUM/AB3C01270_0008.GIF
32. Feng J, Shi Y, Yu Q et al (2016) Effect of emulsifying process on stability of pesticide nanoemulsions. Colloids Surf A Physicochem Eng Asp 497:286–292. https://doi.org/10.1016/J.COLSURFA.2016.03.024
33. Feng J, Zhang Q, Liu Q et al (2018) Application of nanoemulsions in formulation of pesticides. nanoemulsions: formulation, applications, and characterization, pp 379–413. https://doi.org/10.1016/B978-0-12-811838-2.00012-6
34. Fojtová D, Vašíčková J, Grillo R et al (2019) Nanoformulations can significantly affect pesticide degradation and uptake by earthworms and plants. Environ Chem 16:470–481. https://doi.org/10.1071/EN19057
35. Fraceto LF, Grillo R, de Medeiros GA et al (2016) Nanotechnology in agriculture: which innovation potential does it have? Front Environ Sci 4:186737. https://doi.org/10.3389/FENVS.2016.00020/BIBTEX
36. Gao Y, Li X, He L et al (2019) Role of adjuvants in the management of anthracnose—change in the crystal morphology and wetting properties of fungicides. J Agric Food Chem 67:9232–9240. https://doi.org/10.1021/ACS.JAFC.9B02147/SUPPL_FILE/JF9B02147_SI_001.PDF
37. Gao Y, Zhou Z, Tang G et al (2024) Facile fabrication of a fungicide and plant immune inducer co-delivery nanosystem for enhanced control efficacy against plant disease. Chem Eng J 482:148817. https://doi.org/10.1016/J.CEJ.2024.148817
38. Gazzurelli C, Migliori A, Mazzeo PP et al (2020) Making agriculture more sustainable: an environmentally friendly approach to the synthesis of Lignin@Cu pesticides. ACS Sustain Chem Eng 8:14886–14895. https://doi.org/10.1021/ACSSUSCHEMENG.0C04645/ASSET/IMAGES/LARGE/SC0C04645_0007.JPEG
39. Giese B, Klaessig F, Park B et al (2018) Risks, release and concentrations of engineered nanomaterial in the environment. Sci Rep 8:1, 8:1–18. https://doi.org/10.1038/s41598-018-19275-4
40. Gomes SIL, Campos EVR, Fraceto LF et al (2022) High-throughput transcriptomics reveals the mechanisms of nanopesticides—nanoformulation, commercial formulation, active ingredient—finding safe and sustainable-by-design (SSbD) options for the environment. Environ Sci Nano 9:2182–2194. https://doi.org/10.1039/D1EN00735A
41. Gottschalk F, Kost E, Nowack B (2013) Engineered nanomaterials in water and soils: a risk quantification based on probabilistic exposure and effect modeling. Environ Toxicol Chem 32:1278–1287. https://doi.org/10.1002/ETC.2177

42. Gouthami K, Lakshminarayana L, Faniband B et al (2023) Introduction to polymeric nanomaterials. Smart Polym Nanocompos Des Synth Functionalization Prop Appl 3–25. https://doi.org/10.1016/B978-0-323-91611-0.00008-6
43. Grillo R, dos Santos NZP, Maruyama CR et al (2012) Poly(e{open}-caprolactone)nanocapsules as carrier systems for herbicides: physico-chemical characterization and genotoxicity evaluation. J Hazard Mater 231–232:1–9. https://doi.org/10.1016/j.jhazmat.2012.06.019
44. Grillo R, Fraceto LF, Amorim MJB et al (2021) Ecotoxicological and regulatory aspects of environmental sustainability of nanopesticides. J Hazard Mater 404. https://doi.org/10.1016/j.jhazmat.2020.124148
45. Gupta R, Malik P, Rani R et al (2024) Recent progress on nanoemulsions mediated pesticides delivery: insights for agricultural sustainability. Plant Nano Biol 8:100073. https://doi.org/10.1016/J.PLANA.2024.100073
46. Hafeez S, Nazir A (2023) Environmental risk assessment of nanopesticides and nanofertilizers. Impact Nanoparticles Agric Soil 351–362. https://doi.org/10.1016/B978-0-323-91703-2.00009-9
47. Hazafa A, Jahan N, Zia MA et al (2022) Evaluation and optimization of nanosuspensions of chrysanthemum coronarium and azadirachta indica using response surface methodology for pest management. Chemosphere 292:133411. https://doi.org/10.1016/J.CHEMOSPHERE.2021.133411
48. Heydarifard S, Taneja K, Bhanjana G et al (2018) Modification of cellulose foam paper for use as a high-quality biocide disinfectant filter for drinking water. Carbohydr Polym 181:1086–1092. https://doi.org/10.1016/J.CARBPOL.2017.11.038
49. Hofmann T, Lowry GV, Ghoshal S et al (2020) Technology readiness and overcoming barriers to sustainably implement nanotechnology-enabled plant agriculture. Nat Food 1:7, 1:416–425. https://doi.org/10.1038/s43016-020-0110-1
50. Huang Y, Yang Y, Liang B et al (2023) Green nanopesticide: pH-responsive eco-friendly Pillar[5]arene-modified selenium nanoparticles for smart delivery of carbendazim to suppress sclerotinia diseases. ACS Appl Mater Interfaces 15:16448–16459. https://doi.org/10.1021/ACSAMI.2C23241/SUPPL_FILE/AM2C23241_SI_001.PDF
51. Hussain S, Siddique T, Saleem M et al (2009) Chapter 5 Impact of pesticides on soil microbial diversity, enzymes, and biochemical reactions. Adv Agron 102:159–200. https://doi.org/10.1016/S0065-2113(09)01005-0
52. Justo-Hanani R (2024) Governing nanotechnology safety: the politics of risk regulation in Europe and the United States. Governing Nanotechnol Saf Polit Risk Regul Europe US 1–130. https://doi.org/10.4337/9781800372870
53. Kah M, Johnston LJ, Kookana RS et al (2021) Comprehensive framework for human health risk assessment of nanopesticides. Nat Nanotechnol 16:955–964. https://doi.org/10.1038/s41565-021-00964-7
54. Kah M, Kookana R (2020) Emerging investigator series: nanotechnology to develop novel agrochemicals: critical issues to consider in the global agricultural context. Environ Sci Nano 7:1867–1873. https://doi.org/10.1039/D0EN00271B
55. Kah M, Tufenkji N, White JC (2019) Nano-enabled strategies to enhance crop nutrition and protection. Nat Nanotechnol 14:6, 14:532–540. https://doi.org/10.1038/s41565-019-0439-5
56. Kah M, Walch H, Hofmann T (2018) Environmental fate of nanopesticides: durability, sorption and photodegradation of nanoformulated clothianidin. Environ Sci Nano 5:882–889. https://doi.org/10.1039/c8en00038g
57. Keller AA, Huang Y, Nelson J (2018) Detection of nanoparticles in edible plant tissues exposed to nano-copper using single-particle ICP-MS. J Nanopart Res 20:1–13. https://doi.org/10.1007/S11051-018-4192-8/METRICS
58. Khursheed A, Rather MA, Jain V et al (2022) Plant based natural products as potential ecofriendly and safer biopesticides: a comprehensive overview of their advantages over conventional pesticides, limitations and regulatory aspects. Microb Pathog 173:105854. https://doi.org/10.1016/J.MICPATH.2022.105854

59. Kumar A, Gupta K, Dixit S et al (2019) A review on positive and negative impacts of nanotechnology in agriculture. Int J Environ Sci Technol 16:2175–2184. https://doi.org/10.1007/S13762-018-2119-7/METRICS
60. Kumar R, Kumar N, Rajput VD et al (2022) Advances in biopolymeric nanopesticides: a new eco-friendly/eco-protective perspective in precision agriculture. Nanomaterials 12:3964. https://doi.org/10.3390/NANO12223964
61. Kumar S, Nehra M, Dilbaghi N et al (2019) Nano-based smart pesticide formulations: emerging opportunities for agriculture. J Control Release 294:131–153. https://doi.org/10.1016/J.JCONREL.2018.12.012
62. Kumar S, Sarita NM et al (2018) Recent advances and remaining challenges for polymeric nanocomposites in healthcare applications. Prog Polym Sci 80:1–38. https://doi.org/10.1016/J.PROGPOLYMSCI.2018.03.001
63. Li Z, Wang H, An S, Yin X (2021) Nanochitin whisker enhances insecticidal activity of chemical pesticide for pest insect control and toxicity. J Nanobiotechnology 19:1–13. https://doi.org/10.1186/S12951-021-00792-W/TABLES/2
64. Liang W, Xie Z, Cheng J et al (2021) A light-triggered ph-responsive metal-organic framework for smart delivery of fungicide to control sclerotinia diseases of oilseed rape. ACS Nano 15:6987–6997. https://doi.org/10.1021/ACSNANO.0C10877/SUPPL_FILE/NN0C10877_SI_001.PDF
65. Liang W, Zhang J, Wurm FR et al (2022) Lignin-based non-crosslinked nanocarriers: a promising delivery system of pesticide for development of sustainable agriculture. Int J Biol Macromol 220:472–481. https://doi.org/10.1016/J.IJBIOMAC.2022.08.103
66. Lowry GV, Avellan A, Gilbertson LM (2019) Opportunities and challenges for nanotechnology in the agri-tech revolution. Nat Nanotechnol 14:6, 14:517–522. https://doi.org/10.1038/s41565-019-0461-7
67. Lv S, Hong T, Wan M et al (2023) Multifunctional mesoporous silica nanosheets for smart pesticide delivery and enhancing pesticide deposition. Langmuir 39:12807–12816. https://doi.org/10.1021/ACS.LANGMUIR.3C01661/SUPPL_FILE/LA3C01661_SI_001.PDF
68. MacHado TO, Beckers SJ, Fischer J et al (2020) Bio-based lignin nanocarriers loaded with fungicides as a versatile platform for drug delivery in plants. Biomacromol 21:2755–2763. https://doi.org/10.1021/ACS.BIOMAC.0C00487/ASSET/IMAGES/LARGE/BM0C00487_0007.JPEG
69. Machado TO, Grabow J, Sayer C et al (2022) Biopolymer-based nanocarriers for sustained release of agrochemicals: a review on materials and social science perspectives for a sustainable future of agri- and horticulture. Adv Colloid Interface Sci 303:102645. https://doi.org/10.1016/J.CIS.2022.102645
70. Manna S, Roy S, Dolai A et al (2023) Current and future prospects of "all-organic" nanoinsecticides for agricultural insect pest management. Front Nanotechnol 4:1082128. https://doi.org/10.3389/FNANO.2022.1082128/BIBTEX
71. Markus AA, Parsons JR, Roex EWM et al (2015) Modeling aggregation and sedimentation of nanoparticles in the aquatic environment. Sci Total Environ 506–507:323–329. https://doi.org/10.1016/J.SCITOTENV.2014.11.056
72. McKenzie FC, Williams J (2015) Sustainable food production: constraints, challenges and choices by 2050. Food Secur 7:221–233. https://doi.org/10.1007/S12571-015-0441-1/METRICS
73. Mgadi K, Ndaba B, Roopnarain A et al (2024) Nanoparticle applications in agriculture: overview and response of plant-associated microorganisms. Front Microbiol 15:1354440. https://doi.org/10.3389/FMICB.2024.1354440/BIBTEX
74. Miernicki M, Hofmann T, Eisenberger I et al (2019) Legal and practical challenges in classifying nanomaterials according to regulatory definitions. Nat Nanotechnol 14:3, 14:208–216. https://doi.org/10.1038/s41565-019-0396-z

75. Ministério da Agricultura P e A (2013) ATO Nº 15, DE 14-03-2013. In: https://sogi8.sogi.com.br/Manager/texto/arquivo/exibir/arquivo?eyJ0eXAiOiJKV1QiLCJhbGciOiJIUzI1NiJ9AFFIjAvMjg2NTkvU0dfUmVxdWlzaXRvX0xlZ2FsX1RleHRvLzAvMC9BVE8gTsK6IDE1LCBERSAxNC0wMy0yMDEzLmRvYy8wLzAiAFFrYS9IxqDoSOPtMJWtuQKwj12OjShQvywVvnvqIAm7WE
76. Mishra P, Dutta S, Haldar M et al (2019) Enhanced mosquitocidal efficacy of colloidal dispersion of pyrethroid nanometric emulsion with benignity towards non-target species. Ecotoxicol Environ Saf 176:258–269. https://doi.org/10.1016/J.ECOENV.2019.03.096
77. Monteiro RA, Camara MC, de Oliveira JL et al (2021) Zein based-nanoparticles loaded botanical pesticides in pest control: an enzyme stimuli-responsive approach aiming sustainable agriculture. J Hazard Mater 417:126004. https://doi.org/10.1016/J.JHAZMAT.2021.126004
78. Muthukrishnan L (2022) An overview on the nanotechnological expansion, toxicity assessment and remediating approaches in agriculture and food industry. Environ Technol Innov 25:102136. https://doi.org/10.1016/J.ETI.2021.102136
79. Nasser R, Ibrahim E, Fouad H et al (2023) Termiticidal, biochemical, and morpho-histological effects of botanical based nanoemulsion against a subterranean termite. Odontotermes Formosanus Shiraki Front Plant Sci 14:1292272. https://doi.org/10.3389/FPLS.2023.1292272/BIBTEX
80. Neme K, Nafady A, Uddin S, Tola YB (2021) Application of nanotechnology in agriculture, postharvest loss reduction and food processing: food security implication and challenges. Heliyon 7:e08539. https://doi.org/10.1016/J.HELIYON.2021.E08539
81. Neves J, Cardoso DN, Malheiro C et al (2019) Copper toxicity to Folsomia candida in different soils: a comparison between nano and conventional formulations. Environ Chem 16:419–429. https://doi.org/10.1071/EN19061
82. Nielsen MB, Skjolding L, Baun A, Hansen SF (2023) European nanomaterial legislation in the past 20 years—closing the final gaps. NanoImpact 32:100487. https://doi.org/10.1016/J.IMPACT.2023.100487
83. Nogueira DJ, Vaz VP, Neto OS et al (2020) Crystalline phase-dependent toxicity of aluminum oxide nanoparticles toward Daphnia magna and ecological risk assessment. Environ Res 182:108987. https://doi.org/10.1016/J.ENVRES.2019.108987
84. Nsom MV, Etape EP, Tendo JF et al (2019) A green and facile approach for synthesis of starch-pectin magnetite nanoparticles and application by removal of methylene blue from textile effluent. J Nanomater 2019:4576135. https://doi.org/10.1155/2019/4576135
85. Oliveira CR, Garcia TD, Franco-Belussi L et al (2019) Pyrethrum extract encapsulated in nanoparticles: toxicity studies based on genotoxic and hematological effects in bullfrog tadpoles. Environ Pollut 253:1009–1020. https://doi.org/10.1016/J.ENVPOL.2019.07.037
86. Oliveira J, Campos EVR, Pereira AES et al (2018) Zein nanoparticles as eco-friendly carrier systems for botanical repellents aiming sustainable agriculture. J Agric Food Chem 66:1330–1340. https://doi.org/10.1021/acs.jafc.7b05552
87. Peixoto S, Henriques I, Loureiro S (2021) Long-term effects of $Cu(OH)_2$ nanopesticide exposure on soil microbial communities. Environ Pollut 269. https://doi.org/10.1016/J.ENVPOL.2020.116113
88. Peixoto S, Morgado RG, Prodana M et al (2024) Responses of soil microbiome to copper-based materials (nano and bulk) for agricultural applications: an indoor-mesocosm experiment. NanoImpact 34:100506. https://doi.org/10.1016/J.IMPACT.2024.100506
89. Preisler AC, Pereira AES, Campos EVR et al (2020) Atrazine nanoencapsulation improves pre-emergence herbicidal activity against Bidens pilosa without enhancing long-term residual effect on Glycine max. Pest Manag Sci 76:141–149. https://doi.org/10.1002/ps.5482
90. Rasmussen K, Bleeker EAJ, Baker J et al (2023) A roadmap to strengthen standardisation efforts in risk governance of nanotechnology. NanoImpact 32. https://doi.org/10.1016/J.IMPACT.2023.100483
91. Santaella C, Plancot B (2020) Interactions of nanoenabled agrochemicals with soil microbiome. nanopesticides: from research and development to mechanisms of action and sustainable use in agriculture, pp 137–163. https://doi.org/10.1007/978-3-030-44873-8_6

92. Santana I, Jeon SJ, Kim HI et al (2022) Targeted carbon nanostructures for chemical and gene delivery to plant chloroplasts. ACS Nano 16:12156–12173. https://doi.org/10.1021/ACSNANO.2C02714/SUPPL_FILE/NN2C02714_SI_001.PDF
93. Schnoor B, Elhendawy A, Joseph S et al (2018) Engineering atrazine loaded poly (lactic-co-glycolic Acid) nanoparticles to ameliorate environmental challenges. J Agric Food Chem 66:7889–7898. https://doi.org/10.1021/ACS.JAFC.8B01911/SUPPL_FILE/JF8B01911_SI_001.PDF
94. Shan P, Lu Y, Lu W et al (2022) Biodegradable and light-responsive polymeric nanoparticles for environmentally safe herbicide delivery. ACS Appl Mater Interfaces 14:43759–43770. https://doi.org/10.1021/ACSAMI.2C12106/SUPPL_FILE/AM2C12106_SI_001.PDF
95. Shoueir KR, El-Desouky N, Rashad MM et al (2021) Chitosan based-nanoparticles and nanocapsules: overview, physicochemical features, applications of a nanofibrous scaffold, and bioprinting. Int J Biol Macromol 167:1176–1197. https://doi.org/10.1016/J.IJBIOMAC.2020.11.072
96. Simonin M, Colman BP, Tang W et al (2018) Plant and microbial responses to repeated $Cu(OH)_2$ nanopesticide exposures under different fertilization levels in an agro-ecosystem. Front Microbiol 9:406948. https://doi.org/10.3389/FMICB.2018.01769/BIBTEX
97. Sousa GFM, Gomes DG, Campos EVR et al (2018) Post-emergence herbicidal activity of nanoatrazine against susceptible weeds. Front Environ Sci 6:1–6. https://doi.org/10.3389/fenvs.2018.00012
98. Specht A, Ricardo Sosa-Gómez D, Vieira de Paula-Moraes S et al (2013) Identificação morfológica e molecular de Helicoverpa armigera (Lepidoptera: Noctuidae) e ampliação de seu registro de ocorrência no Brasil. Pesqui Agropecu Bras 48:689–692. https://doi.org/10.1590/S0100-204X2013000600015
99. Specht A, Sosa-Gómez DR, Rios DAM et al (2021) Helicoverpa armigera (Hübner) (Lepidoptera: Noctuidae) in Brazil: the big outbreak monitored by light traps. Neotrop Entomol 50:53–67. https://doi.org/10.1007/S13744-020-00836-0/METRICS
100. Su X, Yan S, Zhao W et al (2023) Self-assembled thiophanate-methyl/star polycation complex prevents plant cell-wall penetration and fungal carbon utilization during cotton infection by Verticillium dahliae. Int J Biol Macromol 239:124354. https://doi.org/10.1016/J.IJBIOMAC.2023.124354
101. Takeshita V, Carvalho LB, Galhardi JA et al (2022) Development of a preemergent nanoherbicide: from efficiency evaluation to the assessment of environmental fate and risks to soil microorganisms. ACS Nanoscience Au 2:307–323. https://doi.org/10.1021/ACSNANOSCIENCEAU.1C00055/ASSET/IMAGES/MEDIUM/NG1C00055_M006.GIF
102. Takeshita V, de Sousa BT, Preisler AC et al (2021) Foliar absorption and field herbicidal studies of atrazine-loaded polymeric nanoparticles. J Hazard Mater 418:126350. https://doi.org/10.1016/J.JHAZMAT.2021.126350
103. Takeshita V, Munhoz-Garcia GV, Pereira AES et al (2023) Radiometric strategy to track nanopesticides: an important approach to understand the fate, mechanisms of action and toxicity. TrAC, Trends Anal Chem 165:117156. https://doi.org/10.1016/J.TRAC.2023.117156
104. Tang Q, Xu Z, Hong A et al (2021) Response of soil enzyme activity and bacterial community to copper hydroxide nanofertilizer and its ionic analogue under single versus repeated applications. Sci Total Environ 796:148974. https://doi.org/10.1016/J.SCITOTENV.2021.148974
105. Trump BD, Antunes D, Palma-Oliveira J et al (2024) Safety-by-design and engineered nanomaterials: the need to move from theory to practice. Environ Syst Decis 44:177–188. https://doi.org/10.1007/S10669-023-09927-W/TABLES/1
106. Wang Y, Ma S, Yang X et al (2022) Facile synthesis of the dual pesticide-loaded metal-organic framework hybrid for pH-responsive release. ACS Agric Sci Technol 2:1267–1275. https://doi.org/10.1021/ACSAGSCITECH.2C00222/ASSET/IMAGES/MEDIUM/AS2C00222_0009.GIF

107. Wibowo D, Hui Y, Middelberg APJ, Zhao CX (2016) Interfacial engineering for silica nanocapsules. Adv Colloid Interface Sci 236:83–100. https://doi.org/10.1016/J.CIS.2016.08.001
108. Wu H, Cui H, Fu C et al (2024) Unveiling the crucial role of soil microorganisms in carbon cycling: a review. Sci Total Environ 909:168627. https://doi.org/10.1016/J.SCITOTENV.2023.168627
109. Wu H, Hu P, Xu Y et al (2021) Phloem delivery of fludioxonil by plant amino acid transporter-mediated polysuccinimide nanocarriers for controlling fusarium wilt in banana. J Agric Food Chem. https://doi.org/10.1021/acs.jafc.0c07028
110. Wu J, Zhai Y, Monikh FA et al (2021) The differences between the effects of a nanoformulation and a conventional form of atrazine to lettuce: physiological responses, defense mechanisms, and nutrient displacement. J Agric Food Chem 69:12527–12540. https://doi.org/10.1021/ACS.JAFC.1C01382/ASSET/IMAGES/LARGE/JF1C01382_0005.JPEG
111. Xu W, Shen D, Chen X et al (2023) Rotenone nanoparticles based on mesoporous silica to improve the stability, translocation and insecticidal activity of rotenone. Environ Sci Pollut Res 30:106047–106058. https://doi.org/10.1007/S11356-023-29842-6/METRICS
112. Yadav J, Jasrotia P, Kashyap PL et al (2021) Nanopesticides: current status and scope for their application in agriculture. 58:1–17. https://pps.agriculturejournals.cz/doi/1017221/102/2020-PPS.html, https://doi.org/10.17221/102/2020-PPS
113. Zhang N, Hu W, Hou R et al (2023) Enhanced fungicidal activity and mechanism of pyraclostrobin nanoparticle with reactive oxygen species responsiveness against Rhizoctonia solani. J Clean Prod 421:138494. https://doi.org/10.1016/J.JCLEPRO.2023.138494
114. Zhang X, Xu Z, Qian X et al (2020) Assessing the impacts of $Cu(OH)_2$ nanopesticide and ionic copper on the soil enzyme activity and bacterial community. J Agric Food Chem 68:3372–3381. https://doi.org/10.1021/ACS.JAFC.9B06325/SUPPL_FILE/JF9B06325_SI_001.PDF
115. Zhang X, Xu Z, Wu M et al (2019) Potential environmental risks of nanopesticides: application of $Cu(OH)_2$ nanopesticides to soil mitigates the degradation of neonicotinoid thiacloprid. Environ Int 129:42–50. https://doi.org/10.1016/J.ENVINT.2019.05.022
116. Zhang Y, Goss GG (2022) Nanotechnology in agriculture: comparison of the toxicity between conventional and nano-based agrochemicals on non-target aquatic species. J Hazard Mater 439:129559. https://doi.org/10.1016/J.JHAZMAT.2022.129559
117. Zhang Y, Sheedy C, Nilsson D, Goss GG (2020) Evaluation of interactive effects of UV light and nano encapsulation on the toxicity of azoxystrobin on zebrafish. Nanotoxicology 14:232–249. https://doi.org/10.1080/17435390.2019.1690064
118. Zheng Q, Wu J, Yan W et al (2023) Green synthesis of a chlorfenapyr chitosan nanopesticide for maize root application: reducing environmental pollution and risks to nontarget organisms. Int J Biol Macromol 253:126988. https://doi.org/10.1016/J.IJBIOMAC.2023.126988
119. Zhou Z, Tang G, Liu Y et al (2024) Facile fabrication of pesticide nanocapsules using cinnamaldehyde derived imide polymer as wall material for pH-responsive and ultraviolet shielding properties. Chem Eng J 489:151430. https://doi.org/10.1016/J.CEJ.2024.151430
120. (2008) A report for IRGC Risk Governance of Nanotechnology Applications in Food and Cosmetics A report prepared for IRGC by Antje Grobe, Ortwin Renn and Alexander Jaeger Dialogik GmbH international risk governance council international risk governance council Risk Governance of Nanotechnology Applications in Food and Cosmetics
121. Regulation—2015/2283—EN—EUR-Lex. https://eur-lex.europa.eu/eli/reg/2015/2283/oj. Accessed 3 Jul 2024a
122. Regulation—528/2012—EN—EUR-Lex. https://eur-lex.europa.eu/eli/reg/2012/528/oj. Accessed 3 Jul 2024b

MIX
Papier aus verantwortungsvollen Quellen
Paper from responsible sources
FSC® C105338

If you have any concerns about our products, you can contact us on
ProductSafety@springernature.com

In case Publisher is established outside the EU, the EU authorized representative is:
Springer Nature Customer Service Center GmbH
Europaplatz 3, 69115 Heidelberg, Germany

Printed by Libri Plureos GmbH
in Hamburg, Germany